Recent Developments in General Relativity

Springer-Verlag Italia Srl.

B. Casciaro · D. Fortunato · M. Francaviglia ·
A. Masiello (Eds)

Recent Developments in General Relativity

Springer

B. CASCIARO
D. FORTUNATO
Dipartimento Interuniversitario
di Matematica
Università di Bari, Bari, Italy

A. MASIELLO
Dipartimento Interuniversitario
Politecnico di Bari
Bari, Italy

M. FRANCAVIGLIA
Dipartimento di Matematica
Università di Torino
Torino, Italy

Springer-Verlag is a company in the BertelmannSpringer publishing group.

Originally published by Springer-Verlag, Italia, Milano 2000

ISBN 978-88-470-0068-1 ISBN 978-88-470-2113-6 (eBook)
DOI 10.1007/978-88-470-2113-6

Library of Congress Cataloging-in-Publication Data
Recent developments in general relativity / B. Casciaro ... [et al.], eds.
p. cm.
Includes bibliographical references.
ISBN 978-88-470-0068-1
1. General relativity (Physics)–Congresses. 2.
Gravitation–Congresses. I. Casciaro, B. (Biagio) II. Italian
Conference on General Relativity and Gravitational Physics (13th : 1998
: Monopoli, Italy)
QC173.6 .R44 2000
530.11–cd21
00-938079

Cover design: Simona Colombo, Milan
Typesetting: Bürosoft/Text- und DTP-Service, Berlin/Heidelberg

SPIN: 10735291

Preface

The 13th Italian Conference on General Relativity and Gravitational Physics was held in Cala Corvino–Monopoli (Bari) from September 21 to September 25, 1998.

The Conference, which is held every other year in different Italian locations, has brought together, as in the earlier conferences in this series, those scientists who are interested and actively work in all aspects of general relativity, from both the mathematical and the physical points of view: from classical theories of gravitation to quantum gravity, from relativistic astrophysics and cosmology to experiments in gravitation.

About 70 participants came from Departments of Astronomy and Astrophysics, Departments of Mathematics and Departments of Experimental and Theoretical Physics from all over the Country; in addition a few Italian scientists working abroad kindly accepted invitations from the Scientific Committee. The good wishes of the University and of the Politecnico di Bari were conveyed by the director of Dipartimento Interuniversitario di Matematica, Prof. Franco Altomare.

These proceedings contain the contributions of the two winners of the SIGRAV prizes, the invited talks presented at the Conference and most of the contributed talks. We thank all of our colleagues, who did their best to prepare their manuscripts.

The pleasant atmosphere induced by the beauty of the place was greatly enhanced not only by the participation of so many colleagues, who had lively discussions about science well beyond Conference hours, but also by the feeling of hospitality extended to the participants by the staff of the Cala Corvino Hotel, where the Conference was held.

As editors of the proceedings, we take this opportunity to express our gratitude as well as the gratitude of all the participants to the members of the National Scientific Committee and of the Local Organising Committe for their efforts and dedication.

We sincerely thank our generous sponsors for financing the Conference. Finally we warmly thank Dr. Francesca Bonadei, Dr. Carlotta d'Imporzano and the staff of Springer Italia for their efficient and professional work in the preparation of this book.

We would like to thank: Politecnico di Bari, Università degli Studi di Bari; Dipartimento Interuniversitario di Matematica di Bari; Dipartimento di Fisica, Università di Lecce; Comitato per la Matematica,Consiglio Nazionale delle Ricerche; SIGRAV (Società Italiana di Relatività Generale e Fisica della Gravitazione), for their contributions to the Conference.

Bari, April 2000

B. Casciaro
D. Fortunato
M. Francaviglia
A. Masiello

Contents

Introduction 1

General Perspectives

Special Relativity at Action in the Universe 5
G. Ghisellini

String Theory and General Relativity 19
M. Porrati

Virgo – a Laser Interferometer for Gravitational Wave Detection 25
C. Bradaschia

A Space Test of General Relativity 31
I. Ciufolini

Canonical "Loop" Quantum Gravity and Spin Foam Models 43
R. De Pietri

A Possible Way to Define an Effective Cosmological Constant in Scalar–Tensor Cosmologies 63
S. Capozziello, R. de Ritis, A.A. Marino, C. Rubano, P. Scudellaro

A Microscopic Approach to Nuclear Physics: from Deuteron to Neutron Stars 71
A. Fabrocini

Perspectives in the Research of Gravitational Waves: Omnidirectional Resonant Detectors 87
V. Fafone

The Entropy of Black Holes via Noether's Theorem 105
L. Fatibene, M. Ferraris, M. Francaviglia, M. Raiteri

Dual Lagrangian Theories 119
M. Ferraris, M. Raiteri

The Boundary State Description of D-Branes 137
A. Lerda

The N-body Problem in Tetrad Gravity:
A First Step towards a Unified Description of the Four Interactions 149
L. Lusanna

Applications of Calculus of Variations to General Relativity 173
A. Masiello

The Search for Gravitational Waves with Resonant Detectors 197
G. Pizzella

Cauchy-Perturbative Matching and Outer Boundary Conditions 205
L. Rezzolla

Branes in Anti-de Sitter Space-Time 223
M. Trigiante

Black Hole Entropy 241
L. Vanzo

Gravitational Wave Astronomy from Space 253
A. Vecchio

Bar Mode Instability in Relativistic Rotating Stars 271
S. Zane

RECENT RESULTS

General Relativity and Theory of Gravitation

Theoretical Relativity: Classical Topics 287
F. Giannoni

Why is the Gravitational Mass Equal to the Inertial Mass? 289
V. Benci, D. Fortunato

Multiplicity of Timelike Geodesics in Splitting Lorentzian Manifolds 299
A.M. Candela

Gravitational Collapse of Fluid Bodies
and Cosmic Censorship: Analytic Insights 307
S. Jhingan, G. Magli

Black Holes and Solitons 323
L. Martina

The Prolongation Problem for the Heavenly Equation 337
M. Palese, R.A. Leo, G. Soliani

Time Extremizing Trajectories of Massive and Massless Objects in General Relativity ... 345
P. Piccione

Null Geodesics Joining Two Submanifolds in Stationary Lorentzian Manifolds ... 365
A. Salvatore

Analytic Solution of Regge–Wheeler Differential Equation for Black Hole Perturbations in Radial Coordinate and Time Domains ... 373
A.D.A.M. Spallicci

Quantum Gravity

Problems in Quantum General Relativity and Gravitation ... 381
E. Sorace

A Quantum Implementation of the Cosmic Censorship Conjecture ... 387
M.M. Caldarelli

Foundational Problems in Quantum Gravity ... 395
I.G. Avramidi, G. Esposito

Maximal Acceleration Effects for Photons in Cavity Resonators ... 405
A. Feoli, G. Papini, G. Scarpetta

Review of Four-Dimensional Supersymmetric AdS Black Holes ... 411
D. Klemm

Invariants of Spin Networks with Boundary in Quantum Gravity and TQFTs ... 419
G. Carbone, M. Carfora, A. Marzuoli

A Review of Recent Results of the ζ-Function Regularization Procedure in Curved Spacetime ... 427
V. Moretti

Solitons in Three Space Dimensions as a Model for Relativistic Particles ... 437
L. Pisani

Relativistic Astrophysics and Cosmology

Relativistic Astrophysics and Cosmology ... 447
F. Occhionero

Voids in Open Universes 449
F. Occhionero, L. Amendola

Non-Stationary Accretion onto Black Holes 461
J.C. Miller, E. Szuszkiewicz

Stochastic Backgrounds of Gravitational Waves from Compact Sources . . 469
R. Schneider, V. Ferrari, S. Matarrese

Experimental Gravitation

Sights of Experimental Gravity 483
E. Coccia

Mechanical Isolation of Gravitational Wave Antennae: Present and Future 485
E. Coccia, V. Fafone, S. Merkowitz, G. Raffone, M. Schipilliti, M. Visco

Off-Line Subtraction of Seismic Newtonian Noise 495
G. Cella

de Broglie-like Oscillations of Spacetime 505
A. Feoli, G. Scarpetta

Testing Newton's Inverse Square Law 513
S. Focardi

Recent Developments in General Relativity

National Scientific Committee

V. Benci (Univ. of Pisa)
B. Bertotti (Univ. of Pavia)
R. Cianci (Univ. of Genova) – Secretary
I. Ciufolini (Univ. of Roma I)
E. Coccia (Univ. of Roma II)
V. Ferrari (Univ. of Roma I)
R. Fidecaro (I.N.F.N. Sect. of Pisa)
D. Fortunato (Univ. of Bari) – President
M. Francaviglia (Univ. of Torino)
P. Fre' (Univ. of Torino)
L. Guerriero (Politecnico of Bari)
G. Fogli (Univ. of Bari)
L. Lusanna (I.N.F.N. of Firenze)
G. Marmo (Univ. of Napoli)
G. Pizzella (Univ. of Roma I)
T. Regge (Univ. of Torino)
C. Reina (S.I.S.S.A. Trieste)
A. Sagnotti (Univ. of Roma II)
G. Soliani (Univ. of Lecce)
A. Treves (Univ. of Como)

Local Organizing Committee

F. Altomare (Univ. of Bari) – President
O. Amici (Univ. of Bari)
B. Casciaro (Univ. of Bari)
R. Cianci (Univ. of Genova)
D. Fortunato (Univ. of Bari)
M. Francaviglia (Univ. of Torino)
R. Leo (Univ. of Lecce)
L. Martina (Univ. of Lecce)
M. Maiellaro (Univ. of Bari)
A. Masiello (Univ. of Bari) – Secretary
A. M. Pastore (Univ. of Bari)
A. Salvatore (Univ. of Bari)
A. Strumia (Univ. of Bari)

Workshop Coordinators

F. Giannoni (Univ. of L'Aquila)
E. Sorace (Univ. of Firenze)
F. Occhionero (Univ. of Roma I)
E. Coccia (Univ. of Roma II)

List of Contributors

Amendola L.
Osservatorio Astronomico di Roma, Viale del Parco Mellini 84, 00136 Roma, Italy

Avramidi G.I.
Department of Mathematics, The University of Iowa, 14 MacLean Hall,
Iowa City, IA 52242, USA

Benci V.
Dipartimento di Matematica Applicata, Università di Pisa, Via B. Pisano 25/B,
56127 Pisa, Italy
e-mail: benci@dma.unipi.it

Bradaschia C.
INFN Sezione di Pisa, Via Livornese 1291, 56010 San Piero a Grado (Pisa), Italy
e-mail: carlo.bradaschia@pi.infn.it

Caldarelli M.M.
Dipartimento di Fisica, Università di Trento, and Istituto Nazionale di Fisica Nucleare, Gruppo Collegato di Trento, Via Sommarive 14, 38050 Povo (TN), Italy
e-mail: caldarel@science.unitn.it

Candela A.M.
Dipartimento Interuniversitario di Matematica, Via E. Orabona 4, 70125 Bari, Italy
e-mail: candela@pascal.dm.uniba.it

Capozziello S.
Dipartimento di Science Fisiche "E.R. Caianiello", Università di Salerno, 84081 Baronissi (Salerno), Italy
e-mail: capozziello@vaxsy.csied.unisa.it

Carbone G.
SISSA–ISAS, Via Beirut 2-4, 34013 Trieste, Italy

Carfora M.
Dipartimento di Fisica Nucleare e Teorica, Università degli Studi di Pavia,
Via A. Bassi 6, 27100 Pavia, Italy

Cella G.
Dipartimento di Fisica, Università di Pisa, Piazza Torricelli 2, 56100 Pisa, Italy
e-mail: cella@hawaii.difi.unipi.it

Ciufolini I.
Dipartimentio di Ingegneria dell'Innovazione, Università di Lecce, and Dipartimento Aerospaziale, Università degli Studi di Roma "La Sapienza", Via Eudossiana 16, 00184 Roma, Italy
e-mail: ciufoli@nero.ing.uniroma1.it

Coccia E.
Dipartimento di Fisica, Università di Roma II Tor Vergata, and INFN Sezione di Roma, Via della Ricerca Scientifica 1, 00131 Roma, Italy
e-mail: coccia@roma2infn.it

De Pietri R.
Dipartimento di Fisica, Università di Parma, and INFN, Gruppo Collegato di Parma, Via Parco Area delle Science 7/A, 43100 Parma, Italy
e-mail: depietri@pr.infn.it

de Ritis R.
Dipartimento di Science Fisiche, Università di Napoli, Mostra d'Oltremare pad. 20, 80125 Napoli, and Istituto Nazionale di Fisica Nucleare NF, Sezione di Napoli, Complesso Universitario di Monte Sant'Angelo, Via Cintia, Edificio G, 80126 Napoli, Italy
e-mail: deritis@axpna1.na.infn.it

Esposito G.
Istituto Nazionale di Fisica Nucleare, Sezione di Napoli, Mostra d'Oltremare Padiglione 20, 80125 Napoli, Italy
e-mail: giampiero.esposito@na.infn.it

Fabrocini A.
Dipartimento di Fisica, Università di Pisa, Via Buonarroti 2, 56127 Pisa, and INFN, Sezione di Pisa, Via Livornese 1291, 56010 San Piero a Grado (Pisa), Italy
e-mail: fabro@adelchi1.df.unipi.it

Fafone V.
INFN Laboratori Nazionali di Frascati, Via E. Fermi 40, 00044 Frascati (Roma), Italy
e-mail: fafone@lnf.infn.it

Fatibene L.
Dipartimento di Matematica, Università degli Studi di Torino, Via Carlo Alberto 10, 10123 Torino, Italy
e-mail: fatibene@dm.unito.it

Feoli A.
Facoltà di Ingegneria, Università del Sannio, Palazzo dell'Aquila Bosco Lucarelli, Corso Garibaldi 107, 83100 Benevento, and Dipartimento di Scienze Fisiche "E.R. Caianello", Università di Salerno, Via S. Allende, 84081 Baronissi (SA), Italy
e-mail: Feoli@vaxsa.sa-infn.it

Ferrari V.
Dipartimento di Fisica, Università degli Studi "La Sapienza" di Roma, and Sezione INFN Roma 1, Piazzale Aldo Moro 5, 00185 Roma, Italy
e-mail: valeria@roma1.infn.it

Ferraris M.
Dipartimento di Matematica, Università degli Studi di Torino, Via Carlo Alberto 10, 10123 Torino, Italy
e-mail: ferraris@dm.unito.it

Focardi S.
Dipartimento di Fisica, Università di Bologna, and INFN, Sezione di Bologna, Via Irnerio 46, 40126 Bologna, Italy
e-mail: focardi@bo.infn.it

Fortunato D.
Dipartimento Interuniversitario di Matematica, Università di Bari, Via E. Orabona 4, 70125 Bari, Italy
e-mail: fortunat@pascal.dm.uniba.it

Francaviglia M.
Dipartimento di Matematica, Università degli Studi di Torino, Via Carlo Alberto 10, 10123 Torino, Italy
e-mail: francaviglia@dm.unito.it

Ghisellini G.
Osservatorio Astronomico di Brera, Via Bianchi 46, 23807 Merate, Italy
e-mail: gabriele@merate.mi.astro.it

Giannoni F.
Dipartimento di Matematica e Fisica, Università di Camerino, Via Madonna delle Carceri 60, 62032 Camerino (MC), Italy
e-mail: giannoni@dma.unipi.it

Jhingan S.
Dipartimento di Matematica, Politecnico di Milano, Piazza Leonardo da Vinci 32, 20133 Milano, Italy
e-mail: sanjhi@mate.polimi.it

Klemm D.
Dipartimento di Fisica, Università degli Studi di Milano, Via Celoria 16, 20133 Milano, Italy
e-mail: Dietmar.Klemm@mi.infn.it

Leo R.A.
Dipartimento di Fisica dell'Università di Lecce, Via Arnesano, 73100 Lecce, Italy
e-mail: leora@le.infn.it

Lerda A.
Dipartimento di Scienze e Tecnologie Avanzate, Università del Piemonte Orientale, 15100 Alessandria, and INFN, Sezione di Torino, Via P. Giuria 1, 10125 Torino, Italy
e-mail: lerda@to.infn.it

Lusanna L.
Sezione INFN di Firenze, Largo E. Fermi 2, 50125 Firenze, Italy
e-mail: lusanna@fi.infn.it

Magli G.
Dipartimento di Matematica, Politecnico di Milano, Piazza Leonardo da Vinci 32, 20133 Milano, Italy
e-mail: magli@mate.polimi.it

Marino A.A.
Osservatorio Astronomico di Capodimonte, Via Moiariello 16, 80100 Napoli, Italy
e-mail: marino@astrna.na.astro.it

Martina L.
Dipartimento di Fisica dell'Università di Lecce, and INFN Sezione di Lecce, Via Arnesano, 73100 Lecce, Italy
e-mail: martina@le.infn.it

Marzuoli A.
Istituto Nazionale di Fisica Nucleare, Sezione di Pavia, Via A. Bassi 6, 27100 Pavia, Italy
e-mail: marzuoli@pv.infn.it

Masiello A.
Dipartimento Interuniversitario di Bari, Via E. Orabona 4, 70125 Bari, Italy
e-mail: masiello@pascal.dm.uniba.it

Matarrese S.
Dipartimento di Fisica, Università degli Studi di Padova, and Sezione INFN Padova, Via Marzolo 8, 35131 Padova, Italy
e-mail: matarrese@pd.infn.it

Merkowitz S.
INFN, Laboratori Nazionali di Frascati, Via E. Fermi 40, 00044 Frascati, Italy

Miller J.C.
International School for Advanced Studies, SISSA, Via Beirut 2–4, 34013 Trieste, and Nuclear and Astrophysics Laboratory, University of Oxford, Keble Road, Oxford OX1 3RH, England
e-mail: miller@sissa.it

Moretti V.
Dipartimento di Matematica, Università di Trento, and Istituto Nazionale di Fisica Nucleare, Gruppo Collegato di Trento, 38050 Povo (TN), Italy
e-mail: moretti@science.unitn.it

Occhionero F.
Osservatorio Astronomico di Roma, Viale del Parco Mellini 84, 00136 Roma, Italy

Palese M.
Dipartimento di Matematica, Università di Torino, Via Carlo Alberto 10, 10123 Torino, Italy
e-mail: alese@dm.unito.it

Papini G.
Department of Physics, University of Regina, Regina, Sask, S4S 0A2, Canada

Piccione P.
Departamento de Matemática, Instituto de Matemática e Estatística, Universidade de São Paulo, São Paulo, Brazil
e-mail: piccione@ime.usp.br

Pisani L.
Università Nazionali degli Studi di Bari, Dipartimento Interuniversitario di Matematica, Campus, Via Orabona 4, 70125 Bari, Italy
e-mail: pisani@pascal.dm.uniba.it

Pizzella G.
Università Nazionali di Roma Tor Vergata, Via della Ricerca Scientifica, 00131 Roma, and INFN, Laboratori Nazionali di Frascati, Via E. Fermi 40, 00044 Frascati (Roma), Italy
e-mail: pizzella@roma2.infn.it

Porrati M.
Theory Division, CERN, 1211 Geneva 23, Switzerland
Department of Physics, New York University, 4 Washington Pl., New York, NY 10003, USA
Rockefeller University, 1230 York Avenue, New York, NY 10021-6399, USA
e-mail: massimo.porrati@nyu.edu, massimo.porrati@cern.ch

Raffone G.
INFN, Laboratori Nazionali di Frascati, Via E. Fermi 40, 00044 Frascati, Italy

Raiteri M.
Dipartimento di Matematica, Università degli Studi di Torino, Via Carlo Alberto 10, 10123 Torino, Italy
e-mail: raiteri@dm.unito.it

Rezzolla L.
Department of Physics and NCSA, University of Illinois at Urbana-Champaign, Urbana, Illinois 61801, USA
e-mail: rezzolla@astro.physics.uiuc.edu
Present address: SISSA/ISAS International School for Advanced Studies,
Via Beirut 2–4, 34014 Trieste, Italy

Rubano C.
Dipartimento di Science Fisiche, Università di Napoli, Mostra d'Oltremare pad. 20, 80125 Napoli, and Istituto Nazionale di Fisica Nucleare NF, Sezione di Napoli, Complesso Universitario di Monte Sant'Angelo, Via Cintia, Edificio G, 80126 Napoli, Italy
e-mail: rubano@axpna1.na.infn.it

Salvatore A.
Dipartimento Interuniversitario di Matematica, Via E. Orabona 4, 70125 Bari, Italy
e-mail: salvator@pascal.dm.uniba.it

Scarpetta G.
Dipartimento di Scienze Fisiche "E.R. Caianello", Università di Salerno, Via S. Allende, 84081 Baronissi (SA), and INFN Sezione di Napoli, Complesso Universitario, Via Cintia, 80126 Napoli, Italy

Schipilliti M.
INFN, Laboratori Nazionali di Frascati, Via E. Fermi 40, 00044 Frascati, Italy

Schneider R.
Dipartimento di Fisica, Università degli Studi "La Sapienza" di Roma, and Sezione INFN Roma 1, Piazzale Aldo Moro 5, 00185 Roma, Italy
e-mail: schneider@roma1.infn.it

Scudellaro P.
Dipartimento di Scienze Fisiche, Università di Napoli, Mostra d'Oltremare pad. 20, 80125 Napoli, and Istituto Nazionale di Fisica Nucleare NF, Sezione di Napoli, Complesso Universitario di Monte Sant'Angelo, Via Cintia, Edificio G, 80126 Napoli, Italy
e-mail: scud@axpna1.na.infn.it

Soliani G.
Dipartimento di Fisica, Università di Lecce, Via Arnesano, 73100 Lecce, Italy
e-mail: soliani@le.infn.it

Sorace E.
INFO Sezione di Firenze, Dipartimento di Fisica, Università di Firenze, Largo Enrico Fermi 2, 50125 Firenze, Italy
e-mail: sorace@fi.infn.it

Spallicci A.D.A.M.
Facoltà di Ingegneria, Università del Sannio di Benevento, Palazzo dell'Aquila Bosco Lucarelli, Corso Garibaldi 107, 83100 Benevento, Italy
e-mail: aspallic@mat.uniroma1.it

Szuskiewicz E.
Center of Astronomy, Nicolaus Copernicus University, ul. Gagarina 11, 87–100 Toruń, Poland

Trigiante M.
Department of Physics, University of Wales Swansea, Singleton Park, Swansea SA2 8PP, United Kingdom
e-mail: m.trigiante@swansea.ac.uk

Vanzo L.
Dipartimento di Fisica, Università degli Studi di Trento, and INFN, Gruppo Collegato di Trento, Via Sommarive 14, 38050 Povo, Italy
e-mail: vanzo@science.uniti.it

Vecchio A.
Max Planck Institut für Gravitationsphysik, Albert-Einstein-Institut, Am Mühlenberg 5, 14476 Golm, Germany
e-mail: vecchio@aei-potsdam.mpg.de

Visco M.
CNR-IFSI, Via Fosso del Cavaliere 100, 00133 Roma, and INFN Sezione di Roma, Via della Ricerca Scientifica 1, 00131 Roma, Italy
e-mail: visco@roma2.infn.it

Zane S.
Nuclear and Astrophysics Laboratory, University of Oxford, Keble Road, Oxford, OX 1 3RH, UK
e-mail: zane@astro.ox.ac.uk

Introduction

There is a long established tradition of Italian conferences on general relativity. That held in Bari in September 1998 was the 13th in a series of biennial events. Since the Italian Society of General Relativity and Gravitational Physics (SIGRAV) was founded in 1990, the biennial conference has been organized not only as a general scientific event but also as the official periodic meeting for the members of the Society. This contributes to the special features of this series of conferences which reflect the spirit and aims of SIGRAV, a scientific society of academics both from the mathematics and from the physics communities, both theoreticians and experimentalists, coming both from particle physics and from astrophysics and cosmology oriented institutes. The common ground for SIGRAV members is their interest in gravity and in time–honoured general relativity: the viewpoints on GR are very different but this is the special attraction of these conferences. Every year there are many specialized meetings in string theory, in experimental gravity, in cosmology and astrophysics, in gravitational wave detection and so on. Each of us goes to such meetings to hear about the specialized aspects of one's field. The purpose of attending the biennial SIGRAV conference is different: there one seeks an overview, not of one's own field, but rather of the other fields related to general relativity in which one is not a specialist. Indeed it is probably a unique feature of this surprising theory that it is still so much on the move after almost ninety years after its birth and that it is capable of uniting such a variety of scientific interests, applications and cultural backgrounds.

The Bari 1998 edition of the SIGRAV Conference was completely successful and very well organized in the friendly and beautiful environment of the Puglia coastline. As President of SIGRAV I was very much pleased and, on behalf of the whole Society, I would like to express my gratitude to the Scientific Committee and to the Local Organizing Committee for their excellent work.

Personally I am a string theorist and I can evaluate the success of the conference precisely from what I learned in the fields that are most remote from my expertise but not at all from my curiosity. I think that, in Bari, I had the opportunity to get a very comprehensive and well balanced overview on the prospects of detection of gravitational waves and on satellite tests of GR. The excellent reviews on nuclear matter in neutron stars and on the statistical entropy of black holes deserve a special mention. The talk on the possible detection of scalar gravitational waves by means of a spherical resonant antenna, besides being of the utmost interest, must be singled out as the most representative of the spirit of this conference and of SIGRAV. Indeed it originates from the joint work of experimentalists and of a string theorist curious about investigating possible macroscopic consequences of his *theory*.

Finally I must mention that, at Bari Conference, the 1998 SIGRAV prizes have been awarded to

Prof. Massimo Porrati of New York University for his studies in supersymmetry breaking in string theory;

Dr. Gabriele Ghisellini of Milan Astronomical Observatory for his studies on the astrophysics of superluminal sources.

On the occasion, the two distinguished recipients of the award gave brilliant and most interesting **SIGRAV PRIZE LECTURES**. They are included in the present volume and contribute to its value in an essential way.

April 2000

P. Fré

SIGRAV President

General Perspectives

Special Relativity at Action in the Universe

G. Ghisellini

Abstract. Nature succeeds in accelerating extended and massive objects to relativistic velocities. Jets in active galactic nuclei and in galactic superluminal sources and gamma-ray burst fireballs have bulk Lorentz factors from a few to several hundreds. A variety of effects then arises, such as the beaming of the radiation produced, light aberration, time contraction and the Doppler frequency shift. I will emphasize that special relativity applied to real (i.e., extended) observed objects inevitably must take into account the fact that any piece of information is carried by photons. Being created in different parts of the source, they travel different paths to reach the observer, depending on the viewing angle. The object is seen rotated, not contracted, and at small viewing angles time intervals are observed shorter than intrinsic ones.

1 Introduction

We are used to considering earth's particle accelerators as the realm of special relativity. But nature can do better and bigger, and is able to accelerate extremely large masses to relativistic speeds. Besides cosmic rays, we now know that in the jets of active galactic nuclei (AGN) which are also strong radio emitters, plasma flows at $v \sim 0.99c$, and that in gamma-ray bursts (GRB) the fireball resulting from the release of $\sim 10^{52}$ erg in a volume of radius comparable to the Schwarzschild radius of a solar mass black hole reaches $v \sim 0.999c$. These speeds are so close to the speed of light that it is more convenient to use the corresponding Lorentz factor of the bulk motion, $\Gamma = (1 - \beta^2)^{-1/2}$: for $\Gamma \gg 1$, we have $\beta \sim 1 - 1/(2\Gamma^2)$.

The energetics involved is huge. In AGNs, a fraction of a solar mass per year can be accelerated to $\Gamma \sim 10$, leading to powers of $\sim 10^{46}$ erg s^{-1} in bulk motion. In GRBs, the radiation we see, if isotropically emitted, can reach 10^{54} erg s^{-1}, suggesting even larger values for the bulk motion power. Recently, very interesting sources have been discovered within our Galaxy through their activity in X-rays: they occasionally produce radio jets closely resembling those of radio-loud quasars. During these ejection episodes, the power in bulk motion can exceed 10^{40} erg s^{-1}, a value thought to exceed the Eddington limit for these sources.

The study of these *extended* objects moving close to c requires us to take into account the different travel paths of the photons reaching us. Curiously enough, the resulting effects had not been studied until 1959 when Terrel [1] pointed out that a moving sphere does not *appear* contracted, but *rotated*, contrary to what was generally thought (even by Einstein himself ...). These results, which were "academic" in those years, are now fully applicable to the above mentioned relativistic cosmic objects.

In this paper I will present some of the evidence in support of relativistic bulk motion in astronomical objects, and then discuss how "text-book special relativity" has to be applied when information is carried by photons.

2 Superluminal motion

Rees [2] realized that an efficient way of transporting energy from the vicinity of super-massive black holes to the radio lobes of the recently discovered radio galaxies is through the bulk motion of relativistically moving plasma. If this plasma emits radiation on its way, then we ought to see moving spots of emission in radio maps. One of the most spectacular predictions by Rees was that this motion could appear to exceed the speed of light. This was indeed confirmed in the early seventies when radio-interferometric techniques allowed the linking of radio telescopes thousands of kilometers apart. Among the first few observed targets was 3C 279, a radio-loud quasar at a redshift of $z = 0.538$. Bright spots in radio maps taken at intervals of months were apparently moving at a speed exceeding 10 times c. For obvious reasons, sources presenting this phenomenon are referred to as *superluminal sources*.

This phenomenon can be simply explained as long as the plasma is moving at velocities close to c at small viewing angle (i.e., the angle between the velocity vector and the observer's line of sight). Consider Figure 1: suppose the moving blob emits a photon from position A and then from B.

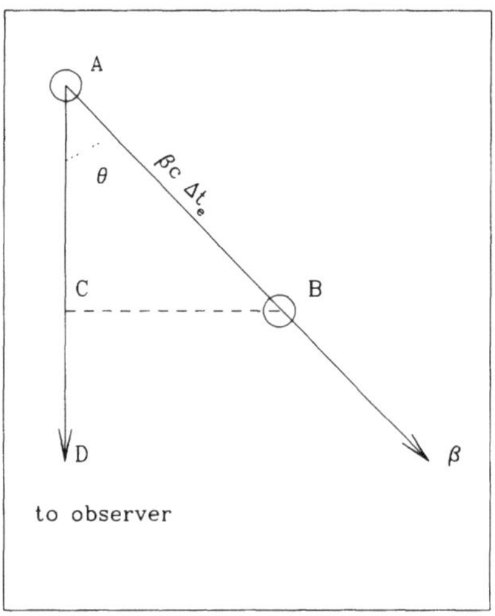

Fig. 1. Explanation of the apparent superluminal speed. A blob emits a photon from A and, after a time Δt_e, from B. If the true velocity is close to the speed of light and the angle θ is sufficiently small, the apparent velocity will exceed the speed of light

The time between the two emissions, as measured by an observer which sees the blob moving, is Δt_e. Therefore the distance AB is equal to $\beta c \Delta t_e$, and $AC = \beta c \Delta t_e \cos\theta$. In the same time interval, the photon emitted at A has reached the point D, and the distance AD is equal to $c\Delta t_e$. Thus the two photons are now separated by $DC = AD - AC = c\Delta t_e(1 - \beta\cos\theta)$. The difference between the arrival times of these two photons is then $\Delta t_a = \Delta t_e(1 - \beta\cos\theta)$, and the projected separation of the blob in the two images is $CB = c\beta\Delta t_e \sin\theta$, leading to an apparent velocity:

$$\beta_{\rm app} = \frac{\beta \sin\theta}{1 - \beta\cos\theta}. \tag{1}$$

It can be readily seen that $\beta_{\rm app} > 1$ for $\beta \to 1$ and small viewing angles θ. The apparent speed is maximized for $\cos\theta = \beta$, where $\beta_{\rm app} = \beta\Gamma$. Notice that this simple derivation *does not require* any Lorentz transformation (no Γ factor involved!). The superluminal effect arises only from the Doppler contraction of the arrival times of photons.

3 Beaming

Let us assume that a source emits isotropically in its rest frame K'. In the frame K, where the source is moving relativistically, the radiation is strongly anisotropic, and three effects occur:

- **Light aberration:** photons emitted at right angles with respect to the velocity vector (in K') are observed in K to make an angle given by $\sin\theta = 1/\Gamma$. This means that in K half of the photons are concentrated in a cone of semi-aperture angle corresponding to $\sin\theta = 1/\Gamma$.
- **Arrival time of the photons**: as discussed above, the emission and arrival time intervals are different. As measured in the same frame K we have, as before, $\Delta t_a = \Delta t_e(1 - \beta\cos\theta)$. If $\Delta t'_e$ is measured in K', $\Delta t_e = \Gamma\Delta t'_e$ leading to

$$\Delta t_a = \Gamma(1 - \beta\cos\theta)\Delta t'_e \equiv \frac{\Delta t'_e}{\delta}. \tag{2}$$

  Here we have introduced the factor δ, referred to as the beaming or Doppler factor. It exceeds unity for small viewing angles, and if so, observed time intervals are *contracted.*
- **Blueshift/redshift of frequencies**: since frequencies are the inverse of times, we just have $\nu = \delta\nu'$.

It can be demonstrated (see, e.g., [3]) that the specific intensity $I(\nu)$ divided by the cube of the frequency is Lorentz invariant, and therefore

$$I(\nu) = \delta^3 I'(\nu') = \delta^3 I'(\nu/\delta). \tag{3}$$

Integration over frequencies yields $I = \delta^4 I'$. The corresponding transformation between bolometric *luminosities* sometimes generates confusion. It is often said that $L = \delta^4 L'$. What this means is: if we estimate the observed luminosity L from the received flux *under the assumption of isotropy*, this is related to L' through the above equation.

But suppose that the photon receiver covers the entire sky, i.e., completely surrounds the emitting source: what is then the relation between the power received in the frames K and K'? In this case we must integrate $dL/d\Omega$ over the solid angle obtaining:

$$L = \int \delta^4 \frac{L'}{4\pi} 2\pi \sin\theta d\theta = \Gamma^2 \frac{\beta^2 + 3}{3} L' \rightarrow \frac{4}{3} \Gamma^2 L'. \quad (4)$$

Note that in this situation *power is not a Lorentz invariant*. This is because here we are concerned with the power *received* in the two frames, not with the *emitted* power (which is indeed Lorentz invariant). This is yet another difference due to "text-book" special relativity not accounting for photons: here the time transformation involves the Doppler term $(1 - \beta \cos\theta)$, causing the difference between *emitted* and *received* power [4].

Because of beaming, relativistically moving objects appear much brighter if their beams point at us, and can therefore be visible up to large distances. Besides being extremely important in order to calculate the intrinsic physical parameters of a moving source, beaming is also crucial for the moving object itself. The observed objects are rarely isolated and more often are part of a jet immersed in a bath of radiation. Just for illustration, let us consider a blob moving close to an accretion disk and surrounded by gas clouds responsible for the emission of the broad lines seen in the spectra of quasars: as the blob moves at relativistic speed in a bath of photons it will see this radiation enhanced. Furthermore, because of aberration, in its frame most of the photons will appear to come from the hemisphere towards which the blob is moving, and be blueshifted. For an observer at rest with respect to the photon bath, the radiation energy density seen by the blob is enhanced by $\sim \Gamma^2$. This increases the rate of interaction between the photons and the electrons in the blob, leading to enhanced inverse Compton emission and possibly even deceleration of the blob by the so-called *Compton drag effect*.

4 Evidence for relativistic motion

Radio-loud AGN with flat radio spectrum

• **Superluminal motion** – The most striking evidence of bulk motion comes from the observation of superluminal sources. Improvements in interferometric techniques have led to the discovery of more than 100 of these sources [5]. The typical bulk Lorentz factors inferred range between 5 and 20.

• **Compton emission** – From radio data (size, flux, spectrum) one can derive, through synchrotron theory, the number density of emitting particles and the radiation energy density [6]. These quantities determine the probability that particles and photons interact through the inverse Compton process, and thus it is possible to predict the amount of high energy radiation (i.e., X-rays) produced. However, this estimated flux is often orders of magnitude larger than what is observed, if beaming is not taken into account. Conversely, the requirement that the radio source emits at most the observed X-ray flux, sets a (lower) limit on the beaming factor δ. Typical values are in agreement with those derived from superluminal motion [7].

• **High brightness temperatures** – This argument is similar to that just presented. The brightness temperature T_b [defined through $I(\nu) \equiv 2kT_b\nu^2/c^2$] is related to the density of particles and photons, and therefore to the probability of Compton scattering: high brightness temperature implies powerful Compton emission. More precisely, if $T_b > 10^{12}$ K (this value is called the *Compton limit*), the luminosity produced by first order Compton scattering is larger than the synchrotron luminosity, and that in the second order exceeds (by the same factor) that in the first order, and so on. Clearly, this can only continue until the typical photon energy reaches the typical electron energy, above which the power has to drop. This increasingly important particle cooling is called the *Compton catastrophe*. To avoid it, we resort to beaming, recalling that T_b transforms according to

$$T_b = \delta T_b' \quad \text{(source size measured directly)}$$
$$T_b = \delta^3 T_b' \quad \text{(source size measured through variability)}. \tag{5}$$

The Compton limit of $T_b > 10^{12}$ K is derived by requiring that the radiation energy density is smaller than the magnetic energy density. A more severe limit can be obtained by imposing the condition of equipartition between particle and magnetic energy densities, as proposed by Readhead [8]. In the latter case one derives larger δ factors, which are in agreement with those obtained by the two previous methods. A note of caution: there is a significant number of sources, called *intraday variables*, whose radio flux changes on a timescale of *hours*. For them, $T_b > 10^{18}$ K, and therefore it is inferred that $\delta > 100$, a value too large to be consistent with those derived in other ways. This is an open issue, and probably effects due to interstellar scintillation and/or coherent radiation (as in pulsars) have to been invoked (for a review see [9]).

• **Gamma-ray emission** – The γ-ray satellite CGRO discovered that radio-loud quasars with flat radio spectrum (FSRQ) can be strong γ-ray emitters, and that often most of their power is radiated in this band. This emission is also strongly variable, on timescales of days or less.

However, photons above $m_ec^2 = 511$ keV can interact with lower energy photons, producing electron–positron pairs. This happens if the optical depth for the photon–photon interaction is greater than unity, and this depends on the density of the target photons and the typical size of the region they occupy. If not beamed, the large power and rapid variability observed imply optical depths largely in excess of unity, and so γ-rays would be absorbed within the source. The condition of transparency to this process leads to lower limits on beaming factors somewhat smaller than in the previous cases. There are also objects (still a few, but increasing in number) observed above a few tenths of a TeV (from the ground by Cherenkov telescopes). The corresponding limits on δ are the most severe.

• **One-sidedness of jets** – Radio-sources are very often characterized by two lobes of extended emission, but in some of them only one jet – starting from the nucleus and pointing to one of the lobes – is visible. If the emitting plasma is moving relativistically, the radiation from the jet approaching us is enhanced, while that from the receding jet is dimmed, explaining the observed asymmetry. This is often referred to as *Doppler favoritism.*

• **Super-Eddington luminosities** – FSRQs show violent activity: large luminosity changes on short timescales. If the variability timescales are associated to the Schwarzschild radius (hence the black hole mass) luminosities exceeding the Eddington limit are often derived (if isotropy is assumed). This difficulty can be easily overcome by beaming (which affects both the variability timescales and the observed power).

• **The Laing–Garrington effect** – The two lobes of radio sources are often differently polarized (or, more precisely, they are differently depolarized): as the (one sided) jet always points towards the more polarized lobe, this convincingly indicates that the visible jet is the one approaching the observer [10, 11].

• **Jet bending** – Jets of AGNs are often significantly curved, posing problems of stability. But this strong bending could be largely apparent if the jet is seen at small viewing angles. This constitutes a further independent hint that these jets are pointing at us (although it is not direct evidence of relativistic motion).

• **Parent population** – For each source whose jet aims directly towards us, there must be $\sim \Gamma^2$ other sources with jets pointing away. If beaming is important, then these objects must be less luminous, not (extremely) superluminal, and not showing violent activity. This *parent population* of sources can be identified with that of radiogalaxies ([12]; for a review see [13]): indeed their number agrees with what is expected from the beaming parameters derived by the other methods.

Galactic superluminal sources

In 1994, the two X-ray transients GRS1915+105 and J1655–40 were monitored with the VLA during a campaign aimed at finding radio jets associated with galactic sources. Surprisingly, superluminal motion was discovered in both of them [14, 15]. The most interesting observational fact is that in these two sources we see *both* the jet and the counter-jet. This is unprecedented among superluminal sources, and it is possible because: i) the viewing angle is large, suppressing the Doppler favoritism effect; ii) the bulk Lorentz factor is small, $\Gamma = 2.5$, leading to moderate beaming and thus an observable flux even at large viewing angles.

Under the assumption that the superluminal blobs move in opposite directions and are characterized by the same (true) velocity βc, we can apply Eq. (1) *twice* to derive *both* β and θ. This in turn determines δ.

Very severe limits can be set on the power carried by the moving blob in the form of both particle bulk motion and Poynting flux. The radio emission observed with VLA comes from a region $\sim 10^{15}$ cm in size, and is believed to be produced by synchrotron emission: it is then possible to calculate the number of particles responsible for the observed radiation (this is a lower limit, since other non-emitting, i.e., sub-relativistic, particles might be present).

The number inferred depends on the value of the magnetic field B: an increase in B decreases the amount of particles required, but increases the implied Poynting flux. *The sum of the power in particle bulk motion and in Poynting flux therefore has a minimum.* For GRS1915+105, this is of the order of 10^{40} erg s^{-1} [16], and exceeds by a factor $\sim$ 10 the power emitted by the accretion disk (in soft and medium energy X-rays). This result itself puts strong constraints on any model for the acceleration of jets, by excluding the possibility that this occurs through radiation pressure.

Gamma-ray bursts

GRBs are flashes of hard X- and γ-rays, lasting for a few seconds. Discovered by the Vela satellites in the late sixties, their origin remained mysterious until the Italian–Dutch satellite *Beppo*SAX succeeded in locating them in the sky with sufficiently small error-boxes: prompt follow-up observations were then possible and led to the discovery that they are at cosmological distances. This in turn allowed the estimation of the energy involved which, if the radiation is emitted isotropically, is in the range 10^{52}–10^{54} erg. The energetics is in itself strong evidence of relativistic bulk motion: the short duration and the even shorter variability timescale (of the order of 1 ms) imply a huge compactness (i.e., the luminosity over size ratio). This resembles the conditions during the Big Bang, and implies a similar evolution: no matter in which form the energy is initially injected, a quasi-thermal equilibrium between matter and radiation is reached, with the formation of electron–positron pairs accelerated to relativistic speeds by the high internal pressure. This is a *fireball.* When the temperature of the radiation (as measured in the comoving frame) drops below $\sim$50 keV the pairs annihilate faster then the rate at which are produced (50 keV, not 511 keV, as a thermal photon distribution has a high energy tail...). But the presence of even a small amount of baryons, corresponding to only $\sim 10^{-6}\, M_{\odot}$, makes the fireball opaque to Thomson scattering: the internal radiation thus continues to accelerate the fireball until most of its initial energy has been converted into bulk motion. The fireball then expands at a constant speed and at some point becomes transparent. If the central engine is not completely impulsive, but works intermittently, it can produce many shells (i.e., many fireballs) with slightly different Lorentz factors. Late but faster shells can catch up with early slower ones, producing shocks which give rise to the observed burst emission. In the meantime, all shells interact with the interstellar medium, and at some point the amount of swept up matter is large enough to decelerate the fireball and produce other radiation which can be identified with the afterglow emission observed at all frequencies.

For GRBs the limits to the required bulk Lorentz factors follow mainly from two arguments (for reviews see [17, 18]):

- **Compactness** – We see high energy (i.e., γ-rays above 100 MeV) emission, varying on short timescales. As for the AGN case (see above) bulk motion is required to lower the implied luminosity and increase the intrinsic size of the emission region in order to decrease the opacity to pair production. Limits to the Lorentz factor in the range 100–300 are derived.
- **Variability** – The radiation we see originates when the fireball has become optically thin to Thomson scattering, i.e., when it has expanded to radii of the order of $R_t =$

10^{13} cm. Yet we see 1 ms variability timescales. Assuming that the radiation is produced while the fireball expands by a factor of two in radius, short timescales are possible if $\Delta t_a = (1 - \beta \cos\theta) R_t / c$. Furthermore, if isotropic, we always see the portion of it pointing at us ($\theta = 0°$), and thus $\Delta t_a \sim 0.01 (R_t / 10^{13})(100/\Gamma)^2$. Also this argument leads to values of Γ in the range 100–300.

5 Rulers and clocks vs photographs and light curves

In special relativity we are used to two fundamentals effects:

- lengths shrink in the direction of motion;
- times get longer.

The Lorentz transformations for a motion along the x axis are (K is the lab frame and K' is the comoving one)

$$\begin{aligned} x' &= \Gamma(x - vt) \\ t' &= \Gamma\left(1 - \frac{v}{c^2} x\right) \end{aligned} \tag{6}$$

with the inverse relations given by

$$\begin{aligned} x &= \Gamma(x' + vt') \\ t &= \Gamma\left(t' + \frac{v}{c^2} x'\right). \end{aligned} \tag{7}$$

The length of a moving ruler has to be measured through the position of its extremes *at the same time* t. Therefore, as $\Delta t = 0$, we have

$$x'_2 - x'_1 = \Gamma(x_2 - x_1) - \Gamma v \Delta t = \Gamma(x_2 - x_1), \tag{8}$$

i.e.,

$$\Delta x = \frac{\Delta x'}{\Gamma} \rightarrow \text{contraction}. \tag{9}$$

Similarly in order to determine a time interval a (lab) clock has to be compared with one in the comoving frame, which has, in this frame, *the same position* x'. It follows that

$$\Delta t = \Gamma \Delta t' + \Gamma \frac{v}{c^2} \Delta x' = \Gamma \Delta t' \rightarrow \text{dilation}. \tag{10}$$

An easy way to remember the transformations is to think of mesons produced in collisions of cosmic rays in the high atmosphere, which can be detected even if their lifetime (in the comoving frame) is much shorter than the time needed to reach the earth's surface. For us, on the ground, relativistic mesons live longer.

All this is correct if we measure lengths by comparing rulers (at the same time) and by comparing clocks (at the same position) - the meson lifetime *is* a clock. In other words, **if we do not use photons** for the measurement process.

The moving bar

If information (about position and time) is carried by photons, we **must** take into account their (different) paths. When we take a picture, we detect photons arriving at the same time at our camera: if the moving body which emitted them is extended, we must consider that these photons have been emitted at different times, when the moving object occupied different locations in space. This may seem quite obvious. And it is. Nevertheless these facts were pointed out in 1959 [1], more than 50 years after the publication of the theory of special relativity.

Let us consider a moving bar, of proper dimension ℓ', moving in the direction of its length at velocity βc and at an angle θ with respect to the line of sight (see Fig. 2). The length of the bar in the frame K (according to relativity "without photons") is $\ell = \ell'/\Gamma$. The photon emitted in A_1 reaches the point H in the time interval Δt_e. After Δt_e the endpoint B_1 has reached the position B_2, and by this time, photons emitted by the other end of the bar can reach the observer simultaneously with the photons emitted by A_1, since the travel paths are equal. The length $B_1B_2 = \beta c \Delta t_e$, while $A_1H = c\Delta t_e$. Therefore

$$A_1H = A_1B_2 \cos\theta \rightarrow \Delta t_e = \frac{\ell' \cos\theta}{\Gamma(1-\beta\cos\theta)}. \tag{11}$$

Note the appearance of the term $\delta = 1/[\Gamma(1-\beta\cos\theta)]$ in the transformation: this accounts for both the relativistic length contraction $(1/\Gamma)$, and the Doppler effect

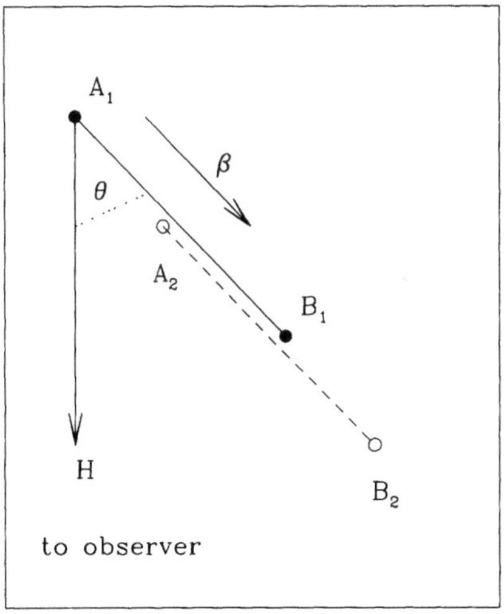

Fig. 2. A bar moving with velocity βc in the direction of its length. The path of the photons emitted by the endpoint A is longer than the path of photons emitted by B. When we make a picture of the bar (or a map), we collect photons reaching the detector simultaneously. Therefore the photons from A have to be emitted before those from B, when the bar occupied another position

$[1/(1-\beta\cos\theta)]$. The length A_1B_2 is then given by

$$A_1B_2 = \frac{A_1H}{\cos\theta} = \frac{\ell'}{\Gamma(1-\beta\cos\theta)} = \delta\ell'. \tag{12}$$

In a real picture, we would see the projection of A_1B_2, i.e.,

$$HB_2 = A_1B_2\sin\theta = \ell'\frac{\sin\theta}{\Gamma(1-\beta\cos\theta)} = \ell'\delta\sin\theta. \tag{13}$$

The maximum observed length is ℓ' for $\cos\theta = \beta$.

The moving square

Now consider a square of size ℓ' in the comoving frame, moving at 90° to the line of sight (Fig. 3). Photons emitted at A, B, C and D have to arrive at the film plate at the same time. But the paths of photons from C and D are longer → they have to be emitted earlier than photons from A and B: when photons from C and D were emitted, the square was in another position. The interval of time between emission

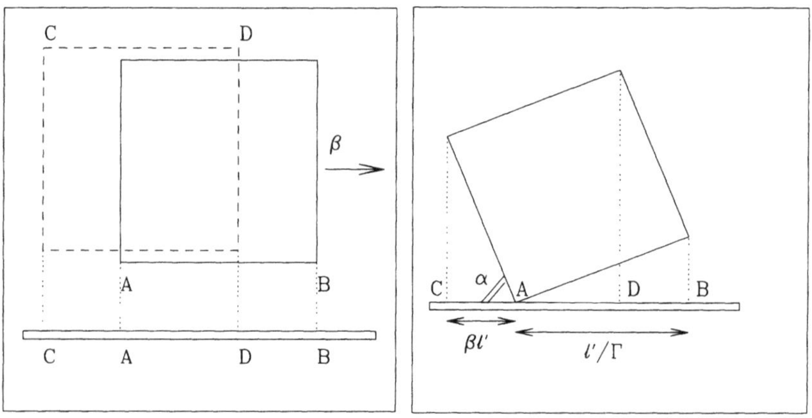

Fig. 3. Left: A square moving with velocity βc seen at 90°. The observer can see the left side (segment CA). Light rays are assumed to be parallel, i.e., the square is assumed to be at large distance from the observer. **Right:** The moving square is seen as *rotated* by an angle α given by $\cos\alpha = \beta$

from C and from A is ℓ'/c. During this time the square moves by $\beta\ell'$, i.e., the length CA. Photons from A and B are emitted and received at the same time and therefore $AB = \ell'/\Gamma$. The total observed length is given by

$$CB = CA + AB = \frac{\ell'}{\Gamma}(1+\Gamma\beta). \tag{14}$$

As β increases, the observer sees the side AB increasingly shortened by the Lorentz contraction, but at the same time the length of the side CA increases. The maximum total length is observed for $\beta = 1/\sqrt{2}$, corresponding to $\Gamma = \sqrt{2}$ and to $CB = \ell'\sqrt{2}$, i.e., equal to *the diagonal* of the square. Note that we have considered the square (and the bar in the previous section) to be at large distances from the observer, so that the emitted light rays are all parallel. If the object is near to the observer, we must take into account the fact that different points of one side of the square (e.g., the side AB in Fig. 3) have different travel paths to reach the observer, producing additional distortions. See Mook and Vargish [19] for some interesting illustrations.

Rotation, not contraction

The next result (taking into account **both** the length contraction **and** the different paths) is an apparent **rotation** of the square, as shown in Fig. 3 (right panel). The rotation angle α can be simply derived (even geometrically) and is given by

$$\cos\alpha = \beta. \tag{15}$$

A few considerations follow.

- If you rotate a sphere you still get a sphere: you **do not** observe a contracted sphere.
- The total length of the projected square, appearing on the film, is $\ell'(\beta + 1/\Gamma)$. It is maximum when the "rotation angle" $\alpha = 45° \rightarrow \beta = 1/\sqrt{2} \rightarrow \Gamma = \sqrt{2}$. This corresponds to the diagonal.
- The appearance of the square *is the same as that seen in a comoving frame for a line of sight making an angle α' with respect to the velocity vector, where α' is the aberrated angle* given by

$$\sin\alpha' = \frac{\sin\alpha}{\Gamma(1 - \beta\cos\alpha)} = \delta\sin\alpha. \tag{16}$$

See Fig. 4 for a schematic illustration.

The last point is particularly important, because it introduces a great simplification in calculating not only the appearance of bodies with a complex shape but also the light curves of varying objects.

Light curves

We have already seen how intrinsic time intervals $\Delta t'_e$ transform in observed Δt_a when taking into account different photon travel paths. The Doppler effect can oppose time expansion and, depending on the viewing angle, Δt_e can be longer or shorter than $\Delta t'_e$. There are however more complex cases where it may be difficult to derive a prescription as simple as Eq. (2). For instance, a relativistically moving blob which also expands relativistically (i.e., "a bomb" exploding in flight). Accounting for the superposition of the two motions is complex, but the introduction of the "aberrated

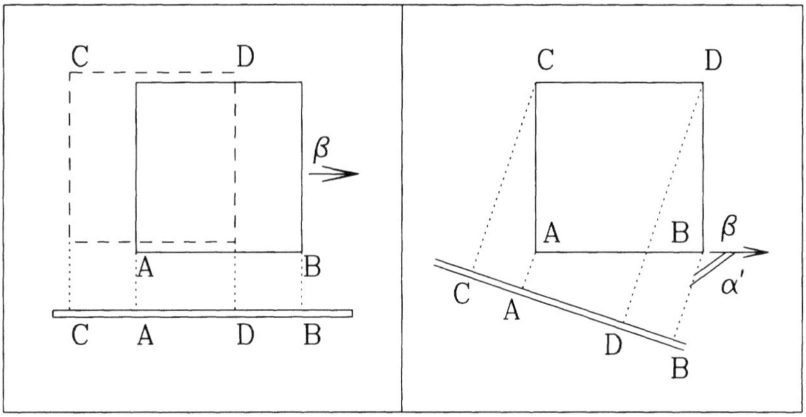

Fig. 4. An observer who sees the object at rest at a viewing angle given by $\sin\alpha' = \delta \sin\alpha$, will take the same picture as the observer who sees the object moving and making an angle α with his/her line of sight. Note that $\sin\alpha' = \sin(2\pi - \alpha')$

angle observer" greatly simplifies this kind of problem: this observer would see the blob without bulk motion and the different travel paths are the geometric ones. Then the observer in the lab frame K just sees the same light curve as the "aberrated angle observer", but with time intervals divided by δ and specific intensities multiplied by δ^3. In fact, in the frame K' the photons emitted at the "de-aberrated angle" α' are the very same ones that reach the observer in K, at the "aberrated angle" α.

Table 1. Useful transformations

$\nu = \delta\nu'$	frequency
$t = t'/\delta$	time
$V = \delta V'$	volume
$\sin\theta = \sin\theta'/\delta$	sine
$\cos\theta = (\cos\theta' + \beta)/(1 + \beta\cos\theta')$	cosine
$I(\nu) = \delta^3 I'(\nu')$	specific intensity
$I = \delta^4 I'$	total intensity
$j(\nu) = \delta^2 j'(\nu')$	specific emissivity
$\kappa(\nu) = \kappa'(\nu')/\delta$	absorption coefficient
$T_B = \delta T'_B$	brightness temperature (size measured directly)
$T_B = \delta^3 T'_B$	brightness temperature (size from variability)

6 Conclusions

In the last 25 years special relativity has become necessary for the understanding of some of the most violent phenomena in our universe, in which large masses are accelerated to relativistic velocities. Instead of dealing with elementary particles in accelerators or with energetic cosmic rays, we can see fractions of a solar mass moving at $0.999c$. These motions give rise to several effects, the most important probably being the collimation of the radiation along the direction of motion, making the source visible to distant observers. This implies that relativistic emitting sources – such as the plasma in jets of AGNs and gamma-ray bursts – can be good probes of the far universe.

As discussed in this contribution, spatial and temporal information are carried by photons, and therefore the differences in their paths to reach the observer must be taken into account. Extended moving objects are seen rotated and therefore spheres remain spheres.

Acknowledgements

It is a pleasure to thank Annalisa Celotti, Laura Maraschi, Aldo Treves and Meg Urry for years of fruitful collaboration.

References

1. Terrel J. (1959): Phys. Rev. **116**, 1041
2. Rees M.J. (1966): Nature **211**, 468
3. Rybicki G.B., Lightman A. P. (1979): Radiative Processes in Astrophysics. Wiley, New York
4. Rybicki G.B., Lightman A. P. (1979): Radiative Processes in Astrophysics. Wiley, New York, p. 141
5. Vermeulen R.C., Cohen M.H. (1994): ApJ **430**, 467
6. Hoyle F.S.R., Burbidge G.R., Sargent W.L.W. (1966): Nature **209**, 751
7. Ghisellini G., Padovani P., Celotti A., Maraschi L. (1993): ApJ **407**, 65
8. Readhead A.C.S. (1994): ApJ **426**, 51
9. Wagner S.J., Witzel A. (1995): ARA&A **33**, 163
10. Laing R.A. (1988): Nature **331**, 149
11. Garrington S.T., Leahy J.P., Conway R.G., Laing R.A. (1988): Nature **331**, 147
12. Blandford R.D., Rees M.J. (1987): in *Pittsburgh Conference on BL Lac objects*, ed. by A.N. Wolfe, University of Pittsburgh Press, Pittsburgh, p. 161
13. Urry M.C., Padovani P. (1995): PASP **107**, 803
14. Mirabel I.F., Rodriguez L.F. (1994): Nature **371**, 46
15. Hjellming R.M., Rupen M.P. (1995): Nature **375**, 464
16. Gliozzi M., Bodo G., Ghisellini G. (1999): MNRAS **303**, L37
17. Piran T. (1999): Phys. Rep. **314**, 575
18. Meszaros P. (1999): Nuclear Phys. B, (Proceedings Supplements), Elsevier Science, in press (astro-ph/9904038)
19. Mook D.E., Vargish T. (1987): Inside Relativity. Princeton University Press, Princeton

String Theory and General Relativity

M. Porrati

Abstract. We give a historical review of some aspects of string theory relevant to our present understanding of General Relativity and connected with Einstein's unification program.

The quest for a unified theory of all fundamental interactions is an old one, dating back to just after the formulation of General Relativity. Its main early advocate was Einstein himself; this is well known to physicists and laymen alike. What is less generally known is that Einstein's view of unification was not the same as the unification program of the 1970s. The latter, indeed, aimed at a unified description of all elementary interactions – through gauge theories, for instance – and at a reduction of the number of free parameters needed to describe them.

For Einstein, a successful unified theory had also to provide a classical derivation of the quantization rules of (old) quantum mechanics, and to allow for regular solutions with finite energy, that could be interpreted as elementary particles [1].

While the idea that quantum mechanics can, to some extent, be reduced to standard, local, classical field theory was exceedingly naive, and was disproved forever by Bell's inequalities [2], the possibility of describing elementary particles as solitons has been resurrected, unexpectedly, in the context of string theory, as we shall discuss in a moment.

First of all, let us recall the main obstacle encountered by any unification program based on field theory, namely, quantum gravity. The Einstein action can be quantized in a more or less standard fashion, at least perturbatively around a given background (as flat Minkowsky space or a black hole); but its coupling constant, G_{Newton}, has negative mass dimension ($G_{\text{Newton}} = M_{\text{Planck}}^{-2}$, $M_{\text{Planck}} \approx 2 \times 10^{19}\, GeV$) and the theory is thus non-renormalizable. To make clear my view of the problem let me discuss another non-renormalizable theory: namely, the Fermi theory of beta decay. Its coupling constant, $G_{\text{Fermi}} = M_{\text{Fermi}}^{-2}$, $M_{\text{Fermi}} \approx 100\,\text{GeV}$, also has negative mass-dimension. Unlike General Relativity, however, this theory is well known to be just an effective low-energy approximation of the true theory of electroweak interactions. Moreover, the Fermi theory is not cloaked in the beautiful geometrical structure of General Relativity, that may lead us astray in our search for a solution to the problem of non-renormalizability: in Fermi's theory the problem appears clearly, as well as its solution.

At its simplest the problem is that a tree-level cross section in the Fermi theory, σ, behaves like $\sigma \sim E^2/M_{\text{Fermi}}^4$ as a function of the c.o.m. energy of a typical process – beta decay, for instance. When $E \gg M_{\text{Fermi}}$, the cross section exceeds the unitarity bound $\sigma \sim M_{\text{Fermi}}^{-2}$. This means that any tree-level result receives enormous radiative corrections, and that a perturbative analysis of the theory breaks down at energies exceeding the Fermi scale. The solution to this problem does not come from a non-perturbative quantization of the theory; rather, it comes from realizing that the pathology of the cross section at energies above the Fermi scale suggests that the true theory of weak interactions contains new degrees of freedom, with mass $O(100\,\mathrm{GeV})$. Nowadays, we know that they are the intermediate vector bosons, $W^{\pm}$ and Z_0, and the neutral Higgs boson.

In a certain sense, non-renormalizability is a signal that a theory is incomplete, since it does not contain all the heavy degrees of freedom necessary to understand its ultraviolet behavior; in other words, it is an effective theory.

Quantum gravity cannot be tamed as easily as the Fermi theory. It is well known that one cannot cure the lack of tree-level unitarity by introducing a finite number of new massive degrees of freedom [3]. An infinite number of new, massive degrees of freedom, of ever-increasing spin, may succeed, nevertheless. Obviously, trying to add by hand these degrees of freedom to Einstein's action is hopeless. Luckily, Scherk and Schwarz pointed out a long time ago [4] that a perturbatively consistent theory, reducing to General Relativity at energies well below M_{Planck} exists, namely, (super)string theory. It does not have the ultraviolet pathologies of General Relativity, and it contains, as it must, infinitely many new particles, with mass $> M_{\mathrm{Planck}}$, and maximum spin proportional to their square mass.

In string theory, the basic interacting objects are no longer point-like, as in field theory, but one-dimensional. More precisely, in field theory, the one-particle states are parametrized by position (or momentum), x^m, (p^m), spin, s, and internal indices, a, taking a finite number of values – they can label a representation of a symmetry group, for instance. In string theory, instead, the one-string states are labelled by a map $X^m(\sigma)$, $\sigma \in [0, \pi]$, giving the position in target space of a one-dimensional object (the string) plus internal indices. One-string states in superstrings are labelled by a superfield whose bosonic component is $X^m(\sigma)$ [5].

A theory of extended objects may seem to contradict a basic requirement we demand of any theory of fundamental interaction, namely, locality. What saves the day here is that, while strings themselves are non-local, their interactions – joining and splitting, basically – are local.

Consistent string theories (superstrings) are naturally formulated in ten dimensions; this means that to make contact with our low-energy world, six dimensions must be compactified. String theory has a fundamental scale, M_S, and a dimensionless coupling constant, g_S. The relation between the string scale and the Planck scale is, in ten dimensions,

$$M_{\mathrm{Planck}}^8 = \frac{1}{g_S^2} M_S^8. \qquad (1)$$

It may be somewhat surprising that a theory with a fundamental scale, which acts as an ultraviolet cutoff, is nevertheless local. This can be explained by noticing that strings have a tension $T = M_S^2$, and that their vibrational eigenmodes are interpreted as particles of mass $O(M_S)$. More precisely, by quantizing the free vibrating (super) string one gets a wealth of massive particles, with degeneracy:

$$d(M) = \exp(\text{const}\, M/M_S), \quad M \gg M_S. \tag{2}$$

In words, the asymptotic multiplicity of particles of mass M grows exponentially with their mass. The infinite tower of new massive states acts as a cutoff in the sense that, by scattering string states of energy larger than M_S, one does not probe shorter and shorter distances, but, rather, one produces new states – i.e., the extra energy pumped into the system increases its entropy instead of probing shorter distances.

Superstring theory does not solve all problems of quantum gravity. Its standard formulation is a perturbative prescription for computing S-matrix elements. Thus, string theory does not easily answer questions about the eventual resolution of General Relativity singularities, or whether quantum gravity induces quantum decoherence. Only very recently have we been able to extract (some) information about these non-perturbative aspects of gravity. On the other hand, many perturbative questions, unanswerable in Einstein's theory, *are* directly answered by strings. Among them: perturbative corrections to particle scattering at Planckian energies [5]; non-minimal corrections to the gravitational coupling of high-spin, massive particles [6].

Let us come back to unification. All unification programs have one aim in common: to explain why the parameters governing fundamental interactions are what they are, possibly by a dynamical mechanism. In string theory *all* physical parameters are vacuum expectation values of some scalar field; even the string coupling constant itself, which is the VEV of the 10-d dilaton. Thus, the possibility exists of a dynamical determination of all fundamental constants of nature. This possibility is still remote, though, because the high-energy (stringy) dynamics that would fix these parameters is still unknown. It is typically non-perturbative.

Even in the absence of a full understanding of string dynamics, the structure of physics below the string scale is nevertheless constrained by approximate symmetries, broken only at scales much smaller than M_S. The most powerful such symmetry is supersymmetry. It is an extension of the Poincaré symmetry that contains, besides the usual Poincaré generators, some fermionic generators, Q^I_α, where α is a spinor index and $I = 1, \ldots, N$ is a replica (extension) index. The supercharges Q^I_α transform as spinors under Lorentz transformations, commute with the momentum P_m and *anticommute* among themselves:

$$\{Q^I_\alpha, Q^J_\beta\} = \delta^{IJ}(C\gamma^m)_{\alpha\beta} P_m + \sum_K (C\gamma^{m_1} \ldots \gamma^{m_K})_{\alpha\beta} Z^{IJ}_{m_1 \ldots m_K}. \tag{3}$$

We will have more to say about the central charges $Z^{IJ}_{m_1 \ldots m_K}$ later on. Here, let us notice that, obviously, the larger the algebra, the more constrained is the low-energy effective action. Unfortunately, while $N = 1$ supersymmetry allows for particles in chiral representations of the gauge group, all other supersymmetries imply that matter is in real representations. This is incompatible with the standard model. Only recently a model was found with $N = 2$ supersymmetry, in which the spontaneous breakdown of supersymmetry also gives mass to the fermions with the "wrong" chirality under $SU(2) \times U(1)$ [7].

Much is still to be done before we will be able to find a true dynamical mechanism that fixes the coupling constants of elementary interactions. Rather than pursuing this problem, let us look at the other aspect of Einstein's unification, not shared by other, more modest programs; namely, the idea that elementary particles should appear as solitons of the unified theory.

Even in this case "reality" is stranger than fiction: string theory contains both elementary particles and solitons, and the distinction between the two depends on the dynamical regime of the theory! What distinguishes an elementary particle from a soliton is that the classical size of a soliton, L_C, is much larger than its Compton wavelength, i.e., its inverse mass: $L_C \gg \lambda_{\text{Compton}} = 1/M$. For an elementary particle, the opposite limit holds. In a certain class of supersymmetric field theories, $N = 4$ super Yang–Mills, obtained as low-energy limit of superstrings, we know *exact* formulae [8] for the soliton mass, valid even at strong coupling. The solitons, in this case, are magnetic monopoles, and their mass and classical size can be related to the VEV of a scalar field, Φ, and to the gauge coupling constant, g, as follows:

$$L_C = \frac{1}{g\langle\Phi\rangle}, \quad \lambda_{\text{Compton}} = \frac{g}{\langle\Phi\rangle}. \tag{4}$$

At weak coupling, $g \ll 1$, $L_C \gg \lambda_{\text{Compton}}$, and the monopole is a classical soliton, while at strong coupling, $g \gg 1$, $L_C \ll \lambda_{\text{Compton}}$, and the monopole is to all effects an elementary particle. It is also the true, light, charged degree of freedom at strong coupling, since its mass is $M_{mag} = \langle\Phi\rangle/g$, while electrically charged "elementary" particles have mass $M_{el} = g\langle\Phi\rangle \gg M_{mag}$. In other words, in going from weak to strong coupling, "elementary" particles and solitons have exchanged roles, and the strong-coupling theory can be re-interpreted as a weakly coupled $N = 4$ theory with monopoles as elementary charged degrees of freedom, and coupling constant $g_{mag} \equiv 1/g$. This is the simplest example of strong-weak coupling duality [9].

In the full superstring theory, the picture is even more interesting. Indeed, besides the elementary string states, of mass $O(M_S)$, there exist additional non-perturbative states, missed by our previous analysis.

To be specific, let us study what happens in the so-called type IIA superstring, starting with the weak-coupling regime first. There, besides the fundamental string, with tension $T = M_S^2$, there exist five-dimensional classical solitons (the five-branes) with tension – i.e., energy per unit area – $T_5 = g_S^{-2} M_S^6$, and other extended objects, with even dimensions ranging from 0 to 6: the D-branes [10]. D-branes have tension $T = g_S^{-1} M_S^{D+1}$; thus, they are heavier than elementary string states but lighter than solitons. A D-brane is charged under the central charge $Z^{IJ}_{m_1 \cdots m_D}$. In the absence of extended objects, the only central charge that can be nonzero is the scalar one, $D = 0$ [11].

At strong coupling, $g_S \gg 1$, the mass of the D0 branes is less than that of the elementary string states; thus, the true "elementary" degrees of freedom for type IIA superstring at strong coupling are not strings, but rather charged particles and the five-brane! A careful analysis of the theory reveals that the effective theory of the type IIA superstring at strong coupling is 11-d supergravity [12], which, indeed, contains no strings among its low-energy degrees of freedom. In this case the strong coupling dual of our theory is a theory that looks completely different from the weak-coupling one, unlike the case of $N = 4$ super Yang–Mills.

Various strong-weak coupling dualities can be performed on all five superstring theories in maximal dimension (d=10), with a very pleasant result [10]. While perturbatively these five theories are distinct, non-perturbatively they can be seen as different (dual) limits of each other. They are just different facets of a yet-to-be-determined underlying theory: M-theory.

A satisfactory description of M-theory is lacking today; the best that we can do is to study some of its limiting forms.

One particularly interesting limit is scattering at c.o.m. energies $E \gg M_{\text{Planck}}$. In this limit General Relativity is classical, in the sense that quantum effects start to be relevant when the colliding particles are within a Compton wavelength of each other, but this region, of radius $\lambda_{\text{Compton}} = 1/E$, is hidden beyond the Schwarzschild horizon, of radius $R \sim E/M^2_{\text{Planck}} \gg \lambda_{\text{Compton}}$ [13]. Now, since the entropy of a black hole is proportional to its area, and since the entropy counts the number of degrees of freedom of a physical system, it seems that our ultra-Planckian scattering process can be described by far fewer degrees of freedom than one would expect from local field theory. Indeed, in field theory, the maximum number of degrees of freedom necessary to describe a process taking place in a volume V is proportional to the volume, while here it seems that a number of degrees of freedom proportional to the area bounding the volume would suffice. This fact has led 't Hooft and Susskind [13] to the *holographic conjecture*, stating that there should exist a formulation of M-theory (or any other complete theory of quantum gravity) in which all the (non-redundant) degrees of freedom necessary to describe a system enclosed in a volume V live on the boundary of that volume[1]. The possibility of completely describing physical processes happening in space-time in terms of a theory living only at the boundary of space-time has been given a precise meaning when space-time has a negative cosmological constant [14]. I shall proceed no further on this avenue but rather refer to the original literature in this exciting and fast-developing field.

Today, string theory is evolving at a fast pace and undoubtedly many new surprises are in store in the near future, surprises that may shed completely new light on long-standing problems that have existed in General Relativity since its appearance, such as the information loss puzzle in quantum black holes, the fate of singularities, the ultimate structure of a unified theory, and the dynamical determination of all fundamental parameters of nature. Some of these problems Einstein himself tried to answer; it is a tribute to the vitality of string theory and theoretical physics today, that many of the as yet partial answers to these questions which we discussed here are much more intriguing and exotic than he thought.

Acknowledgements

This work is supported in part by NSF grant no. PHY-9722083.

[1] Whence the name, as a 2-d hologram completely encodes a 3-d image.

References

1. Pais A. (1982): Subtle is the Lord: The Science and the Life of Albert Einstein, Oxford University Press, Oxford
2. Bell J.S. (1964): Physics (Long Island City, NY) **1**, 195
3. Cornwall J.M., Levin D.N., Tiktopoulos G. (1974): Phys. Rev. D **10**, 1145; Erratum (1975) ibid. D **11**, 972
4. Scherk J., Schwarz J.H. (1974): Phys. Lett. B **52**, 347
5. Schwarz J.H. (1982): Phys. Rep. **89**, 223
6. Giannakis I., Liu J.T., Porrati M. (1999): Phys. Rev. D **59**, 104013 (sic)
7. Girardello L., Porrati M., Zaffaroni A. (1997): Nucl. Phys. B **505**, 272
8. Montonen C., Olive D. (1977): Phys. Lett. B **72**, 117
9. Girardello L., Giveon A., Porrati M., Zaffaroni A. (1995): Nucl. Phys. B **448**, 127
10. Polchinski J.: TASI Lectures on D-Branes, hep-th/9611050
11. Haag R., Lopuszanski J.T., Sohnius M. (1975): Nucl. Phys. B **88**, 257
12. Witten E. (1995): Nucl. Phys. B **443**, 85, hep-th/9503124
13. 't Hooft G. (1993): Dimensional reduction in quantum gravity, in *Salamfestschrift: a collection of talks*, ed. by A. Ali, Y. Ellis, S. Randjabar-Daemi, World Scientific, Singapore; Susskind, S. (1995): J. Math. Phys. **36**, 6377
14. Maldacena J. (1998): Adv. Theor. Math. Phys. **2**, 231, hep-th/9711200;
    Gubser S.S., Klebanov I.R., Polyakov A.M. (1998): Phys. Lett. B **428**, 105, hep-th/9802109;
    Witten E. (1998): Adv. Theor. Math. Phys. **2**, 253, hep-th/9802150

Virgo – a Laser Interferometer for Gravitational Wave Detection

C. Bradaschia

Abstract. The Virgo interferometer is described, together with the status of its preparation.

1 Virgo

Virgo [1] will be a recycled Michelson interferometer, powered by a 25 W Nd:YAG laser ($\lambda = 1.06\,\mu$m) having in each arm one 3 km Fabry-Perot resonant cavity. The Virgo Project is under construction in Italy at Cascina, near Pisa, by CNRS (France) and INFN (Italy). The participating laboratories are: ESPCI-Paris, LAL-Orsay, LAPP-Annecy, IPN-Lyon, Observatoire de la Côte d'Azur-Nice for CNRS and Firenze, Frascati, Napoli, Perugia, Pisa, Roma for INFN. The aim is to attain a sensitivity allowing the detection of gravitational waves emitted by coalescing binary compact stars, gravitational collapses and spinning neutron stars or constituting the stochastic background gravitational radiation. Such sensitivity is at the limits of present technologies, between a few Hz and a few kHz. An introduction to gravitational waves and methods of detection can be found in other contributions to this conference. The Virgo sensitivity curve is shown in Figure 1, as a function of frequency; the vertical scale gives the linear spectral density of the minimum detectable dimensionless gravitational amplitude. Thanks to the exceptional frequency and power stability of the laser and to the quality of the mirrors, all noises of optical origin are kept at a very low level in the frequency region where they dominate (above about 50 Hz). All the optical components will be suspended in vacuum by anti-seismic Super Attenuators (SA); the optical path will be completely under vacuum. The vacuum system will consist of: two orthogonal vacuum tubes, 3 km long, 1.2 m in diameter, containing the arm optical paths; one vacuum tube, 144 m long, 0.3 m in diameter for the mode cleaner cavity; 10 vertical vacuum towers, up to 11 m high, 2 m in diameter, containing the different SA's.

The construction of Virgo is planned in two main steps: the realization of the Central Interferometer by early 2000, and the realization of the 3 km arm Full Interferometer, by early 2002. This strategy is the consequence of technical and scientific choices and of budgetary constraints. The Central Interferometer will consist of all the Virgo final components in the central area, including the 144 m mode cleaner. It will operate as a simple Michelson interferometer where the arm end mirrors will be installed at the place of the input mirrors of the future 3 km Fabry–Perot cavities. The vacuum system, suspensions and electronics will be permanent; only the mirrors will be provisional, for quality and dimensions; the two 3 km vacuum tubes will be

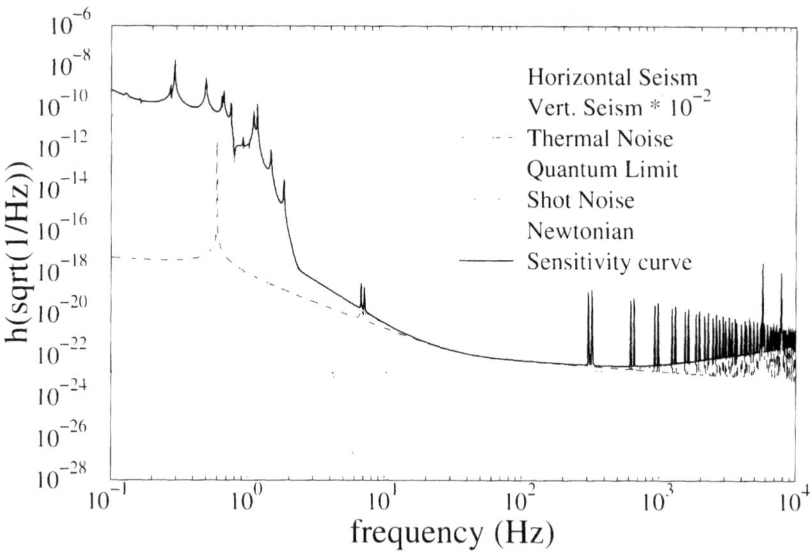

Fig. 1. Virgo sensitivity curve

in construction. The expected results of the first step are: the complete development and test of the methods for aligning and controling a large suspended interferometer; the commissioning of the read-out and data acquisition system; initial experience of the different noises contained in the detected signal. This procedure should allow the Full Interferometer to be put into operation very rapidly, after the completion of the two arms in early 2002. The optical scheme of the Full Interferometer is shown in Figure 2.

2 Virgo description and status of preparation

Infrastructure

This part includes all the buildings and such equipment as power distribution, telephones, white rooms, etc. All the buildings are essentially above ground, including the "tunnels" containing the vacuum tubes. All the buildings of the central area are presently finished; they have been occupied and the installation of the apparatus for the Central Interferometer has been initiated. The central area buildings are:

- Central building, containing the core of the Full Interferometer, hence the whole Central Interferometer (the laser, 6 optical elements and their SA's, the read-out electronics);
- Control building, containing the computer system and the control room, besides a meeting room and some office space;

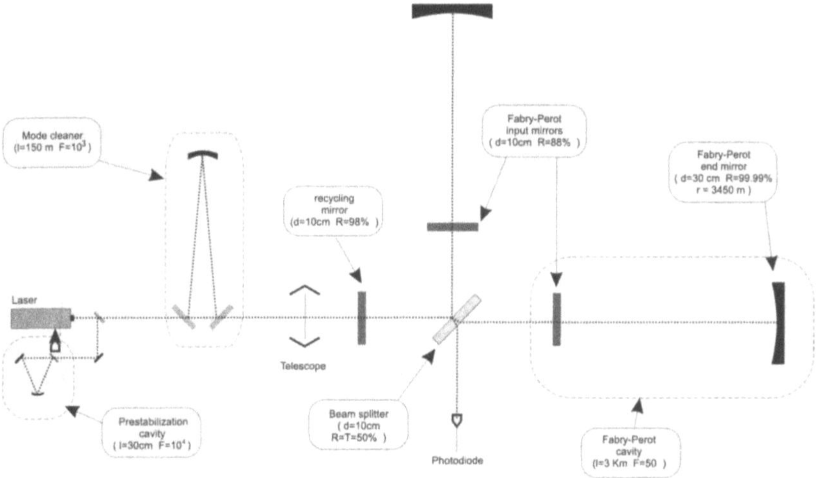

Fig. 2. Virgo optical scheme

- Mode Cleaner building, containing the mode cleaner mirror inside its vacuum tower;
- Mode Cleaner tunnel, containing the mode cleaner vacuum tube;
- Technical building, containing electrical power generators and transformers together with hot and chilled water generators.

All the buildings have been designed so as to avoid any mechanical resonance in their structures; this is necessary to avoid spoiling the attenuation performance of the SA's. The buildings' foundations rest on piles driven into the ground as deep as 30 m in order to reduce the effects of soil subsidence to an acceptable level. The call for tender of the arm infrastructure (3 km tunnels, intermediate and terminal buildings) has been completed; construction started in April 1999; the buildings are expected to be completed in early 2001.

Vacuum system

The most relevant chacteristic of the vacuum system is its dimension; the evacuated volume is one order of magnitude larger than those of the largest particle accelerators in the world. The pressure required is 10^{-9} mbar or less and an extremely low partial pressure of hydrocarbons is required (10^{-14} mbar) in order to protect the optical components from pollution. In order to achieve these performances with a relatively low cost pumping system, Virgo has developed [2] a thermal treatment of the vacuum system components (400°C in air) enabling the attainment of an outgassing rate value as low as 5×10^{-15} mbar l s^{-1} cm^{-2}. This is two orders of magnitude better than typical Ultra High Vacuum systems. The eight towers for the Central Interferometer have been installed; three of them and the Mode Cleaner tube are already under vacuum. The two arm terminal towers are expected to be ready in due course. The vacuum tube realization has been started, its installation is expected to be completed in early 2002, shortly after completion of the last (West) arm infrastructure. Baffles to stop the propagation of light diffused by the mirrors will be installed inside the towers (glass baffles) and the tubes (stainless steel baffles). The pumping system for the Central Interferometer is being installed. The rest will be ordered as needed.

Interferometer

The 10 W laser for the Central Interferometer is running in the laboratory without problems; the stability requirements have been met. To reach the power (25 W) required for the Full Interferometer, two or more 10 W lasers will be added coherently. The mirror requirements have been attained, as far as reflectivity and roughness are concerned, on small diameters (50–80 mm), sufficient for the Central Interferometer. A new facility able to handle larger mirrors has been built at Lyon; as soon as it is in operation, good performances are also expected for the large mirrors needed for the Full Interferometer. A completely new facility has been set up for the metrology of mirror surfaces and substrates, with the required performances on the large diameters (more than 30 cm) required for the Full Interferometer. Input and detection benches, together with special suspension elements, are quite advanced as regards both mechanics and optics. A detailed description of the optical aspects of interferometers for gravitational wave detection and in particular of Virgo can be found in reference [3].

Suspensions

Each full height SA [4] is made of one three-leg (6 m high) inverted pendulum, supporting a chain of pendula, each containing cantilever blades for vertical seismic filtering. The last three elements of each chain are: steering filter, "marionetta" and mirror (or optical table); they are coupled through magnetic actuators to allow alignment and control of the interferometer optical components. The SA's, providing a seismic attenuation of more than a factor of 10^{12} above a few Hz, allow the extension of the gravitational wave detection range down to this frequency region. All the different elements composing the SA's have already been ordered; several elements have already been delivered. The first final full size SA has been assembled in the laboratory, in Pisa, using serial production elements. The measured attenuation performances largely satisfy the requirements. This SA made possible the certification of the assembly procedure and is being used to tune the control system. Later, in parallel to the SA's assembly in Cascina, it will be used to advance developments and to make thermal noise studies. The assembly and installation of SA's on site started in April 1999 and will last until end 1999. This includes all the electronics and cabling for local control of the suspensions.

Electronics and software

The control and read-out electronics is expected to be available in due course. Most of the designs are frozen; final prototypes have been accepted; a few are still under evaluation. Several subsystems are ready and working: Local Readout, Frame Builder, Vacuum Control, Buildings Control, Data Distribution [5]. The lay-out of the computer system has been approved for some time; a first set of workstations has been received and has been put into operation together with data handling and a network of PC's. The data formatting and the corresponding software has been agreed with LIGO (the similar project in the USA) and a first joint release has been made [6]. A simulation program has been developed in the last few years for both the optical part and the suspensions [7]. It is quite satisfactory and is used for setting up the interferometer control system and for preparation of the analysis programs.

3 Conclusions

If the Virgo development strategy and plan are held to, commissioning will start in early 2002, with the aim of attaining the design sensitivity in a few years. It has to be mentioned that vigorous preparation for data analysis has been started recently; it is expected to reach working status as the same time as the completion of the detector.

References

1. The Virgo Collaboration (1997): Final Design Report, Pisa
2. Bernardini M., Braccini S., Bradaschia C. (1998): Journ. Vac. Sci. Technol. A **16**, (1)
3. Hello P. (1998): Optical aspects of interferometric gravitational wave detectors, in *Progress in Optics*, Vol. XXXVIII, Elsevier, Amsterdam, pp. 85–164
4. Braccini S., Bradaschia C., Del Fabbio R. et al. (1996): Rev. Sci. Instrum. **67** (8), 2899–2902;
   Beccaria M., Bernardini M., Bougleux E. et al. (1997): Nucl. Instr. and Meth. in Phys. Res. A **394**, 397–408;
   De Salvo R., Di Virgilio A., Enrico A. et al. (1999): Nucl. Instr. and Meth. in Phys. Res. A **420**, 316–335;
   Losurdo G., Bernardini M., Braccini S. et al. (199): Rev. Sci. Instrum. **70** (5), 2507–2515
5. Barone F., Garufi F., Milano L. et al. (1997): Rev. Sci. Instrum. **68** (10), 3907–3913
6. Mours B. et al. (1996): A common data format for gravitational wave interferometers, in *Proc. of the TAMA International Workshop on Gravitational Wave Detection*, ed. by K. Tsubono, M.K. Fujimoti, K. Kurda, Universal Academy Press, Tokyo
7. Caron B., Derome L., Flaminio R. et al. (1999): SIESTA, a time domain, general purpose simulation program for the VIRGO experiment, Astroparticle Physics **10**, 369–386

A Space Test of General Relativity

I. Ciufolini

Abstract. In this paper we describe a test in space of general relativity and gravitational physics: the measurement of this effect, obtained in 1995–98, using two laser-ranged satellites. We first briefly present the gravitomagnetic field and the Lense–Thirring effect in general relativity. We then review the method proposed to measure this effect by analyzing the orbits of the two laser-ranged satellites LAGEOS and LAGEOS II, and we report on the corresponding measurement of the Lense–Thirring effect by analyzing the data of LAGEOS and LAGEOS II with the orbital programs GEODYN-SOLVE, using the Earth gravitational models JGM-3 and EGM-96 and our method. The first result was obtained in 1995, and the most accurate measurements were obtained in 1998 using EGM-96, with an estimated uncertainty of about 20%–30%.

1 Spin and gravitomagnetism

The contention of Ernst Mach that inertial forces arise from rotations and accelerations with respect to the distant masses in the universe [1, 2] was one of the ideas inspiring Einstein to develop the general theory of relativity. General relativity satisfies this "Mach principle" only in part, nevertheless, Einstein's theory predicts the occurrence of peculiar phenomena in the vicinity of a spinning body, caused by its rotation, which may be considered as weak manifestations of Mach's ideas on the origin of inertia and which have not yet been measured [2, 3][1].

When a clock that co-rotates very slowly around a spinning body returns to its starting point, it finds itself advanced relative to a clock kept there at "rest" (with respect to the "distant stars"). Indeed, synchronization of clocks all around a closed path near a spinning body is not possible, and light co-rotating around a spinning body would take less time to return to a fixed point than light rotating in the opposite direction [2]. Similarly, the orbital period of a particle co-rotating around a spinning body would be longer than the orbital period of a particle counter-rotating on the same orbit. Furthermore, a particle orbiting around a spinning body will have its orbital plane "dragged" around the spinning body in the same sense as the rotation of the body, and small gyroscopes that determine the axes of a local, freely falling, inertial frame, where "locally" the gravitational field is "unobservable", will rotate with respect to the"distant stars" because of the rotation of the body. This phenomenon – called "dragging of inertial frames" or, more simply, "frame dragging", as Einstein

[1] In the famous gedanken experiment of Newton concerning a rotating bucket filled with water, the question was: "what is the origin of the inertial forces that curve the surface of the water in the bucket?" "Are these forces produced by the relative rotation of the bucket with respect to absolute space or with respect to the other masses of the universe?"

called it – is also known as the Lense–Thirring effect [1–3]. In Einstein's general theory of relativity, all of these phenomena are the result of the rotation of the central mass.

Rotation, inertia, and the "fictitious" inertial forces arising in a rotating system have been central issues and problems of mechanics since the time of Galileo and Newton. Mach thought that centrifugal forces were the result of rotation with respect to the masses in the universe, and Einstein's development of the general theory of relativity was influenced by Mach's ideas on the origin of inertia and inertial forces. Today, the level at which general relativity satisfies Mach's ideas on inertia is still debated and discussed. However general relativity satisfies at least a kind of "weak manifestation" of Mach's ideas: the dragging of inertial frames [2]. Indeed, in Einstein's gravitational theory, the concept of an inertial frame has only a local meaning, and a local inertial frame is "rotationally dragged" by mass-energy currents because moving masses influence and change the orientation of the axes of a local inertial frame (that is, the gyroscopes); thus, a current of mass such as the spinning Earth "drags" and changes the orientation of the gyroscopes with respect to the distant stars.

To understand these phenomena of general relativity associated with the rotation of a mass, one may use a formal analogy with the classical theory of electromagnetism. Newton's law of gravitation has a formal counterpart in Coulomb's law of electrostatics; however, Newton's theory has no phenomenon formally analogous to magnetism. On the other hand, Einstein's theory of gravitation predicts that the force generated by a current of electrical charge, described by Ampere's law, should also have a formal counterpart "force" generated by a current of mass. The detection and measurement of this "gravitomagnetic" force is the subject of this paper.

In general relativity, for a localized, stationary, mass-energy distribution, in the weak-field and slow-motion limit, the three "vector" components of the metric tensor are given by: $\mathbf{h} \cong -2(\mathbf{J} \times \mathbf{x})/r^3$, where $\mathbf{h}$ is known as the gravitomagnetic potential. The gravitomagnetic field $\mathbf{H}$ is given by $\mathbf{H} = \nabla \times \mathbf{h} \cong 2\,[\frac{\mathbf{J} - 3(\mathbf{J} \cdot \hat{\mathbf{x}})\,\hat{\mathbf{x}}}{\mathbf{r}^3}]$. By using the geodesic equation, in the weak-field and slow-motion limit, one then has, for a test particle: $m\,\frac{d^2\mathbf{x}}{dt^2} \cong m\,(\mathbf{G} + \frac{d\mathbf{x}}{dt} \times \mathbf{H})$, where $\mathbf{G} \cong -\frac{M}{\mathbf{r}^2}\,\hat{\mathbf{x}}$ is the standard Newtonian acceleration. Thus, a central object with angular momentum $\mathbf{J}$, via the second term in the "force", $\dot{\mathbf{x}} \times \mathbf{H}$, drags the orbital plane (and the orbital angular momentum) of a test particle in the sense of rotation of the central body. This dragging of the whole orbital plane is described by formula (2), below, for the rate of change of the longitude of the nodes. Similarly, by integrating the above equation of motion of the test particle, one finds formula (3), below, for the secular rate of change of the longitude of the pericenter. On a gyroscope with angular momentum $\mathbf{S}$ there is a torque $\tau \cong (1/2)\,\mathbf{S} \times \mathbf{H} = d\mathbf{S}/dt \equiv \dot{\Omega} \times \mathbf{S}$; therefore, the gyroscope precesses with respect to an asymptotic inertial frame with angular velocity $\dot{\Omega}$ given by formula (1) below; on the gyroscope there is also a force exerted by the gravitomagnetic field.

Thus, an external current of mass, such as the spinning Earth, "drags" and changes the orientation of gyroscopes. Indeed, a test gyroscope precesses with respect to "an asymptotic inertial frame" with angular velocity:

$$\dot{\Omega} = -\frac{1}{2}\mathbf{H} = \frac{[-\mathbf{J} + 3(\mathbf{J}\cdot\hat{\mathbf{x}})\hat{\mathbf{x}}]}{|\mathbf{x}|^3}, \tag{1}$$

where **J** is the angular momentum of the central object, **H** its gravitomagnetic field generated by **J**, and **x** the gyroscope's position vector. This formula quantifies the Lense–Thirring effect for a gyroscope [1, 2].

The whole orbital plane of a test particle is itself a kind of enormous gyroscope (for a motion under a central force) dragged by the gravitomagnetic field. Indeed, the orbit of a test particle around a central body with angular momentum **J** has a secular rate of change of the longitude of the line of the nodes (intersection between the orbital plane of the test particle and the equatorial plane of the central object), discovered by Lense–Thirring [3]:

$$\dot{\Omega}^{\text{Lense–Thirring}} = \frac{2\mathbf{J}}{[a^3(1-e^2)^{3/2}]}, \tag{2}$$

where a is the semimajor axis of the test particle, and e its orbital eccentricity. The pericenter of an orbiting test particle is also a kind of enormous gyroscope (for a motion under a central force $\sim 1/r^2$). Indeed, the argument of the pericenter, ω, that is, the angle from the equatorial plane to the pericenter, and thus the longitude of the pericenter $\tilde{\omega} = \Omega + \omega$ (determining the Runge-Lenz vector), have a secular rate of change:

$$\dot{\tilde{\omega}}^{\text{Lense–Thirring}} = \frac{2J\ (\hat{\mathbf{J}} - 3\ \cos I\ \hat{\mathbf{l}})}{[a^3(1-e^2)^{3/2}]}, \tag{3}$$

where $\hat{\mathbf{l}}$ is the orbital angular momentum, unit vector, of the test particle, and I its orbital inclination (angle between the orbital plane and the equatorial plane of the central object): $\cos I \equiv \hat{\mathbf{J}} \cdot \hat{\mathbf{l}}$.

In astrophysics, a supermassive spinning black hole and its gravitomagnetic field might explain the constant orientation of the long jets of matter ejected from the cores of quasars and active galactic nuclei [4]. These jets have emission times that may reach millions of years.

To characterize the gravitomagnetic field generated by the intrinsic angular momentum of a body and the Lense–Thirring effect, and to distinguish it from other relativistic phenomena such as the de Sitter effect – due to the motion of a local inertial frame through a static gravitational field – one may give a description of this gravitomagnetic field in terms of spacetime-curvature invariants [2, 5]. The pseudoinvariant $^*\mathbf{R}\cdot\mathbf{R} = \frac{1}{2}\varepsilon^{\alpha\beta\sigma\rho}R_{\sigma\rho}{}^{\mu\nu}R_{\alpha\beta\mu\nu}$, built from the Riemann tensor **R** and its dual $^*\mathbf{R}$, gives an invariant characterization of this gravitomagnetic phenomenon since it is non-zero in the field of a central body if and only if the body is rotating. Indeed the pseudoinvariant $^*\mathbf{R}\cdot\mathbf{R} \sim \frac{JM}{r^7}\cos\theta$ is proportional to the angular momentum of the central body. Therefore, using this invariant, one can determine whether or not there is a contribution of angular momentum, and mass currents, to spacetime geometry and curvature. Thus, one may describe this gravitomagnetic effect, as that phenomenon of nature which is an "absolute" (coordinate-independent) feature of general relativity such that *spacetime curvature is generated by the spin of a body* [2, 5].

Thus, the direct measurement of the Lense–Thirring effect described in Sect. 2 confirms one of the remaining fundamental predictions of general relativity, that a current of mass-energy, such as a spinning mass, as a result of its mass-energy-motion, changes the geometry of the universe [2] by generating spacetime curvature.

2 Measurement of the Lense–Thirring effect

Our detection and measurement of the Lense–Thirring effect, one of the phenomena due to the gravitomagnetic field, was obtained by using the satellite laser-ranging data of the satellites LAGEOS (LAser GEOdynamics Satellite, NASA) and LAGEOS II (NASA and ASI, the Italian Space Agency) and the Earth gravitational field models, JGM-3 and EGM-96.

The measurement of distances has always been a fundamental issue in astronomy, engineering, and science in general. So far, laser ranging has been the most accurate technique for measuring the distances to the moon and artificial satellites [6, 36]. Short-duration laser pulses are emitted from lasers on Earth, aimed at the target through a telescope, and then reflected back by optical cube-corner retroreflectors on the moon [8] or the artificial satellite [9], such as LAGEOS. By measuring the total round-trip travel time, one can determine the distance to a retroreflector on the moon with an accuracy of about 2 cm and to the LAGEOS satellites with a few millimeters accuracy.

The LAGEOS satellites are heavy brass and aluminum satellites, of about 406 kg weight, completely passive and covered with retroreflectors, orbiting at an altitude of about 6,000 km above the surface of Earth. LAGEOS, launched in 1976 by NASA, and LAGEOS II, launched by NASA and ASI in 1992, have an essentially identical structure but they have different orbits. The semimajor axis of LAGEOS is $a \cong 12,270$ km, the period $P \cong 3.758$ hr, the eccentricity $e \cong 0.004$, and the inclination $I \cong 109.9°$. The semimajor axis of LAGEOS II is $a_{II} \cong 12,163$ km, the eccentricity $e_{II} \cong 0.014$, and the inclination $I_{II} \cong 52.65°$.

We analyzed the laser-ranging data using the principles described in [10] and adopted the IERS conventions [11] in our modeling, except that, in the 1998 analysis, we used the recently released static and tidal EGM-96 model [12]. Error analysis of the LAGEOS orbits indicated that the EGM-96 errors can only contribute periodic root-sum-square errors of 2 to 4 mm radially, and in all three directions they do not exceed 10 to 17 mm. The initial positions and velocities of the LAGEOS satellites were adjusted for each 15-day batch of data, along with a constant and a variable acceleration in the along-track direction (to account for the residual drag at the satellites' altitude) and very small variations in their reflectivities. Solar radiation pressure, Earth albedo, and anisotropic thermal effects were also modeled [13–16]. In modeling the thermal effects, the orientation of the satellite spin axis was obtained from [17]. Lunar, solar and planetary perturbations were also included in the equations of motion, formulated according to Einstein's general theory of relativity with the exception of the Lense–Thirring effect, which was purposely set to zero. All of the tracking station coordinates were adjusted (accounting for tectonic motions) except for those defining the TRF terrestrial reference frame. Polar motion was also adjusted, and Earth's rotation was modeled from the very long baseline interferometry-based series SPACE96 [18]. We analyzed the orbits of the LAGEOS satellites using the orbital analysis and data reduction software GEODYN II [19].

The node and perigee of LAGEOS and LAGEOS II are dragged by the Earth's angular momentum. From the Lense–Thirring formula (2), we get $\dot{\Omega}_I^{\text{Lense–Thirring}} \cong$ 31 mas/yr and $\dot{\Omega}_{II}^{\text{Lense–Thirring}} \cong 31.5$, where mas is a millisecond of arc. The argument of pericenter (perigee in our analysis), ω, also has a Lense–Thirring drag. From formula (3) we get for LAGEOS: $\dot{\omega}_I^{\text{Lense–Thirring}} \cong 32$ mas/yr, and for LAGEOS II: $\dot{\omega}_{II}^{\text{Lense–Thirring}} \cong -57$ mas/yr. The nodal precessions of LAGEOS and LAGEOS II can be determined with an accuracy of the order of 1 mas/yr. Over our total observational period of about 4 years, we obtained a root mean square of the node residuals of about 4 mas for LAGEOS and about 7 mas for LAGEOS II. For the perigee, the observable quantity is the product $e \cdot a \cdot \dot{\omega}$, where e is the orbital eccentricity of the satellite. Thus, the perigee precession $\dot{\omega}$ for LAGEOS is difficult to measure because its orbital eccentricity e is $\sim 4 \times 10^{-3}$. The orbit of LAGEOS II is more eccentric, with $e \sim 0.014$, and the Lense–Thirring drag of the perigee of LAGEOS II is almost twice as large in magnitude as that of LAGEOS. Over about 4 years, we obtained a root mean square of the residuals of the LAGEOS II perigee of about 25 mas, whereas the total Lense–Thirring effect on the perigee is, over 4 years, $\cong -228$ mas.

To precisely quantify and measure the gravitomagnetic effects we have introduced the parameter μ that is by, definition, 1 in general relativity [2] and 0 in Newtonian theory.

The main error in this measurement is due to the uncertainties in the Earth's even zonal harmonics and their time variations. The unmodeled orbital effects due to the harmonics of lower order are comparable to, or larger than, the Lense–Thirring effect. However, by analyzing both the JGM-3 and the EGM-96 models with their uncertainties in the even zonal harmonic coefficients, and by calculating the secular effects of these uncertainties on the orbital elements of LAGEOS and LAGEOS II, we find [20] that the main sources of error in the determination of the Lense–Thirring effect are concentrated in the first two even zonal harmonics, J_2 and J_4. We can, however, use the three observable quantities $\dot{\Omega}_I$, $\dot{\Omega}_{II}$ and $\dot{\omega}_{II}$ to determine μ [20, 21], thereby avoiding the two largest sources of error, those arising from the uncertainties in J_2 and J_4. We do this by solving the system of three equations for $\delta\dot{\Omega}_I$, $\delta\dot{\Omega}_{II}$ and $\delta\dot{\omega}_{II}$ in three unknowns μ, J_2 and J_4, obtaining:

$$\begin{aligned} &\delta\dot{\Omega}_I^{\text{Exp}} + c_1\delta\dot{\Omega}_{II}^{\text{Exp}} + c_2\delta\dot{\omega}_{II}^{\text{Exp}} \\ &= \mu(31 + 31.5c_1 - 57c_2)\ \text{mas/yr} + \text{ other errors } \cong \mu(60.2\ \text{mas/yr}), \end{aligned} \tag{4}$$

where $c_1 = 0.295$ and $c_2 = -0.35$. Equation (4) for μ does not depend on J_2 and J_4 nor on their uncertainties; thus, the value of μ that we obtain is unaffected by the largest errors, which are due to δJ_2 and δJ_4, and is sensitive only to the smaller errors due to δJ_{2n}, with $2n \geq 6$.

Similarly, regarding tidal, secular, and seasonal changes in the geopotential coefficients, the main effects on the nodes and perigee of LAGEOS and LAGEOS II, caused by tidal and other time variations in Earth's gravitational field [22, 23], are due to changes in J_2 and J_4. However, the tidal errors in J_2 and J_4 and the errors resulting from other unmodeled, medium and long period, time variations in J_2 and J_4, including their secular and seasonal variations, are eliminated by our combination of residuals of nodes and perigee. In particular, most of the errors resulting from the 18.6- and 9.3-year tides, associated with the lunar node, are eliminated in our measurement. An extensive discussion of the various error sources that can affect the result is given in [23]; only a brief discussion of the error sources is given here.

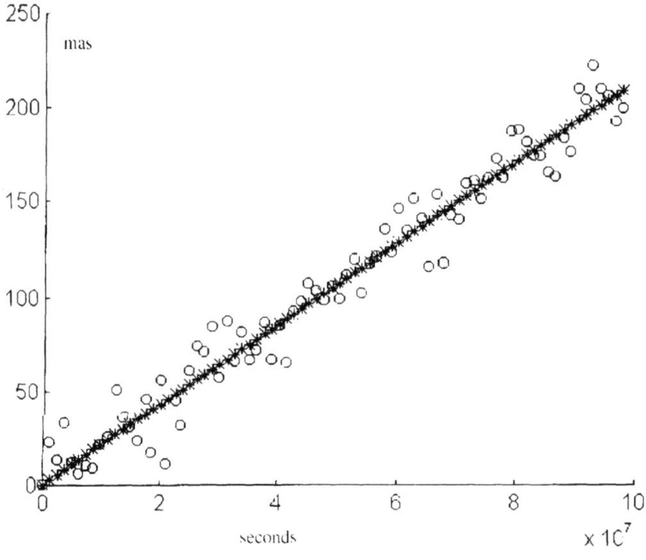

Fig. 1. Sum of the residuals (o) of the nodes of LAGEOS and LAGEOS II and perigee of LAGEOS II from November 1992 to December 1995, using the Earth gravitational model JGM-3 and with the method described in the text. On the vertical axis, we plotted: (node residuals of LAGEOS)+0.295(node residuals of LAGEOS II)- 0.35(perigee residuals of LAGEOS II). The best-fit line (*) shown through these combined residuals has a slope of about $\mu_{\text{Measured}} \cong 1.1$ (compared with $\mu_{\text{GR}} \equiv 1$ in general relativity). Together with the slope we fitted and removed ten periodic signals from the combined residuals. The units on the vertical axis are milliseconds of arc (mas), and on the horizontal axis seconds of time

In Fig. 1, we show the linear combination of the residuals of the nodes of LAGEOS and LAGEOS II and perigee of LAGEOS II according to Eq. (4) to eliminate the δJ_2

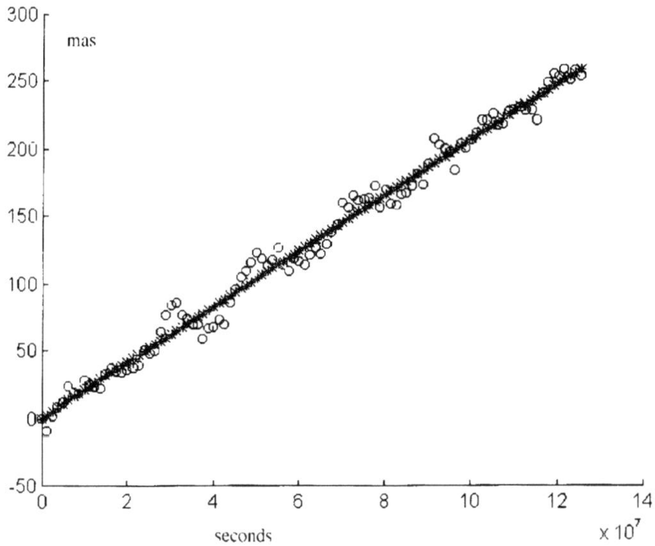

Fig. 2. Sum of the residuals (o) of the nodes of LAGEOS and LAGEOS II and perigee of LAGEOS II from January 1993 to January 1997. On the vertical axis, we plotted: (node residuals of LAGEOS)+0.295(node residuals of LAGEOS II)- 0.35(perigee residuals of LAGEOS II). The best-fit line (*) shown through these combined residuals has a slope of about 66 mas/yr (the total integrated effect corresponds to about 16 m at the LAGEOS altitude); that is, $\mu_{\text{Measured}} \cong 1.1$ (compared with $\mu_{\text{GR}} \equiv 1$ in general relativity). The standard deviation of the slope is about 2 mas/yr, and the root mean square of the post-fit residuals is about 9 mas. Because of modeling errors (secular and periodic) and random errors, we estimate the total error in our measurement of μ to be $\lesssim 20\%$ of μ. Together with the slope we fitted and removed from the combined residuals four small periodic signals, with 1044-, 820-, 569-, and 365.25-day periods, corresponding to, respectively, the nodal period of LAGEOS, the perigee and nodal periods of LAGEOS II, and 1 year. A similar fit (with $\mu_{\text{Measured}} \cong 1.1$) was obtained by removing only three periodic signals with 1044-, 820-, and 569-day periods

and δJ_4 errors, using the Earth gravitational model JGM-3, over about a 3.1-year period, and after removing ten small periodic residual signals and the small observed inclination residuals [23].

In Fig. 2, we display the improved analysis [24] (obtained with the linear combination of the residuals of the nodes of LAGEOS and LAGEOS II and perigee of LAGEOS II according to Eq. (4)) using the more recent static and tidal Earth gravitational model EGM-96; we also refined the non-gravitational perturbations model; the total period of observations was of 4 years, longer by about 1 year with respect to the observational period corresponding to Fig. 1. We only removed four small periodic residual signals and the small observed inclination residuals. The removal of the periodic terms was achieved by a least squares fit of the residuals using a secular trend and four periodic signals with 1044-, 820-, 569-, and 365.25-day periods, corresponding to, respectively, the nodal period of LAGEOS, the perigee and nodal periods of LAGEOS II, and 1 year. A similar fit (with $\mu_{\text{Measured}} \cong 1.1$) was obtained by removing only three periodic signals with 1044-, 820-, and 569-day periods. The 820-day period is the period of the main odd zonal harmonic perturbation of the LAGEOS II perigee; the 1044- and 569-day periods are the periods of the main tidal orbital perturbations, with $l = 2$ and $m = 1$, which were not eliminated using Eq. (4). Some combinations of these frequencies correspond to the main non-gravitational perturbations of the LAGEOS II perigee. We stress that the present analysis using EGM-96 and its accurate tidal model is substantially independent of the signals removed, whereas the previous analysis [23], corresponding to Fig. 1, depended on the periodic terms included in the fit. In other words, our value (Fig. 2) of the secular trend does not significantly change by fitting additional periodic perturbations, and indeed, even the fit of the residuals with only a secular trend, with no periodic terms, just increases the slope by less than 10 %. Nevertheless, in this case, the root mean square of the post-fit residuals increases by about four times with respect to Fig. 2.

Our best-fit straight lines in Figs. 1 and 2, through the combined residuals of nodes and perigee, have respectively slopes $\mu_{\text{Measured}} \cong 1.1$, and $\mu_{\text{Measured}} \cong 1.1 \pm 0.03$, where 0.03 is the standard deviation of the fit corresponding to Fig. 2 using the EGM-96 gravitational model. This combined, measured, gravitomagnetic perturbation of the satellites' orbits corresponds, in a 1-year period, to about 16 m at the LAGEOS altitude, that is, about 265 mas. The root mean square of the post-fit combined residuals corresponding to Fig. 2 is about 9 mas. Our total error is estimated to be of the order of 30 %–40 % of μ_{GR} corresponding to Fig. 1, and of the order of 20 % of μ_{GR} corresponding to Fig. 2. Using the JGM-3 covariance matrix, we found the errors due to the uncertainties in the even zonal harmonics J_{2n}, with $2n \geq 6$, to be: $\delta\mu^{\text{ even zonals: } J_{2n} \geq J_6} \lesssim 17\%$ of μ_{GR}, and using the EGM-96 covariance matrix: $\delta\mu^{\text{ even zonals: } J_{2n} \geq J_6} \lesssim 13\%$ of μ_{GR}. The errors in the modeling of the perigee rate of LAGEOS II due to uncertainties in the odd zonal harmonics J_{2n+1}, are, with EGM-96: $\delta\mu^{\text{odd zonals}} \lesssim 2\%$ of μ_{GR}. Using the EGM-96 tidal model, we estimated the effect of tidal perturbations and other variations of Earth's gravitational field to be $\delta\mu^{\text{tides} +} \lesssim 4\%$ of μ_{GR}. On the basis of analyses [23] of the non-gravitational perturbations – in particular, those on the perigee of LAGEOS II – we found $\delta\mu^{\text{non-gravitational}} \lesssim 13\%$ of μ_{GR}, and the error due to uncertainties in the orbital inclinations of LAGEOS and LAGEOS II was estimated as $\delta\mu^{\text{inclination}} \lesssim 5\%$ of μ_{GR}. Taking into account all of these error sources, we arrived at a total root-sum-square error $\lesssim 20\%$ of μ_{GR}. Therefore, over an observational period of 4 years and using EGM-96, we determined $\mu_{\text{Measured}} = 1.1 \pm 0.2$, where 0.2 is the estimated total uncertainty due to all the error sources.

Based on the analysis of the orbits of the laser-ranged satellites LAGEOS and LAGEOS II, we conclude that the gravitomagnetic or Lense–Thirring effect exists, and its value is within 10 %, plus or minus ~ 20 %, of what is predicted by Einstein's theory of general relativity, that is, $\mu_{\text{GR}} = 1$.

Testing Our Method to Measure the Lense–Thirring Effect and Its Error Budget

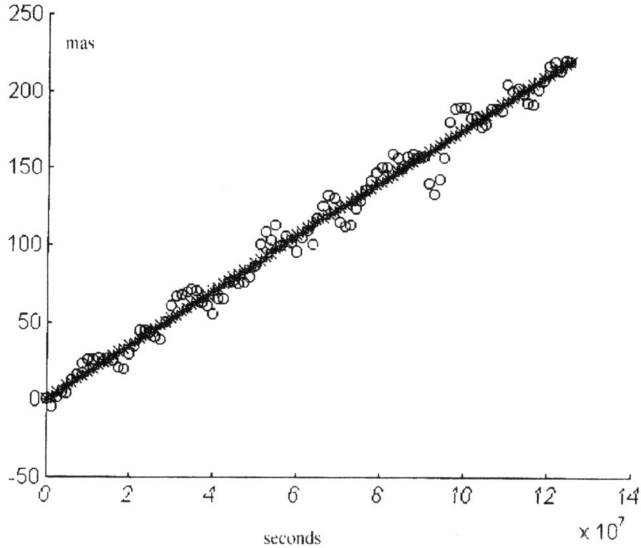

Fig. 3. The fit of the parameter μ using the combined residuals of six orbital elements $\delta\Omega_I$, $\delta\Omega_{II}$, $\delta\omega_{II}$, δe_{II}, δM_{II}, and δa_{II}. This fit and the corresponding value of μ are completely independent of the adjusted along-track accelerations of LAGEOS II and of the errors δJ_2 and δJ_4 (see Sect. 2). Together with the secular trend shown in this figure we have simultaneously fitted four periodic signals with the periods of perigee and node of LAGEOS II, node of LAGEOS and 365-day. The measured value of the Lense–Thirring effect obtained with this method is $\mu_{\text{Measured}} \cong 1$

A basic concern in our analyses was to estimate the total error in our measurement of the Lense–Thirring effect. In order to support our measurement and the corresponding error analysis, we have performed: (a) a test and (b) a preliminary blind-test simulation, explained in the next paragraphs (a) and (b).

(a) Testing our solution for μ using node, perigee, mean anomaly, eccentricity and semimajor axis of LAGEOS II, and node of LAGEOS.
To test our measurement of the Lense–Thirring effect with the residuals of node and perigee of LAGEOS II and node of LAGEOS we have also analyzed the residuals of eccentricity, mean anomaly and semimajor axis of LAGEOS II. Thus, we have produced a solution for μ by using all these orbital elements. Indeed, using the residuals of these orbital elements, we have a system of equations (for $\delta\Omega_I$, $\delta\Omega_{II}$, $\delta\omega_{II}$, δe_{II}, δM_{II}, and δa_{II}) in the unknowns: Lense–Thirring effect, δJ_2 and δJ_4 errors, and a constant and a variable, once-per-revolution, along-track acceleration. Therefore, the solution of this system for μ and the corresponding fit (shown in Fig. 3) are *completely independent* of the adjusted accelerations and of the errors δJ_2 and δJ_4. The result of our solution with the residuals of these orbital elements, and the corresponding fit for μ, is shown in Fig. 3: the measured value of the Lense–Thirring effect is $\mu_{\text{Measured}} \cong 1$. Thus,

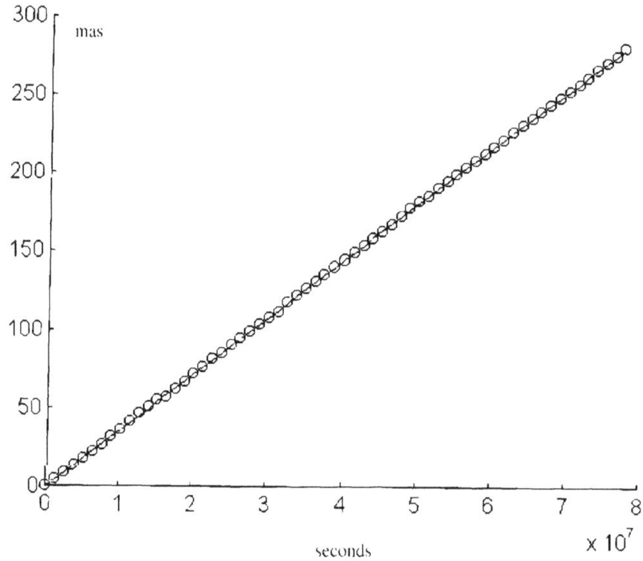

Fig. 4. Result of a *preliminary* simulation and blind-test analysis. The fit (solid line) of the parameter μ using about two and a half years of simulated data (o) of the LAGEOS satellites is shown; their simulated orbits were obtained by a model of perturbations in which the a priori value of the Lense–Thirring effect was *twice* its general relativistic value. The secular trend of the combined post-fit residuals shown is more than 90% of the a priori Lense–Thirring value (twice the general relativistic effect) set up in the simulation of the orbits

this further, new analysis, and the corresponding fit test and confirm our previous result [24].

(b) Preliminary blind-test-simulation. Finally, following a suggestion by the group at CSR of UT-Austin [25], we have performed a *preliminary* simulation and blind-test analysis. Using GEODYN we have simulated about two and a half years of data of the LAGEOS satellites with a model of perturbations in which the a priori value of the Lense–Thirring effect was *twice* its general relativistic value. Next we performed our standard analysis previously explained in this section, using the same model of perturbations but without the Lense–Thirring effect. In Fig. 4 is shown the result of our fit of the residuals of node and perigee of LAGEOS II and node of LAGEOS, obtained with the analysis of the simulated data of the orbits of these two satellites. With our combination (4) of the residuals of the orbital elements, we found a value of μ of about 1.9 times the Lense–Thirring effect, that is, more than 90% of the value a priori set up in our simulation of their orbits and of the corresponding laser-ranging data. We observe that this test was just a preliminary blind-test-simulation; indeed the simulated orbits and the corresponding laser-ranging data have to be simulated by using a model of perturbations including variations in all the relevant parameters (such as J_2 and

J_4) within known uncertainties. Nevertheless, this preliminary simulation was important in that it showed the consistency of our method and of the preliminary error budget.

Acknowledgements

The measurement described in this paper was done in collaboration with Erricos C. Pavlis (University of Maryland and NASA-Goddard). We also appreciate unstinting help from Despina E. Pavlis and D. D. Rowlands.

References

1. Misner C.W., Thorne K.S., Wheeler J.A. (1973): Gravitation. Freeman, San Francisco
2. Ciufolini I., Wheeler J.A. (1995): Gravitation and Inertia. Princeton University Press, Princeton
3. Lense J., Thirring H. (1918): Phys. Z. **19**, 156;
   Mashhoon B., Hehl F.W., Theiss D.S. (1984): Gen. Relativ. Gravit. **16**, 711 (English translation)
4. Thorne K.S., Price R.H., MacDonald D.A., eds. (1986): Black Holes, the Membrane Paradigm. Yale University Press, New Haven
5. Ciufolini I. (1994): Class. and Quantum Grav. A **11**, 73
6. Cohen S.C., Dunn P.F. (1985): LAGEOS Scientific Results. J. Geophys. Res. B **90** (11), 9215
7. Bender P., Goad C.C. (1979): The use of satellites for geodesy and geodynamics, in *Proc. Second International Symposium on the Use of Artificial Satellites for Geodesy and Geodynamics.*, ed. by G. Veis, E. Livieratos, Vol. II, National Technical University of Athens, p. 145
8. Nordtvedt K. (1996): Physics Today **49**, 26
9. Degnan J.J., Pavlis E.C. (1994): GPS World **5**, 62
10. International Earth Rotation Service (IERS) (1997): Annual Report, 1996, Observatoire de Paris, Paris
11. McCarthy D. (1996): The 1996 IERS Conventions. Observatoire de Paris, Paris
12. Lemoine F.G., Pavlis N.K., Cox C.M. et al. (1997): Gravity Field Improvement Activities at NASA GSFC, in *Proc. of the 1997 Institute of Astronomy and Geophysics Scientific Assembly, Rio de Janeiro, Brazil, September 5–9, 1997*
13. Rubincam D.P. (1988): J. Geophys. Res. B **93** (11), 13803
14. Rubincam D.P. (1990): J. Geophys. Res. B **95** (11), 4881
15. Rubincam D.P., Mallama A. (1995): J. Geophys. Res. B **100** (10), 20285
16. Martin C.F., Rubincam D.P. (1996): J. Geophys. Res. B **101** (2), 3215
17. Farinella P., Vokrouhlicky D.V.Y., Barlier F. (1996): J. Geophys. Res. B **101** (8), 17861
18. Gross R.S. (1996): J. Geophys. Res. B **101** (4), 8729
19. Pavlis D.E. et al. (1997): GEODYN II, Operations Manual, 3
20. Ciufolini I. (1986): Phys. Rev. Lett. **56**, 278
21. Ciufolini I. (1996): Nuovo Cimento A **109**, 1709
22. Ciufolini I. (1989): Int Journ. Mod. Phys. A **4**, 3083;
   Tapley B., Ciufolini I., Ries J.C. et al. (1989): NASA-ASI Study on LAGEOS III, CSR-UT publication n. CSR-89-3, Austin, Texas
23. Ciufolini I., Chieppa F., Lucchesi D. et al. (1997): Class. and Quantum Grav. **14**, 2701
24. Ciufolini I., Pavlis E.C. et al. (1998): Science **279**, 2100
25. Eanes R., Ries J. (1998): private communication

Canonical "Loop" Quantum Gravity and Spin Foam Models

R. De Pietri

Abstract. The canonical "loop" formulation of quantum gravity is a mathematically well defined, background independent, non-perturbative standard quantization of Einstein's theory of General Relativity. Some of the most meaningful results of the theory are: 1) the complete calculation of the spectrum of geometric quantities like area and volume and the consequent physical predictions about the structure of space-time at the Planck scale; 2) a microscopical derivation of the Bekenstein–Hawking black-hole entropy formula. Unfortunately, despite recent results, the dynamical aspect of the theory (imposition of the Weller–De Witt constraint) remains elusive.

After a short description of the basic ideas and the main results of *loop* quantum gravity we show in which sense the exponential of the super-Hamiltonian constraint leads to the concept of *spin foam* and to a four dimensional formulation of the theory. Moreover, we show that some topological field theories as the BF theory in 3 and 4 dimensions admits a spin foam formulation. We argue that the spin-foam/spin-network formalism it is the natural framework in which to discuss loop quantum gravity and topological field theory.

1 Introduction

The *loop approach* to quantum gravity [1, 2] has reached a mature status as a physical theory and it is one of the most likely candidates to be a complete consistent quantum theory of gravitational phenomena (for a detailed review and update bibliography we refer to [3] for quantum gravity in general and to [4] for the loop approach in particular). In this approach it was possible:

i) to construct the auxiliary Hilbert space [5, 6] appropriate to the discussion of the kinematics;
ii) to find an explicit basis (the spin-network basis) on this Hilbert space [7, 8];
iii) to solve rigorously the 3-diffeomorphism constraints [9];
iv) to construct operators corresponding to volume and area and determine their spectra [10–13];
v) to derive the Bekenstein–Hawking formula for the entropy of Black Holes [14–17].

The elusive feature of the loop approach to quantum gravity is that in the *frozen-time* canonical picture the relation between operator and physical observable is cumbersome. A space-time picture was lacking until M. Reisenberger and C. Rovelli [18] derived a spin foam model (using the terminology introduced by J. Baez [19]), which they refer to as a *world sheet model*, from the Exponential of the super-Hamiltonian constraint. They also show that this is formally related to the construction of the 4 dimensional path integral for gravity. This work was a first step in the direction of developing a sum over surfaces framework in "loop quantum gravity" whose importance was first noted by Reisenberger in [20]. This recent development indicated a surprising formal analogy between loop quantum gravity and other approaches to the quantization of gravity. The model of dynamical triangulation [21] and the Ponzano–Regge model[1] [30] are exactly models of this kind. The analogy is even deeper if one considers the Ponzano–Regge model [the physicist's version of the celebrated Turaev–Viro state sum model for 3-manifold invariants [31] that can be formulated exactly as a sum over branched surfaces with colored graph observables (spin networks) on the boundary.

The previous consideration, together with the fact that the topological BF field theory in 4 dimensions [32, 33] (see also [34]) admits an analogous formulation as a spin foam model [in this case, the spin foam dual to a simplicial decomposition (triangulation)], strongly suggest that quantum gravity should be naturally formulated using the language of spin foam models. Now, two explicit proposals for the realization of this idea have been put forward. The first (by Barrett and Crane [35]) is related to the discretization of the path integral of the Euclidean Plebanski action [36] (see [37, 38]) and the second (by M. Reisenberger [39]) of self-dual Euclidean gravity [40]. The last model is the explicit quantization of the lattice model [41] that has self-dual Euclidean gravity as its continuum limit. This line of research is deeply connected to the idea of expressing classical gravity as a deformation (by the imposition of additional Lagrange constraints) of a topological BF field theory in 4 dimensions. The idea is that [42] topological BF theory, in the context of diff-invariant theory, can play a role analogous to that of free theory in standard Poincaré invariant quantum field theories. This idea is at the heart of the systematic proposal for deriving *Spin Foam Weight Factors* from a Classical Action Principle that has been recently proposed by Freidel and Krasnov [43].

The discussion here is in the framework of piecewise linear topology and we refer the reader to [44] for details. Intuitively, this setting is equivalent to assuming that the cells, surfaces and edges that we use to decompose a manifold are finite unions of piecewise analytic sub-manifolds. In particular, a graph is a finite collection of analytic paths. This framework has certain advantages and the resulting theory is simpler than those in the analytic or in the smooth setting. The utility of the PL setting in loop quantum gravity was first pointed out by Zapata [45, 46]. All the topological objects (graphs, manifolds, cellular decompositions, ...) are assumed orientable. For the Wigner nJ-symbols we adopt the non-normalized graphical convention of [47, 48] and we restrict to the case of the group SU(2) and use Penrose's binor formalism [49]. Most of the formulas remain valid in the quantum group case, but, in this case, factors related to the braiding can appear.

[1] This model corresponds to three dimensional BF theory [22–24] and it is closely related to Witten's Chern-Simons approach to the quantization of 2+1 dimensional gravity [25–29].

2 Loop quantum gravity

Loop quantum gravity is based on the Hamiltonian formulation of classical general relativity in terms of the canonically conjugated variables $\tilde{E}_i^a(x)$ (densitized triad fields) and $A_a^i(x)$ (Ashtekar's connection) [50]. The dynamics are given, as happens for any diff-invariant theory, only in terms of constraints. In our case:

$$\begin{aligned} C_i(x) &= \mathcal{D}_a \tilde{E}_i^a(x) \simeq 0 && \text{(Gauge constraints)} \\ \mathcal{H}_a(x) &= \tilde{E}_i^b(x) F_{ab}^i(x) \simeq 0 && \text{(Diffeomorphism constraints)} \\ \mathcal{H}(x) &= \frac{\epsilon^{ijk}}{\sqrt{g}} \tilde{E}_i^a(x) \tilde{E}_j^b(x) F_{ab}^k(x) \simeq 0 && \text{(Hamiltonian constraints).} \end{aligned}$$

The loop program efffects the quantization of gravity in a straightforward way. It is based on the assumption that a particular class of classical functions (the $\mathcal{T}$ observables, i.e., the *traces* of the *holonomy* of Ashtekar's connection) becomes one of well-defined quantum operators; i.e., it is assumed that, for any choice of closed loop γ, to the classical functional of the Ashtekar connection:

$$\mathcal{T}_\gamma[A_a^i(x)] = \mathrm{Tr}\left[\exp\left(\int_\gamma A_a(x) dx^a\right)\right] \tag{1}$$

corresponds a well-defined quantum operator on a Hilbert space $\mathcal{H}$. All the other properties should be derived essentialy from this assumption. A consequence of this choice is that the canonical variables $\tilde{E}_i^a(x), A_a^i(x)$ do not become well-defined operators on the Hilbert space $\mathcal{H}$ and the previous definition of the constraint (unless properly regularized) is meaningless on $\mathcal{H}$. This is one of the main technical reasons behind the difficulty of dealing with the super-Hamiltonian constraint.

Given this assumption, there are two possible ways to define the theory. Explicitly construct the vector space on which these operators are well-defined (loop representations, see [11]), in a algebraic way, and then determine the correct scalar product. Alternatively, construct a Hilbert space $\mathcal{H}$ (the connection representation) and show that, on this Hilbert space, these operators are well-defined [9]. These two approaches are equivalent [48]. However, two technical but important decisions must be taken. The first is connected to the properties of the loops γ. The requirement of piecewise analyticity demands a different setting from that of the smooth case since two piecewise analytic loops can have only a finite number of isolated intersection points. The second is that these contractions are well-defined only in the case of a real connection. Now, the standard Ashtekar connection is naturally real only in the case of Euclidean gravity while in the case of Lorentzian gravity it is naturally complex and indeed what follows naturally applies only to Euclidean gravity. However, it is important to note that it is possible to use a real connection in the Lorentzian case as well (the Barbero connection [51,52]) at the price of a more complicated form for the super-Hamiltonian constraint.

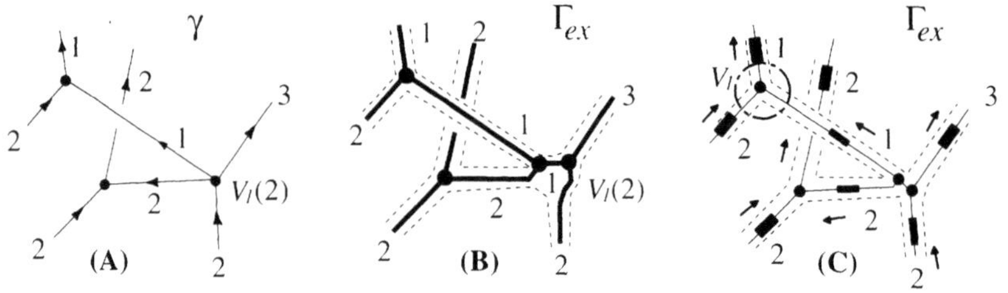

Fig. 1. The graph γ, a possible extended-planar projection Γ_{ex} and the graphical representation of a spin-network cylindrical function

The auxiliary Hilbert space

A full discussion of the kinematical setting is not important here. We refer to [9, 48] for the technical details and for complete references. For the following discussion it is sufficient to recall the basic structure of the Hibert space $\mathcal{H}$. Since a connection associates a group element $g_{e_i} = \mathcal{P}\exp(-\int_{e_i} A)$ to each edge (segment embedded in M^3) of the graph γ. It is natural to consider for each graph γ the Hilbert space ($\mathcal{H}_\gamma = L^2[G^n, d\mu(g)]$) of square integrable functions, with respect to the unique normalized Haar measure $d\mu(g)$, on n (the number of edges e_i in γ) copies of the group $[G = SU(2)]$ G^n. $\mathcal{H}$ can be characterized as the projective limit of the previous family of Hilbert spaces $\mathcal{H}_\gamma$ associated to all possible graphs γ embedded in M^3. The gauge invariant Hilbert space $\mathcal{H}_{\text{aux}}$ is the projective limit of the gauge invariant sub-sector of the Hilbert spaces $\mathcal{H}_\gamma$. Moreover, the Peter-Weyl theorem gives us a natural basis in this space of gauge invariant functions $f_\gamma(g_{e_1}, \ldots, g_{e_n})$, the spin networks basis:

$$\begin{aligned} f_\gamma(g_{e_1}, \ldots, g_{e_n}) &= \sum_{\pi, \iota} f(\gamma, \boldsymbol{\pi}, \boldsymbol{\iota})\ \mathcal{T}_{\gamma,\boldsymbol{\pi},\boldsymbol{\iota}}[A], \\ \mathcal{T}_{\gamma,\boldsymbol{\pi},\boldsymbol{\iota}}[A] &\stackrel{def}{=} \left(\otimes_{i=1}^{\#\text{edge}} \pi_i(g_{e_i})\right) \cdot \left(\otimes_{j=1}^{\#\text{vertex}} \iota_j\right) \end{aligned} \tag{2}$$

where: **(i)** $\boldsymbol{\pi} = (\pi_1, \ldots, \pi_N)$ denotes the labeling of the edges with irreducible representation π_i of G; **(ii)** $\boldsymbol{\iota} = (\iota_1, \ldots, \iota_M)$ a labeling of the vertices with invariant contractors ι_j (the intertwining matrices ι_j, at each of the vertices v_j, represent the invariant coupling of the n_j representations associated to the n_j edges that start or end at v_j). In the case of loop quantum gravity (where A_a^i is an SU(2) connection) the vectors of the basis (a spin network state) are characterized (see Fig. 1) by a colored graph (the irreducible representations of SU(2) are labelled by the associated spin or by an integer, which we call the color of the representations, which is its twice the spin).

The diffeomorphism invariant state

A rigorous definition of diffeomorphism invariant Hilbert space can be obtained [9] using the gauge averaging procedure. The essential idea is that on $\mathcal{H}_{\text{aux}}$ there is a natural definition of the action of a diffeomorphism $\phi \in \text{Diff}$ ($\phi : M \to M$). In fact, in the spin-networks basis we can associated to each diffeomorphism ϕ the operator

$$\hat{U}_\phi|S\rangle = \hat{U}_\phi|\gamma, \pi, \iota\rangle = |\phi(S)\rangle = |S'\rangle = |\phi(\gamma), \pi, \iota\rangle, \tag{3}$$

where $S' = \phi(S)$ is the image under ϕ of the spin network S. At this point one can define the class of spin network knots s as the equivalence class of spin networks under diffeomorphism, i.e., $S, S' \in s$ if there exists a $\phi \in \text{Diff}$ such that $S' = \phi(S)$. In this way, we can define $\mathcal{H}_{\text{Diff}} = \mathcal{H}_{\text{aux}}/\text{Diff}$ with the scalar product:

$$\langle s|s'\rangle = N\int [D\phi]\langle S|\hat{U}_\phi S'\rangle = N\int [D\phi]\langle\phi(S)|S'\rangle \tag{4}$$

where N is an arbitrary, not yet fixed, normalization factor and $S, S' \in \mathcal{H}_{\text{aux}}$ are two arbitrary spin networks in the s-knot class of s and s', respectively. The integration in Eq. (4) is meaningful because the scalar product of two spin networks of $\mathcal{H}_{\text{aux}}$ is different from zero only if they have support on the same graph.

In the previous derivation of the Hilbert space $\mathcal{H}_{\text{Diff}}$ we made an implicit assumption on the kind of diffeomorphism, which are allowed. Having previously assumed that the loops 0 are in the class of the piecewise analytic ones, we find it natural to assume piecewise analytic diffeomorphisms. In this case, the quotient space $\mathcal{H}_{\text{Diff}}$ is a separable Hilbert space. This leads directly to the PL framework. We refer to the work of Zapata for the discussion of the subtleties involved [45, 46].

The super-Hamiltonian constraint

In the first presentation of *loop quantum gravity* [1, 2] it was emphasized that loops without intersection (i.e., spin network states whose graph γ has only bivalent vertices) are automatically in the kernel of the super-Hamiltonian constraint. This brings to light an infinite dimensional subspace of the **physical** Hilbert space $\mathcal{H}_{\text{Phys}}$. Unfortunately, with the introduction of the volume operator [10], it was realized that all the spin network states based on a graph γ with all vertices of valence less than 3 are in the kernel of the volume operator. Consequently, all the previously mentioned states, belonging to the kernel of the super-Hamiltonian constraint, correspond to degenerate metrics and, therefore, cannot correspond to genuine quantum 3 geometries. This suggested defining the super-Hamiltonian constraint $\mathcal{H}(x)$ on the whole diff-invariant Hilbert space $\mathcal{H}_{\text{Diff}}$.

The situation improved in 1996 when Thiemann [53–55] was able to construct an operator with the *naivë* classical limit. Using this definition, it is possibile to investigate the true **physical** Hilbert space $\mathcal{H}_{\text{Phys}}$. For the purpose of the present discussion we do not need a complete account of its definition and its properties. What it is interesting for us is that: 1) it is possible to define such an operator; 2) a rough idea of its action $\mathcal{H}_{\text{Diff}}$. Its main property is that it acts locally around each vertex of the spin network. It adds or deletes an edge of color 1

$$\hat{A}_{vIJK\epsilon\tilde{\epsilon}} =$$

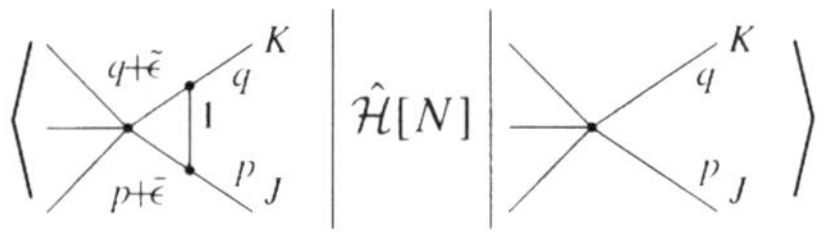

(5)

with well-defined weight factors [$(\hat{A}_{vIJK\epsilon\tilde{\epsilon}})$ matrix elements of the operator] that can be explicitly computed [56].

The hierarchy of Hilbert spaces of loop quantum gravity

In summary, the mathematical definition of the loop quantization of gravity, and the determination of the **physical** Hilbert space $\mathcal{H}_{\mathrm{Phys}}$, involves the analysis of the following hierarchy of Hilbert spaces:

$$\mathcal{H} \overset{C_i(x)}{\Longrightarrow} \mathcal{H}_{\mathrm{aux}} \overset{\mathcal{H}_a(x)}{\Longrightarrow} \mathcal{H}_{\mathrm{Diff}}\ (?) \overset{\mathcal{H}(x)}{\Longrightarrow} (?)\ \mathcal{H}_{\mathrm{Phys}} \tag{6}$$

where an arrow means imposition of a constraint and the question marks emphasize the fact that the last step is still not completely understood. In fact: 1) the physical correctness of the Thiemann's proposal for the regularization of the super-Hamiltonian constraint has been questioned; 2) more than one variant of this operator can be constructed and indeed the problem is determine the correct one; 3) despite [57] there is not a clear understanding of the kernel of the super-Hamiltonian constraint nor of the class of operators that are well-defined on this Hilbert space.

3 The exponential of the Hamiltonian constraint

Reisenberger and Rovelli in [18, 58] suggested (taking notice of the existence of a well-defined version of the Hamiltonian constraint [53–55] on $\mathcal{H}_{\mathrm{Diff}}$) that a better way to address this last remaining problem is to use a 4-dimensional picture. In the case of classical general relativity the lapse $N(x, \tau)$ and shift function $N^a(x, \tau)$ ($N^\mu(x, \tau) = d\Sigma^\mu(x, \tau)/d\tau$) generate the evolution along foliations ($\Sigma(\tau)$) between $\Sigma(0)$ and $\Sigma(1)$. In an analogous way, in the quantum case, we can say that the lapse and shift function generate the evolution from a spin network state $|s, \Sigma(0)\rangle \in \mathcal{H}_{\mathrm{aux}}$ associated to the boundary $\Sigma(0)$ to the spin network state $|s', \Sigma(1)\rangle \in \mathcal{H}_{\mathrm{aux}}$ associated to the boundary $\Sigma(1)$. Formally[2] we have

$$\frac{d}{dt}|S', \Sigma(1)\rangle = \left(N(x, \tau)\hat{\mathcal{H}}(x) + N^a(x, \tau)\hat{\mathcal{H}}_a(x) \right)|S, \Sigma(t)\rangle \tag{7}$$

[2] One should keep in mind that the following formulae do not have a well-defined meaning: there are no rigorous or unambiguous definitions of the 3-diff constraints nor of the Hamiltonian constraints in the Hilbert space $\mathcal{H}_{\mathrm{aux}}$.

$$|S', \Sigma(1)\rangle = \mathcal{T}\exp\Big[-\mathrm{i}\int_0^T d\tau \int d^3x \Big(N(x,\tau)\hat{\mathcal{H}}(x) + N^a(x,\tau)\hat{\mathcal{H}}_a(x)\Big)\Big] \cdot |S, \Sigma(0)\rangle. \tag{8}$$

The last equation suggests the definition of the transition amplitude, averaged over all the possible embeddings (bordisms) interpolating from $\Sigma(0)$ and $\Sigma(1)$:

$$\begin{aligned} Z[S', S; 1] &= \langle S', \Sigma(1)|S, \Sigma(0)\rangle = \\ &= \int [dN(x,\tau)][dN^i(x,\tau)] \\ &\cdot \langle s'|\mathcal{T}\exp\Big[-\mathrm{i}\int_0^1 d\tau \int d^3x \Big(N(x,\tau)\hat{\mathcal{H}}(x) + N^a(x,\tau)\hat{\mathcal{H}}_a(x)\Big)\Big]|S\rangle, \end{aligned} \tag{9}$$

where the integration $[dN(x,\tau)][dN^i(x,\tau)]$ is over all the lapse and shift functions that represent a bordism between $\Sigma(0)$ and $\Sigma(1)$. They note that it is possible to give a well-defined meaning to integration over the lapse of the exponential of infinitesimal diffeomorphisms in term of the gauge averaging procedure used in [9] to solve the diffeomorphism constraints:

$$\hat{U}_{\phi_{N^i}} =: \exp\Big[-\mathrm{i}\int d^3x\, N^a(x)\hat{\mathcal{H}}_a(x)\Big] \tag{10}$$

$$|s\rangle := \int [D\phi]\, \hat{U}_\phi|S\rangle =: \int [dN^a(x)]\, \hat{U}_\phi|S\rangle, \tag{11}$$

and then, by use of a series of formal arguments, that definition (9) depends only on the s-knot classes s and s' of S and S', respectively. Morover, they show that it can be expressed in terms of a diff-invariant definition (such as the one proposed by Thiemann) of the super-Hamiltonian constraint and of its associated proper time propagator:

$$\begin{aligned} Z_{RR}[s', s; 1] &= \int dT < s'|e^{-\mathrm{i}\,T\,\mathcal{H}}|s > \\ &= \int dT \sum_\sigma \frac{(\mathrm{i}\,T)^{n(\sigma)}}{n(\sigma)} \prod_{(\text{coloring of } \sigma)} \prod_{v\in[\sigma]} A_v(\sigma). \end{aligned} \tag{12}$$

In the previous equation, $\mathcal{H}$ is a diff-invariant definition of the Hamiltonian constraints and σ is a 2-dimensional branched surface embedded over $\Sigma \times I[0, T]$. In Fig. 2 we give an example of a branched surface, and of the corresponding spin-network transition. The kind of surfaces that appear on the sum is explicitly determined by the kind of non-null matrix elements of the super-Hamiltonian constraints. We want to emphasize, once more, that this derivation is based on formal argument. The use of formal arguments is of great help in understanding the general picture but one should stop here. The result of this formal derivation is that the functional integral associated to quantum gravity is defined (or, better, should be) as a sum over branched surfaces interpolating between two spin networks. Moreover, it shows that one has to

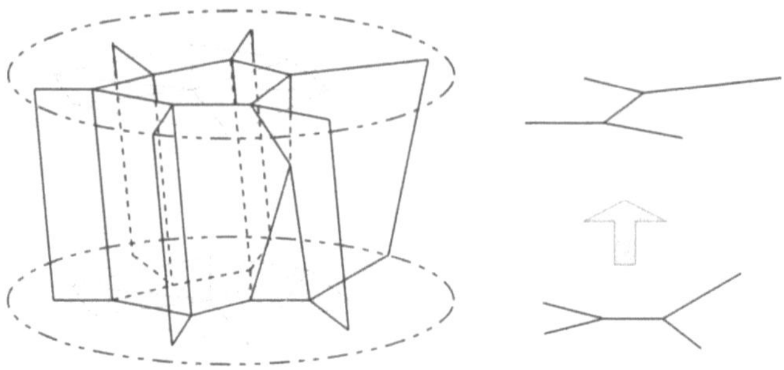

Fig. 2. A potential first order contribution to the partition function $Z_{RR}[s', s; \sigma; 1]$ of Eq. (3) and the corresponding spin-network transition. A three-dimensional situation is depicted. From the point of view of a 4-dimensional theory this contribution corresponds to a branched-polyhedron dual to a singular triangulation

consider the sum over all the possible (not just one) such branched surfaces. This is a major breakthrough towards the understanding of loop quantum gravity and it shows a strict relationship between loop quantum gravity and topological field theories of BF type opening the possibility of using the techniques developed for such theories.

The first occurrence (to our knowledge) of the idea of formulating the transition amplitude between two Hilbert spaces (spanned by embedded colored graphs) associated to two boundary component of a manifold, as a weight associated to the colored branched surface interpolating between the two graphs, first appeared in Turaev and Viro [31]. These authors used the term: two dimensional polyhedral quantum field theory. On that construction is based the idea of *spin foam* models. The next section will be dedicated to the exposition of this idea in the case of 2+1 dimensional gravity or, more precisely, of the Ponzano–Regge–Turaev–Viro model.

4 The Ponzano–Regge–Turaev–Viro model

The Ponzano–Regge model and its quantum group analogue, the Turaev–Viro partition function [it is known that its evaluation gives a quantum invariant of 3-manifolds], are the best understood example of a theory formulated on a triangulation Δ of a manifold M. From the physical point of view it is known (for details see [24, 41, 43]) that it corresponds to the quantization of Euclidean 2+1 dimensional gravity, i.e., to the quantization of the SU(2) BF theory in 3 dimensions:

$$Z_{BF}^{(3)} = \int [\mathcal{D}e(x)][\mathcal{D}\omega(x)] e^{\mathrm{i}\int \mathrm{Tr}[e\wedge R[\omega]] + \Lambda\ \mathrm{Tr}[e\wedge e\wedge e]} \tag{13}$$

$$= \int [\mathcal{D}A(x)][\mathcal{D}B(x)] e^{\frac{\mathrm{i}k}{4\pi}\int [\mathrm{Tr}[A\wedge F[A]] + \frac{2}{3}\mathrm{Tr}[A\wedge A\wedge A]] - [\mathrm{Tr}[B\wedge F[B]] + \frac{2}{3}\mathrm{Tr}[B\wedge B\wedge B]]}$$

$$= \left| \int [\mathcal{D}A(x)] e^{\frac{\mathrm{i}k}{4\pi}\int + \mathrm{Tr}[A\wedge F[A]] + \frac{2}{3}\mathrm{Tr}[A\wedge A\wedge A]} \right|^2 = |Z_{CS}|^2.$$

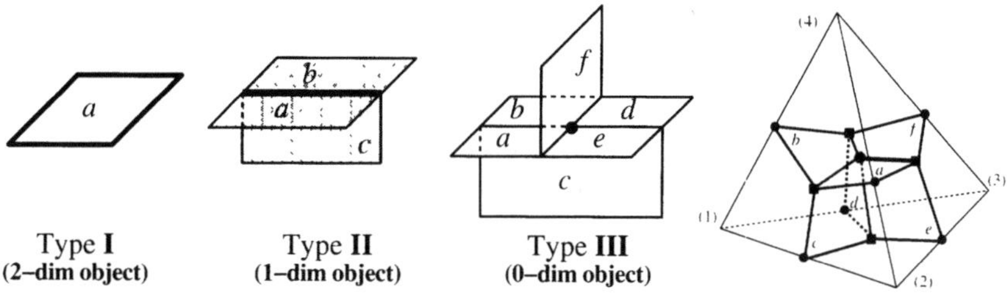

Fig. 3. The description of the standard polyhedron dual to a triangulation and a schematic representation of the duality. Note that in 3 dimensions, a colorating of the edges of the triangulation determines a unique colorating of the faces of its dual 2-skeleton, and *vice versa*

where

$$\omega = \frac{1}{2}[A + B], \quad e = \frac{k}{8\pi}[A - B], \quad \Lambda = \left(\frac{4\pi}{k}\right)^2 .$$

The recipe for the construction of this model (to which we will refer as the PRTV-model) is the following. Color all the edges (1-dimensional objects of the triangulation) with irreducible representations of the SU(2) group and assign the following factors to the various objects of the triangulation: 1) to each edge the dimension of the representation Δ_{c_e}; 2) to each triangle the inverse of the 3J symbol made out of the three colors associated to each of its edges $\theta(a_t, b_t, c_t)^{-1}$; 3) to each tetrahedron the associated 6J symbol (a tetrahedron has 6 edges) Tet$[abed; cf]$.

In the case of the PRTV-model, it is useful to consider its dual formulation, i.e., to take advantage of the duality between a triangulation and the standard polyhedron[3] suggested by Fig. 3. In the dual cellular complex Δ^* of the triangulation Δ we have that, to each n-dimensional cell of Δ, a $3n$-dimensional cell of Δ^* is uniquely assigned. In three dimensions, the object dual to an edge (1-dimensional object) is a surface (2=3-1 dimensional object). This implies that to each coloring of the edges of the triangulation Δ is associated a unique coloring of the faces (two-dimensional objects) of its dual standard polyhedron Δ^*. In turn, it is possible to consider the PRTV-model as a model defined over a two dimensional branched surface, i.e., using this duality, we can interpret the PRTV model both as a state sum model associated to a triangulation but also as a state sum model associated to a branched surface:

$$TV_G[M^3, \Delta] = TV_G[S_2^{\Delta^*}] \tag{14}$$

where by $S_2^{\Delta^*}$ we denote the collection of all the faces, edges and vertices (the 2-skeleton) of the cellular decomposition of M dual to the triangulation Δ. We will

[3] A compact connected two dimensional polyhedron is called quasi-standard if each of its points has a neighborhood homeomorphic either to a plane (type I), to the union of three half-planes with common boundary (type II), or the union of the four half-planes dual to a tetrahedron in its barycenter (type III). We say that it is standard if all its two-dimensional component are disks.

also use the notation $\mathcal{F}(\mathcal{S}_2)$, $\mathcal{E}(\mathcal{S}_2)$ and $\mathcal{V}(\mathcal{S}_2)$ to denote the collection of its faces, edges and vertices, respectively.

This expresses the fact that the definition of the PRTV-model can be given as a series of rules with respect to the objects of the triangulation or as an equivalent set of rules with respect to the objects to its dual standard polyhedron:

Δ	Δ^*		Factor in $TV[M^3, \Delta] = TV_G[\mathcal{S}_2(\Delta^*)]$
vertex	3-cell	Type 0	$\Lambda^{-1} = \left[\sum_a \Delta_a\right]^{(-1)}$
edge	face	Type **I**	$\Delta_a = \dim(a) =$ 
face	edge	Type **II**	$\theta(a, b, c)^{-1} =$ [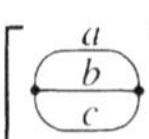]$^{(-1)}$
tetrahedron	vertex	Type **III**	Tet$[abed; cf] =$ 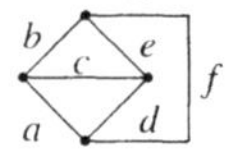

One of the nice characteristics of the dual formulation is that it involves only the coloring of a 2-dimensional standard polyhedron embedded in the manifold. By completing the transition to the dual world it is possible to forget the original triangulation and to associate directly a state-sum to any standard polyhedron of the (M^3) manifold.

$$TV_G[\mathcal{S}_2] = \Lambda^{-\#(3-cells)} \sum_{c \in D(\mathcal{S}_2, G)} \left[\prod_{f \in \mathcal{F}(\mathcal{S}_2)} \left(\cdot \right)^{\chi(f)} \right]$$

$$\cdot \left[\prod_{e \in \mathcal{E}(\mathcal{S}_2)} \left(\cdot \right)^{-\chi(e)} \right] \left[\prod_{v \in \mathcal{V}(\mathcal{S}_2)} \cdot \right] \qquad (15)$$

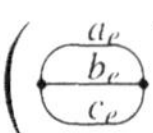

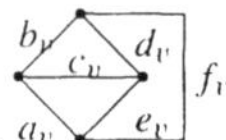

where $\chi(f)$ and $\chi(e)$ are the Euler characteristics of the face f and of the edge e, respectively. For the purpose of the present discussion the introduction of the Euler characteristics is completely irrelevant. All the edges and faces of a standard 2-polyhedron dual to a triangulation of M^3 have Euler-characteristic equal to 1! We explicitly wrote the Euler characteristic because one of the interesting features of the PRTV partition function is that it is defined for any stratification of the manifolds M whose two skeleton is a standard polyhedron. That means that it is defined on the larger class of singular triangulations [31]. For the rest of the discussion we will alway deal with the regular polyhedron dual to a triangulation and we will always replace the Euler characteristic with the appropriate value for polyhedrons belonging to the dual of a triangulation, namely, 1.

This model manifests its topological nature from the fact that it is possible to prove that $TV_G[\mathcal{S}_2]$ it is independent of the chosen standard polyhedron dual to a triangulation (pseudo-triangulation) of M. In fact, it is possible to prove that, if

$\mathcal{S}_2$ and $\tilde{\mathcal{S}}_2$ are two different standard polyhedra dual to triangulations of the same manifold M, then

$$TV_G[M] = TV_G[\mathcal{S}_2] = TV_G[\tilde{\mathcal{S}}_2]. \tag{16}$$

Decorating a 2-dimensional standard polyhedron (spin foam)

In the previous section we used the notation $D(\mathcal{S}_2, G)$ to denote all the possible decorations (colorings) of the 2-surfaces of a standard polyhedron with inequivalent irreducible representations of a group G. The admissibility condition amounts to the existence of an intertwining matrix (in the case of the SU(2) group, if it exists, it is unique up to normalization) between the three-representation associated to the three faces that meet at an edge (a type II object of Fig. 3). It is natural to extend this concept of decoration to the more general situation where there are n-faces that meet at a given edge. In this case, we have to assign to each edge an invariant contractor on the tensor product of the spaces of all the representations associated to the faces that meet at the edge. In this way, we have a decoration of the two- and one-dimensional objects of the 2-polyhedron completely analogous to the decoration (coloring) of the edges and vertices of the graph γ that we used to label the spin networks basis of equation (2), which, in the notation $D(\mathcal{S}_2, G)$, can be rewritten as:

$$f_\gamma(g_{e_1}, \ldots, g_{e_n}) = \sum_{\boldsymbol{\pi}, \iota \in D(\gamma, SU(2))} f(\gamma, \boldsymbol{\pi}, \iota)\ \mathcal{T}_{\gamma, \boldsymbol{\pi}, \iota}[A]. \tag{17}$$

This completes the definition of the concept of a group-decorated 2-dimensional oriented polyhedron. The extension to non-simple polyhedra of this concept (implictly contained in [20]) of group decoration of 2-polyhedra with boundary is due to Baez. He used the now generally accepted term *spin foam* to denote a single group-decorated 2-polyhedron and the term *spin foam model* to denote a theory based on the sum on all the possible decorations (colorings). We refer the reader to Baez' work [19] for more details and for a discussion of the additional care needed in the case of a general group.

The Hilbert space on the boundary and the functorial nature of the invariant

Following Turaev and Viro [31], we note that it is possibile to generalize the PRTV-model to include the case of a manifold M with boundary with a decorated (colored) graph sitting on its boundary ∂M. The PRTV partition function (15) in the case of a decorated polyhedron $\mathcal{S}_2$ with a graph γ on the boundary (i.e., $\partial \mathcal{S}_2 = \mathcal{S}_2 M^3 = \gamma$) is defined as

$$TV_G[\mathcal{S}_2; (\gamma, \boldsymbol{c})] = \frac{1}{N(\gamma, \boldsymbol{c})} \Lambda^{\frac{\#(2-cells\ of\ \gamma)}{2}} \Lambda^{-\#(3-cells)} \times \tag{18}$$

$$\times \sum_{\boldsymbol{c} \in D_{in}(\mathcal{S}_2, G)} \left[\prod_{f \in \mathcal{F}_{in}(\mathcal{S}_2)} \left(\right. \right.$$

$$\left. \left. \right) \right] \times$$

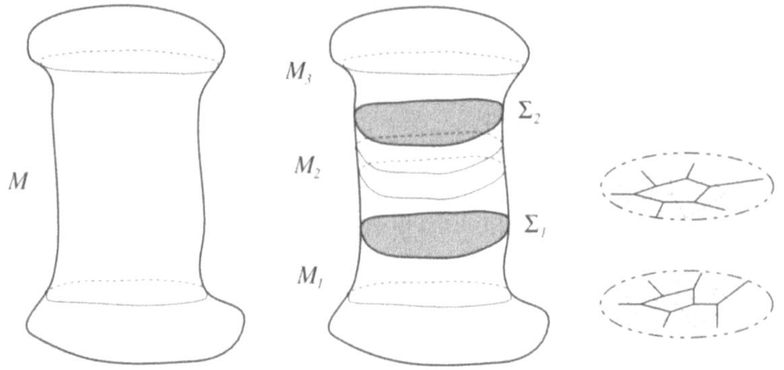

Fig. 4. A possible decomposition of an orientable manifold in three components. This drawing has only a schematic purpose since we suppose that $\partial M_3 = -\Sigma_2$, $\partial M_1 = \Sigma_1$ and $\partial M_2 = \Sigma_2 \cup -\Sigma_1$

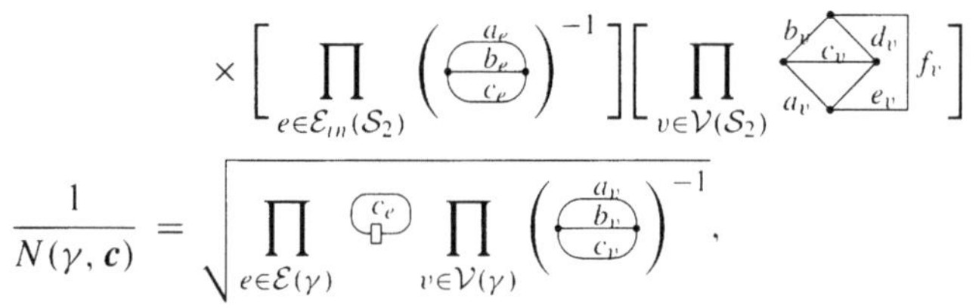

(19)

where, by $\mathcal{F}_{in}(\mathcal{S}_2)$ and $\mathcal{E}_{in}(\mathcal{S}_2)$, we have denoted the faces and edges of $\mathcal{S}_2$ that do not end or belong to the boundary of $\mathcal{S}_2$. Moreover $D_{in}(\mathcal{S}_2, G)$ denotes a decoration of the internal faces $\mathcal{F}_{in}(\mathcal{S}_2)$ compatible with the decoration of the graph γ on the boundary[4] ($c_f = c_e$ if $f \in \mathcal{F}(\mathcal{S}_2)$, $e \in \mathcal{E}(\gamma)$ and $e \subset \partial f$).

Using the previous generalization of the PRTV-model we can reinterpret the state sum model of Eq. (14) in the following way. As suggested by Fig. 4 we can imagine splitting the manifold M into three disjoint components glued together along the common boundary and rewriting the partition function associated to any given triangulation as:

$$TV_G[M^3, \Delta] = \sum_{c' \in D(\gamma, G)} \sum_{c \in D(\gamma, G)} TV_G[\mathcal{S}_2^{\Delta_1^*}; (\gamma', c')] \frac{1}{N(\gamma', c')^2} \quad (20)$$
$$TV_G[\mathcal{S}_2^{\Delta_2^*}; (\gamma' \cup \gamma, c + c')] \frac{1}{N(\gamma, c)^2} TV_G[\mathcal{S}_2^{\Delta_3^*}; (\gamma, c)]$$

where $\gamma \in \Sigma_1$ and $\gamma' \in \Sigma_2$ are the graphs determined by the intersection of the 2-bone of the dual cellular decomposition Δ^* with Σ_1 and Σ_2, respectively. Now, forming the intersection of Δ^* with a 2-surface Σ, we find that the 3-cells are mapped into 2-cells, the 2-cells are mapped into 1-cells (the edges of γ) and the 1-cells into vertices of γ. In this operation we are indeed inducing a coloring of γ from the

[4] A similar costruction was used by G. Carbone, M. Carfora and A. Marzuoli to construct a true invariant for manifold with non-empty boundary [59].

coloring of the branched polyhedron. Moreover, if Δ is a triangulation of M then the resulting graph is trivalent. Equation (20) has a nice interpretation in terms of an operator equation. Consider the Hilbert spaces $\mathcal{H}_\gamma$ of Sect. 2.1 associated to graphs embedded on the two Cauchy surfaces Σ_1 ($\mathcal{H}_{\gamma_1}$) and Σ_2 ($\mathcal{H}_{\gamma_2}$). In the normalization of [11] and [48] the completeness of $\mathcal{H}_\gamma$ is given by

$$\hat{1}_\gamma = \sum_{c \in D(\gamma, G)} \frac{1}{N(\gamma, \boldsymbol{c})^2} |\Sigma; \gamma, \boldsymbol{c}\rangle \langle \Sigma; \gamma, \boldsymbol{c}|. \tag{21}$$

Moreover, we can define an operator $\hat{P}^{(M_2)}_{\gamma_2, \gamma_1}$

$$\hat{P}^{M_2}_{\gamma_2, \gamma_1} : \mathcal{H}_{\gamma_1} \Longrightarrow \mathcal{H}_{\gamma_2} \tag{22}$$

$$\langle \gamma', c' | \hat{P}^{M_2}_{\gamma', \gamma} | \gamma, c \rangle = TV_G[\mathcal{S}_2^{\Delta_2^*}; (\gamma' \cup \gamma, c + c')]. \tag{23}$$

The operator $\hat{P}^{M_2}_{\gamma', \gamma}$ does not depend on the interior of the simple polyhedron $\mathcal{S}_2^{\Delta_2^*}$ used to compute the transition amplitude [31]. Using this notation we can rewrite Eq. (20) as

$$TV_G[M^3, \Delta] = \langle \emptyset | \hat{P}^{M_3}_{\emptyset, \gamma_2} \hat{P}^{M_2}_{\gamma_2, \gamma_1} \hat{P}^{M_1}_{\gamma_1, \emptyset} | \emptyset \rangle \ . \tag{24}$$

In the previous notation we have implicitly assumed that the boundary of each of the manifolds M_i can be viewed naturally as comprised of two disjoint components ($\partial M = (\partial M)^+ (\partial M)^-$), one associated to the *future* $(\partial M)^+$ and the other to the *past* $(\partial M)^-$. $\hat{P}^M_{\gamma', \gamma}$ is the operator of equation (23) between the Hilbert space associated to the graph $\gamma \subset -(\partial M)^-$ and the Hilbert space associated to the graph $\gamma' \subset (\partial M)^+$. With this convention it is natural to define the composition of two manifolds with boundary M_1 and M_2 when $(\partial M)_2^+ = -(\partial M)_1^-$ as the manifold $M = M_2 \circ M_1$ that one obtains by glueing M_1 and M_2 along the isomorphic boundary $(\partial M)_2^+ = -(\partial M)_1^-$. At this point we are ready to discuss the functorial nature of the invariant. From the independence of the PRTV state sum from the chosen polyhedron we see that $\hat{P}$ defines a semifunctor on the space of manifolds with a fixed graph on their boundary:

$$\hat{P}^{M_2 \circ M_1}_{\gamma', \gamma} = \hat{P}^{M_2}_{\gamma', \gamma''} \hat{P}^{M_1}_{\gamma'', \gamma}. \tag{25}$$

This $\hat{P}$ does not properly define a topological field theory since it is not a functor ($\hat{P}^{\Sigma \times I}_{\gamma, \gamma} \neq 1_\gamma$). There is, however, a standard procedure that allows us to define a functor from a semifunctor. In fact, if we consider the Hilbert space $\overline{\mathcal{H}_\gamma} = \mathcal{H}_\gamma / Ker(\hat{P}^{\Sigma \times I}_{\gamma, \gamma})$ from $\hat{P}^M_{\gamma', \gamma} = \hat{P}^M_{\gamma', \gamma} \circ \hat{P}^{\Sigma \times I}_{\gamma, \gamma}$, it follows that its restriction to the family of Hilbert spaces $\overline{\mathcal{H}_\gamma}$ becomes a functor, i.e., we used the operator $\hat{P}^{\Sigma \times I}_{\gamma, \gamma}$ as the projector on the physical Hilbert space of the theory. In this way the PRTV-model defines a 2+1 dimensional quantum field theory [31] in Atiyah's terminology [60].

From the physical point of view this discussion can be seen as a computation of the Hartle–Hawking wave function and of the topology change amplitude of the

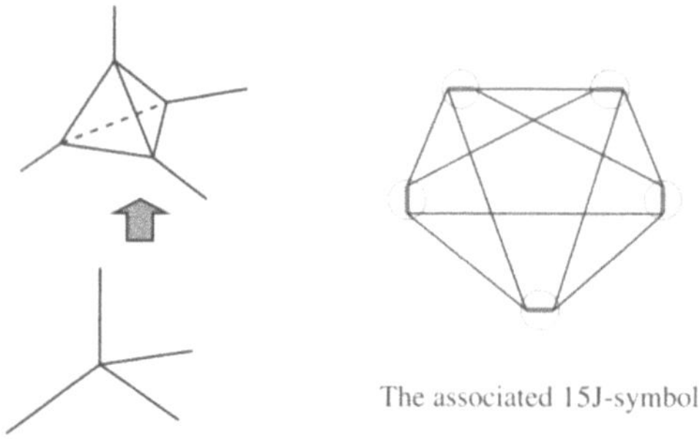

Fig. 5. The elementary vertex of the BF theory. There are 5 nodes (with the associated intertwiner) and 10 links (with the associated representation). To the elementary vertex there are indeed 15 associated colors

3-dimensional lattice gravity of Ponzano and Regge [23]. From the point of view of a Hamiltonian (initial values) formulation in terms of loop-variables this is particularly nice. In fact, in dimension n the Hamiltonian formulation is defined on an $(n-1)$-dimensional Cauchy surface and in any dimension the intersection of an $(n-1)$-surface and any 2-dimensional shadow is a 1-dimensional graph embedded on the Cauchy surface. Even more interesting is the fact that the Hilbert space structure associated to the BF theory is very similar to that of loop quantum gravity.

5 Four-dimensional models

In the previous section we showed how the Ponzano–Regge model corresponds to the BF theory in 3 dimensions. In this section we discuss an example of a 4-dimensional theory that can be expressed within the spin foam formalism: the Crane–Yetter–Ooguri model of BF theory and the spin foam model of Barrett and Crane. We do not discuss here the model of Reisenberger [39, 41]. All these models share the characteristic of being based on a theory of BF type. This led Freidel and Krasnov [43] to propose a way of systematically deriving the *spin foam model* associated to any theory that admits a formulation in terms of an action of BF type plus constraints imposed by Lagrange multipliers.

The Crane–Yetter–Ooguri model

The state sum formulation of the topological BF theory in 4 dimensions,

$$Z_{BF}^{(4)} = \int [\mathcal{D}A(x)][\mathcal{D}B(x)] e^{\mathrm{i}\int \mathrm{Tr}[B\wedge F[A]] + \Lambda\ \mathrm{Tr}[B\wedge B]} \qquad (26)$$
$$= \int [\mathcal{D}A(x)] e^{\frac{1}{2\Lambda}\int \mathrm{Tr}[F[A]\wedge F[A]]}$$

$$= \int [\mathcal{D}A(x)] e^{\frac{1}{2\Lambda}\int d\left(\mathrm{Tr}[A\wedge F[A]]+\frac{2}{3}\mathrm{Tr}[A\wedge A\wedge A]\right)},$$

with cosmological constant is given by the Crane–Yetter model [33] or, without cosmological constant, by the Ooguri model [32]. The partition function of the model associated to a triangulation Δ of a manifold M can be written as

$$Z_{BF}(\Delta, \Lambda) = \sum_{j_f, i_t} \prod_f \dim_q(j_f) \prod_v \phi_{q,v}(\boldsymbol{j}, \boldsymbol{i}), \tag{27}$$

where $q = \exp(\mathrm{i}\Lambda)$, j_f denotes a coloring of the faces of Δ by irreducible representations of $U_q(su(2))$, i_t denotes a coloring of the tetrahedra of Δ by intertwiners and the sum is over all such colorings with $j < \frac{\pi}{\Lambda}$. Moreover $\dim_q(j)$ denotes the quantum dimension of the representation of spin j and $\phi_v(\boldsymbol{j}, \boldsymbol{i})$ denotes the quantum 15-$\boldsymbol{J}$ symbol associated with the 4-simplex v. More precisely, associated to a 4-simplex v we consider a graph Γ_v given by the intersection between the 2-skeleton of the complex which is dual to the triangulation and the boundary of the 4-simplex v (see Fig. 5). Γ_v corresponds to the pentagram graph and we color its 10 edges by $\boldsymbol{j}$ and its 5 vertices by $\boldsymbol{i}$. $\phi_{q,v}(\boldsymbol{j}, \boldsymbol{i})$ corresponds to the Reshetikhin–Turaev evaluation of the colored graph Γ_v. This state sum does not depend on the triangulation when correctly normalized by a factor $N^{n_0 - n_1}$, where n_0, n_1 are the numbers of 0-,1-simplices, respectively and it corresponds to the evaluation of the BF partition function of equation (26).

The Barrett–Crane model

The Relativistic-Spin-Foam model is a modification of the state sum (26) in the case of the $so(4)$ gauge group in which the conditions corresponding to the quantum 4-simplex conditions (when $\Lambda = 0$) are imposed by hand. Yetter generalized this construction to the case of non-vanishing Λ [61]. Yetter emphasized that the quantum group that should be considered is $U_q(su(2)) \otimes U_{q^{-1}}(su(2))$. This state sum model can be written

$$Z_{BC}(\Delta, \Lambda) = \sum_{j_f, j'_f, i_t, i'_t} \prod_f \dim_q(j_f) \dim_{q^{-1}}(j'_f) \prod_v \phi_{q,v}(\boldsymbol{j}, \boldsymbol{i}) \phi_{q^{-1},v}(\boldsymbol{j}', \boldsymbol{i}') \delta_{\boldsymbol{j},\boldsymbol{j}'} \delta_{\boldsymbol{i},\boldsymbol{i}'}. \tag{28}$$

Thus this model corresponds to two copies of $su(2)$ BF theories (or an $so(4)$ BF theory) together with 15 constraints imposed for each 4-simplex. The presence of $\delta_{\boldsymbol{j},\boldsymbol{j}'}$ and $\delta_{\boldsymbol{i},\boldsymbol{i}'}$ corresponds to the geometricity constraint. In [29] it is shown that the above spin foam model can be associated to a well-defined field action: the $so(4)$ Plebanski action [36]. This is a deformed BF theory depending on an $so(4)$ connection $\omega = \omega_\mu^{IJ} X_{IJ} dx^\mu$, a two-form valued into $so(4)$ $B = B_{\mu\nu}^{IJ} X_{IJ} dx^\mu \wedge dx^\nu$, and a scalar symmetric traceless matrix $\phi_{[IJ][KL]}$ ($\epsilon^{IJKL}\phi_{IJKL} = 0$) that acts as

Lagrange multiplier fields. This action reads:

$$\mathcal{S}[\omega; B; \phi] = \int_M \Bigg[B^{IJ} \wedge F_{IJ}(\omega) - \frac{\Lambda}{4} \epsilon_{IJKL} B^{IJ} \wedge B^{KL} - \frac{1}{2} \phi(B)_{IJ} \wedge B^{IJ} \Bigg], \tag{29}$$

where $\phi(B)_{IJ} = \phi_{IJKL} B^{KL}$. In this context, the geometry constraints are imposed, at the quantum level, by integration over Lagrange multiplier fields $\phi_{[IJ][KL]}$. This action has two nice features: one of its sectors of solution is closely connected to gravity and, secondly, it is a deformation of a quantizable topological theory (the BF theory) that can play a role analogous to that of the free theory in standard quantum field theory.

6 Conclusions

In this work we showed that the spin foam formalism gives an unified framework for the discussion of an important class of diff-invariant quantum theories. Between them: topological BF theory in 3 and 4 dimensions, the four-dimensional formulation of loop quantization of gravity, the Barrett–Crane model. The main unsolved problems of the general picture we described here are:

1. the state sum models we described are properly defined only on a single given triangulation (or its dual spin-foam). If this is valid in the case of a topological field theory without local degrees of freedom, it is questionable in the case of gravity. Work is now in progress [62] in the direction of dealing with all possible triangulations. This is somehow analogous to considering generalizations (where the triangulation is not the only dynamical variable) of the model of dynamical triangulations [21].
2. the connection between the spin foam models and the exponential of the constraints needs to be improved with a rigorous mathematical analysis of the derivation.
3. a systematic study of the physical implications of taking different weight factors is missing.

Acknowledgements

This work was greatly influenced by numerous discussions with M. Carfora, L. Freidel, J. Lewandowski, L. Lusanna, A. Marzuoli, M. Pauri, C. Rovelli and J. Zapata, to all of whom I am extremely grateful. The work of R.D.P. at CPT Marseille is supported by a Dalla Riccia Fellowship.

References

1. Rovelli C., Smolin L. (1988): Knot theory and quantum gravity, Phys. Rev. Lett. **61**, 1155
2. Rovelli C., Smolin L. (1990): Loop space representation of quantum general relativity. Nucl. Phys. B **331**, 80
3. Rovelli C. (1998): Strings, Loops and Others: A Critical Survey of the Present Approaches in to Quantum Gravity, in *Gravitation and Relativity: At the Turn of the Millennium (Pune, India)*, ed. by N. Dadhich, J. Narlikar, Proc. GR15 Conference, I.U.C.A.A., December 16–21, p. 281
4. Rovelli C. (1998): Loop Quantum Gravity in Living Reviews in Relativity, 1998-1. Eleconic publications of the Max-Planck-Institut für Gravitationsphysik, Albert-Einstein-Institut, Potsdam, http://www.livingreviews.org//Articles/Volume1/1998-1rovelli/
5. Ashtekar A., Isham C. J. (1992): Inequivalent observable algebras: another ambiguity in field quantization. Phys. Lett. B **274**, 393–398
6. Ashtekar A., Lewandowski J. (1995): Differential geometry on the space of connections via graphs and projective limits. J. Geom. Phys. **17**, 191
7. Baez J.C. (1994): Generalized measures in gauge theory. Lett. Math. Phys. **31**, 213–224
8. Rovelli C., Smolin L. (1995): Spin networks and quantum gravity. Phys. Rev. D **52**, 5743–5759
9. Ashtekar A., Lewandowski J., Marolf D., Mourao J., Thiemann T. (1995): Quantization of diffeomorphism invariant theories of connections with local degrees of freedom. J. Math. Phys. **36**, 6456–6493
10. Rovelli C., Smolin L. (1995): Discreteness of area and volume in quantum gravity. Nucl. Phys. B **442**, 593–622, (1995) erratum, (1995): Nucl. Phys. B **442**
11. De Pietri R., Rovelli C. (1996): Geometry eigenvalues and scalar product from recoupling theory in loop quantum gravity. Phys. Rev. D **54**, 2664–2690
12. Ashtekar A., Lewandowski J. (1997): Quantum theory of geometry. 1: Area operators. Class. Quant. Grav. A **14**, 55–82
13. Ashtekar A., Lewandowski J. (1998): Quantum theory of geometry. 2. Volume operators. Adv. Theor. Math. Phys. **1**, 388
14. Rovelli C. (1996): Black hole entropy from loop quantum gravity. Phys. Rev. Lett. **77**, 3288
15. Barreira M., Carfora M., Rovelli C. (1996): Physics with nonperturbative quantum gravity: radiation from a quantum black hole. Gen. Rel. Grav. **28**, 1293
16. Krasnov K.V. (1997): Geometrical entropy from loop quantum gravity. Phys. Rev. D **55**, 3505–3513
17. Ashtekar A., Baez J., Corichi A., Krasnov K. (1998): Quantum geometry and black hole entropy. Phys. Rev. Lett. **80**, 904–907
18. Reisenberger M.P., Rovelli C. (1997): 'Sum over surfaces' form of loop quantum gravity. Phys. Rev. D **56**, 3490–3508
19. Baez J.C. (1998): Spin foam models. Class. Quant. Grav. **15**, 1827
20. Reisenberger M.P. (1994): World sheet formulations of gauge theories and gravity, gr-qc/9412035
21. Ambjorn J., Carfora M., Marzuoli A. (1997): The geometry of Dynamical Triangulations. Lecture Notes in Physics, Vol. **50**, Springer Berlin Heidelberg New York
22. Ooguri H., Sasakura N. (1991): Discrete and continuum approaches to three-dimensional quantum gravity. Mod. Phys. Lett. A **6**, 591–3600
23. Ooguri H. (1992): Partition functions and topology changing amplitudes in the 3-d lattice gravity of Ponzano and Regge. Nucl. Phys. B **382**, 276–304

24. Archer F., Williams R.M. (1991): The Tauraev–Viro state sum model and three-dimensional quantum gravity. Phys. Lett. B **273**, 438–444
25. Witten E. (1988): Topological quantum field theory. Commun. Math. Phys. **117**, 353
26. Witten E. (1988): (2+1)-dimensional gravity as an exactly soluble system. Nucl. Phys. B **311**, 46
27. Witten E. (1988): Topological gravity. Phys. Lett. B **206**, 601
28. Witten E. (1989): Quantum field theory and the Jones polynomial. Commun. Math. Phys. **121**, 351
29. Witten E. (1989): Topology changing amplitudes in (2+1)-dimensional gravity. Nucl. Phys. B **323**, 113
30. Ponzano G., Regge T. (1968) Semiclassical Limit of Racach Coefficients, in *Spetroscopy and group theoretical methods in Physics*, ed. by F. Bloch, North-Holland, Amsterdam, pp. 1–58
31. Turaev V.G., Viro O.Y. (1992): State sum invariant of 3-manifolds and quantum 6j-symbols. Topology **31** (4), 865–902
32. Ooguri H. (1992): Topological lattice models in four-dimensions. Mod. Phys. Lett. A **7**, 2799–2810
33. Crane L., Kauffman L.H., Yetter D.N. (1997): State-sum invariants of 4-manifolds. J. Knot Theory Ramifications **6** (2), 177–234, hep-th/9409167
34. Carter J.S., Kauffman L.H., Saito M. (1998): Structures and diagrammatics of four-dimensional topological lattice field theories. math.GT/9806023
35. Barrett J.W., Crane L. (1998): Relativistic spin networks and quantum gravity. J. Math. Phys. **39**, 3296–3302
36. Plebanski J. (1977): On the separations of Einteinian substructure. J. Math. Phys. **18**, 2511
37. De Pietri R., Freidel L., (1999): SO(4) Plebanski action and relativistic spin foam model. Class. Quant. Grav. **16**, 2187–2196
38. Reisenberger M.P. (1999): Classical Euclidean general relativity from 'lefthanded area = righthanded area'. Class. Quant. Grav. **16**, 1357–1371
39. Reisenberger M.P. (1997): A lattice world sheet sum for 4-d Euclidean general relativity, gr-qc/9711052
40. Reisenberger M.P. (1995): New constraints for canonical general relativity. Nucl. Phys. B **457**, 643–687
41. Reisenberger M.P. (1997): A lefthanded simplicial action for Euclidean general relativity. Class. Quant. Grav. **14**, 1753
42. Horowitz G.T. (1989): Exactly soluble diffeomorphism invariant theories. Comm. Math. Phys. **125**, 417
43. Freidel L., Krasnov K. (1999): Spin foam models and the classical action principle. Adv. Theor. Math. Phys. **2**, 1183–1247
44. Rourke C.P., Sanderson B.J. (1972): Introduction to Picewise-Linear Topology. Springer, Berlin Heidelberg New York
45. Zapata J.A. (1998): Combinatorial space from loop quantum gravity. Gen. Rel. Grav. **30**, 1229
46. Zapata J.A. (1997): A combinatorial approach to diffeomorphism invariant quantum gauge theories. J. Math. Phys. **38**, 5663–5681
47. Kauffman L.H., Lins S. L. (1994): Temperley-Lieb Recoupling Theory and Invariants of 3-Manifolds. Princeton University Press, Princeton
48. De Pietri R. (1997): On the relation between the connection and the loop representation of quantum gravity. Class. Quant. Grav. **14**, 53–70

49. Penrose R. (1971): Application of Negative Dimensional Tensors, in *Combinatorial Mathematics and its Applications*, ed. by D.J. Welsh, Academic Press, London, pp. 221–243
50. Ashtekar A. (1986): New variables for classical and quantum gravity. Phys. Rev. Lett. **57**, 2244
51. Barbero J.F. (1995): Real Ashtekar variables for Lorentzian signature space times. Phys. Rev. D **51**, 5507–5510
52. Barbero J.F. (1996): From Euclidean to Lorentzian general relativity: the real way. Phys. Rev. D **54**, 1492–1499
53. Thiemann T. (1996): Anomaly – free formulation of nonperturbative, four-dimensional Lorentzian quantum gravity. Phys. Lett. B **380**, 257–264
54. Thiemann T. (1998): Quantum spin dynamics (QSD). Class. Quant. Grav. **15**, 839
55. Thiemann T. (1998): Quantum spin dynamics (QSD). II. Class. Quant. Grav. **15**, 875
56. Borissov R., De Pietri R., Rovelli C. (1997): Matrix elements of Thiemann's Hamiltonian Constraint in Loop Quantum Gravity, Class. Quant. Grav. **14**, 2793
57. Thiemann T. (1998): Quantum spin dynamics (QSD). III. Quantum constraint algebra and physical scalar product in quantum general relativity. Class. Quant. Grav. **15** , 1207
58. Rovelli C. (1998): The projector on physical states in loop quantum gravity. Phys. Rev. D **59**, 104015
59. Carbone G., Carfora M., Marzuoli A. (1998): Wigner symbols and combinatorial invariants of 3-manifolds with boundary. SISSA Preprint 118/98/FM
60. Atiyah M. (1989): Topological quantum field theories. Publ. Math. IHES **68**, 175–186
61. Yetter D.N. (1998): Generalised Barrett–Crane vertices and invariants of embedded graphs. math.QA/9801131
62. De Pietri R., Freidel L., Krasnov K., Rovelli C. (1999): Barrett–Crane model from a Boulatov-Ooguri field theory over a homogeneous space. Nucl. Phys. B, in press, hep-th/9907154

A Possible Way to Define an Effective Cosmological Constant in Scalar–Tensor Cosmologies

S. Capozziello, R. de Ritis, A.A. Marino, C. Rubano, P. Scudellaro

Abstract. The cosmological constant has attained a leading role in recent researches in cosmology. We extend the cosmic no–hair theorem to non-minimally coupled theories of gravity where ordinary matter is also present in the form of a perfect fluid. To achieve this goal we give a set of conditions for obtaining the asymptotic de Sitter expansion independently of any initial data (no fine–tuning problem), that is, we introduce a time–dependent (effective) cosmological constant. Finally, we apply the results to some specific models.

There are many reasons today to believe that several cosmological models have to incorporate a cosmological "constant" term. In fact, if we consider, e.g., gravitational lensing, there are at least three important motivations for taking into account a cosmological "constant": the first is connected to lensing phenomena from distant sources; the second is related to the statistics of the lensed quasars; finally there is the question of "weak lensing" and the formation of multiple images (arclets) of distant galaxies (for a detailed analysis of these three topics and an exhaustive list of references, see [1]). The value of Ω_Λ which comes from such observations seems to be close to $\Omega_\Lambda = 0.6$, the so-called Kochanek limit. The very recent observations of two teams, the *High-z Supernova Search Team* [2] and the *Berkeley Supernova Cosmology Project* [3], have reached the same conclusions concerning the necessity of a nonzero Λ; they also agree with the Kochanek limit given above. The same considerations can be made in studying the cosmological age problem [4]. However, from this last perspective, we have to deal with the problem of introducing Λ from the theoretical side.

In this talk we seek to find a more satisfactory way of introducing a Λ–term which can also be connected to the inflationary paradigm as well to current observations. What we are going to do is to follow the so-called "*cosmological no–hair theorem*" [5], extending it to the case of non-minimally coupled scalar–tensor theories of gravity which are playing an important role in inflationary cosmology (see, e.g., [6],[7],[8] and references therein). In such a context it could be interesting to analyse the meaning of a cosmological "constant" reconstructed using only the "scalar–field aspect" of such theories. From this point of view, the real problem is how to define this quantity since it is for several reasons "unpleasant" to introduce it by hand.

In the context of general relativity, the so-called *no–hair conjecture* was introduced by Hoyle and Narlikar [9]. They claimed that, if there is a positive cosmological constant, all the expanding universe models would approach the de Sitter behaviour. That is, if a cosmological constant is present, no matter what the initial conditions, the universe will become homogeneous–isotropic– de Sitter. However, there is something vague in such a formulation. The meaning of *"expanding universe"* as well as of *"approach the de Sitter behaviour"* is unclear. Furthermore, there is no general proof (or disproof) of such a conjecture; on the contrary, there are counter–examples of initially expanding and then recollapsing universe models which never become de Sitter [10].

In 1983, Wald gave a proof of a simplified version of the conjecture. He proved that: *All Bianchi cosmologies (except IX), in the presence of a positive cosmological constant, asymptotically approach the de Sitter behaviour* [5].

In all these discussions (in Wald's paper too), the cosmological constant is a true constant and it is put by hand in the gravitational arena. It is relevant that, in Wald's proof, the Bianchi identities are not used, and then the proof is independent of the dynamical evolution of the material sources. As already said, here we discuss how to introduce a time–dependent "cosmological constant" in the context of scalar–tensor theories of gravity in which a scalar field ϕ is non-minimally coupled to geometry, a standard perfect fluid matter (noninteracting with ϕ) being also present (see [11] also). So doing, we introduce a more general set of conditions (with respect to those given by Wald), not based on the "energy conditions", by which it is possible to get a *de Sitter asymptotic* behaviour for the scale factor of the universe. In other words, we introduce an *"effective cosmological constant"* via the asymptotic de Sitter expansion. Furthermore, it is shown that the de Sitter asymptotic behaviour does not depend on the (asymptotic) sign, as well as on the value, of the effective gravitational "constant", which is given by $(8\pi G_N = \hbar = c = 1)$

$$G_{\text{eff}} = -\frac{1}{2F(\phi)}, \tag{1}$$

where $F(\phi)$ is the (non-minimal) gravitational coupling.

Before entering into a detailed exposition, it could be useful to present briefly how it is possible to generalize Wald's conditions. For the sake of simplicity, a Friedman–Robertson–Walker (FRW)–flat cosmology will be used. However, the results can be easily extended to any Bianchi model [12].

First of all, let us recall Wald's conditions: *i)* $(H - \sqrt{\Lambda/3})(H + \sqrt{\Lambda/3}) \geq 0$, and *ii)* $\dot{H} \leq \Lambda/3 - H^2$; here Λ is the cosmological constant, $a(t)$ is the scale factor and $H = \dot{a}/a$ is the Hubble parameter. Instead of conditions $i)$, $ii)$, let us assume the weaker ones: $A)$ $(H - \Lambda_1)(H + \Lambda_2) \geq 0$, and $B)$ $\dot{H} \leq 0$. In general, the constants $\Lambda_{1,2}$ of condition $A)$ need not coincide with the true cosmological constant Λ. They coincide, as we shall see below, if the Bianchi identities are taken into account. Condition $B)$ is more general than $ii)$ since $B) \Longrightarrow ii)$, but the converse is not true. Condition $i)$ and $A)$ are essentially the same, apart from the identification of thecosmological constant. Under $A)$ and $B)$, $H(t)$ is a decreasing function bounded from below. In other words $H(t)$ has a horizontal asymptote, or, equivalently, H goes to a constant. This can be seen in the following way: conditions $A)$, $B)$ imply that the following sequence can be constructed: $\sigma_n = H(t_0 + t_n)$, where t_0 is an arbitrarily chosen large time and t_n is an ordered increasing sequence of instants greater than 0 for each (integer) n. Then the given sequence σ_n is such that

$$\sigma_{n+1} \leq \sigma_n \text{ for any } n \geq 0, \quad (\text{as } \dot{H} \leq 0 \text{ by hypothesis } (B)) \tag{2}$$

and

$$\sigma_n \geq \sqrt{\frac{\Lambda}{3}} \text{ for any } n \geq 0, \quad (\text{as } H \text{ is bounded from below by hypothesis } (A)). \tag{3}$$

This means that σ_n is a monotone decreasing sequence; thus it is convergent, that is, $\lim_{t\to\infty} H(t) = H_0$. Therefore, H has a horizontal asymptote and $a(t)$ asymptotically becomes like de Sitter, independent of any initial condition, i.e., $a(t) \to \exp(\alpha t)$, where α is an unknown constant (the details of the proof are given in [14]). In practice, hypotheses A), B) imply a de Sitter behaviour satisfying the statement given by Wald.

Let us now consider the Bianchi identity as well (we have not used it so far in obtaining the asymptotic de Sitter behaviour): with standard matter present, the matter–energy density (in an FRW background) is $\rho_m = Da^{-3\gamma}$, where $\gamma\epsilon[1, 2]$ as usual. Then, considering the (0, 0) Einstein equation, we see that

$$H^2 = \frac{\Lambda}{3} + \frac{\rho_m}{3}, \tag{4}$$

or equivalently

$$\left(H - \sqrt{\frac{\Lambda}{3}}\right)\left(H + \sqrt{\frac{\Lambda}{3}}\right) = \frac{D}{3}e^{-3\gamma\alpha t}, \tag{5}$$

and so $H \to \sqrt{\frac{\Lambda}{3}}$, i.e., $\alpha = \sqrt{\frac{\Lambda}{3}}$. By proving the no–hair theorem in this way, we have shown the different role that the Bianchi identity actually plays: among many possible asymptotic de Sitter behaviours, it fixes one of them (that is, the cosmological constant) without introducing any initial data dependence. Of course, the first issue to be addressed is: where do hypotheses A), B) come from? Wald's conditions i), ii) are strictly related to the energy conditions so that they are plausible. What about conditions A), B)?

Let us now go into more detail about the theories which we are using. The FRW–Einstein cosmological equations are

$$H^2 + \frac{\dot{F}}{F}H + \frac{\rho_\phi}{6F} + \frac{\rho_m}{6F} = 0, \tag{6}$$

$$\dot{H} = -\left(H^2 + \frac{V}{6F}\right) - H\frac{\dot{F}}{2F} + \frac{\dot{\phi}^2}{6F} - -\frac{1}{2}\frac{\ddot{F}}{F} + \frac{3p_m + \rho_m}{12F}, \tag{7}$$

where $H = \dot{a}/a$, $\rho_\phi = \frac{1}{2}\dot{\phi}^2 + V(\phi)$, ρ_m, p_m are, respectively, the Hubble parameter, the energy density of the scalar field (the potential is generic), the energy density and the pressure of standard matter. (Actually such equations come from the general field action

$$\mathcal{A} = \int d^4x\sqrt{-g}\left[F(\phi)R + \frac{1}{2}g^{\mu\nu}\phi_{;\mu}\phi_{;\nu} - V(\phi) + \mathcal{L}_m\right], \tag{8}$$

rewritten in the cosmological case. $V(\phi)$ and $F(\phi)$ are generic functions of the field ϕ and $\mathcal{L}_m$ is the ordinary matter contribution to the total Lagrangian density [13].) Equation (6) can be rewritten using the second degree polynomial $\mathcal{P}(H)$, as

$$\mathcal{P}(H) \equiv \left(H - \Lambda_{\text{eff},1}\right)\left(H - \Lambda_{\text{eff},2}\right) = -\frac{\rho_m}{6F}, \tag{9}$$

where

$$\Lambda_{\text{eff},1,2} = -\frac{\dot{F}}{2F} \pm \sqrt{\left(\frac{\dot{F}}{2F}\right)^2 - \frac{\rho_\phi}{6F}}. \tag{10}$$

Here "1" corresponds to the plus sign and "2" to the minus sign. The two values $\Lambda_{\text{eff},1,2}$ have to be real, and so the restriction $\left(\frac{\dot{F}}{2F}\right)^2 \geq \frac{\rho_\phi}{6F}$ has to be satisfied.

It is easy to see that

$$\Lambda_{\text{eff},1} + \Lambda_{\text{eff},2} = -\frac{\dot{F}}{F},$$

and

$$\Lambda_{\text{eff},1} - \Lambda_{\text{eff},2} = 2\sqrt{\left(\frac{\dot{F}}{2F}\right)^2 - \frac{\rho_\phi}{6F}} \geq 0.$$

In general: $\Lambda_{\text{eff},1} \geq \Lambda_{\text{eff},2}$. Of course, the equation for $\dot{H}$ can be also written using $\mathcal{P}(H)$. We now make the following hypotheses, namely, for $t \gg 0$, we suppose that: **i)** $\frac{\dot{F}}{F} \longrightarrow \Sigma_0$, and **ii)** $\frac{\rho_\phi}{6F(\phi)} \longrightarrow \Sigma_1$, where $\Sigma_{0,1}$ are two constants depending, respectively, on the parameters of the coupling, and on the parameters of the coupling and of the potential. Under these two hypotheses, we see that the two quantities $\Lambda_{\text{eff},1,2}$ asymptotically go to constants. Conversely, if it is assumed that $\Lambda_{\text{eff},i} \rightarrow \Lambda_i$ (constants), then $\dot{F}/F$ and $\rho_\phi/6F$ become constants for large t. Hypotheses **i)** and **ii)** are necessary and sufficient conditions since the two Λ's are asymptotically constants.

We will also assume that asymptotically the sign of $F(\phi)$ is constant (this is the third, quite natural, assumption); then two cases have to be considered: $F(t \gg 0) \leq 0$ and $F(t \gg 0) \geq 0$. Since we are supposing that, asymptotically, $\dot{F}/F$ is constant, each of the above cases has two subcases related to the sign of $\dot{F}$. Only the case $F(t \gg 0) \leq 0$ is physically relevant. The other case (repulsive gravity) can be interesting if related to the possibility of recovering the asymptotic de Sitter behaviour for $a(t)$; this implies that recovering such a behaviour for $a(t)$ in not connected with recovering the standard sign of gravity. Let us now consider the case $F(t \gg 0) \leq 0$ and $\dot{F}(t \gg 0) \leq 0$; from hypothesis **i)** we see that $\Sigma_0 \geq 0$. Then the reality condition on the discriminant connected to the Λ_{eff}'s is (asymptotically) satisfied. For $\mathcal{P}(H)$ the inequality $\mathcal{P}(H) \geq 0$ is obtained, and then we have $H \geq \Lambda_1$, $H \leq \Lambda_2$. For the Λ_i, there is then the asymptotic expression $\Lambda_{1,2} = -\frac{\Sigma_0}{2} + \sqrt{\left(\frac{\Sigma_0}{2}\right)^2 + |\Sigma_1|} \geq 0$. For $\dot{H}$, we get $\dot{H} = -\left(H^2 - \frac{V}{6|F|}\right) - H\frac{\dot{F}}{2F} - \frac{\dot{\phi}^2}{6|F|} - \frac{1}{2}(\frac{\ddot{F}}{F}) - \frac{3p_m + \rho_m}{12|F|}$. If (this is the last hypothesis) **iii)** $H^2 \geq \frac{V}{6|F|}$, then $\dot{H} \leq 0$. In other words, $H(t)$ has a horizontal asymptote, or, equivalently, H goes to a constant. Then the universe, for large t, has a de Sitter behaviour (i.e., $a(t) \sim \exp(\alpha t)$, where α is an unknown constant). The universe, for large t, expands as de Sitter, even if the parameter which specifies such an expansion, i.e., the effective cosmological constant, is not fixed.

If we compare Wald's conditions with those presented here, we have:

(Wald's)		(new conditions)
$\left(H - \sqrt{\frac{\Lambda}{3}}\right)\left(H + \sqrt{\frac{\Lambda}{3}}\right) \geq 0$	$\Longleftrightarrow$	$(H - \Lambda_1)(H - \Lambda_2) \geq 0,$
$\dot{H} \leq \frac{\Lambda}{3} - H^2 \leq 0$	$\Longrightarrow$	$\dot{H} \leq 0.$

The equations involving H are the same in both cases. The true difference concerns the equation for $\dot{H}$; our condition ($\dot{H} \leq 0$) is more general than $\dot{H} \leq (\Lambda/3 - H^2) \leq 0$. Hypothesis **iii)**, when $\phi \to$ const., is nothing else but $H^2 \geq \frac{\Lambda}{3}$, that is, we recover the standard case in which $V =$ const is interpreted as the cosmological constant. By some algebra, it is easy to show that such a hypothesis is equivalent to

$$\frac{1}{12F}\frac{\dot{\phi}^2}{2V} \geq \left(\frac{F'}{F}\right)^2 = \left(\frac{G'_{\mathrm{eff}}}{G_{\mathrm{eff}}}\right)^2 . \tag{11}$$

Having shown that $a(t)$ behaves like de Sitter for large t, we may now fix α in order to recover the effective cosmological constant. For this purpose, the Bianchi contracted identity for matter is needed (it is important to stress that we have not used any Bianchi contracted identity to find the asymptotic behaviour of $a(t)$). As usual, we get $\rho_m = Da^{-3\gamma}$ (here the standard matter state equation has been used; D is the integration constant giving the matter content of the universe). Introducing this expression for the matter in Eq. (7), for large t, we have $(H - \Lambda_1)(H + |\Lambda_2|) = \frac{D}{|F_0|}e^{-(3\gamma\alpha + \Sigma_0)t}$, where $3\gamma\alpha + \Sigma_0 \geq 0$. Then we get $(H - \Lambda_1)(H + |\Lambda_2|) \to 0$, i.e., $H \to \Lambda_1$. The (effective) matter content, $\rho_m/6F(\phi)$, indicates how far H is "distant" from the de Sitter behaviour given by the cosmological constant Λ_1. In other words, the Bianchi identity is not used for finding the type of expansion; we use it only to (asymptotically) select the specific value of what we call "the cosmological constant". Essentially, we have introduced the (effective) cosmological constant in a "pragmatic" way, through the (asymptotic) de Sitter behaviour for $a(t)$. In a certain sense, the approach in [5] is reversed: there, Λ (constant) is introduced *a priori* and this leads, under certain hypotheses, to a de Sitter expansion. Here, the de Sitter expansion is recovered under completely different hypotheses, and this selects the effective cosmological constant. Moreover, we have obtained such a result without assuming to recover the standard gravity (i.e., that $G_{\rm eff} \to G_N$ is not needed). In this way, we believe that we have answered the question: how can one recover Wald's hypotheses from A), B)?

Also taking into account the Klein–Gordon equation $\ddot{\phi} + 6\left(\dot{H} + 2H^2\right)F'(\phi) + 3H\dot{\phi} + V'(\phi) = 0$, we get, for large t, $\dfrac{\dot{\phi}^2}{F(\phi)} = \text{const} = C_1(\Sigma_0, \Sigma_1)$, which implies that $|\Sigma_1| \geq 2\Sigma_0^2$. We also get $\dfrac{V}{6F} = C_2(\Sigma_0, \Sigma_1)$, as well as $\frac{\dot{\phi}^2}{V} = C_3(\Sigma_0, \Sigma_1)$. In the other case, $\dot{F}(\phi(t \gg 0)) \geq 0$, everything proceeds as in the Wald case.

Finally, let us consider the case $F(\phi(t \gg 0)) \geq 0$. Here also there are two subcases, $\dot{F}(\phi(t \gg 0)) \leq 0$ and $\dot{F}(\phi(t \gg 0)) \geq 0$. As already stressed, even if this situation seems unphysical, it gives a better understanding of the non-necessary correlation between the (asymptotic) de Sitter behaviour (i.e., between the *no–hair* theorem) and the recovering of standard gravity. Of course, now the condition on the reality of Λ_i has to be carefully considered. The most interesting case is $\dot{F} \leq 0$. Here we have two (asymptotic) positive cosmological constants, that is, $\Lambda_{\rm eff\,1,2} \to \Lambda_{1,2} \geq 0$. We now have $\Lambda_1 \leq H \leq \Lambda_2$. Now, it is crucial to know the sign of $\dot{H}$: if $\dot{H} \geq 0$, the effective Λ is given by max (Λ_1, Λ_2); conversely, if $\dot{H} \leq 0$, Λ is given by their minimum.

Examples

We restrict the analysis to a dust–dominated universe since we are interested in asymptotic regimes. We discuss two cases: in the first, the sign of the function describing the coupling is negative (standard gravity); in the second the opposite case is discussed (repulsive gravity).

First case: $F(\phi) = k_0\phi^2$, $V(\phi) = \lambda\phi^2$, $\gamma = 1$, where $k_0 < 0$ and $\lambda > 0$ are free parameters; the de Sitter regime is recovered even if solutions do not converge towards standard gravity. The coupling $F(\phi)$ is always negative, whereas $V(\phi)$ is always positive and $\dot{F}(\phi(t \gg 0)) < 0$. In fact, the general solution [8],[14] is de Sitter for $t \gg 0$ as $a(t) \sim \exp(\Lambda t)$, $\phi(t) \sim \exp(\Lambda_1 t)$, where $\Lambda = \sqrt{\dfrac{\lambda(1-8k_0)^2}{2k_0(12k_0-1)(3-32k_0)}} > 0$, is exactly the asymptotic behaviour of $\Lambda_{\rm eff,\,1}$. Λ_1 is another constant depending on λ and k_0. It is relevant to stress that: $F(\phi(t \gg 0)) < 0$ and $F(\phi(t))$ diverges. We do not recover asymptotically the standard G_N. Actually we have (plus infinity) asymptotic gravitational freedom [15]: nevertheless we have de Sitter behaviour at infinity for $a(t)$. Furthermore, condition **ii)** is always satisfied.

The second case we take into account is: $F(\phi) = k_0\phi^2, \quad V(\phi) = \lambda\phi^2, \quad \gamma = 1$, with $k_0 > 0$ (specifically, $1/12 < k_0 < 3/32$). The solutions are asymptotically the same as in the case $k_0 < 0$ [8]. Now $\phi(t)$ is a decreasing function of time, which implies that $\dot{F}(\phi(t \gg t)) \leq 0$ and $\Sigma_0 \leq 0$. We see that $\dot{H} \leq 0$ and then the effective cosmological constant is given by min (Λ_1, Λ_2). By some algebra, it is possible to verify that the true cosmological constant is Λ_2 (which is always less than Λ_1). This example is useful to emphasize the fact that the de Sitter asymptotic behaviour, connected with the presence of a cosmological constant, is independent of the sign of gravitational coupling. (For a more detailed discussion, and for a more extended class of examples, see [14].)

In conclusion, it is possible to formulate the cosmic no–hair theorem in the framework of non-minimally coupled scalar–tensor theories by introducing a time dependent cosmological "constant", not using the "geometrical side" of such theories (i.e., $\Lambda g_{\mu\nu}$, as usual) but the "scalar side". Under the hypotheses used, the de Sitter asymptotic regime is obtained and this is not necessarily connected with recovering standard gravity. It is interesting to stress that, by this mechanism, the "amount of Λ" is strictly related to the "amount" of matter content of the universe. We believe that this has some interest for a better theoretical understanding of the role played by the Ω parameter as we mentioned at the beginning of our talk.

References

1. Fort B., Mellier Y. (2000): Gravitational Lensing anf the Test of the Cosmological Constant, in *Proc. of the VI Colloque de Cosmologie*, June 4–6, 1997, in press
2. Riess A.G., Filippenko A.V., Challis P. (1998): Astron. J. **116**, 1009
3. Pelmutter S., Alderling G., Goldhaber G. (1999): Astrophys. J. **517**, 565
4. Krauss L.M., Turner M.S. (1995): General Relativity and Gravitation **27**, 1137
5. Wald R.M. (1983): Phys. Rev. D **28**, 2118
6. Bergmann P.G. (1968): Int. Journ. Theor. Phys. **1**, 25
7. Sciama D.W. (1953): Mon. Not. R. Astrom. Soc. **113**, 34;
   Brans C., Dicke R.H. (1961): Phys. Rev. **124**, 925;
   Zee A. (1979): Phys. Rev. Lett. **42**, 417;
   Smolin L. (1979): Nucl. Phys. B **160**, 253;
   Adler S. (1980): Phys. Rev. Lett. **44**, 1567
8. Capozziello S., de Ritis R., Rubano C., Scudellaro P. (1996): La Rivista del Nuovo Cimento **4**, 1
9. Hoyle F., Narlikar J.V. (1963): Proc. R. Soc. A **273**, 1
10. Cotsakis S., Flessas G. (1993): Phys. Lett. B **319**, 69;
    Burd A.B., Barrow J.D. (1988): Nucl. Phys. B **308**, 929;
    Yokogawa J., Maeda K. (1988): Phys. Lett. B **207**, 31;
    Barrow J.D., Götz, G. (1989): Phys. Lett. B **231**, 228
11. Pimentel L.O., Stein-Schabes J. (1989): Phys. Lett. B **216**, 27
12. Capozziello S., de Ritis R. (1996): Int. Journ. Mod. Phys. D **5**, 209
13. Capozziello S., de Ritis, R. (1993): Phys. Lett. A **117**, 1;
    Capozziello S., de Ritis, R. (1994): Class. Quantum Grav. **11**, 107;
    Capozziello S., de Ritis R., Scudellaro P. (1994): Phys. Lett. A **188**, 130

14. Capozziello S., de Ritis R. (1997): Gen. Relativ. Grav. **29**, 1425;
Capozziello S., de Ritis R., Marino A. A. (1998): Gen. Relativ. Grav. **30**, 1247
15. Capozziello S., de Ritis R. (1995): Phys. Lett. A **208**, 181

A Microscopic Approach to Nuclear Physics: from Deuteron to Neutron Stars

A. Fabrocini

Abstract. The latest attempts to understand the properties of all nuclei in terms of microscopic interactions between individual nucleons are described. This contribution tries to give a consistent picture of nuclear systems, from deuteron up to the heaviest nuclei (including neutron stars), based on modern, non-relativistic Hamiltonians containing two- and three-body potentials which fit, in an extremely accurate way, the large amount of experimental data on ground and scattering states of two or more nucleons. The resolution of the Schrödinger equation for light nuclei provides, for the first time, clear evidence of a shell structure without the need of introducing an *ad hoc* mean field. The microscopic equation of state of dense nucleon matter, when used for deriving neutron star structure, gives results in good agreement with the present observational data.

1 Introduction

One of the many, and still unaccomplished, dreams of theoretical nuclear physicists is the description of the whole zoo of nuclear systems in terms of the same microscopic nucleon–nucleon (NN) interaction. It was not a long time ago that the deuteron was the only nucleus whose Schrödinger equation, with a realistic interaction, we knew how to solve. Luckily enough the situation has improved with the advent of powerful and fast computers and the development of sophisticated and accurate semianalytical or purely numerical techniques. By now the ground and low lying states of nuclei with mass number $A \leq 9$ are currently studied starting from a NN interaction directly derived from the NN scattering data. Even for heavier systems the perspectives are changing and, at least for the ground states, the shell model is no longer the only feasible tool for examining these nuclei. Variational methods are opening the way for the study of the nuclear *correlations*, e.g., the modifications induced in the many-body nuclear wave function by the strong interaction.

The mass difference between a nucleus, with Z protons and N neutrons, and free nucleons gives the nuclear binding energy curve

$$B(Z, N) = \left[Zm_p + Nm_n - m_{nuc}(Z, N)\right] c^2 \tag{1.1}$$

which is the first and most relevant information in nuclear physics. The curve of $B(Z, N)/A$ for stable nuclei as a function of A rises rapidly up to its maximum of 8.8 MeV in the region $A \sim 60$ and gradually descends to 7.9 Mev at $A = 208$ because of the Coulomb repulsion. As is well-known, the curve exhibits local spikes in correspondence with the magic numbers 2, 8, 20, 28, 59, 82 and 126 for N or Z. The *saturation* of the binding energy per nucleon was the first evidence of the short range of the nuclear forces [1] and the existence of magic numbers was at the origin of the shell model [2].

A simple parametrization of $B(Z, N)$ is given by the *semiempirical mass formula*

$$B(Z, N) = E_v A - E_s A^{2/3} - E_\beta \frac{(N-Z)^2}{A} - E_c \frac{Z^2}{A^{1/3}} \tag{1.2}$$

where the coefficients correspond to the volume ($E_v = 15.8$ MeV), the surface ($E_s = 18.3$ MeV), the symmetry ($E_\beta = 23.2$ MeV for $N \neq Z$ and $\beta = (N-Z)/A$) and the Coulomb ($E_c = 0.71$ MeV) energies. The nuclear radius behaves like $R_{nuc} \sim r_0 A^{1/3}$ with $r_0 \sim 1.12$ fm $= 1.12 \times 10^{-13}$ cm.

The correct theoretical description of $B(Z, N)$ in finite nuclei and of the *Equation of State* (EOS) of infinite nucleonic matter is the first goal which a consistent microscopic theory must pursue. This work is mostly concerned with the description of the latest results in this field when a realistic interaction is adopted. A nuclear potential is defined as realistic when it gives a good fit to data from experiments involving two nucleons. One of the main outcomes from this analysis is that the interaction must strongly (and in a complicated way) depend on the relative state of the nucleons. In addition, there is evidence of the existence of many–body forces. In fact, at least three nucleon (NNN) potentials are required to obtain nuclear masses in quantitative agreement with the experiments.

The plan of this paper is as follows. Section 2 gives a brief description of modern nuclear interactions. Section 3 is devoted to outlining the methods for computing the ground state expectation value of the Hamiltonian in finite nuclei with a presentation of the latest results. Infinite nuclear and neutron matters are discussed in Sect. 4, together with their EOS and its implication for neutron star structure. Some conclusions are drawn in Sect. 5.

2 The nuclear Hamiltonian

The starting point of a non-relativistic approach to nuclear physics is the derivation (or construction) of a suitable microscopic potential between nucleons to be included in the Hamiltonian. QCD, despite being presently recognized as the theory of strong interactions, cannot be realistically used in the non-perturbative nuclear regime, so mesonic field theories have been largely exploited either to directly address some particular features of the NN potential (as with the large distances behavior) or to serve as a guide for developing models. This is an extremely successful approach, since *modern* potentials, derived along these lines, describe with excellent accuracy NN scattering data up to $E_{lab} \leq 350$ MeV [3, 4, 5]. They simultaneously fit proton–proton and neutron–proton scattering and, to this aim, contain a detailed description of the electromagnetic interaction as well as terms violating the isospin symmetry (as a consequence of the small mass difference between the nucleons and between the exchanged mesons).

In nuclear many–body theory the system is described by a Hamiltonian

$$H = -\frac{\hbar^2}{2m}\sum_i \nabla_i^2 + \sum_{i<j} V_{ij} + \sum_{i<j<k} V_{ijk} \tag{2.1}$$

where the two–body potential can be separated as

$$V_{ij} = V_{ij}^{em} + V_{ij}^{\pi} + V_{ij}^{R}. \tag{2.2}$$

The different parts of V_{ij} come from:

(i) electromagnetic interactions (V^{em}) as Coulomb, Darwin–Foldy term, vacuum polarization and magnetic moments;

(ii) One Pion Exchange Potential (V^{π}, OPEP), due to the exchange of one pion, the lightest meson ($m_\pi \sim 140$ MeV/c^2), and so dominant at large NN distances; the explicit expression of OPEP is

$$V_{ij}^{\pi} = \frac{f_{\pi NN}^2}{4\pi}\frac{m_\pi}{3} X_{ij} \tau_i \cdot \tau_j \tag{2.3}$$

where

$$X_{ij} = Y_\pi(r_{ij})\sigma_i \cdot \sigma_j + T_\pi(r_{ij})S_{ij} \tag{2.4}$$

$$S_{ij} = 3\hat{r}_{ij} \cdot \sigma_i \hat{r}_{ij} \cdot \sigma_j - \sigma_i \cdot \sigma_j \tag{2.5}$$

$$Y_\pi(r_{ij}) \rightarrow e^{-\mu r_{ij}}/(\mu r_{ij}) \tag{2.6}$$

$$T_\pi(r_{ij}) \rightarrow e^{-\mu r_{ij}}/(\mu r_{ij}) \left[1 + \frac{3}{\mu r_{ij}} + \frac{3}{(\mu r_{ij})^2}\right] \tag{2.7}$$

with $\mu = m_\pi/\hbar c = .7\ \text{fm}^{-1}$;

(iii) the exchange of more pions or of heavier mesons, such as ρ and ω, which are at the base of the short- and intermediate-range potential V^R; alternatively, the very short distance behavior might be thought of as caused by the quark substructure of the nucleons; however, V^R is mostly phenomenological because of the large theoretical uncertainties in a microscopic description of its origin.

The interaction is found to depend strongly on the total spin and isospin ($S, T = 0, 1$) of the pair. In fact, only the state $S, T = (1, 0)$, corresponding to the deuteron, is bound. Moreover, the deuteron has a non-zero quadrupole moment pointing to a non-spherical shape and, consequently, to the presence of an $l = 2$ tensor force. These features are also found in OPEP, which is the only truly microscopic part of the theoretical interaction.

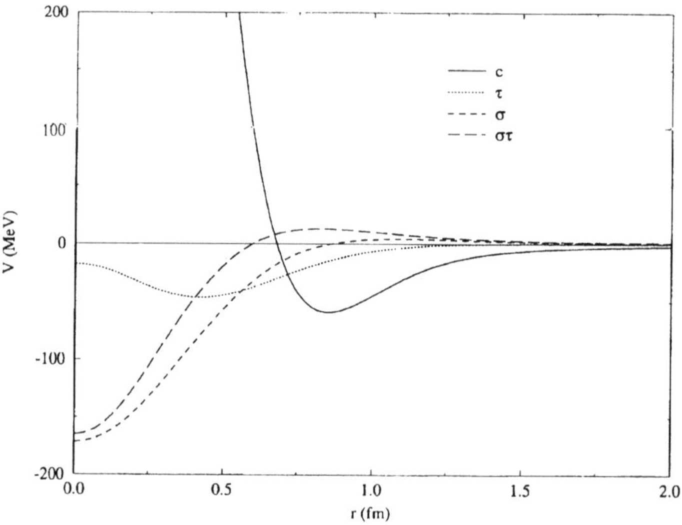

Fig. 1. The first four components of the Argonne v_{18} potential

Realistic NN interactions may be represented as

$$V_{ij} = \sum_{p=1,18} v_p(r_{ij}) O^p_{ij} \tag{2.8}$$

where the first fourteen components are isoscalar,

$$O^{p=1,14}_{ij} = \left[1, \sigma_i \cdot \sigma_j, S_{ij}, \mathbf{L} \cdot \mathbf{S}, \mathbf{L}^2, \mathbf{L}^2 \sigma_i \cdot \sigma_j, (\mathbf{L} \cdot \mathbf{S})^2\right] \otimes \left[1, \tau_i \cdot \tau_j\right], \tag{2.9}$$

and the remaining components,

$$O^{p=15,18}_{ij} = \left[1, \sigma_i \cdot \sigma_j, S_{ij}\right]\left[3\tau_{iz}\tau_{jz} - \tau_i \cdot \tau_j\right], \left[\tau_{iz} + \tau_{jz}\right], \tag{2.10}$$

break both charge dependence and symmetry. The most accurate potential of this kind is the so-called Argonne v_{18} (A18) [4] which fits the NN scattering data with χ^2/datum $\sim$ 1. Other representations are possible [3,5], but, as long as they reproduce the same set of experimental data with a similar accuracy, they provide predictions in very good agreement with each other (it has to be stressed that this was not the case for older, less accurate potentials).

Some characteristics of Argonne v_{18} are: it contains 43 parameters, $f^2_{\pi NN}/4\pi = 0.075$ and it fits 1787 pp and 2514 np diffusion data with $\chi^2/$ datum $= 1.09$. The older Argonne v_{14} potential (A14), without the 15–18 components [6], had 32 parameters, $f^2_{\pi NN}/4\pi = 0.081$ and fitted 1762 np diffusion data with $\chi^2/$ datum $= 1.67$. Figure 1 shows the behavior of A18 in the first four (c, σ, τ and $\sigma\tau$) channels.

A two-body interaction, by itself, is not able to reproduce the binding energy of the $A = 3, 4$ nuclei and the saturation density of nuclear matter. Thus, a three-nucleon potential, V_{ijk}, is needed. Again the NNN potential can be partially constructed on the basis of meson theory and its long range part is due to two-pion exchanges between three nucleons with an intermediate non-nucleonic state. The most important of such states is the $\Delta(1232)$, leading to ([7])

$$V_{ijk}^{2\pi} = A_{2\pi} \sum_{cycl.} \{X_{ij}, Xik\}\{\tau_i \cdot \tau_j, \tau_i \cdot \tau_k\} + \frac{1}{4}\left[X_{ij}, Xik\right]\left[\tau_i \cdot \tau_j, \tau_i \cdot \tau_k\right] \tag{2.11}$$

where $A_{2\pi}$ is a factor related to the πNN and πNΔ coupling constants and the sum is taken over the indices i, j and k. The full V_{ijk} is then written as

$$V_{ijk} = V_{ijk}^{2\pi} + V_{ijk}^{R} \tag{2.12}$$

where V_{ijk}^{R} is a short–range repulsive term, having no spin–isospin dependence, of the form

$$V_{ijk}^{R} = U_0 \sum_{cycl.} T_\pi^2(r_{ij}) T_\pi^2(r_{ik}), \tag{2.13}$$

U_0 being an additional parameter [8]. This type of repulsive NNN potential is expected on account of more complicated meson exchanges and of relativistic corrections. In the Urbana family of V_{ijk} the parameters $A_{2\pi}$ and U_0 are fixed in such a way as to reproduce correctly the light nuclei binding energy and the saturation point of nuclear matter.

3 The many-body methods

Solving the Schrödinger equation

The exact solution of the Schrödinger equation for the ground state of A$>$ 2 nucleons

$$H\Psi_0(A) = E_0\Psi_0(A) \tag{3.1}$$

with realistic Hamiltonians, as (2.1), has been an elusive goal for a long time. However, in the eighties a few methods were developed which have allowed researchers to study light nuclei accurately. The Green's function Monte Carlo (GFMC) [9] technique is the technique taken to the highest mass number, A$=$ 8 [10].

In GFMC one stochastically solves the imaginary time Schrödinger equation

$$-\frac{\partial H}{\partial \Psi_\tau} = (H - E)\Psi_\tau \tag{3.2}$$

whose formal solution is:

$$\Psi_\tau = e^{-(H-E)\tau}\Psi_t\,. \tag{3.3}$$

The exact ground state, Ψ_0, can be extracted from the trial function, Ψ_t, (if it has non-zero projection on Ψ_0) through the limit

$$\Psi_0 = \lim_{\tau\to\infty} e^{-(H-E)\tau}\,\Psi_t\,. \tag{3.4}$$

In practice, a transient energy estimator is constructed

$$E_\tau = \frac{\langle\Psi_t|H|\Psi_\tau\rangle}{\langle\Psi_t|\Psi_\tau\rangle} = \frac{\langle\Psi_{\tau/2}|H|\Psi_{\tau/2}\rangle}{\langle\Psi_{\tau/2}|\Psi_{\tau/2}\rangle} \tag{3.5}$$

and $E_0 = E_{\tau\to\infty}$.

Table 1 gives the GFMC ground state energies from the deuteron (^{2}H) to A= 8 nuclei (^{8}Be, ^{8}Li and ^{8}He) using the A18 NN potential and the Urbana model IX (UIX) [10] three–nucleon interaction. The Table reports the expectation values of the various terms of the interaction as well as the differences with the experimental binding energies ($\Delta E_{\rm expt}$). Most of the nuclear binding comes from V^π_{ij} whereas V^R_{ij} becomes increasingly important along A. $\langle V^{2\pi}_{ijk}\rangle$ is an order of magnitude smaller than $\langle V^\pi_{ij}\rangle$ but it gives a significant fraction of the energy. The total energy comes from a large cancellation between kinetic and potential energies.

Table 1. GFMC ground state energies and expectation values of the potential (in MeV) for light nuclei with the A18+UIX interaction. From [10]

Nucleus	$\langle V^\pi_{ij}\rangle$	$\langle V^{2\pi}_{ijk}\rangle$	$\langle V^R_{ij}\rangle$	$\langle V^R_{ijk}\rangle$	E_{GFMC}	$\langle\Delta E_{\rm expt}\rangle$
^{2}H	−21.3	0	−0.8	0	−2.22	0.0
^{3}H	−43.8	−2.2	−14.6	1.0	−8.47	−0.01(1)
^{4}He	−99.4	−11.7	−36	5.3	−28.30	0.0
^{6}He	−109	−13.6	−56	6.4	−27.64	−1.63(14)
^{6}Li	−129	−13.5	−50	6.3	−31.25	−0.74(11)
^{7}He	−110	−14.1	−61	6.7	−25.2	−3.7(2)
^{7}Li	−153	−17.1	−68	8.2	−37.4	−1.8(3)
^{8}He	−121	−15.8	−74	7.5	−25.8	−5.6(6)
^{8}Li	−157	−22.2	−104	11.0	−38.3	−3.0(6)
^{8}Be	−224	−28.1	−72	13.3	−54.7	−1.8(6)

The A>5 nuclei are all underbound and $\Delta E_{\rm expt}$ is comparable to $\langle V^R_{ijk}\rangle$ in neutron rich nuclei (^{7}He and ^{8}He) indicating that the UIX model might not be adequate in describing the three-neutron interaction. In addition, the underbinding is larger in nuclei with higher isospin. These data are presently used to develop new models of three-nucleon interactions. Figure 2 shows the energies for the ground and first excited states for the same nuclei. The description of the low lying states given by GFMC is a little worse than that for the ground state; however, it has to be remarked that a nuclear shell structure naturally originates from the solution of the many–body Schrödinger equation without the need of introducing an apposite shell model.

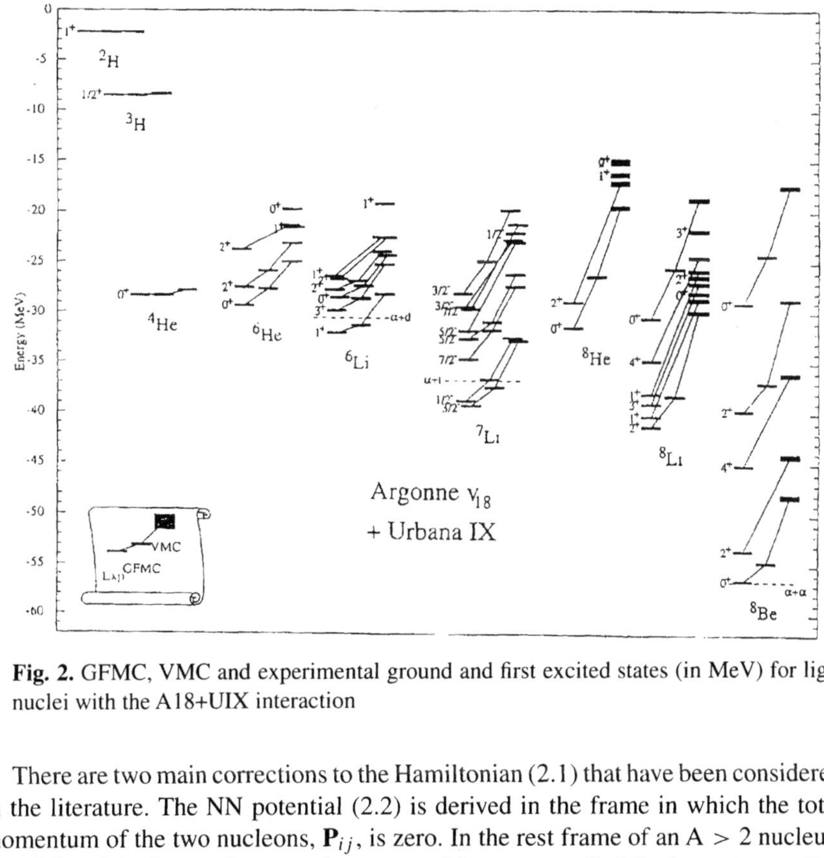

Fig. 2. GFMC, VMC and experimental ground and first excited states (in MeV) for light nuclei with the A18+UIX interaction

There are two main corrections to the Hamiltonian (2.1) that have been considered in the literature. The NN potential (2.2) is derived in the frame in which the total momentum of the two nucleons, $\mathbf{P}_{ij}$, is zero. In the rest frame of an A $>$ 2 nucleus, $\mathbf{P}_{ij} \neq 0$ and the interaction must be corrected by a term called the *boost interaction* [11]. The other correction arises from the use of the relativistic kinetic energy operator $T_R = \sum_i \sqrt{p_i^2 + m^2}$. In fact, even though non-interacting nucleons have small momenta (of the order of $m/4$) and so relativistic corrections are expected to be negligible, a pair of interacting nucleons may, however, acquire large relative momenta at small distances. The results for light nuclei with a Hamiltonian containing both the boost interaction and the relativistic corrections are similar to those presented above [12].

Other techniques for solving the few–body Schrödinger equation, besides GFMC, are on the market. The Faddeev [13], Faddeev–Yakubovsky [14] and Hyperspherical Functions expansion [15] methods deserve, in my opinion, particular mention because not only do they have the same degree of accuracy as GFMC in the description of the ground state properties but they also have permitted a quantitative study of a few low energy reactions, involving A = 3 and 4 nuclei, which are of particular relevance in nuclear astrophysics. These powerful methods appear to be more versatile than GFMC but they cannot easily be extended to heavier systems.

The Correlated Basis Function approach

Solving the Schrödinger equations becomes a prohibitive task when the mass number increases. In fact, the many-nucleon wave function is made up of Slater determinants (at least one per each spin-isospin state, in the non-interacting case) and the stochastic sample per determinant grows as A^3. Moreover, when fermions are present, inevitably the *sign problem* arises. In a few words, the fermion wave function is not positive everywhere and it cannot be interpreted as a probability distribution to guide the Monte Carlo sample.

An alternative and accurate way of studying heavy nuclei and nuclear matter is the Correlated Basis Function (CBF) theory [16]. An independent particle basis, such as the Fermi gas, is not adequate for describing the short-range behavior of the nuclear wave function, which is dramatically affected by the strong NN interaction. The idea behind CBF consists of including these non-perturbative effects directly into the basis.

The CBF theory is based on a set of correlated A-body wave functions

$$\Psi_n^{(CBF)}(1,2\cdots \mathrm{A}) = F(1,2\cdots \mathrm{A})\Phi_n(1,2\cdots \mathrm{A}), \tag{3.6}$$

where $\Phi_n(1,2\cdots \mathrm{A})$ is a generic mean field state (a Slater determinant of single particle wave functions) and $F(1,2\cdots \mathrm{A})$ is an A–body correlation operator acting on Φ_n and reflecting the operator structure of the interaction. The zeroth order of the CBF based perturbation theory is denoted as the variational level, in the sense that the correlation operator is chosen in such a way as to minimize

$$E_v = \frac{\langle \Psi_0^{(CBF)} | H | \Psi_0^{(CBF)} \rangle}{\langle \Psi_0^{(CBF)} | \Psi_0^{(CBF)} \rangle} \geq E_0 \,. \tag{3.7}$$

The richer the correlation operator, the more accurate are the CBF estimates. Thus a realistic and flexible ansatz for $F(1,\cdots,\mathrm{A})$ is the key to a successful application of CBF theory. Moreover, the CBF states can be used to evaluate other matrix elements than ground state expectation values. The simplest correlation operator has the Jastrow form [17]

$$F_J(1,\cdots,A) = \prod_{i<j} f_J(r_{ij}) \tag{3.8}$$

where the two-body correlation function, $f_J(r)$, depends only on the interparticle distance. More realistic studies use a state dependent correlation operator modeled on the interaction:

$$F(1,\cdots,A) = \mathcal{S}\left[\prod_{i<j} f_{ij}^{(2)}\right] \mathcal{S}\left[\prod_{i<j<k} f_{ijk}^{(3)}\right], \tag{3.9}$$

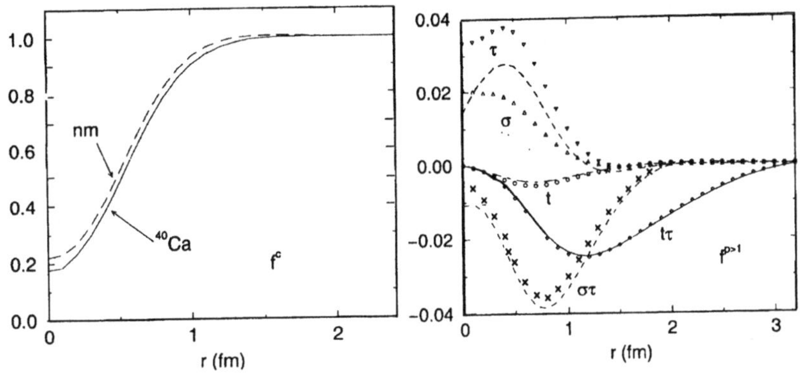

Fig. 3. Two–body correlation functions in ^{40}Ca (lines) and nuclear matter (symbols)

e.g., a symmetrized product of two- and three-body correlation operators, whose spin-isospin dependence is

$$f_{ij}^{(2)} = \sum_{n=1,8} f^{(n)}(r_{ij}) O_{ij}^{n} \tag{3.10}$$

$$f_{ijk}^{(3)} = -\epsilon V_{ijk}^{2\pi} \,. \tag{3.11}$$

Typical correlation functions are given in Fig. 3 for ^{40}Ca and for nuclear matter. The Jastrow component ($n = 1$) is small at short distances and reduces the importance of the repulsive core of the interaction. The other components are smaller, the most important being the tensor–isospin component, $f^{(t\tau,n=6)}(r)$, directly related to the long range OPEP. It is remarkable how short range correlations do not vary sensibly in going from finite nuclei to infinite matter.

The evaluation of the energy, E_v, (or in general of diagonal and off-diagonal CBF matrix elements) can be performed in a number of ways.

[a] Variational Monte Carlo (VMC) [18]. For A<8 nuclei the expectation value (3.7) is directly computed by Monte Carlo sampling of the 3A dimensional integral. The calculation has only a statistical error that can be made arbitrarily small.

[b] Cluster Monte Carlo (CMC) [19]. For 15 <A< 41 nuclei an expansion in the number n of nucleons connected by non-Jastrow correlations is performed. The n-th term, fully antisymmetrized and Jastrow correlated to all the remaining nucleons, is evaluated by Monte Carlo. Up to now, terms up to $n = 4, 5$ ($CMC_{4,5}$) have been computed.

[c] Fermi Hypernetted Chain (FHNC) equations [20]. In all doubly closed shell nuclei and infinite matter the matrix element is expanded in powers of $h(r) = [f^{(1)}(r)]^2 - 1$ and $f^{(n>1)}(r)$. The expansion is analogous to that of the partition function of a classical gas in Mayer diagrams ($h_{cl}(r) = \exp[-U(r)/KT] - 1$). Infinite classes of diagrams are summed via the FHNC equations and the Single

Operator Chain (SOC) [21,22] approximation. In FHNC/SOC, 29 coupled integral equations are solved iteratively. Some classes of diagrams (the *elementary diagrams*) cannot be summed in a closed form. However, they become important only at high density and are not relevant at nuclear densities (their inclusion is instead essential in denser systems such as atomic liquid Helium).

The variational results obtained for nuclei, from A = 3 to A = 208, are reported in Table 2. In the light systems the comparison with the exact GFMC results shows that the error introduced by the variational approach may be taken to be about 1 MeV per nucleon. The two ^{16}O results (CMC and FHNC/SOC) provide a check on the accuracy of the integral equation method which is the only one reliable when A increases. The results in heavier systems have been obtained with either truncated or simplified interactions and work is proceeding quickly towards the use of a complete interaction there also.

Table 2. Variational ground state energies per nucleon (in MeV) for several nuclei. Details on the calculations and interactions are given in the references

Nucleus	Method	Interaction	Reference	E_v/A	E_{expt}/A
^{4}He	VMC	$A18+UIX$	[10]	−6.95	−7.08
^{6}He	VMC	$A18+UIX$	[10]	−4.14	−4.89
^{8}He	VMC	$A18+UIX$	[10]	−2.46	−3.93
^{16}O	CMC$_5$	$A14+UVII$	[19]	−7.7	−7.72
^{16}O	FHNC/SOC	$A14+UVII$	[22]	−8.1	−7.72
^{40}Ca	FHNC/SOC	$U6$	[22]	−7.9	−8.30
^{208}Pb	FHNC	$S3$	[23]	−8.5	−7.87

4 Nuclear and neutron matter and neutron stars

The Equation of State

Nuclear matter (NM) is an infinite system of nucleons, at constant density $\rho = A/V$ $(A, V = \infty)$, without Coulomb interaction. The value of the symmetry parameter, β, measures the concentrations of the nucleonic species, going from symmetrical NM (SNM, $\beta = 0$, N=Z) to pure neutron matter (PNM, $\beta = 1$, Z=0). In the $A = \infty$ limit the mass formula (1.2) simplifies to $B(Z,N)/A = E_v - \beta^2 E_\beta$ so providing information on the properties of NM. From this analysis it is possible to extract the SNM energy (corresponding to the volume energy, $E_{SNM} = -16 \pm 1$ MeV), while the SNM saturation density, $\rho_{SNM} = 0.16 \pm 0.01$ fm^{-3} is obtained from charge-density distribution measures. CBF has been used extensively to study ground state properties and the responses of NM to external probes. Here, I will briefly review the results obtained for the EOS with realistic potentials.

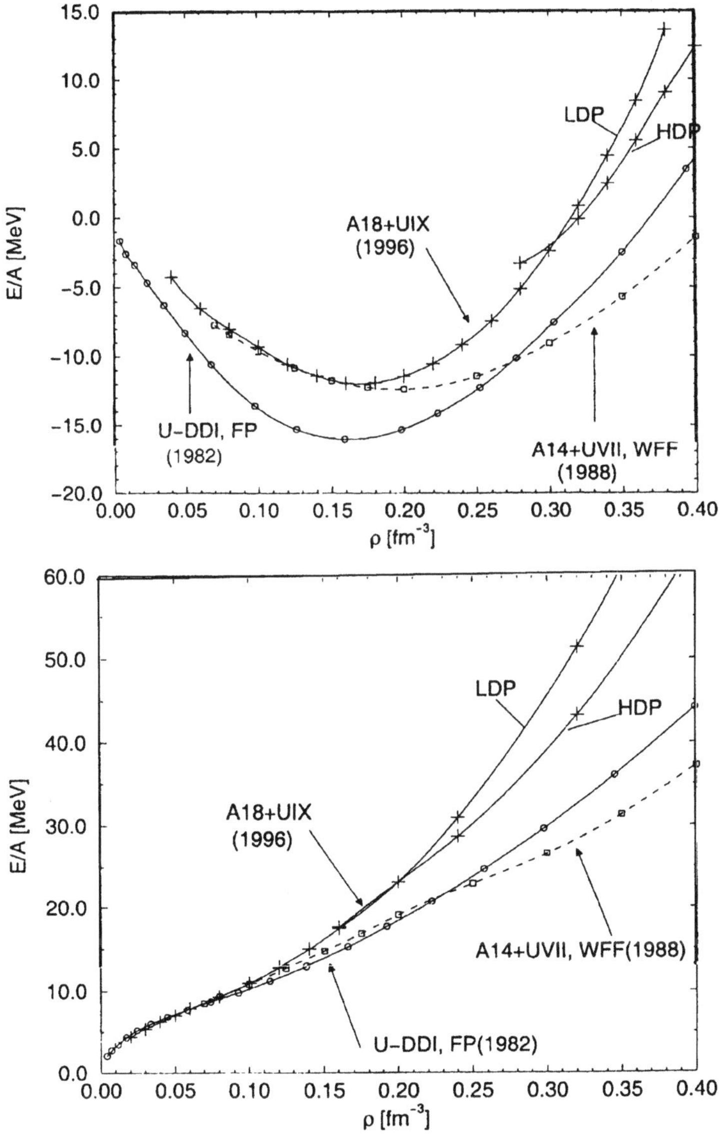

Fig. 4. $E(\rho)$ in SNM (upper panel) and PNM (lower panel). The two sets of variational minima beyond the phase transition point are labeled LDP and HDP for the low and high density phases

The most reliable calculations have been performed within the FHNC/SOC scheme [24] using the correlation operator (3.9) without three-body correlations ($f^{(3)}_{ijk} = 1$). Realistic Hamiltonians having two- and three-nucleon potentials have been considered.

The SNM $E(\rho)$ is depicted in Fig. 4 for the A18+UIX and the A14+UVII models. The U–DDI (Urbana–density dependent interaction) [25] was set in such a way as to reproduce the empirical saturation properties of SNM, so its minimum can be regarded as experimental data. The calculated energies are higher than the empirical value by $\sim 3-4$ MeV. This fact must be ascribed to the simplicity of the correlation function. In fact, CBF perturbative corrections lower the energy by $\sim 2-3$ MeV [26].

The SNM curve points to a phase transition at $\rho \sim 0.3$ fm^{-3}. The phase at higher density has spin–isospin order due to neutral pion condensation [27]. The variational wave function is appropriate for isotropic matter and the phase transition is signaled by a change in the range and strength of the tensor correlations. The transition is very sensitive to the short range behavior of the spin and spin–isospin interactions and $V^{2\pi}_{ijk}$ is essential for its occurrence in SNM.

Figure 4 also gives the PNM energies. The phase transition occurs at $\rho \sim 0.2$ fm^{-3} and at much higher density in absence of $V^{2\pi}_{ijk}$.

Asymmetric NM (N $\neq$ Z) calculations at a fixed proton fraction, $x = Z/A$, are difficult to perform because of the isospin flip induced by the interaction and the correlation. However, in an expansion in powers of β, quadratic terms are by far the most important which allows us to write

$$\frac{E(\rho,x)}{A} = T_F(\rho,x) + V_0(\rho) + \beta^2 V_2(\rho) \tag{4.1}$$

where $T_F(\rho,x)$ is the Fermi gas kinetic energy

$$T_F(\rho,x) = \frac{3\hbar^2}{10m}(3\pi^2\rho)^{2/3}\left[x^{5/3} + (1-x)^{5/3}\right] \tag{4.2}$$

and $V_{0,2}(\rho)$ are obtained from the SNM and PNM results.

At a given density the symmetry energy is then

$$E_\beta(\rho) = \frac{1}{8}\left(\frac{\partial^2 E(\rho,x)}{\partial x^2}\right)_{x=1/2} = \frac{5}{9}T_F(\rho,\frac{1}{2}) + V_2(\rho)\,. \tag{4.3}$$

At saturation density the A18+UVII model gives $E_\beta(\rho) = 29$ Mev/A, close enough to the empirical value.

The approximation (4.1) is particularly useful in estimating the energy and the proton fraction for β-stable matter in the interior of neutron stars, where protons and neutrons are in equilibrium with electrons and muons via the reactions $n \to p + e^- \to p + \mu^-$.

The balance of the chemical potentials, $\mu_n - \mu_p = \mu_e = \mu_\mu$, with

$$\mu_n - \mu_p = \frac{\hbar^2}{2m}(3\pi\rho)^{2/3}\left[(1-x)^{2/3} - x^{2/3}\right] + 4(1-2x)\,V_2(\rho) \tag{4.4}$$

and

$$\mu_{e,\mu} = \left[m_{e,\mu}^2 + \hbar^2(3\pi^2 x_{e,\mu})^{2/3}\right]^{1/2}, \tag{4.5}$$

and the charge neutrality, $x = x_e + x_\mu$, must be imposed and the two equations are solved numerically to obtain $x(\rho)$ and $E(\rho, x)$.

Realistic potentials and FHNC/SOC provide a β-stable matter proton fraction increasing along the density. However, it remains below the value of 0.148 needed for direct Urca cooling in the range of the neutron star densities.

Moreover, the collapse of a massive star, producing type II Supernovae, occurs at $x \sim 1/3$, where matter saturation properties are significantly different from those of SNM. The variations may be easily computed within the parabolic approximation (4.1). The possibilty of a prompt shock mechanism is very sensitive to these changes. With the A18+UVII model the compressibility modulus, K_0, is reduced by $\sim$25%, so favouring such a process during the collapse. However, the curve of pressure as a function of density gives an adiabatic index $\gamma = (\rho_m/P)(\partial P/\partial\rho_m) \sim 3.5$ (ρ_m is the mass density), apparently against the prompt shock mechanism.

Neutron star structure

The EOS of β-stable matter governs the structure of a neutron star. It gives $P(\epsilon)$, i.e., the pressure P that the matter can support at a given mass-energy density ϵ. Once $E(\rho, x)$ is known, $P(\epsilon)$ is derived from $\epsilon = \rho\left[E(\rho) + mc^2\right]$ and $P = \rho^2\left[\partial E(\rho)/\partial\rho\right]$.

The stellar structure is determined by the general relativistic equation of hydrostatic balance (the Tolman–Oppenheimer–Volkoff equation, TOV) [28]

$$\frac{\partial P(r)}{\partial r} = -G\frac{\left[\epsilon(r) + P(r)/c^2\right]\left[m(r) + 4\pi r^3 P(r)/c^2\right]}{r^2\left[1 - 2Gm(r)/c^2\right]}, \tag{4.6}$$

$$m(r) = \int_0^r 4\pi s^2 \epsilon(s) ds\,. \tag{4.7}$$

Given the central density, $\epsilon_c = \epsilon(r = 0)$ and the corresponding central pressure P_c, the TOV equation is integrated outward up to a radius R where the pressure vanishes. R is the stellar radius and $M_G = m(R)$ is the gravitational mass. Other quantities can be calculated such as the moment of inertia I, the surface redshift z and the amu mass M_A

$$M_A = m_A \int_0^R \frac{4\pi r^2 \rho(r)}{\left[1 - 2Gm(r)/rc^2\right]^{1/2}} dr \tag{4.8}$$

where m_a is one atomic mass unit (1.66×10^{-24} g). The difference $M_A - M_G$ is the stellar binding energy. The prediction of the A14+UVII model for the properties of a typical neutron star are given in Table 3. The gravitational mass is shown in Fig. 5.

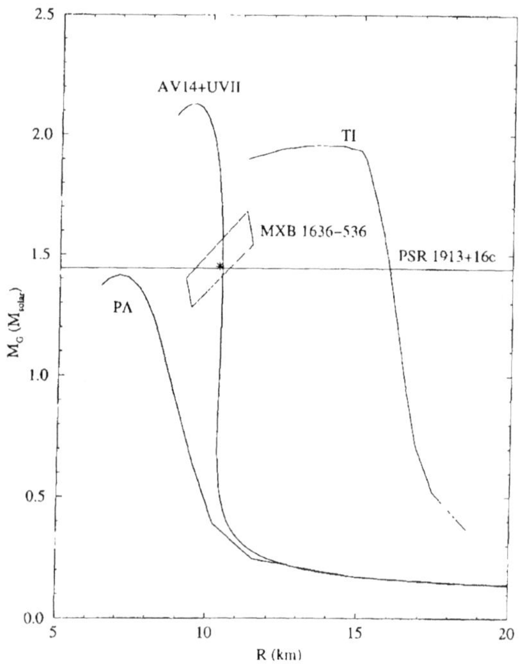

Fig. 5. Neutron star gravitational mass for the A14+UVII interaction. See text

Table 3. Properties of a neutron star with the A14+UVII model

ρ_c	$0.66\,\mathrm{fm}^{-3}$
ϵ_c	$1.2 \times 10^{15}\,\mathrm{g\,cm}^{-3}$
P_c	$2.1 \times 10^{35}\,\mathrm{dyn\,cm}^{-2}$
M_G	$1.4\,M_\odot$
$M_A - M_G$	$0.18\,M_\odot$
R	$10.4\,\mathrm{km}$
I	$1.2 \times 10^{45}\,\mathrm{g\,cm}^2$
z	0.29

Figure 5 also shows the results of other models: the hyperon (Λ) model [29] and the tensor interaction (TI) model [30]. The first gives a soft EOS and the second a stiff one. In addition, the measured mass for the radio pulsar PSR 1916+13 ($M_G = 1.44 M_\odot$) [31] is reported. Also given, as a cross with an error box, is the result of an analysis of data from the x–ray burst source MXB 1636–536, giving both the mass and the radius of the star [32]. The A14+UVII EOS goes through the point, but the analysis is still controversial. The soft EOS does not seem to support the observed neutron star mass and the stiff EOS appears to give too large a radius. Moreover, the non-interacting EOS supports a star with a maximum mass of just 0.7 $M_\odot$, which underlines the relevance of the nuclear interaction for neutron stars.

5 Conclusions

I have described a series of exact and variational calculations of binding energies of finite and infinite nuclear systems. The approaches outlined here have been used to study many other properties, such as density and momentum distributions, sum rules, electromagnetic form factors and cross sections. The general outcome is that two-body interactions alone do not provide quantitative agreement with experiments, as far as the ground state is concerned. Many-body forces (at present, only three-body forces) lead to significant improvements. The same microscopic Hamiltonian appears to be able to describe light and heavy nuclei as well as nuclear matter. The development of better many-body interactions is one of our most important goals.

The interplay with astrophysics is becoming tighter and tighter. Low energy few-body reactions, of major astrophysical interest, are being studied with increasing accuracy and the microscopic Equation of State of nucleonic matter provides a description of static properties of neutron stars in agreement with present knowledge. These subjects surely represent one of the most c hallenging (and rewarding) fields of application of old-fashioned nuclear physics.

Acknowledgements

The results presented here are also the fruit of collaboration over several years with many co-workers and friends, particularly O. Benhar (Roma), J. Carlson (Los Alamos), S. Fantoni (Trieste), R. Schiavilla (TJNAF), S. Pieper and R. B. Wiringa (Argonne) and V. R. Pandharipande (Urbana).

References

1. Bethe H.A., Bacher R.F. (1936): Rev. Mod. Phys. **8**, 82
2. Mayer M.G., Jensen H.D. (1955): Elementary Theory of Nuclear Shell Structure. Wiley, New York
3. Stoks V.G.J., Klomp R.M.A., Terheggen C.P.F.T., De Swart J.J. (1994): Phys. Rev. C **49**, 2950
4. Wiringa R.B., Stoks V.G.J., Schiavilla R. (1995): Phys. Rev. C **51**, 38

5. Machleidt R., Sammarruca F., Song Y. (1996): Phys. Rev. C **53**, R1483
6. Wiringa R.B., Smith R.A., Ainsworth T.L. (1984): Phys. Rev. C **29**, 1207
7. Fujita J., Miyazawa H. (1957): Progr. Theor. Phys. **17**, 360
8. Carlson J., Pandharipande V.R., Wiringa R.B. (1983): Nucl. Phys. A **401**, 59
9. Carlson J. (1988): Phys. Rev. C **38**, 1879
10. Pudliner B.S., Pandharipande V.R., Carlson J. (1997): Phys. Rev. C **56**, 1720
11. Forest J.L., Pandharipande V.R., Friar J.L. (1995): Phys. Rev. C **52**, 568
12. Forest J.L., Pandharipande V.R., Friar J.L. (1995): Phys. Rev. C **52**, 576
13. Chen C. R., Payne G.L., Friar J.L. Gibson B.F. (1986): Phys. Rev. C **33**, 1740; Stadler A., Glöckle W., Sauer, P.U. (1991): Phys. Rev. C **44**, 2319
14. Glöckle W., Kamada H. (1993): Phys. Rev. Lett. **71**, 971
15. Kievsky A., Viviani M., Rosati S. (1993): Nucl. Phys. A **551**, 241; (1994): A **577**, 511
16. Feenberg E. (1969): Theory of quantum fuids. Academic Press, New York
17. Jastrow R. (1955): Phys. Rev. **98**, 1479
18. Carlson J., Pandharipande V.R. (1981): Nucl. Phys. A **371**, 301
19. Pieper S.C., Wiringa R.B., Pandharipande V.R. (1992): Phys. Rev. C **46**, 1741
20. Rosati S. (1982): From Nuclei to Particles, in *Proc. Int. School E. Fermi, Course* LXXIX, ed. by A. Molinari, North Holland, Amsterdam, p. 73
21. Pandharipande V.R., Wiringa R.B. (1979): Rev. Mod. Phys. **51**, 821
22. Fabrocini A., Arias de Saavedra F., Co G.P. (1998): Phys. Rev. C **57**, 1668
23. Arias de Saavedra F., Co G.P., Fabrocini A., Fantoni S. (1996): Nucl. Phys. A **605**, 359
24. Wiringa R.B., Ficks V., Fabrocini A. (1988): Phys. Rev. C **38**, 1010; Akmal A., Pandharipande V.R., Ravenhall D.G. (1998): Phys. Rev. C **58**, 1804
25. Friedman B., Pandharipande V.R. (1981): Nucl. Phys. A **361**, 501
26. Fabrocini A., Fantoni S. (1993): Phys. Lett. B **298**, 263
27. Migdal A.B. (1978): Rev. Mod. Phys. **50**, 50
28. Oppenheimer J.R.,Volkoff G.M. (1939): Phys. Rev. **55**, 374
29. Pandharipande V. R. (1971): Nucl. Phys. A **178**, 213
30. Pandharipande V.R., Smith R.A. (1975): Nucl. Phys. A **237**, 507
31. Taylor J.H., Weisberg J.M. (1989): Astrophys. J. **345**, 434
32. Fujimoto M.Y., Taam R. (1989): Astrophys. J. **305**, 246

Perspectives in the Research of Gravitational Waves: Omnidirectional Resonant Detectors

V. Fafone

Abstract. Spherical gravitational wave detectors offer a wealth of opportunities to detect and characterize gravitational radiation. They allow isotropic sky coverage and determination of the source direction and wave polarizations. The cross section is high at two frequencies, allowing the determination of the features of coalescing binaries. These detectors also appear to be powerful instruments for measuring a stochastic background of gravitational radiation. We describe the features of spherical detectors, their sensitivity to various types of gravitational wave signal and the experimental state of the art.

1 Introduction

For over 40 years, physicists have endeavoured to obtain direct evidence for gravitational waves (GW). GW experiments started with Joseph Weber [1, 2] at the University of Maryland in the early sixties with the development of the first GW detector: a resonant bar operating at room temperature and equipped with a piezoelectric transducer. Since then, cryogenic antennas have been developed, the first of which was operational at Stanford in 1980 [3]. Subsequently improved cryogenic antennas have been operated at low noise temperatures. In 1990 the Rome group obtained continuous data from a similar detector EXPLORER, located at CERN [4], and finally, in May 1991, two large cryogenic antennas entered simultaneous and continuous operation with high sensitivity: EXPLORER and ALLEGRO [5] of the Louisiana State University. They were joined in 1993 by NIOBE [6] of the University of Western Australia, in 1995 by NAUTILUS [7] operated at the INFN Laboratories in Frascati by the Rome group, and in 1997 by AURIGA [8] located at the INFN Legnaro Laboratories.

Now, for the first time, five cryogenic resonant bar detectors – ALLEGRO, AURIGA, EXPLORER, NAUTILUS, NIOBE – are on the air. They are working in collaboration under the IGEC (International Gravitational Events Collaboration) agreement and are already searching for coincident impulsive GW events. In the near future they will extend the searches to continuous sources from specific astronomical objects. Their sensitivity can be conveniently described by the power spectral amplitude of noise at input $\tilde{h} = S_h^{1/2}$ expressed as equivalent metric perturbation. The sensitivity of the resonant GW detectors in operation is $\simeq 5\ 10^{-22}\ \mathrm{Hz}^{-1/2}$ over a bandwidth of fractions of Hz around 1 kHz [9] (see Fig. 1).

This sensitivity allows the detection at SNR= 1 of a continuous GW signal around 1 kHz of amplitude $h = 5\ 10^{-26}$ with an observation time of 100 days and of an impulsive GW signal of duration 1 ms at amplitudes $h = 5\ 10^{-19}$. These values, over

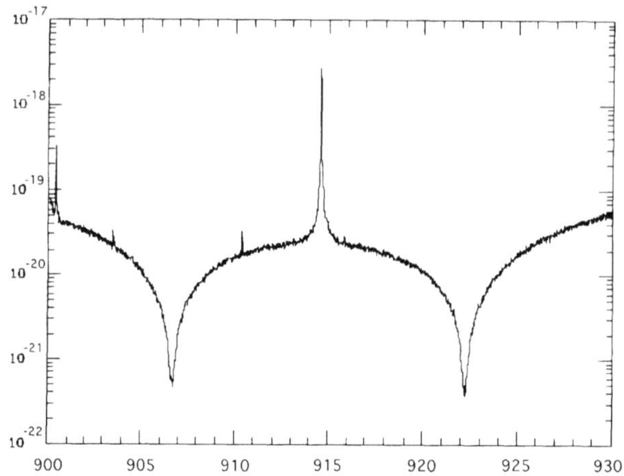

Fig. 1. Experimental strain sensitivity of NAUTILUS (input noise spectral amplitude in units of $Hz^{-1/2}$) versus frequency. The sensitivity at the two resonances is about $4\ 10^{-22} Hz^{-1/2}$. Note that the spectral amplitude is better than $2\ 10^{-20}\ Hz^{-1/2}$ in a total bandwidth of about 25 Hz. The peak at 914.6 Hz is a calibration reference signal

1000 times better than Weber's first antennas, make it possible to monitor the strongest potential sources of GW in our Galaxy and in the local group (i.e., at distances up to 1 Mpc). The final goal for all groups which are engaged in this research is to reach such a sensitivity as to be able to detect GW events occurring in the Virgo Cluster. In order to improve the sensitivity, more advanced transducers and amplifiers are under development [10–13]. Furthermore a huge effort is under way to build large laser interferometers. It is widely believed that, in the near future, sensitivities of the order of $10^{-23}\ Hz^{-1/2}$ over a bandwidth of several tens of Hz for resonant bar detectors and of several hundreds of Hz for interferometric detectors will be attained allowing the observation of GW sources up to distances of the order of 100 Mpc [14].

Projects for spherical resonant detectors have emerged in the last few years in the scientific community [15–18], due to their remarkable capabilities in characterizing a GW.

In what follows a description of the features of spherical detectors and of the experimental state of the art is given.

2 Features of spherical detectors

Direction information and omnidirectionality

In a cylindrical bar, only the first longitudinal mode of vibration interacts strongly with a GW (the cross section of the higher frequency longitudinal odd modes decreases as $1/n^2$ where n is the order of the mode). Consequently only one wave parameter can be measured: the amplitude of a combination of the two polarization states. This deficiency can be overcome only by constructing 6 antennas with different orientations, which is the minimum needed to recover all components of the field with isotropic sensitivity [19].

Each quadrupole mode of a spherical mass is five-fold degenerate (its angular dependence can be described in terms of the five spherical harmonics $Y_{lm}(\theta,\varphi)$ with $l = 2$ and $m = -2 \ldots + 2$). A GW is described as a traveling, time-dependent deviation of the metric tensor, $h_{\mu\nu}$, which can be taken to be transverse and traceless and with only the spatial components h_{ij} different from zero. Because the tensor h_{ij} is traceless, its angular expansion can be determined completely with the five ordinary spherical harmonics of order 2. We call the resulting 5 time-dependent expansion coefficients, denoted by $h_m(t)$, the "spherical gravitational amplitudes." They are a complete and orthogonal representation of the cartesian metric deviation tensor $h_{ij}(t)$. They depend only on the two wave-frame amplitudes (one for each polarization) and the direction of propagation. We now see why a sphere is an ideal GW detector: the 5 quadrupole modes of sphere are in a one-to-one correspondence with the 5 GW amplitudes. Therefore, by observing the quadrupole modes of the sphere, one can obtain all the tensorial components of an incident gravitational wave, with the *same sensitivity* to GW of *all* polarizations and directions.

The method, first outlined by Forward [20] and specified by Wagoner and Paik [21], consists of measuring the vibrations of the sphere in at least five independent locations on the surface of the sphere so as to determine the vibration amplitude of each of the five degenerate modes. The Fourier components of the GW amplitudes at the lowest order quadrupole frequency and the two angles defining the source direction can be obtained as proper combinations of these five outputs [16, 17, 22, 23].

The signal deconvolution is based on the assumption that in the wave reference frame (the frame in which the z axis is aligned along the wave propagation direction) only the modes with $l = 2$ and $m = +2$ and -2 are excited by the GW, as the helicity of a GW is 2 in general relativity. One can take advantage of this to deconvolve the Euler angles between the laboratory frame and the wave frame, thus finding the GW amplitudes in the wave-frame and the wave propagation direction. There will remain a 180° ambiguity for a single sphere; however, this ambiguity can be removed by measuring the time delay between two well separated detectors operated in coincidence.

Many different configurations have been discussed [16, 17, 22, 23] where explicit formulas can be found. In particular, Johnson and Merkowitz proposed a configuration of motion sensors that solved the major practical problems in deconvolving a GW signal [16] (see Fig. 2).

One remarkable thing about a sphere is that only a single sphere is necessary to complete the matrix h_{ij} and thus determine the direction of a gravitational wave. In addition, Zhou and Michelson [17] indicated that a *single* spherical detector, with a sightly *smaller* signal-to-noise ratio, has *less* error in determining source direction than a three interferometer network. Figure 3 shows the results of a numerical simulation similar to that of Zhou and Michelson describing the error in a source direction estimate due to a finite signal-to-noise ratio.

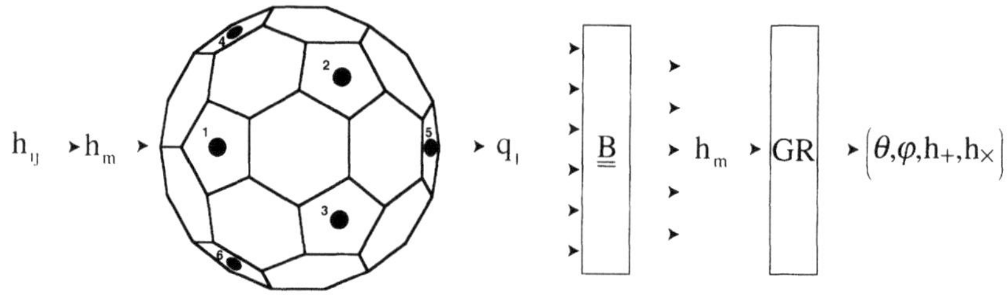

Fig. 2. The technique proposed by Johnson and Merkowitz for detecting GW. The transducers are placed on the surface of a truncated icosahedron at the centers of six non-antipodal pentagonal faces. By transforming to a 5-dimensional abstract vector space, based upon the 5 spherical harmonics of order 2, a one-to-one relationship is found between the force of a gravitational wave, the 5 quadrupole modes of a sphere, and a special linear combination of the resonant transducer responses. This transducer set up gives a simple expression for the matrix B. Through these relationships and a theory of gravitation, the direction, polarization, and amplitude of a gravitational wave can be found with linear algebra using the observable transducers responses

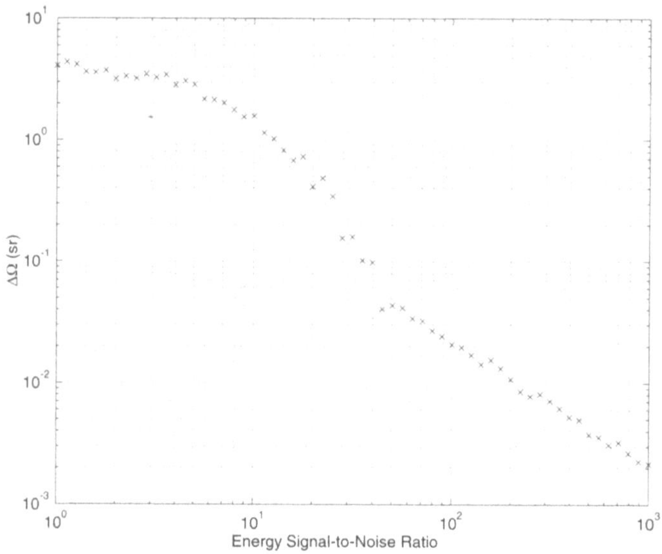

Fig. 3. The results of a numerical simulation (similar to that of Zhou and Michelson) describing the solid angle direction estimation error $\Delta\Omega$ in a source direction measurement due to a finite in signal-to-noise ratio. Each point represents the results of a 200 trial Monte Carlo simulation of a direction measurement with the corresponding signal-to-noise ratio

Large cross section

The GW energy absorbed by a resonant detector can be expressed in terms of the integrated cross section which, for the quadrupole modes of a spherical detector, can be written as [18]

$$\sigma_n = F_n \frac{G}{c^3} M_s v_s^2, \tag{1}$$

where n is the order of the quadrupole mode, M_s is the mass of the sphere, v_s is the speed of sound and F_n is a dimensionless coefficient which is characteristic of each quadrupole mode. Numerically it has been found that $F_1 = 2.98$, $F_2 = 1.14$, $F_3 = 0.107$. It is remarkable that the second order ($n = 2$) quadrupole mode cross section is only a factor 2.61 lower than that of the first order ($n = 1$) quadrupole mode. This means that this detector can be used to advantage at two frequencies.

The cross section of the lowest order mode is larger than that of a cylindrical antenna made of the same material and with the same resonant frequency by a factor of about $0.8(R_s/R_b)^2$ [17, 18], where R_s and R_b are the radii of the sphere and of the bar, respectively, and the impinging GW is assumed to have the best orientation and polarization for the bar. This means a factor of 18 with respect to the present bars (see Fig. 4).

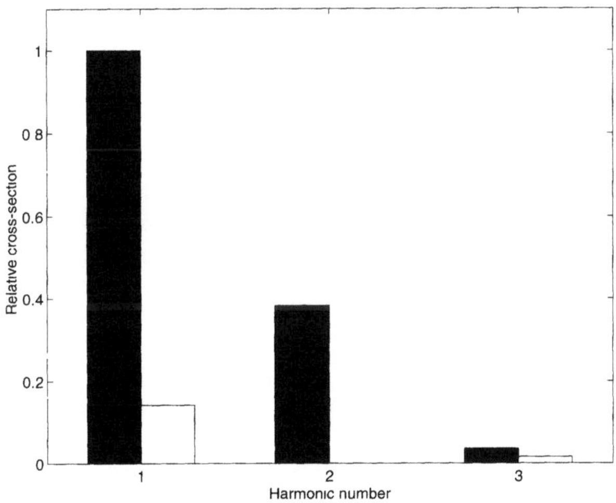

Fig. 4. A comparison of the cross sections for the first few harmonics of a sphere (black) with those of an equivalent optimally oriented bar normalized with the cross section of the first quadrupole mode of the sphere

Averaging over source direction and polarization leads to a loss of energy resolution, compared to the optimum, by a factor of $15/4 = 3.7$. Thus the net result is that the angle-averaged energy resolution of a sphere is $3.7 \times 18 = 67$ times better than the equivalent bar detector (or about 8 times better in h).

The sphere quadrupole eigenfrequencies can be written as [21, 24]

$$\omega_n = \frac{c_n}{R_s} v_s . \tag{2}$$

The dimensionless coefficients c_n are numerically found to be $c_1 = 1.62$ and $c_2 = 3.12$. A Poisson ratio value of 1/3, common to most materials including the ones we are interested in, has been assumed in the numerical results reported.

Values of the first and second quadrupole resonant frequencies for spheres of different diameter and materials are reported in Table 1. The materials reported upon are Al 5056 (the high mechanical quality factor material used for the present bars), and a copper aluminium alloy, indicated as CuAl, investigated at Kamerlingh Onnes Laboratory [25]. The speed of sound is 5400 m/s for Al 5056 at low temperature, and 4700 m/s for CuAl. The masses range from 38 tons (3 m dia Al 5056) up to 250 tons (4 m dia CuAl). The corresponding integrated cross section values of the lowest order quadrupole mode are 9 10^{-22} m^2 Hz, and 4.5 10^{-21} m^2 Hz. The cross section of the typical 3 m long bars is 4.5 10^{-23} m^2 Hz for the best orientation and most favourable polarization.

In the following we will often refer to the sensitivity of the lowest order mode of a 3 m dia, CuAl spherical detector (whose mass is about 100 tons), as that of a typical large spherical detector.

Table 1. Values of the first and second order quadrupole resonance frequencies for spheres of different materials and diameters

	diameter (m)	f_1 (Hz)	f_2 (Hz)
Al 5056	3	950	1826
	3.5	814	1566
	4	712	1370
CuAl	3	725	1395
	3.5	622	1196
	4	544	1046

3 Sensitivity of spherical detectors

The mechanical oscillation induced in the antenna by interaction with the GW is transformed into electrical signals by a set of transducers and then amplified by electrical amplifiers. Unavoidably, Brownian motion noise associated with dissipations in the antenna and in the transducer, and electronic noise from the amplifiers limit the sensitivity of the detector.

The total noise at the output of each resonant mode can be seen as due to an input noise generator having power spectrum of strain $S_h(f)$, acting on a noiseless oscillator. In a resonant-mass detector, this function is a resonant curve [26] and can be characterized by its value at resonance $S_h(f_n)$ and by its half height width. $S_h(f_n)$ can be written as:

$$S_h(f_n) = \frac{G}{c^3}\frac{4kT}{\sigma_n Q_n f_n}. \tag{3}$$

Here T is the thermodynamic temperature of the detector (plus a back-action contribution from the amplifiers, which is very small for the present readout systems), and Q_n is the quality factor of the mode. The predicted sensitivity of the lowest mode of a typical large spherical detector, cooled to 20 mK, and made of a high Q material ($\simeq 10^7$), is in the range of $[S_h(f_1)]^{1/2} \simeq 10^{-24}\,\mathrm{Hz}^{-1/2}$.

The half height width of $S_h(f)$ gives the bandwidth of the resonant mode [27]:

$$\Delta f_n = \frac{f_n}{Q_n}\Gamma_n^{-1/2}. \tag{4}$$

Γ_n is the ratio of the wideband noise in the n-th resonance bandwidth to the narrowband noise. In practice $\Gamma_n \ll 1$ and the bandwidth is much larger than the pure resonance linewidth f_n/Q_n. In the limit $\Gamma_n \to 0$, the bandwidth becomes infinite. If a quantum limited readout system were available, values of the order of 200 Hz could be reached (see Fig. 5).

The last two equations characterize the sensitivity of a resonant-mass detector. The optimum performance of a detector is obtained by filtering the output with a filter matched to the signal. The energy signal-to-noise ratio (SNR) of the filter output is given by the well-known formula

$$\mathrm{SNR} = \int_{-\infty}^{+\infty} \frac{|H(f)|^2}{S_h(f)}\,df, \tag{5}$$

where $H(f)$ is the Fourier transform of $h(t)$.

We will consider the SNR of a spherical detector for various GW signals.

Burst

We model the burst signal as a featureless waveform, rising quickly to an amplitude h_0 and lasting for a time τ_g much shorter then the detector optimum integration time $\Delta t = \Delta f_n^{-1}$. Its Fourier transform will be considered constant within the detector bandwidth: $H(f) \simeq H(f_n) = H_0$.

From Eq. (5) we get [26]

$$\mathrm{SNR} = \frac{\pi \Delta f_n H_0^2}{2S_h(f_n)}. \tag{6}$$

Using the equation $H_0^{\min} = h_0^{\min}\tau_g$, we find that $\mathrm{SNR} = 1$:

$$h_0^{\min} = \tau_g^{-1}\left[\frac{2S_h(f_n)}{\pi \Delta f_n}\right]^{1/2}. \tag{7}$$

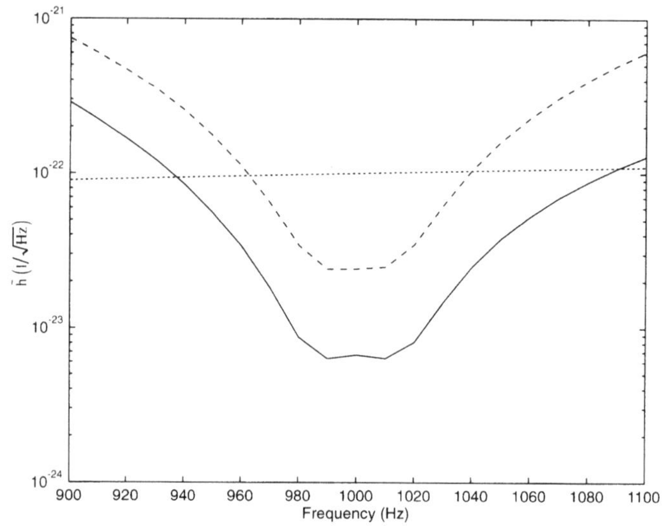

Fig. 5. The calculated strain noise spectrum for various detectors. The solid line is a 1 kHz Aluminum spherical antenna with quantum limited sensor noise, for a single channel (i.e., a single linear polarization GW arriving from an *arbitrary* direction). The dashed line is an equivalent bar antenna with quantum limited sensor noise, for the *optimum* orientation of the wave. The dotted line is the first generation LIGO detector, for the *optimum* orientation of the wave [28]

The requirement of a large bandwidth makes a strong coupling of the transducer to the antenna desirable. The range $h_0^{\min} \simeq 10^{-22}$ could be reached by the lowest order mode of a typical large spherical detector with bandwidth of the order of 10 Hz. This sensitivity would allow the detection of the GW collapses occurring in the Virgo cluster at the level of an energy release of $10^{-4} M_{\odot}$.

Monochromatic signal

We consider a sinusoidal wave of amplitude h_0 and frequency f_s constant for the entire observation time t_m. The Fourier transform amplitude at f_n is $\frac{1}{2} h_0 t_m$ with a bandwidth given by t_m^{-1}. For SNR $= 1$ we obtain the minimum detectable value of h_0 which, at $f_s = f_n$, is

$$h_0^{\min} = \left[\frac{2 S_h(f_n)}{t_m} \right]^{1/2} . \tag{8}$$

The minimization of the factor T/Q maximizes the SNR of the detector over a narrow bandwidth centered on the resonant frequency f_n.

For instance, the nearby pulsar [29] PSR J0437-4715 at a distance of 150 pc should emit at 347 Hz a GW amplitude (optimistically) of 2×10^{-26}. This would give an SNR $= 200$ on a spherical detector of $M = 100$ tons (we may think with a hollow sphere to get such a low resonant frequency; see Sect. 5), $Q = 10^7$, $T = 10\,\text{mK}$, after integrating the signal for 1 year.

In Figure 6 we show the upper limits to the GW amplitude h emitted by known pulsars observed for one year. Assuming a reasonable neutron star distribution in space and in frequency, we can expect to detect several sources in one year's observation time with the spherical antenna [30].

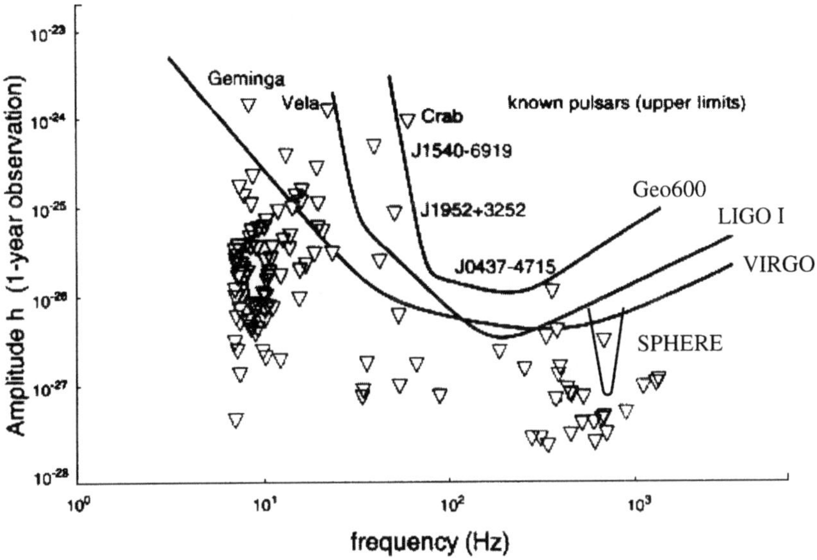

Fig. 6. Upper limits to the gravitational wave amplitude at the Earth from the measured spin-down of known pulsars

Chirp

We consider here the interaction of the spherical detector with the waveform emitted by a binary system, consisting of either neutron stars or black holes, in the inspiral phase. The system, in the Newtonian regime, has a clean analytic behaviour, and emits a waveform of increasing amplitude and frequency that can sweep up to the kHz range of frequency.

From point of view of the resonant-mass detector, the chirp signal can be treated as a transient GW, depositing energy in a time-scale short with respect to the detector damping time [31]. We can then use (6) to evaluate the SNR, where the Fourier transform $H(f_n)$ at the resonant frequency f_n can be written explicitly as

$$H(f_n) = \left\{ \left[\int h(t) cos(2\pi f_n t) dt \right]^2 + \left[\int h(t) sin(2\pi f_n t) dt \right]^2 \right\}^{1/2} \tag{9}$$

with $h(t)$ indicating $h_+(t)$ or $h_\times(t)$. Substituting in (9) the well-known chirp waveforms for optimally oriented orbits of zero eccentricity [32], the SNR for chirp detection is [33]:

$$\mathrm{SNR} = \frac{2^{1/3}5}{48} G^{2/3} \frac{\sigma_n Q_n 2\pi \Delta f_n}{4kT} \frac{1}{r^2} M_c^{5/3} (2\pi f_n)^{-4/3}. \tag{10}$$

M_c is the chirp mass defined as $M_c = (m_1 m_2)^{3/5}(m_1 + m_2)^{-1/5}$, m_1 and m_2 are the masses of the two compact objects (neutron stars or black holes) and r is the distance to the source. The chirp mass is the only parameter that determines the overall frequency acceleration of the chirp signal. It follows that the measurement of the time delay $\tau_2 - \tau_1$ between excitations of the first and second quadrupole modes on a spherical detector will determine the frequency acceleration of the chirp signal and allow the chirp mass to be measured automatically. An explicit formula has been found [33]:

$$M_c = 2^{8/5} \left(\frac{5}{256}\right)^{3/5} \frac{c^3}{G} \left(\frac{\omega_2^{-8/3} - \omega_1^{-8/3}}{\tau_2 - \tau_1}\right)^{3/5}, \tag{11}$$

where ω_1 and ω_2 are the angular frequencies of the first and second quadrupole modes. In Fig. 7 we report the distances at which binaries of the indicated masses will give SNR $= 1$ at the detector second quadrupole mode. For example, a system of two neutron stars or black holes having $m_1 = m_2 = 1.4\ M_\odot$ at 100 Mpc distance will give SNR $= 24$ in the $n = 1$ mode and SNR $= 4$ in the $n = 2$ mode of a CuAl spherical detector of 4 m diameter, 250 tons of weight, having quantum limited sensitivity. Time delays between the excitation of the two modes are of the order of some hundredths of seconds.

Other consequences of the multimode and multifrequency nature of a spherical detector are the possibilities of determining the orbit orientation by the measurement of the relative proportion of the two polarization amplitudes, and the source distance r by the measurement of the intrinsic GW amplitudes [33].

These capabilities of a spherical detector can provide a fundamental astrophysical measurement. In fact, if the binary source position is determined with sufficient precision to locate the binary in a particular galaxy, then the comparison of the galaxy redshift with the GW distance determines the Hubble constant with high accuracy [34].

Stochastic background

In this case $h(t)$ is a random function and we assume that its power spectrum, indicated by $S_{gw}(f)$, is flat and its energy density per unit logarithmic frequency is a fraction $\Omega_{gw}(f)$ of the closure density ρ_c of the universe:

$$\frac{d\rho_{gw}}{d\ln f} = \Omega_{gw}\rho_c. \tag{12}$$

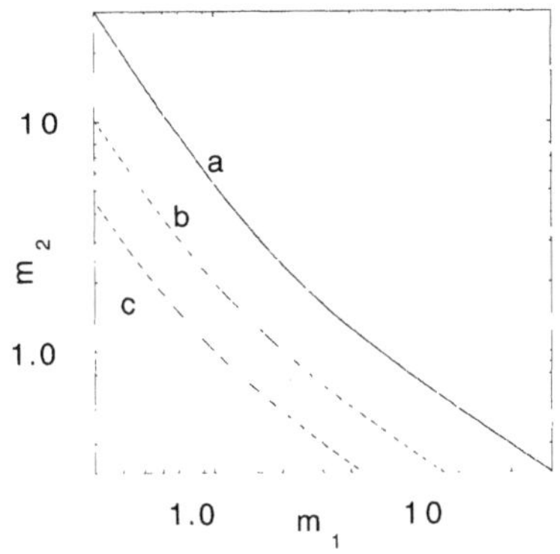

Fig. 7. Contours of constant chirp mass M_c in m_1, m_2 space. Masses are in units of solar mass. To each contour corresponds the maximum distance at which the given binary can be observed with an $SNR = 1$ by a quantum limited, 100 ton, CuAl sphere, 3 m in diameter, at its first order resonance frequency $f_1 = 725\,\mathrm{Hz}$. The corresponding maximum distances are: $a = 204\,\mathrm{Mpc}$, $b = 141\,\mathrm{Mpc}$, and $c = 107\,\mathrm{Mpc}$

$S_{gw}(f)$ is given by

$$S_{gw}(f) = \frac{2G}{\pi} f^{-3} \Omega_{gw}(f) \rho_c. \tag{13}$$

The measured noise spectrum $S_h(f)$ of a single resonant-mass detector automatically gives an upper limit to $S_{gw}(f)$ (and hence to $\Omega_{gw}(f)$).

Two different detectors with overlapping bandwidth Δf will respond to the background in a correlated way. The SNR of a GW background in a cross-correlation experiment between two spherical detectors located near one another and having power spectral densities of noise $S_h^1(f)$ and $S_h^2(f)$ is [35]

$$\mathrm{SNR} = \left(\frac{S_{gw}^2}{S_h^1 S_h^2} \Delta f t_m \right)^{1/4}, \tag{14}$$

where t_m is the total measuring time.

Detectors that are separated by some distance are not as well correlated, because GWs coming from within a certain cone about the line joining the detectors will reach one detector well before the other. The fall-off in the correlation with separation is a function of the ratio of the wavelength to the separation, and has been studied for pairs of bars, pairs of interferometers [36, 37] and pairs of spherical detectors [38].

If we suppose that two identical large spherical detectors are co-located for optimum correlation, the background will reach a SNR = 1 if Ω_{gw} is

$$\Omega_{gw} \simeq 2\ 10^{-7} \left(\frac{f_n}{10^3\ \mathrm{Hz}}\right)^3 \left(\frac{\sqrt{S_h^1(f_n)}}{10^{-24}\sqrt{\mathrm{Hz}}}\right) \left(\frac{\sqrt{S_h^2(f_n)}}{10^{-24}\sqrt{\mathrm{Hz}}}\right) \left(\frac{10\ \mathrm{Hz}}{\Delta f_n}\right)^{1/2} \left(\frac{10^7 s}{t_m}\right)^{1/2}, \quad (15)$$

where the Hubble constant is assumed to be 100 $\mathrm{km\,s^{-1}\,Mpc^{-1}}$.

Spherical detectors can set very interesting limits on the GW background at kHz frequencies. In particular, in view of recent estimations based on cosmological string models [39], it emerges that experimental measurements performed at the level of sensitivity attainable with large spherical detectors would be true tests of Planck-scale physics.

Equations (14) and (15) hold for any cross-correlation experiment between two GW detectors adjacent and aligned for optimum correlation. An interesting consequence is that it may be worthwhile in the near future to move an advanced resonant mass detector very near to a large interferometer, like VIRGO, to perform stochastic searches near 1 kHz [40].

4 Experimental results

- To verify the mechanical behavior of a spherical antenna Merkowitz and Johnson constructed a room temperature prototype in the shape of a truncated icosahedron (TIGA) at LSU [41]. They tested their direction finding algorithm by applying an impulse excitation to the surface of the prototype and measuring the response of the 6 resonant transducers mounted on the surface [42]. The amplitude and phase information of the 5 quadrupole modes of the sphere were successfully found from the transducer data. The strain tensor was then constructed from this information. The impulse location was calculated by properly interpretating the eigenvalues and eigenvectors of the strain tensor for an impulsive excitation. As shown in Fig. 8, Merkowitz and Johnson found the calculated location of the impulses to be consistent with those measured geometrically, thus verifying their technique. In the course of testing their model, they also verified the *in situ* measurement technique based on a generalized mechanical model including deviations from perfect symmetry in the actual detector, such as non-degenerate quadrupole modes, imprecise placement of the mechanical resonators and non-perfect matching of the resonator frequencies [43].
- An experimental study of the eigenfrequencies and quality factors of vibration of aluminum alloy 5056 spherical resonators at ultra-low temperatures was performed at the Kamerlingh Onnes Laboratory in Leiden in collaboration with the Rome group [25]. Three important points were investigated:

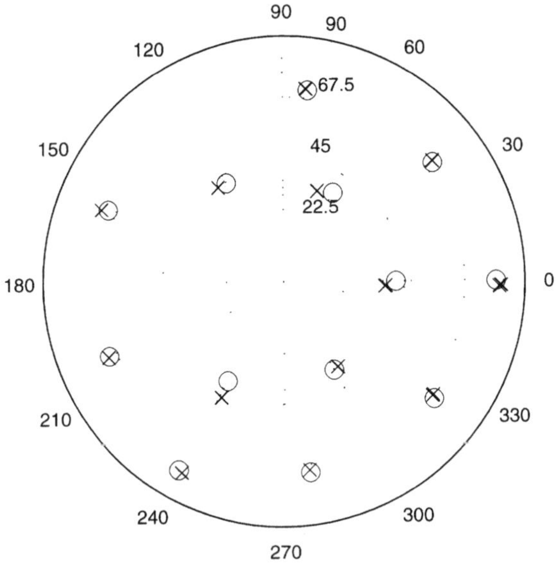

Fig. 8. Location of several impulses applied to the LSU prototype spherical antenna, calculated using the TIGA technique. The x's mark the locations calculated from the motion sensor data, and the o's mark the locations of the mechanical impulse excitation masured geometrically [42]

1. the quality factor of the vibrational modes of a solid sphere at very low temperatures. Previous measurements have in fact investigated the Q of flexural modes of disks, longitudinal modes of bars and torsional modes of resonators;
2. the independence of the 5 degenerate quadrupole modes. The transfer of energy at an appreciable level between these modes could make a potential detector useless;
3. the fabrication technique to produce a large sample while preserving the high inherent Q of the material. It has been proposed [44] that an explosive welding method be used to produce large-diameter spheres out of explosively-bonded plates. This technique produces high-strength bonds between similar and dissimilar metals and appears capable of preserving the inherent attributes of each parent metal.

The experimental study on several 15 cm diameter Al 5056 spheres, suspended at the center of mass, answered these questions [25].

Q-values of spheroidal modes ranging from 3 10^6 to 2 10^7 in a bulk sample and from 1.7 10^6 to 8 10^6 in an explosively-bonded sample were measured. A Q-value of 1.2 10^8 was found in a higher order mode of the bulk sample as shown in Fig. 9. Independence of the quadrupole modes was measured within one part in 10^5 in amplitude.

After these experiments, other materials were tested in Leiden, in particular copper alloys [45]. It was found that CuAl is a promising material for use as a large

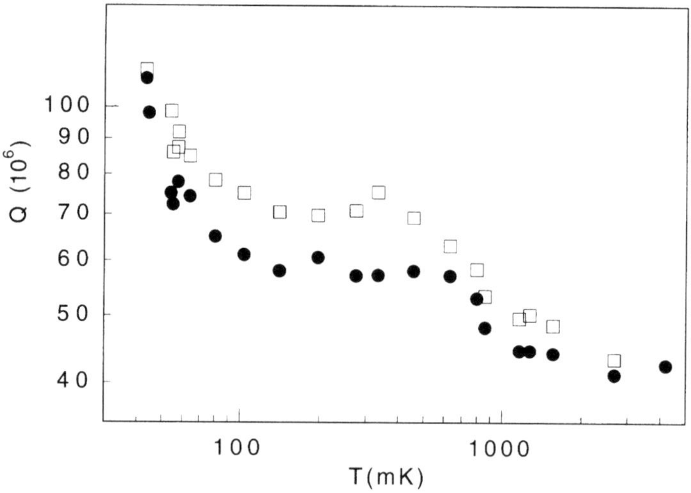

Fig. 9. Q-values of two modes (circles = 35354 Hz, squares = 35359 Hz) of a 15 cm diameter Al 5056 sphere, down to very low temperatures [25]

spherical antenna[1]. Tests on small pieces have demonstrated that high mechanical quality factors are possible with the casting technique. The ability to cast in large pieces, along with high thermal conductivity (lower thermal gradients, thus lower surface temperatures) and relatively high Young modulus (high GW cross section) makes a copper alloy an appealing material for a large detector.

- A study at room temperature on a suspension system for a spherical detector has been developed at the INFN National Laboratories in Frascati with an Al 5056 prototype similar to the sample at LSU. The measurements, in agreement with the predictions obtained with finite element analysis, show the possibility of reaching, in the frequency range of interest, the mechanical isolation of about 200 dB required for a GW resonant detector [46].

5 The hollow sphere

Most of the nice properties of a spherical GW detector depend on its being *spherically symmetric*. A spherical shell, or *hollow sphere*, obviously maintains that symmetry, so it can be considered as an interesting alternative to the usual *solid* sphere [47]. In order to have a large cross section, a resonant GW detector must be made of a high speed-of-sound material, and must have a large mass. The actual construction of a massive spherical body may be technically difficult. In fact, fabricating a large hollow sphere is a different task from fabricating a solid one. Casting a hollow half sphere is a nearly two-dimensional cast, at odds with casting a solid sphere, which requires

[1] LIPS, Drunen (The Netherlands), European company fabricating large propellers, is capable of casting large, high quality pieces of this material in a spherical shape

rather special moulds. As an example of the feasibility of large two-dimensional casting we can mention the fabrication of propellers of more than 10 meters in size and masses of the order of 100 tons[2]. Two hollow hemispheres could then be welded together with electron beam techniques.

However, while it is known that these welding techniques preserve most of the properties of the original material, their effect on the acoustic quality factor (a relevant parameter in resonant mass detectors; see Eq. (3)) must be further studied.

An interesting feature of the hollow sphere is that its quadrupole frequencies are below those of an equally massive solid sphere, thus making the low frequency range accessible to this antenna with good sensitivity. In [47] the system response to the classical GW signal sources (bursts, chirps, continuous and stochastic) for several sizes and materials has been studied: interesting signal-to-noise ratios are attainable with such a detector. Also, the bandwidth of these detectors partly overlaps with that of the projected large interferometers [48, 49], so potentially both kinds of detectors can be operated simultaneously to make *hybrid* GW observatories of unprecedented sensitivity and signal characterisation power. In Table 2 the features of different hollow spheres are given. Strain sensitivities of the order of $10^{-24}\,\mathrm{Hz}^{-1/2}$ over a bandwidth of tens of Hz around the two resonance frequencies can be achieved. They correspond to a sensitivity of $h_0 \sim 10^{-22}$ for an impulsive signal and of $h_0 \sim 10^{-27}$ for a monochromatic signal (integrated for 1 year).

Table 2. Main features for several hypothetical hollow spheres of two different materials (R and a are the outer and inner radii, respectively)

	M (ton)	$2R$ (m)	$(R-a)$ (cm)	f_1 (Hz)	f_2 (Hz)
CuAl	200	4	81	395	1188
	200	6	25	191	753
	100	4	31	302	1161
	100	6	12	185	738
	40	3	22	399	1543
	40	4	11	281	1115
Al 5056	200	6	90	273	935
	100	6	37	230	896
	40	4	35	361	1370
	40	6	14	218	866

As regards the sensitivity to a GW stochastic background, if we assume two identical large hollow spherical detectors co-located for optimum correlation, the background will reach an SNR = 1 if Ω_{gw} is

[2] See footnote at page 100

$$\Omega_{gw} \simeq 10^{-9} \times \left(\frac{f_n}{200\,\text{Hz}}\right)^3 \left(\frac{\sqrt{S_h^1(f_n)}}{10^{-24}\text{Hz}^{\frac{1}{2}}}\right) \left(\frac{\sqrt{S_h^2(f_n)}}{10^{-24}\text{Hz}^{\frac{1}{2}}}\right) \left(\frac{20\text{Hz}}{\Delta f_n}\right)^{\frac{1}{2}} \left(\frac{10^7\text{sec}}{t_m}\right)^{\frac{1}{2}}, \tag{16}$$

where the Hubble constant has been taken as $100\,\text{km}\,\text{s}^{-1}\,\text{Mpc}^{-1}$.

Acknowledgements

I wish to thank Massimo Bassan, Eugenio Coccia, Ivo Modena and Guido Pizzella for useful discussions and suggestions.

References

1. Weber J. (1961): General Relativity and Gravitational Waves. Interscience, New York
2. Weber J. (1960): Phys. Rev. **117** (1), 306–313
3. Boughn S.P., Fairbank W.M., Giffard R.P. (1982): Astrophys. J. L **261**, 19–22
4. Astone P., Bassan M., Bonifazi P. (1993): Phys. Rev. D **47** (2), 362–375
5. Mauceli E., Geng Z.K., Hamilton W.O. (1996): Phys. Rev. D **54**, 1264
6. Blair D.G., Ivanov E.N., Tobar M.E. (1995): Phys. Rev. Lett. **74**, 1908
7. Astone P., Bassan M., Bonifazi P. (1997): Astroparticle Phys. **7**, 231–243
8. Prodi G.A., Conti L., Mezzena R. (1998): Initial operation of the gravitational wave detector AURIGA, in *Proc. of the Second Edoardo Amaldi Conference on Gravitational Wave Experiments (CERN 1997)*, ed. by E. Coccia, G. Pizzella, G. Veneziano, World Scientific, Singapore, pp. 148–158
9. Coccia E. (1998): Resonant-mass Gravitational-Wave Detectors in Operation, in *Proc. of the Second International LISA Symposium (Pasadena, California, 1998)*, ed. by W.M. Folkner, AIP Conference Proceedings, p. 456
10. Bassan M., Minenkov Y., Simonetti R. (1997): Advances in Linear Transducers for Resonant Gravitational Wave Antennas, in *Proc. of the International Conference on Gravitational Waves (Cascina, Pisa, Italy 1996)*, ed. by I. Ciufolini, F. Fidecaro, World Scientific, Singapore, pp. 225–228
11. Tobar M.E., Ivanov E.N., Qi D.K.L (1997): Appl. Phys. B **64**, 153
12. Carelli P., Castellano M.G., Lioni R., Toruoli G. (1998): Appl. Phys. Lett. **72**, 115
13. Conti L., Cerdonio M., Taffarello L. (1998): Rev. Sci. Instr. **69**, 554
14. Coccia E., Pizzella G., Veneziano G., eds. (1998): Proceedings of the Second Edoardo Amaldi Conference on Gravitational Wave Experiments (CERN 1997). World Scientific, Singapore
15. Frossati G., Coccia E. (1994): Cryogenics, ICEC Supplement **34**, 9
16. Johnson W.W., Merkowitz S.M. (1993): Phys. Rev. Lett. **70**, 2367
17. Zhou C.Z., Michelson P.F. (1995): Phys. Rev. D **51**, 2517
18. Coccia E., Lobo J.A., Ortega J.A. (1995): Phys. Rev. D **52**, 3735

19. Cerdonio M. et al. (1993): Phys. Rev. Lett. **71**, 4107
20. Forward R. (1971): Gen. Rel. and Grav. **2**, 149
21. Wagoner R.V., Paik H.J. (1976): Multi-mode detection of gravitational waves by a sphere, in *Proc. of the Int. Symposium on Experimental Gravitation*. Accademia Nazionale dei Lincei, Rome, pp. 257–266
22. Magalhaes N.S., Johnson W.W., Frajuca C., Aguiar O.D. (1995): Mon. Not. R. Astron. Soc. **274**, 670
23. Lobo J.A., Serrano M. (1996): Europhys. Lett. **35**, 253
24. Lobo, J.A. (1995): Phys. Rev. D **52**, 591
25. Coccia E., Fafone V., Frossati G. (1996): Phys. Lett. A **219**, 263–270
26. Astone P., Pallottino G.V. , Pizzella G. (1997): Class. Quantum Grav. **14**, 2019–2030
27. Pallottino G.V., Pizzella G. (1981): Nuovo Cimento C **4**, 237
28. Abramovici A., Althouse W.E., Drever R.W.P. (1992): Science **256**, 325
29. Johnston S., Lorinier D.R., Harrison P.A. (1993): Nature **361**, 613
30. Finn L.S. (1997): Gravitational radiation sources for Acustic Detectors, in *Proceedings of the OMNI-1 Workshop (Sao Josè dos Campos, Sao Paulo, Brazil, 1996)*, ed. by W.F. Velloso Jr., O.D. Aguiar, N.S. Magalhaes, World Scientific, Singapore, pp. 3–12
31. Dewey D. (1987): Phys. Rev. D **36**, 1577
32. Thorne K.S. (1987): Gravitational radiation, in *Three Hundred Years of Gravitation*, ed. by S.W. Hawking, W. Israel, Cambridge University Press, pp. 330–458
33. Coccia E., Fafone V. (1996): Phys. Lett. A **213**, 16
34. Schutz B. (1986): Nature **323**, 310
35. Bendat J.S., Piersol A.G. (1966): Measurement and Analysis of Random Data. Wiley & Sons, New York
36. Michelson P.F. (1987): Mon. Not. R. Astr. Soc. **227**, 933
37. Flanagan E.E. (1987): Phys. Review D **48**, 2389
38. Vitale S., Cerdonio M., Coccia E., Ortolan A. (1997): Phys. Rev. D **55**, 1741
39. Brustein R., Gasperini M., Giovannini M., Veneziano G. (1995): Phys. Lett. B **36**, 45
40. Astone P., Lobo J.A., Schutz B.F. (1994): Class. Quantum Grav. **11**, 2093
41. Merkowitz S.M., Johnson W.W. (1996): Phys. Rev. D **53**, 5377
42. Merkowitz S.M., Johnson W.W. (1998): Europhys. Lett. **41** (4), 355–360
43. Merkowitz S.M., Johnson W.W. (1997): Real life TIGA measurements: results from the LSU prototype, in *Proc. of the OMNI-1 Workshop, (Sao Josè dos Campos, Sao Paulo, Brazil, 1996)*, ed. by W.F. Velloso Jr., O.D. Aguiar, N.S. Magalhaes, World Scientific, Singapore, pp. 190–193
44. Coccia E., Fafone V., Frossati G. (1995): On the Design of Ultrasow Temperature Spherical Gravitational Wave Detectors, in *Proc. of the First Edoardo Amaldi Conference on Gravitational Wave Experiments*, ed. by E. Coccia, G. Pizzella, F. Ronga, World Scientific, Singapore, pp. 463–473
45. Frossati G. (1997): A 100 TON 10 mk spheriacl gravitational wave detector, in *Proc. of the OMNI-1 Workshop, (Sao Josè dos Campos, Sao Paulo, Brazil, 1996)* , ed. by W.F. Velloso Jr., O.D. Aguiar, N.S. Magalhaes, World Scientific, Singapore, pp. 179–189
46. Merkowitz S.M., Coccia E. Fafone V. (1999): Rev. Sci. Instr. **70** (2), 1553
47. Coccia E., Fafone V., Frossati G. (1998): Phys. Rev. D **57** (4), 2051–2060
48. Brillet A., Babusci D., Fang W. (1997): VIRGO-Status Report, pp. 86–96
49. Fritschel P., The LIGO project: progress & plans, pp. 74–85

The Entropy of Black Holes via Noether's Theorem

L. Fatibene, M. Ferraris, M. Francaviglia, M. Raiteri

Abstract. We generalise some properties of Noether charges that have been related recently to the entropy of black holes. We define the variation of the entropy of black holes starting from suitable variations of superpotentials. Superpotentials are associated to Noether currents and are properly defined in any gauge-natural field theory. In this covariant framework the entropy is defined in such a way that it automatically satisfies the first principle of thermodynamics. We consider the BTZ black hole solution as a test model. Entropy is explicitly computed both by our method and by the prescription given by Wald in order to compare results.

1 Introduction

The problem of the definition of the entropy of black holes is one of the intriguing problems in *recent* developments of General Relativity. It is very important in itself because it seems to reveal a deep (and, under some perspectives, still mysterious) relation between macroscopic/classical physics of gravity and microscopic/quantum gravity. Some years ago Wald et al. [1–3] proposed a prescription for the computation of entropy from a geometrical viewpoint. It is based on previously known formulae for conserved quantities, obtained via Noether's theorem, which are obviously related to entropy owing to the first principle of black-hole thermodynamics

$$\delta m = \Omega\, \delta J + T\, \delta S, \tag{1.1}$$

where m is the *total mass*, J is the *total angular momentum*, S is the *entropy*, Ω is the *angular velocity of the horizon* and T is the *temperature* of the Hawking radiation irradiated by the black hole under consideration. We remark that T is an external parameter which can be provided by various techniques (for example, Euclidean path integral, on a curved background).

In general, when gauge symmetries are present, other conserved quantities Q_A, called *gauge charges*, are related to the additional gauge symmetries of the theory. In these cases the first principle of thermodynamics has the more general form

$$\delta m = \Omega\, \delta J + T\, \delta S + b^A\, \delta Q_A, \tag{1.2}$$

where $b^A\, \delta Q_A$ are the *work terms* corresponding to gauge symmetries. We remark that gauge theories are gauge-covariant, i.e., covariant with respect to the group $\mathrm{Aut}(P)$ of automorphisms of a suitable principal bundle P, called the *structure group of the theory*. In general, one cannot identify canonically the group $\mathrm{Diff}(M)$ of spacetime diffeomorphisms as a subgroup of $\mathrm{Aut}(P)$; this latter group instead acts canonically on dynamical fields. As a consequence, the group $\mathrm{Diff}(M)$ does not act naturally on dynamical fields and thence Lie derivatives of fields with respect to spacetime vector fields are undefined. Since Lie derivatives of fields enter Noether currents, a general and systematic framework to deal with conservation laws in gauge covariant theories has to be developed. We shall here introduce *gauge-natural theories* which provide such a framework and also encompass *natural theories*, i.e., the Diff(M)-covariant field theories.

Entropy may be defined in this more general framework, in a global and geometrical setting, which also illuminates the relation between various proposals. This framework clarifies exactly at what stage of the theory the required hypotheses are relevant. The geometrical analysis already allowed us to gauge away a set of irrelevant hypotheses originally required to apply Wald's original prescription, which is however encompassed in ours.

Here we shall present a short though self-contained introduction to the geometrical definition of entropy via Noether's theorem and we shall use the BTZ black hole solution in dimension $(2+1)$ as a test model for applications. As presented here, the BTZ solution is a solution of Einstein's equations with (negative) cosmological constant, i.e., a solution of a natural theory. Both our general framework and Wald's, as well as other proposals [5], apply to BTZ solutions. As a consequence one can compare various techniques directly in this example. Here we compute in detail BTZ entropy, following both our prescription and Wald's.

We remark that there is an equivalent setting which presents BTZ spacetime as a solution of an SO(2, 1)-gauge covariant theory; it is called *BCEA theory*. This is a frame-affine theory coupled to *topological matter* in which gravity is described by means of a frame on spacetime and an independent SO(2, 1)-connection. Since such a theory is a non-natural theory, Wald's prescription [1, 2] does not apply in the very basic sense that most of the concepts used in it are undefined in this setting. We refer to [6] for further details about the application of our prescription to BCEA theory.

We finally stress that there are also other families of solutions (e.g. Taub-NUT solutions) to which, even if they belong to a natural setting, Wald's prescription does not apply due to failure of hypotheses, while our prescriptions readily apply and give the same results as found by other methods (see, e.g., [7]).

2 Notation

We hereafter summarise standard bundle theoretic notation in as far as we need it in the present paper. Further details may be found in [8, 9].

We denote by $\mathcal{C} = (C, M, \pi, F)$ the *configuration bundle* of a field theory; C is the *total space*, M is the *base manifold*, π is the *projection* and F is the *standard fiber*. Let us denote by (x^μ, y^i) a set of fibered coordinates on $\mathcal{C}$. The base manifold M, which is assumed to be orientable, paracompact, connected and of dimension n, is identified with spacetime and the standard fiber is the manifold describing the set where fields take their values. Then a *configuration* is a (global) section σ, represented by the (local) equations $y^i = \sigma^i(x)$. Giving a section amounts to the specification of fields at any point (x^μ) of spacetime.

A projectable vector field over C is written (locally) as

$$\hat{\Xi} = \xi^\mu(x)\,\partial_\mu + \xi^i(x, y)\,\partial_i \qquad \begin{cases} \partial_\mu = \frac{\partial}{\partial x^\mu} \\ \partial_i = \frac{\partial}{\partial y^i} \end{cases} \tag{2.1}$$

and it projects over the field $\xi = \xi^\mu(x)\partial_\mu$ over the base M. [In general there is no canonical way to associate a vector field $\hat{\Xi}$ on the bundle to an arbitrary vector field ξ on the base manifold. Natural bundles, defined here in Sect. 3, are exactly those bundles which do allow a canonical lifting procedure]. If $\hat{\Xi}$ is a projectable vector field, we say that

$$\pounds_{\hat{\Xi}}\sigma = T\sigma(\xi) - \hat{\Xi}\circ\sigma = (\xi^\mu\partial_\mu\sigma^i - \xi^i)\partial_i \tag{2.2}$$

is the *Lie-derivative of the section* σ *with respect to* $\hat{\Xi}$.

Vector fields over C which are tangent to the fibers are called *vertical vector fields*. They are sections of the so-called *vertical bundle* $V(C)$ which is a subbundle of TC. In fibered coordinates a vertical vector field reads as

$$X = \xi^i(x, y)\partial_i. \tag{2.3}$$

A 1-form over the bundle C may be written as:

$$\Theta = \theta_\mu(x, y)\mathrm{d}x^\mu + \theta_i(x, y)\mathrm{d}y^i. \tag{2.4}$$

One can consider the pull-back of Θ along a section σ of C, obtaining

$$(\sigma^*)\Theta = \left(\theta_\mu(x, \sigma(x)) + \theta_i(x, \sigma(x))\frac{\partial\sigma^i}{\partial x^\mu}\right)\mathrm{d}x^\mu \tag{2.5}$$

which is a form over the base M.

We denote by $V^*(C)$ the *dual vertical bundle* which is the vector bundle dual to $V(C)$. We remark that $V^*(C)$ is not a subbundle of T^*C. In the sequel, we denote the dual basis of $\{\partial_i\}$ by $\{\overline{\mathrm{d}y}^{\,i}\}$. The contraction of an element $\Theta = \theta_i\,\overline{\mathrm{d}y}^{\,i}$ of $V^*(C)$ with a vertical vector $X = X^i\,\partial_i$, both projecting to the same point $p \in C$, will be denoted by brackets: $< \Theta \mid X > = \theta_i\; X^i$.

We denote the *k-jet bundle* by J^kC; it is the bundle of all equivalence classes of (local) sections of C, where two (local) sections are equivalent iff they have a contact of order (at least) k. In other words, they have the same Taylor expansion at a point up to order k. *Natural fibered coordinates* over J^kC are denoted by $(x^\mu, y^i, y^i_\mu, \ldots, y^i_{\mu_1\ldots\mu_k})$, where the $y^i_{\mu_1\ldots\mu_l}$ represent the lth-order derivatives of local sections with respect to world coordinates and they are thence understood to be symmetric in lower indices. If σ is a (global) section of C then we use $j^k\sigma$ to denote its kth*-order prolongation* which is immediately seen to be defined and is a (global) section of J^kC.

The *total derivative operators* are defined as follows:

$$\mathrm{d}_\mu = \partial_\mu + y^i_\mu\,\partial_i + \cdots + y^i_{\mu\mu_1\ldots\mu_l}\,\partial_i^{\mu_1\ldots\mu_l} + \ldots. \tag{2.6}$$

The kth*-order lift of a vector field* $\Xi = \xi^\mu \partial_\mu + \xi^i \partial_i$ over C is the vector field $j^k\Xi = \xi^\mu \partial_\mu + \xi^i \partial_i + \xi^i_\mu \partial^\mu_i + \cdots + \xi^i_{\mu_1\ldots\mu_k} \partial^{\mu_1\ldots\mu_k}_i$ over J^kC, where

$$\begin{cases} \xi^i_\mu = \mathrm{d}_\mu \xi^i - \mathrm{d}_\mu \xi^\nu y^i_\nu \\ \ldots \\ \xi^i_{\mu_1\ldots\mu_k} = \mathrm{d}_{\mu_k} \xi^i_{\mu_1\ldots\mu_{k-1}} - y^i_{\mu_1\ldots\mu_{k-1}\nu} \, \mathrm{d}_{\mu_k} \xi^\nu . \end{cases} \tag{2.7}$$

By an abuse of language we still denote by $< \cdot \mid \cdot >$ the duality between $V(J^kC)$ and its dual $V^*(J^kC)$, for any order k of prolongation.

A Lagrangian of order k is a morphism $L : J^kC \to A_n(M)$ from J^kC into the bundle $A_n(M)$ of n-forms over M. One can define the (global) bundle morphism $\delta L : J^kC \to V^*(J^kC) \otimes A_n(M)$, intrinsically identified by

$$< \delta L \circ j^k\sigma \mid j^kX > = \frac{d}{dt}\left(L \circ J^k\hat{\Psi}_t \circ j^k\sigma\right)\Big|_{t=0}, \tag{2.8}$$

where X is any vertical vector field on C, $\hat{\Psi}_t$ is its flow and σ is a configuration. It is well-known that for each Lagrangian L there exists a unique (global) morphism, called *the Euler–Lagrange morphism*,

$$\mathbb{E}(L) = J^{2k}C \to V^*(C) \otimes A_n(M) \tag{2.9}$$

together with a family of (global) morphisms (which depend on the Lagrangian and possibly on a connection γ on M; [9, 10]) called *Poincaré–Cartan morphisms*

$$\mathbb{F}(L, \gamma) = J^{2k-1}C \to V^*(J^{k-1}C) \otimes A_{n-1}(M) \tag{2.10}$$

defined so that the following *first-variation formula* holds for any vertical vector field X on C:

$$< \delta L \mid j^kX > = < \mathbb{E}(L) \mid X > + \operatorname{Div} < \mathbb{F}(L, \gamma) \mid j^{k-1}X > . \tag{2.11}$$

The *formal divergence operator* Div is defined by

$$\operatorname{Div}(\omega) \circ j^{k+1}\sigma = \mathrm{d}(\omega \circ j^k\sigma), \quad \omega : J^kC \to A_l(M), \tag{2.12}$$

$\mathrm{d}(\cdot)$ being the exterior differential operator on forms.

We remark that the first-variation formula (2.11) is written on the bundle C and it encompasses the whole class of *first-variation formulae* that can be written on the base M by pull-back along sections of the configuration bundle C (not necessarily solutions of field equations, i.e., of the Euler–Lagrange equations $\mathbb{E}(L) \circ j^{2k}\sigma = 0$). We shall see that performing calculations in this *off-shell fashion* (i.e., along arbitrary, possibly non-critical sections) turns out to be important; certainly it clarifies relations among intrinsic quantities that can be hidden by direct evaluation along solutions (i.e., in the so-called *on-shell* fashion).

3 Gauge-natural theories and Noether currents

Gauge-natural theories provide the approppriate framework for a systematic treatment of gauge symmetries. Of course any gauge theory can just as well be investigated on its own without introducing further geometric structures. However, once one analyses the general properties of such field theories, the gauge-natural framework spontaneously arises. Furthermore, gauge-natural theories include in a unifying scheme all natural theories such as General Relativity as well as gauge theories, Bosonic and Fermionic matter together with their mutual interactions [11].

The starting point is the observation that a gauge theory has a preferred group of symmetries, the *gauge-transformation group*, which preserves its structure. In the field theories which are relevant to fundamental physics, there exists a Lie group G acting at each point of the configuration bundle.

Mathematically, this amounts to the assumption that there exists a G-principal bundle $\mathcal{P}$, called the *structure bundle*, whose group $\mathrm{Aut}(\mathcal{P})$ of principal automorphisms is required to act on configuration fields precisely by means of elements of the Lie group G at any point.

A *gauge-natural bundle* is a fiber bundle $\mathcal{C} = (C, M, \pi, F)$ together with a canonical (functorial) rule that associates a bundle automorphism of $\mathcal{C}$ to each automorphism of $\mathcal{P}$. This rule is also required to preserve the composition of automorphisms, thence inducing a natural action of $\mathrm{Aut}(\mathcal{P})$ on $\mathcal{C}$. Recalling that sections of $\mathcal{C}$ represent dynamical fields, we see that a natural action of $\mathrm{Aut}(\mathcal{P})$ (i.e., gauge-transformations) is then defined on configurations, which is encountered in physical phenomenology, as noted above (see [8, 11] for further details).

Let us denote the *s-frame bundle* over M by $L^s(M)$. For any integer pair (r, s) such that $r \leq s$, let $W^{(r,s)}\mathcal{P} = J^r\mathcal{P} \times_M L^s(M)$ be the *gauge-natural prolongation of order* (r, s) of $\mathcal{P}$, where we have denoted the fiber product over M by $\times_M$. The bundle $W^{(r,s)}\mathcal{P}$ is a principal bundle whose structure group is denoted by $W^{(r,s)}G$ and it is the semi-direct product of the groups J^rG and $\mathrm{GL}^s(n)$. It can be proved that any gauge-natural bundle is associated to $W^{(r,s)}\mathcal{P}$ for some (finite) order (r, s) (see the first item of [8]). Now, for any manifold F and any representation $\lambda : W^{(r,s)}G \times F \to F$, any automorphism $\Phi \in \mathrm{Aut}(\mathcal{P})$ naturally induces an automorphism $\Phi^{(r,s)}$ over $W^{(r,s)}\mathcal{P}$, which in turn induces a transformation Φ^λ on the gauge-natural bundle $\mathcal{C} = W^{(r,s)}\mathcal{P} \times_\lambda F$ associated to $W^{(r,s)}\mathcal{P}$ through λ given by

$$\Phi^\lambda([q, f]) = [\Phi^{(r,s)}(q), f] \quad q \in W^{(r,s)}\mathcal{P}\,,\ f \in F\,,\ [q, f] \in \mathcal{C} \tag{3.1}$$

where $[\,,\,]$ denotes the orbit with respect to the left action λ. Then the correspondence $\Phi \mapsto \Phi^\lambda$ provides the canonical rule according to which gauge-transformations act on configuration fields.

Let us now consider the infinitesimal generator Ξ of a 1-parameter subgroup of principal automorphisms Φ_t of the structure bundle $\mathcal{P}$. The vector field Ξ is projectable and we denote by ξ its projection over the base manifold M. If $\{\rho_A\}$, $A = 1 \dots k = \dim(G)$, forms a local basis for vertical right-invariant vector fields over $\mathcal{P}$, then the infinitesimal generator Ξ is of the form

$$\Xi = \xi^\mu(x)\,\partial_\mu + \xi^A(x)\,\rho_A. \tag{3.2}$$

Then a 1-parameter subgroup of automorphisms Φ_t^λ of the configuration bundle $\mathcal{C}$ is induced; let us denote by $\hat{\Xi}$ its infinitesimal generator which also projects over ξ. The *Lie derivative of a section* σ with respect to Ξ can be defined by specializing the general definition of Lie derivative with respect to a projectable vector field over $\mathcal{C}$, namely

$$\pounds_\Xi \sigma := \pounds_{\hat{\Xi}} \sigma = T\sigma(\xi) - \hat{\Xi} \circ \sigma. \tag{3.3}$$

We remark that the Lie derivative $\pounds_\Xi \sigma$ defined here is locally a linear combination of the ξ^μ and their derivatives up to order s and the ξ^A and their derivatives up to order r.

As a particular example one can consider the trivial gauge group $G = \{e\}$. Accordingly, any *trivial* gauge-natural bundle $\mathcal{C}$ is of order $(0, s)$, i.e., it is just associated to $L^s(M)$; in this way we recover the class of *natural bundles*, for example, the bundles of tensors and linear connections over M. Owing to the triviality of the gauge group $\{e\}$, the diffeomorphisms on the base manifold M act naturally on the configuration bundle $\mathcal{C}$. Then any vector field ξ over M induces a unique vector field over $\mathcal{C}$ and Lie derivatives of sections of $\mathcal{C}$ with respect to ξ are well-defined according to their "classical" definition. This natural structure is not encountered in *non-trivial* gauge theories, where the configuration field is a principal connection over a structure bundle $\mathcal{P}$ with $G \neq \{e\}$. The bundle of principal connections over $\mathcal{P}$ is gauge-natural of order $(1, 1)$ and thence it is not natural; in particular Lie derivatives with respect to a generic spacetime vector field ξ are not defined.

A Lagrangian L is called *gauge-natural* if it is Aut($\mathcal{P}$)-covariant, i.e., for any generator Ξ of automorphisms of $\mathcal{P}$ it satisfies

$$\pounds_\xi (L \circ j^k \sigma) = < \delta L \circ j^k \sigma \mid j^k \pounds_\Xi \sigma > \quad \forall \sigma : M \to \mathcal{C}. \tag{3.4}$$

We then say that any infinitesimal generator Ξ of automorphisms of the structure bundle $\mathcal{P}$ is an *infinitesimal symmetry*. A field theory is said to be a *gauge-natural theory* if both the configuration bundle and the Lagrangian are gauge-natural. Furthermore, we require that two connections (γ, A), called *dynamical connections*, on $L(M)$ and $\mathcal{P}$, respectively, can be built out of the dynamical fields together with their derivatives up to some (finite) order. This is a reasonable requirement. For example, in General Relativity, γ can be chosen to be the Levi–Civita connection of the dynamical metric, while in gauge theories the dynamical field itself is the dynamical connection A.

The first-variation formula (2.11) and the gauge covariance condition (3.4) alone provide a complete treatment of conserved quantities in gauge-natural field theories. In fact, these two formulae provide an intrinsic and covariant version of Noether's theorem by simply noticing that the Lie derivative of a section σ is vertical. Hence one can apply the first-variation formula in (3.4):

$$\begin{aligned} \pounds_\xi (L \circ j^k \sigma) &= < \delta L \circ j^k \sigma \mid j^k \pounds_\Xi \sigma > = \\ &= < \mathbb{E}(L) \circ j^{2k} \sigma \mid \pounds_\Xi \sigma > + \operatorname{Div} < \mathbb{F}(L, \gamma) \circ j^{2k-1} \sigma \mid j^{k-1} \pounds_\Xi \sigma > \end{aligned} \tag{3.5}$$

which, by expanding the l.h.s. Lie-derivative by the usual rule $\pounds_\xi = i_\xi\, d + d\, i_\xi$, allows the following currents to be defined:

$$\begin{aligned} \mathcal{E}(L, \Xi) &= < \mathbb{F}(L, \gamma) \mid j^{k-1}\Xi_\pounds > -i_\xi\, L \\ \mathcal{W}(L, \Xi) &= - < \mathbb{E}(L) \mid \Xi_\pounds > \end{aligned} \tag{3.6}$$

called, respectively, *Noether current* and the *work current*. We remark that these currents are well-defined fibered morphisms $\mathcal{E}(L, \Xi) : J^{2k-1}\mathcal{C} \to A_{n-1}(M)$ and $\mathcal{W}(L, \Xi) : J^{2k}\mathcal{C} \to A_n(M)$, respectively. Here $\Xi_\pounds : J^1\mathcal{C} \to V(\mathcal{C})$ is the global bundle morphism intrinsically defined by

$$\Xi_\pounds \circ j^1\sigma = \pounds_\Xi \sigma, \tag{3.7}$$

where $\pounds_\Xi\sigma$ is the Lie derivative of a section σ defined by (3.3).

Because of (3.5), these two currents obey a conservation law

$$\text{Div}\, \mathcal{E}(L, \Xi) = \mathcal{W}(L, \Xi). \tag{3.8}$$

Whenever we have a current $\mathcal{J}(L, \Xi)$ over $\mathcal{C}$ we can consider its pull-back along a section σ obtaining a form over M denoted by $\mathcal{J}(L, \Xi, \sigma)$. Let us then denote by $\mathcal{W}(L, \Xi, \sigma)$ and $\mathcal{E}(L, \Xi, \sigma)$ the pull-backs of $\mathcal{W}(L, \Xi)$ and $\mathcal{E}(L, \Xi)$, respectively, along a section σ of $\mathcal{C}$. Since $\mathcal{W}(L, \Xi, \sigma)$ vanishes because of (3.6) whenever σ is a solution of field equations $\mathbb{E}(L) \circ j^{2k}\sigma = 0$, $\mathcal{E}(L, \Xi, \sigma)$ is conserved *on-shell*, i.e., $d\, \mathcal{E}(L, \Xi, \sigma) = 0$ if σ is critical. In other words, one can pull-back Eq. (3.8) along solutions producing a whole class of currents $\mathcal{E}(L, \Xi, \sigma)$ (one for each solution) which are closed forms on M. We remark that Eq. (3.8) is again written on $\mathcal{C}$ (and not on M), a fact which will turn out to be important below when discussing superpotentials.

In any gauge-natural theory Lie derivatives of configuration fields are linear combinations of

$$\xi^\mu,\ \nabla_{\rho_1}\xi^\mu,\ \ldots,\ \nabla_{\rho_1\ldots\rho_s}\xi^\mu, \xi^A,\ \nabla_{\rho_1}\xi^A,\ \ldots,\ \nabla_{\rho_1\ldots\rho_r}\xi^A, \tag{3.9}$$

where *symmetrized covariant derivatives* $\nabla_{\rho_1\ldots\rho_k}\xi^\mu$ and $\nabla_{\rho_1\ldots\rho_k}\xi^A$ are built out of the dynamical connections (γ, A).

By a standard covariant integration by parts, the quantities (3.6) can be recasted [12] as

$$\begin{aligned} \mathcal{W}(L, \Xi) &= \mathcal{B}(L, \Xi) + \text{Div}\, \tilde{\mathcal{E}}(L, \Xi), \\ \mathcal{E}(L, \Xi) &= \tilde{\mathcal{E}}(L, \Xi) + \text{Div}\, \mathcal{U}(L, \Xi). \end{aligned} \tag{3.10}$$

The current $\mathcal{B}(L, \Xi)$ turns out to vanish identically along any section (i.e., off-shell) because of covariance, the *reduced current* $\tilde{\mathcal{E}}(L, \Xi)$ vanishes identically on shell, and the current $\mathcal{U}(L, \Xi)$ is called the *superpotential*.

We stress that both the reduced current $\tilde{\mathcal{E}}(L, \Xi)$ and the superpotential $\mathcal{U}(L, \Xi)$ are calculated on the bundle $\mathcal{C}$ and can be pulled back on the base M along any section obtaining $\tilde{\mathcal{E}}(L, \Xi, \sigma)$ and $\mathcal{U}(L, \Xi, \sigma)$. Reduced currents and superpotentials are easily shown to be global, thanks to their tensorial character; if they are pulled back along a solution σ the reduced current vanishes and the current $\mathcal{E}(L, \Xi, \sigma)$ is not only a closed form, but it is also exact on-shell [12].

In other words, while the current $\mathcal{E}(L, \Xi)$ is conserved just along solutions, the quantity $\mathcal{E}(L, \Xi) - \tilde{\mathcal{E}}(L, \Xi) = \mathrm{Div}\,\mathcal{U}(L, \Xi)$ is conserved along any section of $\mathcal{C}$. We express this different behaviour by saying that $\mathcal{E}(L, \Xi)$ is *weakly conserved* while Div $\mathcal{U}(L, \Xi)$ is said to be *strongly conserved*. As it is well-known, gauge-natural theories always allow (global) superpotentials [12, 13].

Let us finally consider a *region* $D \subset M$ of spacetime M, i.e., a compact $(n-1)$-submanifold with a boundary ∂D which is a compact $(n-2)$-submanifold; we define the *conserved quantities* associated to Ξ along a section σ as

$$Q_D(L, \Xi, \sigma) = \int_D \mathcal{E}(L, \Xi, \sigma) = \int_D \tilde{\mathcal{E}}(L, \Xi, \sigma) + \int_{\partial D} \mathcal{U}(L, \Xi, \sigma). \tag{3.11}$$

If σ is a solution of field equations the reduced current $\tilde{\mathcal{E}}(L, \Xi, \sigma)$ vanishes and $Q_D(L, \Xi, \sigma)$ is just given by the integral of the superpotential on the boundary ∂D of the region D.

Specializing to General Relativity, the superpotential $\mathcal{U}(L, \Xi)$ associated to the natural lift Ξ of a spacetime vector field ξ is the Komar potential [14]. Though the conserved quantities are geometrically well-defined objects, they are affected by the well-known problems of *anomalous factors* when we are interested in defining *mass* and *angular momentum* [15]. Furthermore, integrals of superpotentials may diverge when dealing with non-asymptotically flat solutions [6, 7].

A solution to these problems is achieved by a suitable redefinition of conserved quantities relative to a background solution [6, 16]. The *variation of modified conserved quantities* $\hat{Q}_D(L, \Xi, \sigma)$ is given by

$$\begin{aligned} \delta_X \hat{Q}_D(L, \Xi, \sigma) = \int_{\partial D} & \delta_X \mathcal{U}(L, \Xi, \sigma) \\ & - i_\xi < \mathbb{F}(L, \gamma) \circ j^{2k-1}\sigma \mid j^{k-1} X >, \end{aligned} \tag{3.12}$$

where X is a vertical vector field over $\mathcal{C}$ and $\mathbb{F}(L, \gamma)$ is the bundle morphism (2.10) specialized by setting γ to be the dynamical connection. Since all quantities on the r.h.s. of Eq. (3.12) can be computed, the equation may be regarded as a variational equation for $\hat{Q}_D(L, \Xi, \sigma)$ which determines, once integrated, the *modified conserved quantity* associated to Ξ relative to the fixed background. Notice that, in the first principle of thermodynamics (1.2), only variations of conserved quantities enter, so that (3.12) is particularly suitable for us.

The integrand $\delta_X \mathcal{U}(L, \Xi, \sigma) - i_\xi < \mathbb{F}(L, \gamma) \circ j^{2k-1}\sigma \mid j^{k-1} X >$ is independent on the addition of pure divergence terms to the Lagrangian and once integrated on spatial infinity ∞ (in place of ∂D in (3.12)), it provides the expected conserved quantities in all General Relativity examples examined so far.

4 Entropy

We are now going to define *entropy* as the quantity whose variation satisfied the first principle of thermodynamics (1.2). This can be solved for $T\delta S$ to obtain

$$T\delta S = \delta m - \Omega\, \delta J - b^A\, \delta Q_A. \tag{4.1}$$

With ∂_t and $-\partial_\phi$ denoting the vector fields which produce the mass and the angular momentum, respectively, the r.h.s. of (4.1) is the variation of the conserved quantity associated to $\Xi = \partial_t + \Omega\, \partial_\phi + b^A\, \rho_A$ so that one has for the *variation of the entropy*

$$\delta S = \frac{1}{T} \int_\infty \Big(\delta_X \mathcal{U}(L, \Xi, \sigma) - i_\xi < \mathbb{F}(L, \gamma) \circ j^{2k-1}\sigma \mid j^{k-1} X > \Big). \tag{4.2}$$

Again the quantities appearing on the r.h.s. of Eq. (4.2) are known and Eq. (4.2) may be regarded as a variational equation for S, which, when it is integrable, provides a definition for entropy. In this way entropy is a quantity related to spatial infinity, just as conserved quantities are, and no other additional hypotheses are required on the vector field Ξ or the manifold M.

However, in any gauge-natural theory one can prove the following theorem.

Theorem. *If σ is a solution of field equations, X is a solution of linearized field equations and Ξ is a Killing vector (i.e., $\pounds_\Xi \sigma = 0$), then the following holds:*

$$\mathrm{d}\Big(\delta_X \mathcal{U}(L, \Xi, \sigma) - i_\xi < \mathbb{F}(L, \gamma) \circ j^{2k-1}\sigma \mid j^{k-1} X > \Big) = 0. \tag{4.3}$$

In this way entropy can be obtained *also* as a quantity integrated on the border of a finite region since the integral in (4.2) just depends on the homology of the integration domain ∂D. In other words, under these hypotheses one can recast δS into the form

$$\delta S = \frac{1}{T} \int_\Sigma \Big(\delta_X \mathcal{U}(L, \Xi, \sigma) - i_\xi < \mathbb{F}(L, \gamma) \circ j^{2k-1}\sigma \mid j^{k-1} X > \Big) \tag{4.4}$$

for any Σ such that $\infty - \Sigma$ is a homological boundary in M. We remark that in this way entropy is still *a priori* unrelated to any horizon of spacetime.

The original prescription for entropy due to Wald et al. [1–3] was given for a generally covariant metric theory which admits a stationary, asymptotically flat, black hole solution. Then ∂_t is the Killing vector corresponding to asymptotic time translations while $-\partial_\phi$ is an asymptotic rotation so that Eq. (3.12) provides the variation of mass and angular momentum. Under these hypotheses one can consider Ξ to be the natural lift on C of the spacetime vector $\xi = \partial_t + \Omega\partial_\phi$, where Ω is the *angular velocity of the horizon* of the black hole solution under consideration. Moreover, the horizon of the black hole solution is assumed to be a Killing horizon for ξ and it is assumed to extend to a bifurcate Killing horizon [4], i.e., the union of two Killing horizons which intesect on a spacelike $(n-2)$-surface $\bar{\Sigma}$, called the *bifurcation surface*, on which ξ vanishes. If the surface gravity κ does not vanish on the horizon an explicit recipe to construct *Kruskal-like coordinates* around the horizon is produced [4]. Now by specializing Eq. (4.4) on $\bar{\Sigma}$, one obtains

$$\delta S = \frac{1}{T} \int_{\bar{\Sigma}} \delta_X \mathcal{U}(L, \Xi, \sigma) \tag{4.5}$$

due to the fact that $i_\xi < \mathbb{F}(L, \gamma) \circ j^{2k-1}\sigma \mid j^{k-1}X >$ is linear in ξ and thence it vanishes on $\bar{\Sigma}$.

Although formula (4.5) seems to be simpler than our formula (4.4) we remark that the former holds only on $\bar{\Sigma}$ (provided it exists at all) while (4.4) holds on *any* surface Σ without any further requirements about maximality of the solution or the nature of the horizon. The existence of such a surface $\bar{\Sigma}$, in fact, is garanteed only in the maximal extension of the solution. For example, for the Schwarzschild solution, the surface $\bar{\Sigma}$ corresponds to the surface $U = 0$, $V = 0$ in Kruskal coordinates. Thus Kruskal coordinates are needed in an "essential" way to apply Wald's recipe. For these reasons this earlier prescription turns out to be more difficult to use to perform direct computations (as we shall show in next section) as well as completely useless at the theoretical level. Furthermore, due to some unessential hypotheses which are required, it does not apply when the Kruskal-like extension cannot be made (e.g., in Taub-NUT solutions; [7]).

5 The entropy of BTZ black holes

As observed in [5] Einstein equations with (negative) cosmological constant $\Lambda = -1/l^2$ allow a $(1+2)$-dimensional black hole solution. Let us consider the Lagrangian

$$L = \alpha\Big(R + \frac{2}{l^2}\Big)\sqrt{g}\,\mathbf{ds}. \tag{5.1}$$

Considering its variation along a vertical vector field $X = \delta g^{\mu\nu}\partial/\partial g^{\mu\nu}$ one finds that

$$\begin{aligned}
&< \mathbb{E}(L) \mid X > = \alpha\sqrt{g}\Big[R_{\mu\nu} - \Big(\frac{1}{2}R + \frac{1}{l^2}\Big)g_{\mu\nu}\Big]\delta g^{\mu\nu}\,\mathbf{ds},\\
&< \mathbb{F}(L, \gamma) \mid j^1X > = \alpha\sqrt{g}\Big[g^{\lambda\sigma}g_{\alpha\beta} - \delta^\lambda_{(\alpha}\delta^\sigma_{\beta)}\Big]\nabla_\sigma\delta g^{\alpha\beta}\,\mathbf{ds}_\lambda,\\
&\tilde{\mathcal{E}}(L, \xi) = 2\alpha\sqrt{g}\Big[R^\lambda_{\cdot\nu} - \Big(\frac{1}{2}R + \frac{1}{l^2}\Big)\delta^\lambda_{\cdot\nu}\Big]\xi^\nu\,\mathbf{ds}_\lambda,\\
&\mathcal{U}(L, \xi) = \alpha\sqrt{g}\,\nabla^{[\beta}\xi^{\alpha]}\,\mathbf{ds}_{\alpha\beta}.
\end{aligned} \tag{5.2}$$

This model allows the following black hole solution (BTZ)

$$g_{\mathrm{BTZ}} = -N^2\,\mathrm{d}t^2 + N^{-2}\mathrm{d}r^2 + r^2(N_\phi\mathrm{d}t + \mathrm{d}\phi)^2$$

where we have set

$$\begin{aligned}
N^2 &= -m + \tfrac{r^2}{l^2} + \tfrac{J^2}{4r^2}, & m &= \tfrac{r_+^2 + r_-^2}{l^2},\\
N_\phi &= -\tfrac{J}{2r^2}, & J &= \tfrac{2r_+r_-}{l}.
\end{aligned} \tag{5.3}$$

Now let $\kappa = (r_+^2 - r_-^2)/(r_+ l^2)$ denote the surface gravity so that $T = \kappa/(2\pi)$ is the temperature of the Hawking radiation and let $\Omega = -N_\phi(r_+)$ be the angular velocity of the BTZ horizon. If one computes the entropy using our prescription (4.4) directly in the given coordinate system by choosing any spacelike surface Σ with $t =$ constant and $r =$ constant, one easily finds that

$$\delta\mathcal{S} = 8\pi^2\,\alpha\delta r_+ \quad \Rightarrow \quad \mathcal{S} = 4\pi^2\,\alpha r_+, \tag{5.4}$$

which is exactly the entropy expected [5].

The Wald's original recipe for the entropy forces one to first choose a particular $\bar{\Sigma}$ (the bifurcation surface) on which the Killing vector ξ, and thence the whole term $i_\xi < \mathbb{F}(L,\gamma)\,|\, j^{k-1}X >$ in (4.4), vanishes. Of course the existence of such a surface is ensured just in the maximal extension of the solution.

Thence, in order to apply formula (4.5) to the BTZ solution, one should first construct Kruskal-type coordinates as shown in [4]. Once one has verified that κ is a non-vanishing constant on the horizon, this can be done in two steps. First of all we define Eddington–Finkelstein coordinates (u, ρ, φ) which are the parameters along the flows of the vector fields (ξ, ζ, X) we are going to define. The vector field $\xi = \partial_t + \Omega\partial_\phi$ is the Killing vector whose Killing horizon $\mathcal{N}$ has to be extended; u is the parameter along its flow; X is a vector field tangent to $\mathcal{N}$ such that $X^\lambda\nabla_\lambda u = 0$. Finally ζ is a vector field such that the following conditions have to be satisfied

$$\begin{cases} \xi^\lambda\nabla_\lambda u = 1 \\ g_{\mathrm{BTZ}}(\zeta,\xi) = 1 \qquad |\,\zeta\,|^2 = 0 \text{ on } \mathcal{N}\,. \\ g_{\mathrm{BTZ}}(\zeta, X) = 0 \end{cases} \tag{5.5}$$

In these new coordinates the vector fields (ξ, ζ, X) read as

$$\begin{cases} \xi = \frac{\partial}{\partial u} \\ \zeta = \frac{\partial}{\partial \rho} \\ X = \frac{\partial}{\partial \varphi} \end{cases} \quad . \tag{5.6}$$

Notice that even in Eddington–Finkelstein coordinates the vector field ξ does not vanish anywhere in the chart domain. The BTZ metric reads as

$$g_{\mathrm{BTZ}} = -f(r(\rho))\mathrm{d}u^2 + 2\,\mathrm{d}u\;\mathrm{d}\rho + \Phi(r(\rho))\;\mathrm{d}u\;\mathrm{d}\varphi + \Psi(r(\rho))\mathrm{d}\varphi^2, \tag{5.7}$$

where we set

$$\begin{cases} f(r) = N^2 - r^2(N_\phi + \Omega)^2 \\ \Phi(r) = \frac{N^2 - r^2(N_\phi^2 - \Omega^2)}{\Omega} \\ \Psi(r) = -\frac{N^2 - r^2(N_\phi - \Omega)^2}{4\Omega^2} \end{cases} \quad . \tag{5.8}$$

Here $r(\rho)$ is obtained by inverting the change of coordinates, namely

$$\begin{cases} \rho(r) = \int_{r_+}^{r} \frac{\mathrm{dr}}{F(\mathrm{r})} \\ F(r) = \sqrt{\frac{\Psi(r)}{r^2}} \end{cases} \quad . \tag{5.9}$$

We remark that $F(r)$ is well-defined in a neighbourhood of the horizon $\mathcal{N}$. Now we can define Kruskal-type coordinates (U, V, φ) as

$$\begin{cases} U = e^{\kappa u} \\ V = -\rho\, e^{-\kappa u} \exp\left(2\kappa \int_0^\rho H(\rho)\mathrm{d}\rho\right) \end{cases} \qquad H(\rho) = \frac{1}{f(r(\rho))} - \frac{1}{2\kappa\rho}\,. \tag{5.10}$$

The product UV depends only on ρ and can be regarded as an implicit definition of ρ as a function $\tilde{\rho}(UV)$. Then the BTZ metric reads as

$$\begin{aligned} g_{\mathrm{BTZ}} &= G(UV)\mathrm{d}U\mathrm{d}V + \frac{\Phi(r(\tilde{\rho}))}{\kappa\, U}\mathrm{d}U\mathrm{d}\varphi + \Psi(r(\tilde{\rho}))\mathrm{d}\varphi^2, \\ G(UV) &= \frac{f(r(\tilde{\rho}))}{\kappa^2\, UV}. \end{aligned} \tag{5.11}$$

In these coordinates the Killing vector ξ reads as

$$\xi = \kappa\left(U\frac{\partial}{\partial U} - V\frac{\partial}{\partial V}\right), \tag{5.12}$$

which finally vanishes for $\bar{\Sigma} : (U = V = 0)$.

Thus we have extended the Killing horizon $\mathcal{N}$ to a bifurcate Killing horizon and we have identified the bifurcate surface $\bar{\Sigma}$. Now one can "easily" compute the entropy:

$$\begin{aligned} \delta_X S = \frac{1}{T}\int \delta_X\, \mathcal{U}(L, \xi)_3 \mathrm{d}\varphi \ &\underset{U=V=0}{\overset{\rho=0}{\Longrightarrow}}\ \delta_X S = 8\pi^2\, \alpha\, \delta r_+ \\ &\Longrightarrow\ S = 8\pi^2\, \alpha\, r_+, \end{aligned} \tag{5.13}$$

where $\mathcal{U}(\mathcal{L}, \xi)_\ni$ is the third component of the superpotential 1-form $\mathcal{U}(L, \xi)$. This, of course, reproduces result (5.4).

We stress that the first expression in Eq. (5.13) for $\delta_X S$ is meaningful only on the bifurcate surface where we can ignore the contribution of the term $i_\xi < \mathbb{F}(L, \gamma) \mid j^{k-1}X >$. We also stress that the above method fails in the extreme case. In fact, in this case $\kappa = 0$ and the derivation of Kruskal-type coordinates fails at Eqs. (5.10) and (5.11) (notice that in this case $f(r)$ vanishes identically). On the contrary, as we proved (see Eq. (5.4)) the second expression in (5.13) for $\delta_X S$ is correct on *any* surface Σ. In this way entropy is not related directly to a quantity computed on the horizon (see [7] for a discussion of entropy of Taub-NUT solutions).

Comparing the amount of calculation performed in this section to compute entropy according to the prescriptions (4.4) and (4.5) shows that (4.4) is by far the best suited for applications.

The computations of this Section have been carried out by using MapleV and its Tensor package [17].

Acknowledgements

We are grateful to I. Volovich for having long ago drawn our attention to the entropy formula of Wald, as well as to R. Mann for having drawn our attention to the BTZ solution and for useful discussions about it.

References

1. Iyer V., Wald, R. (1994): Phys. Rev. D **50**, 846
2. Wald R.M. (1993): J. Math. Phys. **31**, 2378
3. Kay B.S., Wald R.M. (1991): Physics Reports **207** (2), 49
4. Racz I., Wald R.M. (1992): Class. Quantum Grav. **9**, 2643
5. Carlip S., Gegenberg J. (1991): Phys. Rev. D **2** (44), 424;
   Bañados M., Teitelboim C., Zanelli J. (1992): Phys. Rev. Lett. **69**, 1849;
   Bañados M., Hnaux M., Teitelboim C., Zanelli J. (1993): Phys. Rev. D **48** (4), 1506;
   Cangemi D., Leblanc M., Mann R.B. (1993): Phys. Rev. D **48**, 3606;
   Brown J.D., Creighton J., Mann R.B. (1994): Phys. Rev. D **50**, 6394;
   Carlip S., Teitelboim C. (1995): Phys. Rev. D **501** (2), 622;
   Carlip S., Gegenberg J., Mann R.B. (1995): Phys. Rev. D **51** (12), 6854
6. Fatibene L., Ferraris M., Francaviglia M., Raiteri M. (1999): Ann. Phys. **275**, 27. hep-th/9810039;
   Fatibene L., Ferraris M., Francaviglia M., Raiteri M. (1999): Phys. Rev. D **60**, 124012;
   Fatibene L., Ferraris M., Francaviglia M., Raiteri M. (1999): Phys. Rev. D **60**, 124013
7. Hawking S.W., Hunter C.J. (1999): Phys. Rev. D **59**, 044025;
   Hunter C.J. (1999): Phys. Rev. D **59**, 024009;
   Hawking S.W., Hunter C.J., Page D.N. (1999): Phys. Rev. D **59**, 044033
8. Kolář I., Michor P.W., Slovák J. (1993): Natural Operations in Differential Geometry. Springer, New York;
   Saunders D.J. (1989): The Geometry of Jet Bundles. Cambridge University Press, Cambridge;
   Trautman A. (1962): Conservation Laws in General Relativity, in *Gravitation: An Introduction to Current Research*, ed. by L. Witten, Wiley, New York, pp. 168 ff.
9. Ferraris M., Francaviglia M. (1991): The Lagrangian Approach to Conserved Quantities, in *Mechanics, Analysis and Geometry: 200 Years after Lagrange*, ed. by M. Francaviglia, Elsevier, Amsterdam, pp. 451 ff.
10. Trautman A. (1967): Comm. Math. Phys. **6**, 248;
    Kolář I. (1984): J. Geom. Phys., **1**, 127
11. Fatibene L., Francaviglia M. (1998): Deformations of Spin Structures and Gravity, in *Gauge Theories of Gravitation*. (Jadwisin, Sept. 1997), Acta Physica Polonica B **29** (4), 915;
    Fatibene L., Ferraris M., Francaviglia M., Godina M. (1998): Gen. Rel. Grav. **30** (9), 1371
12. Ferraris M., Francaviglia M., Robutti O. (1987): On the notion of energy and the existence of superpotentials in gravitational theories, in *Géométrie et Physique, Proceedings of the Journées Relativistes 1985*, (Marseille, 1985), ed. by Y. Choquet-Bruhat, B. Coll, R. Kerner, A. Lichnerowicz, Hermann, Paris, pp. 112 ff.
13. Fatibene L., Gauge-natural formalism for classical field theories. Ph.D. Thesis Univ. di Torino (to appear)
14. Komar A. (1959): Phys. Rev. **113**, 934
15. Katz J. (1985): Class. Quantum Grav. **2**, 423
16. Ferraris M., Francaviglia M. (1990): Gen. Rel. Grav. **22** (9), 965
17. Chu K. Farell C., Fee G. McLenaghan R. (1997): Fields Institute Comm. **15**, 195

Dual Lagrangian Theories

M. Ferraris, M. Raiteri

Abstract. The procedure for constructing dual Lagrangians depending only on canonical momenta is described. The theory is applied to explain the equivalence between certain relativistic theories of gravitation and general relativity. The theory of dual Lagrangians for gauge theories is studied in detail. As an example, we describe SO(3) and SO(2,1) Yang–Mills theories on 3-dimensional and 4-dimensional manifolds.

1 Introduction

In this paper we present some recent results which we obtained about Lagrangian field theories. The results concern the possibility of comparing Lagrangian field theories through a process which, in classical mechanics, would consist in the following chain of transformations: a Legendre transformation $\left(q^i, \dot{q}^i\right) \longmapsto \left(q^i, p_i = \partial L / \partial \dot{q}^i\right)$, followed by a trivial canonical transformation $\left(q^k, p_i\right) \longmapsto \left(q'_i = p_i, p'^k = -q^k\right)$ which, in turn, is followed by an inverse Legendre transformation $\left(q'_k, p'^k\right) \longmapsto \left(q'_k, \dot{q}'_k = \partial H' / \partial p'^k\right) \equiv (p_k, \dot{p}_k)$. Of course this can be done only when the Hamiltonian $H\left(q^k, p_i\right)$ is sufficiently regular. Namely, the Hamiltonian must satisfy the following two regularity conditions:

$$\det \left\| \frac{\partial^2 H}{\partial p_i \partial p_k} \right\| \neq 0 \quad \text{and} \quad \det \left\| \frac{\partial^2 H}{\partial q^i \partial q^k} \right\| \neq 0.$$

The starting point and the end point of the chain can be obtained by eliminating canonical momenta (p_k) from Hamilton's equations or by eliminating configuration variables $\left(q^i\right)$ from Hamilton's equations. The interesting thing in this process is that we start from a system of Euler–Lagrange equations for a Lagrangian $L\left(q^i, \dot{q}^i\right)$, which are second-order equations for configuration variables $\left(q^i\right)$, then we obtain Hamilton's equations for a Hamiltonian $H\left(q^k, p_i\right)$, which are first-order equations for configuration variables $\left(q^i\right)$ and for canonical momenta (p_k), and we end up in a system of second-order equations for canonical momenta (p_k), which are Euler–Lagrange equations for a "dual" Lagrangian $L'(p_k, \dot{p}_k)$. Moreover we are able to provide, in a natural way, a second-order "dual" Lagrangian $L^*(p_k, \dot{p}_k, \ddot{p}_k)$ which differs from the Lagrangian L' by a total time derivative. The procedure is straightforward and is based on the Helmholtz Lagrangian $L_H = p_i \dot{q}^i - H(q^a, p_k)$. We know that, by solving the first half $\dot{q}^i = \partial H / \partial p_i$ of Hamilton's equations with respect to canonical momenta $(p_k)$and substituting the resulting functions $p_k = \bar{p}_k\left(q^i, \dot{q}^i\right)$ into the Helmholtz Lagrangian, we obtain the original Lagrangian $L\left(q^i, \dot{q}^i\right)$, i.e., we have

$$\left[p_i\dot{q}^i - H(q^a, p_k)\right]\Big|_{p=\bar{p}(q,\dot{q})} = L\left(q^i, \dot{q}^i\right).$$

If instead, we solve the other half $\dot{p}_k = -\partial H/\partial q^k$ of Hamilton's equations with respect to the configuration variables $(q^i)$and then we substitute the resulting functions $q^i = \bar{q}^i(p_k, \dot{p}_k)$ into the Helmholtz Lagrangian, we obtain the second-order "dual" Lagrangian

$$L^*(p_k, \dot{p}_k, \ddot{p}_k) = \left[p_i\dot{q}^i - H(q^a, p_k)\right]\Big|_{q=\bar{q}(p,\dot{p})}.$$

On the other hand, we have $L_H = \frac{d}{dt}(p_i q^i) - \dot{p}_k q^k - H(q^a, p_k)$ and

$$L'(p_k, \dot{p}_k) = \left[-\dot{p}_k q^k - H(q^a, p_k)\right]\Big|_{q=\bar{q}(p,\dot{p})}$$

so that the Lagrangians $L^*(p_k, \dot{p}_k, \ddot{p}_k)$ and $L'(p_k, \dot{p}_k)$ differ by the total time derivative $\frac{d}{dt}\left[p_i\bar{q}^i(p_k, \dot{p}_k)\right]$.

In Sect. 2 we shall explain how this procedure can be applied to first-order Lagrangian field theories. In Sect. 3 we briefly review how the theory applies to relativistic theories of gravitation. In Sect. 4 we discuss how the theory applies to gauge-theories and we present recent results for two examples of Yang-Mills theories.

2 First-order Lagrangian field theories

We assume the reader to be familiar with differential geometry, fibered manifolds, jet prolongations and the calculus of variations on fibered manifolds.

Let $\eta : Y \longrightarrow M$ be a fibered manifold, with basis manifold M, of dimension m, and total space Y, of dimension $m+n$, and projection η. The set of local sections of Y will be denoted by $\Gamma(Y)$. When dealing with the fibered manifold Y we shall use only fibered charts (U, x^λ, y^i), $\lambda = 1, \ldots, m$, $i = 1, \ldots, n$, where U is an open subset of Y projecting onto the domain W of a coordinate chart (W, x^λ) of M. The r-th order jet prolongation of the fibered manifold Y, will be denoted by $\eta^r : J^r(Y) \longrightarrow M$. The natural fibered chart on $J^1(Y)$, induced by the fibered chart (U, x^λ, y^i) of Y will be denoted by $(J^1U, x^\lambda, y^i, y^i_\mu)$. The corresponding "natural fibered chart" induced on $J^2(Y)$, will be denoted by $(J^2U, x^\lambda, y^i, y^i_\mu, y^i_{\mu\nu})$ (with $y^i_{\mu\nu} = y^i_{\nu\mu}$).

According to the geometric approach to Lagrangian field theories, a first-order Lagrangian is a base-preserving fibered morphism

$$L : J^1(Y) \longrightarrow A^0_m(M) \tag{1}$$

where $A^0_m(M)$ is the vector bundle of m-forms (scalar densities of weight 1) over M. In local fibered coordinates the Lagrangian L is expressed by

$$L = \mathcal{L}(x^\lambda, y^i, y^i_\mu)ds \tag{2}$$

where $ds = dx^1 \wedge dx^2 \wedge \ldots \wedge dx^m$. The Lagrangian L can be assumed equivalently to be a horizontal m-form over $J^1(Y)$. For any local section $\sigma \in \Gamma(Y)$ an m-form $(j^1\sigma)^* L = (\mathcal{L} \circ j^1\sigma)\, ds$ is induced on M. The action functionals are then defined by

$$A_D(\sigma) = \int_D (j^1\sigma)^* L = \int_D \left(\mathcal{L} \circ j^1\sigma\right) ds \tag{3}$$

where $D \subseteq M$ is a compact domain. Critical sections for the Lagrangian L are local sections $\sigma \in \Gamma(Y)$ which are stationary points for all the action functionals A_D as D ranges through all compact domains of M.

We associate to a given first-order Lagrangian L a further globally defined m-form Θ_L over $J^1(Y)$, which is called the *Poincaré–Cartan* form associated to L. In local fibered coordinates the m-form Θ_L has the following expression:

$$\Theta_L = \partial_i^\lambda \mathcal{L}\, \omega^i \wedge ds_\lambda + \mathcal{L}\, ds \tag{4}$$

where $\omega^i = dy^i - y^i_\mu dx^\mu$ and $ds_\lambda = \partial_\lambda \rfloor ds$. Since for any local section $\sigma \in \Gamma(Y)$ we have $(j^1\sigma)^* L = (j^1\sigma)^* \Theta_L$, the Poincaré-Cartan form gives rise to the same action integrals of the Lagrangian L. Moreover, it allows us to single out all the critical sections of L by means of the following prescription: a local section $\sigma \in \Gamma(Y)$ is a critical section of L if and only if

$$(j^1\sigma)^* \left(\Xi \rfloor d\Theta_L\right) = 0 \tag{5}$$

for any vector field Ξ on $J^1 Y$ (the space of vector fields can be restricted to projectable vector fields on $J^1(Y)$, or to vertical vector fields on $J^1(Y)$ or even to prolongations of vertical vector fields on Y). Equation (5) is equivalent to requiring that the image of the local section $j^2\sigma$ be contained in the kernel of the *Euler–Lagrange morphism* $e(L)$ of L. The Euler–Lagrange morphism is a global second order operator

$$e(L) : J^2 Y \longrightarrow V^*(Y) \bigotimes_{J^2(Y)} A^0_m(M)$$

over Y (where $V^*(Y)$ denotes the dual of the vertical tangent bundle of Y), and it can be represented equivalently by an $(m+1)$-form $e(L) = e_i(L) dy^i \wedge ds$ over $J^2(Y)$, where the coefficients $e_i(L)$ are defined by

$$e_i(L) = \partial_i \mathcal{L} - d_\lambda \left(\partial_i^\lambda \mathcal{L}\right) = \partial_i \mathcal{L} - \left(\partial_\lambda + y^j_\lambda \partial_j + y^j_{\mu\lambda} \partial^\mu_j\right) \partial_i^\lambda \mathcal{L}. \tag{6}$$

Equation (5) is equivalent to the well-known second-order Euler–Lagrange equations $e_i(L) \circ j^2\sigma = 0$ for the local section σ.

We now define the *Legendre bundle* of the fibered manifold Y as the vector bundle

$$L(Y) = A^0_{m-1}(M) \bigotimes_Y V^*(Y) \cong V^*(Y) \bigotimes_Y T(M) \bigotimes_Y A^0_m(M).$$

Fibered coordinates (x^λ, y^i) of Y induce natural fibered coordinates $(x^\lambda, y^i, p^\mu_j)$ on the Legendre bundle $L(Y)$. The contact part $\partial^\lambda_i \mathcal{L}\,\omega^i \wedge ds_\lambda$ of the Poincaré–Cartan form Θ_L allows us to define a global base-preserving fibered morphism $\phi_L : J^1(Y) \longrightarrow L(Y)$ which is called the *Legendre transformation*. In natural fibered coordinates the Legendre transformation is represented by:

$$\phi_L : (x^\alpha, y^a, y^a_\alpha) \longmapsto (x^\alpha, y^a, p^\mu_i = \partial^\mu_i \mathcal{L}(x^\alpha, y^a, y^a_\alpha)) \tag{7}$$

A Lagrangian L is regular in an open subset $U \subseteq J^1 Y$ if the Legendre transformation is invertible when restricted to U. If the open subset U is sufficiently small, this condition is equivalent to

$$\det \left\| \frac{\partial^2 \mathcal{L}}{\partial y^i_\mu \partial y^j_\nu} \right\| \neq 0. \tag{8}$$

A weaker regularity condition consists of requiring the Legendre transformation, or, equivalently, the matrix $\|\partial^\mu_i \partial^\nu_j \mathcal{L}\|$, to have constant rank on U.

Regular theories

When the Lagrangian L is regular in the domain J^1U of a natural fibered chart $(J^1U, x^\lambda, y^i, y^i_\mu)$, the Legendre transformation (7) defines a coordinate transformation in J^1U with inverse map:

$$\phi_L^{-1} : (x^\alpha, y^i, p^\mu_i) \longmapsto (x^\alpha, y^a, y^b_\nu = \bar{y}^b_\nu(x^\alpha, y^i, p^\mu_i)). \tag{9}$$

A real-valued function $\mathcal{H}$ can be defined in J^1U by setting

$$\mathcal{H}(x^\mu, y^i, p^\mu_i) = \bar{y}^i_\mu(x^\sigma, y^j, p^\sigma_j) p^\mu_i - \mathcal{L}(x^\alpha, y^k, \bar{y}^k_\alpha(x^\sigma, y^j, p^\sigma_j)). \tag{10}$$

Using this coordinate system the Euler-Lagrange equations (6) can be rewritten as follows:

$$d_\mu y^i - \partial^i_\mu \mathcal{H} = 0 \tag{11}$$

$$d_\mu p^\mu_i + \partial_i \mathcal{H} = 0. \tag{12}$$

The second-order Euler–Lagrange equations in the variables (y^i) are then reduced to an equivalent system of first-order equations (in normal form) in the variables y^i, p^μ_i. The function $\mathcal{H}$ defined in (10) is the *Hamiltonian* and Eqs. (11) and (12) are *Hamilton's equations*. Equation (11) can be considered as the definition of the inverse Legendre transformation (9). In fact we have

$$\bar{y}^b_\nu(x^\alpha, y^i, p^\mu_i) = \partial^b_\nu \mathcal{H}(x^\alpha, y^i, p^\mu_i) \tag{13}$$

and the regularity condition (8) for the Lagrangian density induces the regularity condition

$$\det \left\| \frac{\partial^2 \mathcal{H}}{\partial p^\mu_i \partial p^\nu_j} \right\| \neq 0 \tag{14}$$

for the Hamiltonian $\mathcal{H}$ which, in turn, allows us to solve Eq. (11) with respect to the variables p_i^μ, obtaining, $p_i^\mu = \partial_i^\mu \mathcal{L}(x^\alpha, y^a, y_\alpha^a)$.

We now assume that, in addition to the regularity condition (14), the Hamiltonian $\mathcal{H}$ also satisfies the "dual" regularity condition

$$\det \left\| \frac{\partial^2 \mathcal{H}}{\partial y^i \partial y^j} \right\| \neq 0. \tag{15}$$

We can, therefore, locally solve Eq. (12) with respect to the variables y^i, obtaining

$$y^i = \bar{y}^i(x^\sigma, p_j^\mu, p_{j\mu}^\mu). \tag{16}$$

Substituting these expressions into Eq. (11), we are able to eliminate the original configuration variables y^i from Hamilton's equations, thus obtaining a system of "dual" second-order equations

$$d_\nu \left[\bar{y}^i(x^\lambda, p_j^\mu, p_{j\mu}^\mu) \right] = \partial_\nu^i \mathcal{H}\left(x^\lambda, \bar{y}^i(x^\lambda, p_j^\mu, p_{j\mu}^\mu), p_k^\alpha \right) \tag{17}$$

for canonical momenta p_i^μ. Euler–Lagrange equations, Hamilton's equations and dual equations are completely equivalent: we can move between the three theories by means of formula (16) and of the Legendre transformation (7).

It is well-known that Hamilton's equations (11) and (12) can be deduced from a variational principle. In fact, they can be obtained by taking independent variations of the "Helmholtz Lagrangian density"

$$\mathcal{L}_H(x^\mu, y^j, p_j^\nu, y_\nu^j, p_{j\sigma}^\nu) = p_i^\nu y_\nu^i - \mathcal{H}(x^\mu, y^j, p_j^\nu) \tag{18}$$

with respect to y^i and p_i^ν. The Helmholtz Lagrangian density $\mathcal{L}_H$ is globally well-defined on $J^1(L(Y))$. Remarkably enough, the "dual" second-order equations (17) can be deduced from a variational principle. In order to verify this assertion it also is sufficient to consider the following "dual" second-order Lagrangian density

$$\mathcal{L}^*(x^\alpha, p_a^\alpha, p_{a\mu}^\alpha, p_{a\mu\nu}^\alpha) = \left[p_i^\nu d_\nu \bar{y}^i - \mathcal{H}(x^\nu, \bar{y}^i, p_j^\nu) \right]_{\bar{y}^i = \bar{y}^i(x^\sigma, p_a^\alpha, p_{a\alpha}^\alpha)} \tag{19}$$

obtained by substituting Eq. (16) into the Helmholtz Lagrangian density (18). The dual Lagrangian density $\mathcal{L}^*$ is a fake second-order Lagrangian density since second-order derivatives $p_{a\mu\nu}^\alpha$ can be collected into a divergence term. In fact we have

$$\mathcal{L}^*(x^\alpha, p_a^\alpha, p_{a\mu}^\alpha, p_{a\mu\nu}^\alpha) = d_\nu \left[p_i^\nu \bar{y}^i(x^\lambda, p_j^\sigma, p_{j\sigma}^\sigma) \right] - \mathcal{U}(x^\lambda, p_i^\mu, p_{i\mu}^\mu) \tag{20}$$

where the local first-order Lagrangian density $\mathcal{U}$ is defined by

$$\mathcal{U}(x^\lambda, p_i^\mu, p_{i\mu}^\mu) = \left[\bar{y}^i p_{i\nu}^\nu + \mathcal{H}(x^\lambda, \bar{y}^i, p_j^\nu) \right]_{\bar{y}^i = \bar{y}^i(x^\lambda, p_j^\sigma, p_{j\sigma}^\sigma)}. \tag{21}$$

Remark 1. We stress that a local Hamiltonian formalism always exists for regular Lagrangians. If we want a global version we have to consider hyperregular Lagrangians, i.e., Lagrangians such that the Legendre map ϕ_L is a diffeomorphism of $J^1(Y)$ onto its image $\mathrm{Im}(\phi_L)$, which is an open subset of $L(Y)$. The Helmholtz Lagrangian density $\mathcal{L}_H$ is then globally well-defined in the open subset $J^1(\mathrm{Im}(\phi_L))$ of $J^1(L(Y))$.

Remark 2. Even in the case of hyperregular Lagrangians it is not always possible to globalize the "dual" second-order Lagrangian density $\mathcal{L}^*$. A sufficient condition for achieving the globalization of the Lagrangian density $\mathcal{L}^*$ consists in restricting our attention to fibered manifolds Y such that the Legendre bundle $L(Y)$ is the fibered product $Y \times_M P$ of Y and of a vector bundle $P \longrightarrow M$. In this case the Lagrangian density $\mathcal{L}^*$ is globally well-defined on $J^2(P)$. The "dual" first-order Lagrangian density $\mathcal{U}$ will instead always give rise to a family of local objects on $J^1(P)$.

Non-regular theories

When the regularity condition (8) does not hold, the Hamiltonian theory described above fails to work. However, if the Legendre map ϕ_L has constant rank $m+n+k$ (with $1 \le k < m \cdot n$), we may proceed as follows.

First of all we remark that, whenever a vertical vector $X = X^i_\alpha \partial^\alpha_i$ of the fibre bundle $\eta^1_0 : J^1Y \longrightarrow Y$ annihilates the momenta $p^\mu_i(x^\alpha, y^a, y^a_\alpha) = \partial^\mu_i \mathcal{L}(x^\alpha, y^a, y^a_\alpha)$, it also annihilates the Hamiltonian

$$\mathcal{H}(x^\alpha, y^a, y^a_\alpha) = y^i_\mu\, \partial^\mu_i \mathcal{L}(x^\alpha, y^a, y^a_\alpha) - \mathcal{L}(x^\alpha, y^a, y^a_\alpha)\,.$$

We have, in fact

$$X^a_\alpha \partial^\alpha_a \mathcal{H} = y^i_\mu\, X^a_\alpha \partial^\alpha_a \partial^\mu_i \mathcal{L} = 0. \tag{22}$$

The rank theorem allows us to replace coordinates $(x^\alpha, y^a, y^a_\alpha)$ with new coordinates $(x^\alpha, y^a, \pi_A, v^\Sigma)$ in J^1Y (with $1 \le A \le k$ and $1 \le \Sigma \le m \cdot n - k$), in such a way that the momenta p^μ_i and the Hamiltonian $\mathcal{H}$ are independent of the $m \cdot n - k$ coordinates v^Σ (for example, we may replace k coordinates y^a_α with k independent momenta p^μ_i, which we denote by π_A, and then we denote by v^Σ the remaining $m \cdot n - k$ coordinates y^a_α). With this prescription the vertical vectors X of the type described above have necessarily vanishing components along the basis vectors $\partial^A = \partial/\partial\pi_A$. Accordingly, we may assume that the Legendre map ϕ_L is obtained as the composition of two maps of constant rank: a surjective submersion

$$\bar{\phi}_L : (x^\alpha, y^a, y^a_\alpha) \longmapsto (x^\alpha, y^a, \pi_A = \bar{\pi}_A(x^\alpha, y^a, y^a_\alpha)),$$

and an embedding

$$\iota_C : (x^\alpha, y^a, \pi_A) \longmapsto (x^\alpha, y^a, p^\mu_i = \bar{p}^\mu_i(x^\alpha, y^a, \pi_A)).$$

The functions $\bar{\phi}_L$ and ι_C have the following properties:

$$t^A \partial_j^\nu \bar{\pi}_A(x^\mu, y^i, y^i_\mu) = 0 \Longleftrightarrow t^A = 0 \tag{23}$$

$$s_A \partial^A \bar{p}^\mu_i(x^\mu, y^i, \pi_B) = 0 \Longleftrightarrow s_A = 0. \tag{24}$$

The Hamiltonian $\mathcal{H}(x^\lambda, y^i, y^i_\lambda)$ is then the composition of the surjective submersion $\bar{\phi}_L$ with a scalar density $\bar{H}(x^\alpha, y^a, \pi_A)$, and the Poincaré–Cartan form Θ_L can be written as $\Theta_L = (\bar{\phi}_L)^* \bar{\Theta}$ with $\bar{\Theta}$ defined by

$$\bar{\Theta} = \bar{p}^\mu_i(x^\alpha, y^a, \pi_A) dy^i \wedge ds_\mu - \bar{H}(x^\alpha, y^a, \pi_A) ds. \tag{25}$$

Owing to (25) we can rewrite the original Lagrangian density $\mathcal{L}$ as follows:

$$\mathcal{L} = \bar{p}^\mu_i(x^\lambda, y^i, \bar{\pi}_A(x^\sigma, y^j, y^j_\sigma))\, y^i_\mu - \bar{H}(x^\lambda, y^i, \bar{\pi}_A(x^\sigma, y^j, y^j_\sigma)). \tag{26}$$

Once condition (23) is taken into account, the variation of $\mathcal{L}$ allows us to rewrite the Euler-Lagrange equations (6) in terms of the scalar density $\bar{H}$:

$$y^i_\mu \partial^A \bar{p}^\mu_i - \partial^A \bar{H} = 0 \tag{27}$$

$$\left(\partial_\mu + \pi_{A\mu} \partial^A\right) \bar{p}^\mu_j + y^k_\mu \left(\partial_k \bar{p}^\mu_j - \partial_j \bar{p}^\mu_k\right) + \partial_j \bar{H} = 0 \tag{28}$$

where we have to substitute $\pi_A = \bar{\pi}_A(x^\sigma, y^k, y^k_\sigma)$ and $\pi_{A\mu} = d_\mu \bar{\pi}_A(x^\sigma, y^k, y^k_\sigma)$. Equations (27) are just the definition of the Legendre map $p^\mu_i = \partial^\mu_i \mathcal{L}$ while Eqs. (28) are the expression of the equations $d_\mu(\partial^\mu_i \mathcal{L}) = \partial_i \mathcal{L}$.

If we do not make the substitution above, Eqs. (27) and (28) may be considered as a reduction of the original second-order Euler–Lagrange equations on the configuration variables y^i to a system of first-order differential equations in the configuration variables (y^i, π_A), which can be deduced from the Helmholtz Lagrangian density

$$\mathcal{L}_H(x^\mu, y^j, \pi_A, y^j_\mu) \doteq \bar{p}^\lambda_i(x^\mu, y^j, \pi_A) y^i_\lambda - \bar{H}(x^\mu, y^j, \pi_A), \tag{29}$$

and they are equivalent to the original Euler–Lagrange equations only if Eq. (27) can be solved uniquely for the variables π_A or, equivalently, if we have

$$\det \left\| y^i_\mu \partial^B \partial^A \bar{p}^\mu_i - \partial^B \partial^A \bar{H} \right\| \neq 0. \tag{30}$$

To give a dual description of the theory we first require the regularity condition (30) to be satisfied. Next we eliminate the dependence of (28) on the variables y^k_μ by assuming that their coefficients vanish identically, i.e., we assume that

$$\partial_k \bar{p}^\mu_j - \partial_j \bar{p}^\mu_k = 0. \tag{31}$$

Then we require that Eq. (28) can be solved uniquely with respect to the variables y^i, by assuming that

$$\det \left\| \left(\partial_\mu + \pi_{A\mu} \partial^A\right) \partial_i \bar{p}^\mu_j + \partial_i \partial_j \bar{H} \right\| \neq 0. \tag{32}$$

The solution will be denoted by $y^i = \bar{y}^i(x^\alpha, \pi_A, \pi_{A\mu})$ and, when substituted back into Eq. (27), produces a second-order equation in the "canonical momenta" π_A:

$$d_\mu(\bar{y}^i(x^\alpha, \pi_A, \pi_{A\sigma}))\, \partial^A \bar{p}_i^\mu - \partial^A \bar{H}(x^\alpha, \bar{y}^i(x^\alpha, \pi_A, \pi_{A\sigma}), \pi_B) = 0. \tag{33}$$

This "dual" second-order equation can be deduced from a "dual" second-order Lagrangian and also from a family of "dual" first-order Lagrangians. We may, in fact, consider the second-order Lagrangian density $\mathcal{L}^*$ obtained by substituting $y^i = \bar{y}^i(x^\alpha, \pi_A, \pi_{A\mu})$ into the Helmholtz Lagrangian density $\mathcal{L}_H(x^\mu, y^j, \pi_A, y^j_\mu)$, i.e.,

$$\mathcal{L}^* = \left[\bar{p}_k^\lambda(x^\mu, \bar{y}^i, \pi_A) d_\lambda \bar{y}^k - \bar{H}(x^\mu, \bar{y}^i, \pi_A)\right]\Big|_{\bar{y}^i = \bar{y}^i(x^\alpha, \pi_A, \pi_{A\mu})}. \tag{34}$$

In order to obtain the first-order Lagrangian density $\mathcal{U}$ we have to take into account the fact that Eq. (31) implies the possibility of finding functions $\bar{h}^\lambda(x^\mu, y^i, \pi_A)$ such that $\bar{p}_k^\lambda(x^\mu, y^i, \pi_A) = \partial_k \bar{h}^\lambda(x^\mu, y^i, \pi_A)$. With this assumption we have

$$\mathcal{L}^* = d_\lambda\left[\bar{h}^\lambda(x^\mu, \bar{y}^i(x^\alpha, \pi_A, \pi_{A\mu}), \pi_A)\right] + \mathcal{U}(x^\alpha, \pi_A, \pi_{A\sigma}) \tag{35}$$

where $\mathcal{U}(x^\alpha, \pi_A, \pi_{A\sigma})$ is defined by

$$\mathcal{U} = -\left[\left(\partial_\lambda + \pi_{B\lambda}\partial^B\right)\bar{h}^\lambda(x, y, \pi) + \bar{H}(x, y, \pi)\right]\Big|_{y = \bar{y}(x^\alpha, \pi_A, \pi_{A\mu})}. \tag{36}$$

If we want to globalize this theory we first have to assume that the Hamiltonian constraint $\mathcal{C} = \mathrm{Im}(\phi_L) \subset L(Y)$ be a fibered submanifold of the vector bundle $L(Y)$. The Legendre transformation is then the composition of a surjective map $\bar{\phi}_L : J^1(Y) \longrightarrow \mathcal{C}$ and of the embedding $\iota_\mathcal{C} : \mathcal{C} \longrightarrow L(Y)$, and we need to assume that the surjective map $\bar{\phi}_L$ be of constant rank, i.e., it has to be a surjective submersion. Finally, we need to assume that the Hamiltonian constraint $\mathcal{C}$ be the fibered product $\mathcal{C} = Y \times_M \bar{P}$ of the fibered manifold Y and of a further fibered manifold $\bar{P}$ over M. For the details we refer the reader to [1] and its references.

3 Relativistic theories of gravitation

An extremely important set of examples, which is actually the one from which all the theory we present in this paper derives, consists in *relativistic theories of gravitation*. As is well-known, three kind of gravitational theories were introduced by Einstein between 1916 and 1925 to describe the interaction of the gravitational field with other matter fields. These theories are known under the names of *purely-metric theory*, *purely-affine theory* and *metric-affine theory*. The purely-metric theory is the standard theory for general relativity while the other two types of theory were introduced while trying to find unified theories or to generalize Einstein's theory. For details on their equivalence we refer the reader to [2] and its references.

Purely-metric theories

These theories were first described by Hilbert [3] and by Einstein [4] and they were later studied in detail by a huge number of people. Purely-metric theories are based on the assumption that the gravitational field is described by a Lorentzian metric $g_{\mu\nu}$ on a four-dimensional space-time manifold M_4, and that matter-fields are described by tensorial objects on the same manifold. According to Einstein, the gravitational part of the field equations is of the form

$$r_{\mu\nu}(j^2 g) - \tfrac{1}{2} r(j^2 g) g_{\mu\nu} = \kappa T_{\mu\nu} \tag{37}$$

where $r_{\mu\nu}(j^2 g)$ denotes the Ricci tensor of the metric field $g_{\mu\nu}$ and $r(j^2 g) = g^{\alpha\beta} r_{\alpha\beta}(j^2 g)$ denotes Ricci's scalar curvature, $\kappa = 8\pi G/c^4$ denotes Einstein's gravitational constant and $T_{\mu\nu}$ denotes the *stress-energy tensor* of the external-matter fields. As was shown by Hilbert, Einstein's equations (37) can be deduced from a generally covariant second-order Lagrangian density of the type

$$\mathcal{L}_{\text{PM}}(j^2 g, j^1 \varphi) = -\tfrac{1}{2\kappa} r(j^2 g)\sqrt{-g} + \mathcal{V}(g_{\alpha\beta}, \varphi^i, D_\mu \varphi^i). \tag{38}$$

The external tensorial matter fields are collectively denoted by φ^i, while $D_\mu \varphi^i$ denotes their covariant derivatives with respect to Christoffel symbols of the metric $g_{\mu\nu}$. The matter Lagrangian $\mathcal{V}(g_{\alpha\beta}, \varphi^i, D_\mu \varphi^i)$ is a generally covariant scalar density and, therefore, it is the product of the metric volume element $\sqrt{-g}$ and of an invariant scalar function $L_{\text{mat}}(g_{\alpha\beta}, \varphi^i, D_\mu \varphi^i)$.

Shortly after, Einstein discovered that Eqs. (37) can also be deduced locally from a first-order Lagrangian density which, however, has the drawback of not being generally covariant. This property is due to the fact that the scalar density $r\sqrt{-g}$ can be rewritten as follows:

$$r\sqrt{-g} = d_\alpha \left(\hat{g}^{\sigma\nu} \gamma^\alpha_{\sigma\nu} - \hat{g}^{\alpha\nu} \gamma^\sigma_{\sigma\nu}\right) + \left(\gamma^\alpha_{\sigma\beta} \gamma^\sigma_{\alpha\nu} - \gamma^\alpha_{\alpha\sigma} \gamma^\sigma_{\beta\nu}\right) \hat{g}^{\beta\nu} \tag{39}$$

where $\gamma^\alpha_{\beta\mu} = \gamma^\alpha_{\beta\mu}(j^1 g)$ denote Christoffel symbols of the metric tensor $g_{\mu\nu}$, while $\hat{g}^{\beta\nu}$ is the tensor density defined by $\hat{g}^{\beta\nu} = g^{\beta\nu}\sqrt{-g}$. Hilbert's second-order Lagrangian density (38) can then be rewritten as follows:

$$\mathcal{L}_{\text{PM}}(j^2 g, j^1 \varphi) = -\tfrac{1}{2\kappa} d_\alpha \left(\hat{g}^{\sigma\nu} \gamma^\alpha_{\sigma\nu} - \hat{g}^{\alpha\nu} \gamma^\sigma_{\sigma\nu}\right) + \mathcal{U}(j^1 g, j^1 \varphi) \tag{40}$$

where Einstein's non-covariant first-order Lagrangian density $\mathcal{U}(j^1 g, j^1 \varphi)$ is defined by

$$\mathcal{U}(j^1 g, j^1 \varphi) = \tfrac{1}{2\kappa} \left(\gamma^\alpha_{\alpha\sigma} \gamma^\sigma_{\beta\nu} - \gamma^\alpha_{\sigma\beta} \gamma^\sigma_{\alpha\nu}\right) \hat{g}^{\beta\nu} + \mathcal{V}(g, \varphi, D\varphi). \tag{41}$$

Since the Lagrangian densities $\mathcal{L}_{\text{PM}}(j^2 g, j^1 \varphi)$ and $\mathcal{U}(j^1 g, j^1 \varphi)$ differ by a divergence, they give rise to the same Euler–Lagrange equations.

Purely-affine theories

These theories were first introduced by Einstein [5, 6] in 1923, and they were extensively studied by Schrödinger [7–11] and later by Kijowski and one of us [2, 12–15]. Purely-affine theories are based on the assumption that the gravitational field is described by a torsionless linear connection $\Gamma^{\alpha}_{\beta\mu}$ on a four-dimensional space-time manifold M_4, that matter-fields are described by tensorial objects on the same manifold and that field equations should be deduced from a variational principle based on a first-order covariant Lagrangian density. Because of the covariance requirement, it turns out that the Lagrangian density must of the form

$$\mathcal{L}_{\mathrm{PA}}(j^1\Gamma, j^1\varphi) = \mathcal{L}(R^{\alpha}_{\beta\mu\nu}, \varphi^i, \nabla_\mu\varphi^i)$$

where

$$R^{\alpha}_{\beta\mu\nu} = d_\mu\Gamma^{\alpha}_{\beta\nu} - d_\nu\Gamma^{\alpha}_{\beta\mu} + \Gamma^{\alpha}_{\sigma\mu}\Gamma^{\sigma}_{\beta\nu} - \Gamma^{\alpha}_{\sigma\nu}\Gamma^{\sigma}_{\beta\mu}$$

denotes the curvature tensor of the connection $\Gamma^{\alpha}_{\beta\mu}$ and φ^i denotes all the *external* matter-fields, while $\nabla_\mu\varphi^i$ denote covariant derivatives of the matter-fields with respect to the connection $\Gamma^{\alpha}_{\beta\mu}$. The Lagrangian density $\mathcal{L}(R^{\alpha}_{\beta\mu\nu}, \varphi^i, \nabla_\mu\varphi^i)$ has to be a generally covariant function of its tensorial arguments.

Instead of considering the general case $\mathcal{L}(R^{\alpha}_{\beta\mu\nu}, \varphi^i, \nabla_\mu\varphi^i)$, we may restrict our attention to Lagrangian densities depending only on the traces of the Riemann tensor $R^{\alpha}_{\beta\mu\nu}$, that is, depending on the Ricci tensor $R_{\beta\nu} = R^{\alpha}_{\beta\alpha\nu}$ and on the first trace $F_{\mu\nu} = R^{\alpha}_{\alpha\mu\nu}$ of the Riemann tensor. Since the Ricci tensor $R_{\beta\nu}$ is not necessarily symmetric, we may further split $R_{\beta\nu}$ into its symmetric part $G_{\beta\nu} = R_{(\beta\nu)}$ and its antisymmetric part which is proportional to $F_{\beta\nu}$ (for symmetric connections we have in fact $F_{\beta\nu} = 2R_{[\beta\nu]}$). We may thus consider Lagrangian densities of the type $\mathcal{L}(G_{\beta\nu}, F_{\mu\nu}, \varphi^i, \nabla_\mu\varphi^i)$; see, e.g., [15]. If we furthermore restrict our attention to Lagrangian densities which do not depend on the tensor $F_{\mu\nu}$, we obtain Lagrangian densities of the form

$$\mathcal{L}_{\mathrm{PA}}(j^1\Gamma, j^1\varphi) = \mathcal{L}(G_{\beta\nu}, \varphi^i, \nabla_\mu\varphi^i) \tag{42}$$

which are the kind of Lagrangian densities that we consider for purely-affine gravitational theories.

Metric-affine theories

The very first example of a metric-affine theory was introduced by Einstein in 1925. Metric-affine theories are based on the assumption that the gravitational field is described by two configuration variables: a Lorentzian metric $g_{\mu\nu}$ and a torsionless linear connection $\Gamma^{\alpha}_{\beta\mu}$ on a four-dimensional space-time manifold M_4.

Also in this case, matter-fields are assumed to be described by tensorial objects on the manifold and field equations should be deduced from a variational principle based on a first-order covariant Lagrangian density. The Lagrangian density is of the type

$$\mathcal{L}_{\text{MA}}(j^1\Gamma, j^1 g, j^1\varphi) = -\tfrac{1}{2\kappa}\sqrt{-g}g^{\beta\nu}G_{\beta\nu}(j^1\Gamma) + \mathcal{W}(g_{\alpha\beta}, \varphi^i, \nabla_\mu\varphi^i) \tag{43}$$

where $G_{\beta\nu}(j^1\Gamma)$ denotes the symmetric part of the Ricci tensor of the connection $\Gamma^\alpha_{\beta\mu}$ and ∇ denotes covariant derivative with respect to Γ.

The three kinds of gravitational theories are apparently very different, but they are related by the same kind of relation we explained in the previous section. If we start from the first-order Lagrangian density $\mathcal{L}_{\text{PA}}(j^1\Gamma, j^1\varphi)$, the role of the Helmholtz Lagrangian density will be played by $\mathcal{L}_{\text{MA}}(j^1\Gamma, j^1 g, j^1\varphi)$, the "dual" second-order Lagrangian density will be $\mathcal{L}_{\text{PM}}(j^2 g, j^1\varphi)$, while $\mathcal{U}(j^1 g, j^1\varphi)$ will be the "dual" first-order Lagrangian density. We refer the reader to [2] for details.

4 Dual formulation of gauge theories

Let us consider a manifold M, of dimension m, endowed with a pseudo-Riemannian metric g and let G be a Lie group of dimension n and Lie algebra $\mathfrak{g}$. A *gauge-theory* with gauge group G on the fixed background (M, g) is usually defined by assigning a gauge-covariant first-order Lagrangian density which describes the interaction between a gauge potential and *external* matter fields. The dynamical variables of the theory are then locally described by the components A^i_μ of the gauge potential 1-form, together with the components φ^B of the external fields (with $1 \leq B \leq r$). Because of gauge-covariance, and the additional requirement of being invariant under isometries of the background metric g, the first-order Lagrangian density necessarily has an expression of the form

$$\mathcal{L}(x, j^1 A, j^1\varphi) = \mathcal{L}(F^i_{\mu\nu}, \varphi^A, D_\mu\varphi^A; g_{\alpha\beta}(x)) \tag{44}$$

where $F^i_{\mu\nu}$ are the components of the field strength 2-form

$$F^i_{\mu\nu} = \partial_\mu A^i_\nu - \partial_\nu A^i_\mu + C^i_{jk}\, A^j_\mu\, A^k_\nu \tag{45}$$

while $D_\mu\,\varphi^A$ denote the components of the covariant derivatives

$$D_\mu\,\varphi^A = \partial_\mu\varphi^A + Z^A_{iB}\, A^i_\mu\, \varphi^B. \tag{46}$$

The constant parameters Z^A_{iB} are the components of the linear map which defines the linear representation of the Lie algebra $\mathfrak{g}$ on the vector space in which the fields φ^A are assumed to live.

The variation of (44) has an expression of the type

$$\delta\mathcal{L} = -\tfrac{1}{2}\tilde{\mathcal{F}}^{\mu\nu}_i\delta F^i_{\mu\nu} + \tilde{f}_A\delta\varphi^A + \tilde{p}^\mu_A\delta(D_\mu\varphi^A), \tag{47}$$

where the coefficients are defined by:

$$\tilde{\mathcal{F}}^{\mu\nu}_i(F;\varphi, D\varphi) \doteq -2\frac{\partial\mathcal{L}}{\partial F^i_{\mu\nu}} \tag{48}$$

$$\tilde{f}_A(F;\varphi, D\varphi) \doteq \frac{\partial \mathcal{L}}{\partial \varphi^A} \tag{49}$$

$$\tilde{p}^{\mu}_A(F;\varphi, D\varphi) \doteq \frac{\partial \mathcal{L}}{\partial (D_\mu \varphi^A)}. \tag{50}$$

We note that the skew-symmetric functions $\tilde{\mathcal{F}}^{\mu\nu}_i$ are the coefficients of a $\mathfrak{g}^*$-valued $(m-2)$-form $\mathcal{F}$. Taking (45) and (46) into account, we see that the second-order Euler–Lagrange equations of the theory are

$$d_\nu \tilde{\mathcal{F}}^{\mu\nu}_i + \tilde{\mathcal{F}}^{\mu\nu}_j C^j_{ik} A^k_\nu = \tilde{p}^{\mu}_A Z^A_{iB}\varphi^B \tag{51}$$

$$d_\mu \tilde{p}^{\mu}_A - \tilde{p}^{\mu}_B Z^B_{iA} A^i_\mu = \tilde{f}_A. \tag{52}$$

We remark that the right-hand side of Eq. (51) vanishes not only in absence of external matter fields, but also if matter fields are present but the Lagrangian does not depend on their covariant derivatives. In both cases the Lagrangian depends on the gauge potential only through the field strength.

Our purpose is now to perform a partial Legendre transformation only with respect to the gauge potential A. The Lagrangian depends on derivatives $\partial_\mu A^i_\nu$ only through the coefficients of the field strength so that we can consider expression (48) as the definition of the (partial) Legendre transformation. The coefficients $\mathcal{F}^{\mu\nu}_i = \tilde{\mathcal{F}}^{\mu\nu}_i(F,\varphi, D\varphi)$ are then the *covariant momenta* conjugated to the gauge potential A. Under the regularity condition

$$\det \left| \frac{\partial^2 \mathcal{L}}{\partial F^i_{\mu\nu}\, \partial F^j_{\alpha\beta}} \right| \neq 0, \tag{53}$$

the Legendre map (48) can be solved explicitly with respect to the fields $F^i_{\mu\nu}$:

$$F^i_{\mu\nu} = \tilde{F}^i_{\mu\nu}(\mathcal{F};\varphi, D\varphi). \tag{54}$$

The Hamiltonian density is defined by

$$H(\mathcal{F}, A, j^1\varphi; g) = H(\mathcal{F}, \varphi, D\varphi; g) \doteq \frac{1}{2}\mathcal{F}^{\mu\nu}_i \tilde{F}^i_{\mu\nu} + \mathcal{L}(\tilde{F}, \varphi, D\varphi; g), \tag{55}$$

while the Helmholtz Lagrangian density $\mathcal{L}_H$ reads as follows:

$$\mathcal{L}_H(\mathcal{F}, j^1 A, j^1\varphi; g) = -\frac{1}{2}\mathcal{F}^{\mu\nu}_i F^i_{\mu\nu}(j^1 A) + H(\mathcal{F}, A, j^1\varphi; g). \tag{56}$$

Taking (55) and (47) into account, we see that the variation of $\mathcal{L}_H$ turns out to be

$$\begin{aligned} \delta\mathcal{L}_H = {} & d_\nu \left(\mathcal{F}^{\mu\nu}_i \delta A^i_\mu + \tilde{p}^{\nu}_A \delta\varphi^A\right) + \frac{1}{2}\left(\tilde{F}^i_{\mu\nu} - F^i_{\mu\nu}\right)\delta\mathcal{F}^{\mu\nu}_i + \\ & - \left(d_\nu \mathcal{F}^{\mu\nu}_i + \mathcal{F}^{\mu\nu}_j C^j_{ik} A^k_\nu - \tilde{p}^{\mu}_A Z^A_{iB}\varphi^B\right)\delta A^i_\mu + \\ & - \left(d_\mu \tilde{p}^{\mu}_A - \tilde{p}^{\mu}_B Z^B_{iA} A^i_\mu - \tilde{f}_A\right)\delta\varphi^A. \end{aligned} \tag{57}$$

Then, the Euler–Lagrange equations ensuing from $\mathcal{L}_H$ are

$$\begin{cases} F^i_{\mu\nu}(j^1 A) = \tilde{F}^i_{\mu\nu}(\mathcal{F}; \varphi, D\varphi) \doteq 2\,\frac{\partial H}{\partial \mathcal{F}^{\mu\nu}_i} \\ d_\nu \mathcal{F}^{\mu\nu}_i + \mathcal{F}^{\mu\nu}_j C^j_{ik} A^k_\nu - \tilde{p}^\mu_A Z^A_{iB}\varphi^B = 0 \\ d_\mu \tilde{p}^\mu_A - \tilde{p}^\mu_B Z^B_{iA} A^i_\mu - \tilde{f}_A = 0. \end{cases} \tag{58}$$

Solving the first equation in (58) with respect to momenta $\mathcal{F} = \tilde{\mathcal{F}}(F; \varphi, D\varphi)$, and inserting this expression in the second and in the third equation of (58), one recovers the original Euler–Lagrangian equations (51) and (52) .

We now want, through a dual Legendre transformation, to construct explicitly the dual Lagrangian for a gauge theory (see [16, 17] for further details). If we assume that the second equation in (58) has a unique solution

$$A^i_\mu = \tilde{A}^i_\mu(j^1\mathcal{F}, j^1\varphi) \tag{59}$$

and if we substitute this solution into the Helmholtz Lagrangian density (56), we obtain the dual second-order Lagrangian density $\mathcal{L}^*$:

$$\mathcal{L}^*(j^2\mathcal{F}, j^2\varphi; g) = -\frac{1}{2}\mathcal{F}^{\mu\nu}_i F^i_{\mu\nu}(j^1\tilde{A}) + H(\mathcal{F}, \tilde{A}, j^1\varphi; g). \tag{60}$$

In agreement with (45) , it can be rewritten as

$$\mathcal{L}^*(j^2\mathcal{F}, j^2\varphi; g) = d_\nu(\mathcal{F}^{\mu\nu}_i \tilde{A}^i_\mu) + \mathcal{U}(j^1\mathcal{F}, j^1\varphi; g) \tag{61}$$

with

$$\mathcal{U}(j^1\mathcal{F}, j^1\varphi; g) = -(d_\nu \mathcal{F}^{\mu\nu}_i)\tilde{A}^i_\mu - \frac{1}{2}\mathcal{F}^{\mu\nu}_i C^i_{jk}\tilde{A}^j_\mu \tilde{A}^k_\nu + H(\mathcal{F}, \tilde{A}, j^1\varphi; g). \tag{62}$$

The variation of (62) turns out to be

$$\begin{aligned} \delta\mathcal{U} = {} & \frac{1}{2}(\tilde{F}^i_{\mu\nu} - C^i_{jk}\tilde{A}^j_\mu \tilde{A}^k_\nu)\delta\mathcal{F}^{\mu\nu}_i - \tilde{A}^i_\mu \delta(d_\nu \mathcal{F}^{\mu\nu}_i) + \\ & + \tilde{f}_A \delta\varphi^A + \tilde{p}^\mu_A(d_\mu \delta\varphi^A + Z^A_{iB}\tilde{A}^i_\mu \delta\varphi^B). \end{aligned} \tag{63}$$

In the dual formulation, the functions $\mathcal{F}^{\mu\nu}_i$ play the role of dynamical variables and their conjugate momenta are

$$\tilde{A}^i_\mu(j^1\mathcal{F}, j^1\varphi) = -\frac{\partial\mathcal{U}}{\partial(d_\nu \mathcal{F}^{\mu\nu}_i)}. \tag{64}$$

The latter expression reproduces formula (59) and it defines the dual Legendre map. The Euler–Lagrange equations for the $\mathcal{F}$ field thus take the form

$$\begin{aligned} \tilde{F}^i_{\mu\nu}(j^1\mathcal{F}, j^1\varphi) = {} & d_\mu\left[\tilde{A}^i_\nu(j^1\mathcal{F}, j^1\varphi)\right] - d_\nu\left[\tilde{A}^i_\mu(j^1\mathcal{F}, j^1\varphi)\right] \\ & + C^i_{jk}\tilde{A}^j_\mu(j^1\mathcal{F}, j^1\varphi)\tilde{A}^k_\nu(j^1\mathcal{F}, j^1\varphi). \end{aligned} \tag{65}$$

In the case when the second equation of (58) reduces to

$$d_\nu \mathcal{F}^{\mu\nu}_i + \mathcal{F}^{\mu\nu}_j C^j_{ik} A^k_\nu = 0, \tag{66}$$

the dual Lagrangian density $\mathcal{L}^*(j^2\mathcal{F}, j^1\varphi; g)$ depends linearly on second-order derivatives of $\mathcal{F}$ and quadratically on first-order derivatives of $\mathcal{F}$ and does not depend on second-order derivatives of fields φ. The possibility of explicitly solving the $p = \dim G \cdot \dim M$ equations (66) is related to the choice of the gauge group G as well as to the dimension of the space-time M. For example, if the Lie algebra $\mathfrak{g}$ is commutative or it admits some non-trivial ideal $\mathfrak{h} \subset \mathfrak{g}$, the components A^i_μ of the gauge potential are not uniquely defined in terms of the $\mathcal{F}$ variables since too many structure constants are equal to zero. Thence, if Eq. (66) does not contain further terms due to the matter fields, as happens in the general situation (see the second equation of (58)), only semi-simple Lie algebras have to be taken into account.

Yang–Mills 3 + 3

We now briefly summarize from [16] the case in which $\dim(M) = 3$ and $\dim(G) = 3$, with $G = \mathrm{SO}(3)$ or $G = \mathrm{SO}(2, 1)$. Under these hypotheses we can write

$$C^i_{jm} = \pm\frac{1}{2}\sqrt{k}k^{il}\epsilon_{ljm} \tag{67}$$

where k_{il} is an Ad-invariant metric on the Lie algebra SO(3) or SO(2, 1) with inverse k^{il}, while $\sqrt{k} \doteq \sqrt{|\det(k_{il})|}$. From now on, the $+$ sign [or the $-$ sign] refers to the choice $G = \mathrm{SO}(3)$ [$G = \mathrm{SO}(2, 1)$ respectively].

The functions $\mathcal{F}$ appearing in (66) can be rewritten as

$$\mathcal{F}^{\mu\nu}_i = k_{ij}\phi^j_\alpha \epsilon^{\alpha\mu\nu}. \tag{68}$$

Assuming for simplicity that $\det(\phi^j_\alpha) \neq 0$, we can define a metric

$$\mathbf{g}_{\alpha\beta} = k_{ij}\phi^i_\alpha \phi^j_\beta \tag{69}$$

on M, which has the same signature as k_{ij}, i.e., it is an Euclidean metric in the SO(3) case while it has signature $+1$ for SO(2, 1). With this notation, Eq. (66) can be rewritten as follows:

$$k_{ki}\epsilon^{\alpha\mu\nu} d_\nu \phi^i_\alpha \pm \frac{1}{2}\sqrt{k}\epsilon_{hkj} A^j_\nu \epsilon^{\alpha\mu\nu}\phi^h_\alpha = 0 \tag{70}$$

which can be solved with respect to the gauge potential. In fact, we stress that the regularity condition of the dual Legendre transformation (70) amounts to the requirement that $\det(\phi^j_\alpha) \neq 0$. After some manipulation (see [16]), we obtain

$$\tilde{A}^a_\mu(j^1\mathcal{F}) = \pm\frac{1}{\sqrt{k}}\epsilon^{aj}{}_i\,\omega^i{}_{j\mu}(j^1\phi) \tag{71}$$

where $\omega^{i}_{j\mu}(j^1\phi)$ is the Levi–Civita connection of the metric $\mathbf{g}$ in the anholonomic basis ϕ^i_α. From (71) we obtain, in turn, the expression for the gauge field strength

$$F^i_{\mu\nu}(j^1\tilde{A}) = \pm\frac{1}{\sqrt{k}}\epsilon^{ip\cdot}{}_q\mathbf{R}^q_{p\mu\nu}, \tag{72}$$

where $\mathbf{R}^q_{p\mu\nu}$ is the Riemann tensor of $\mathbf{g}$.

The first term which appears in the dual Lagrangian density (60) turns out to be

$$-\frac{1}{2}\mathcal{F}^{\mu\nu}_i F^i_{\mu\nu} = \sqrt{\mathbf{g}}\mathbf{g}^{\mu\nu}\,\mathbf{R}_{\mu\nu}(j^2\mathbf{g}) \quad (\sqrt{\mathbf{g}} \doteq \sqrt{|\det(\mathbf{g}_{\mu\nu})|}),$$

and the Lagrangian density (60) takes the form

$$\mathcal{L}^*(j^2\mathbf{g}, j^1\varphi; g) = \sqrt{\mathbf{g}}\,\mathbf{g}^{\mu\nu}\,\mathbf{R}_{\mu\nu}(j^2\mathbf{g}) + H(\mathbf{g}, j^1\varphi; g). \tag{73}$$

The result obtained shows that the dual theory of an $SO(3-\alpha, \alpha)$ gauge theory ($\alpha = 0, 1$) on a three-dimensional manifold is a metric theory: the gauge potential and the metric are variables conjugated to each other (see [16, 19]). So we have exchanged the original 9 configuration variables A^i_μ with the 6 dual configuration variables $\mathbf{g}_{\mu\nu}$: the covariant Legendre transformation allows us to gauge away the gauge degrees of freedom. The dual theory is rewritten in terms of gauge invariant variables.

For example, specializing (44) for the Yang–Mills Lagrangian density $\mathcal{L}_{YM}$:

$$\mathcal{L}_{YM} = -\frac{1}{4}\sqrt{g(x)}\,g^{\alpha\mu}(x)\,g^{\beta\nu}(x)\,F^i_{\alpha\beta}\,F^j_{\mu\nu}\,k_{ij} \tag{74}$$

(with $\sqrt{g(x)} \doteq \sqrt{|\det(g_{\rho\sigma}(x))|}$), we see that the momenta $\mathcal{F}$ turn out to be

$$\mathcal{F}^{\mu\nu}_i = \sqrt{g(x)}\,g^{\alpha\mu}(x)\,g^{\beta\nu}(x)F^j_{\alpha\beta}\,k_{ij}. \tag{75}$$

Namely, they are the Hodge dual of the field strength F, where the Hodge operator is defined with respect to the background metric $g(x)$. The map $(g_{\alpha\beta}, F^i_{\mu\nu}) \mapsto (g_{\alpha\beta}, \mathcal{F}^{\mu\nu}_i(F; g))$ defined by (75) is invertible, with inverse defined by

$$F^i_{\mu\nu} = \tilde{F}^i_{\mu\nu}(\mathcal{F}, g(x)) = \frac{1}{\sqrt{g(x)}}\,g_{\alpha\mu}(x)g_{\beta\nu}(x)\mathcal{F}^{\alpha\beta}_j\,k^{ij}. \tag{76}$$

so that the dual Lagrangian density $\mathcal{L}^*_{YM}$ turns out to be

$$\mathcal{L}^*_{YM} = \sqrt{\mathbf{g}}\,\mathbf{g}^{\mu\nu}\,\mathbf{R}_{\mu\nu}(j^2\mathbf{g}) + \sigma(g)\frac{\sqrt{g(x)}}{2}\,g^{\mu\nu}(x)\,\mathbf{g}_{\mu\nu}. \tag{77}$$

Its Euler–Lagrange equations become the Einstein-like equations

$$\mathbf{R}^{\mu\nu} - \tfrac{1}{2}\mathbf{g}^{\mu\nu}\,\mathbf{R} = \frac{\sigma(g)}{2}\,\frac{\sqrt{g(x)}}{\sqrt{\mathbf{g}}}\,g^{\mu\nu}(x) \tag{78}$$

(indices in the Ricci tensor are raised with the dynamical metric $\mathbf{g}$).

Just as gauge theories in three dimensions can be reformulated as gravitational-like theories, on the opposite side, general relativity in three dimensions can be recast as a gauge theory of orthogonal groups (and also four-dimensional gravity if we assume the existence of a non-null Killing vector; see [16]).

Yang–Mills 3 + 4

If $\dim(M) = 4$ and $\dim(G) = 3$, Eq. (66) is much more difficult to handle. In this case the momenta $\mathcal{F}_i^{\mu\nu}$ can be used to define the $\mathfrak{g}$-valued 2 -form $\Phi^i = \frac{1}{2}\,\Phi^i_{\mu\nu}\,dx^\mu \wedge dx^\nu$ whose coefficients are given by

$$\Phi^i_{\mu\nu} \doteq \frac{1}{2}\,\epsilon_{\mu\nu\alpha\beta}\,\mathcal{F}^{i\alpha\beta} \tag{79}$$

Starting from the configuration variables $\mathcal{F}$ and Φ, we define the following symmetric object (with respect to internal, i.e., Latin, indices):

$$Z^{ij} \doteq \frac{1}{2}\,\Phi^i_{\mu\nu}\,\mathcal{F}^{j\mu\nu}. \tag{80}$$

Moreover, assuming $\det(Z^{ij}) \neq 0$, we also introduce the symmetric tensor (with respect to Greek indices)

$$\mathbf{g}_{\alpha\beta} = \left(\frac{2}{\det Z}\right)^{\frac{1}{3}} \frac{1}{3}\,\epsilon_{ijk}\,\Phi^i_{\alpha\mu}\,\mathcal{F}^{j\mu\nu}\,\Phi^k_{\nu\beta} \tag{81}$$

which can again be considered as a dynamical metric (of Euclidean or Kleinian signature; see [18, 20] and their references). The relevant result is that Eq. (66) can be solved with respect to the gauge potential A^i_μ if $\det\mathbf{g} = \left(\frac{1}{2}\det Z\right)^{2/3} \neq 0$. The explicit form of the dual Lagrangian turns out to be very cumbersome; nevertheless it can again be written in terms of the gauge invariant variables (80) and (81) or of some algebraic combination of them. Detailed computations for this type of theory can be found, for example, in [17, 20, 21].

We stress that the metric tensor (81), to our knowledge, was first introduced by Urbantke [18]. It was also deeply investigated in [22–25] in the context of Ashtekar variables (see [26]). We refer the interested reader to these papers for further details.

In conclusion, we hope that the framework developed in this paper for the construction of *dual theories* (see [1] for deeper insight) may be used to better understand the analogies between different theories, in particular, between gauge and metric theories. Once the relations (and differences) are analyzed we may gain some deeper insight toward a better understanding of a physical theory. For example, in gauge theories (as well as for any gauge-natural theory), conserved quantities associated to space-time symmetries may not be well-defined. This is related mainly to the fact that the Lie derivatives of the dynamical fields, which enter in the definition of Noether currents, may not be uniquely defined because we may not know how space-time diffeomorphisms act on the configuration variables. Instead conserved quantities are well-defined in the dual metric formulation, where Lie derivatives are uniquely defined. For a deeper discussion of this topic we refer the reader to [27] and its references.

References

1. Ferraris M., Francaviglia M., Raiteri M. (2000): Dual Lagrangian field theories. J. Math. Phys., in press
2. Ferraris M., Kijowski J. (1982): On the equivalence of relativistic theories of gravitation. Gen. Rel. Grav. **14** (1), 165–180
3. Hilbert D. (1915): Die Grundlagen der Physik. Königl. Gesel. Wiss., Göttingen Nachr., Math. Phys. Kl., 394–407
4. Einstein A. (1916): Hamiltonsches Prinzip und allgemeine Relativitätstheorie. Sitzungsber. Preuss. Akad. Wiss. (Berlin), 1111–1116
5. Einstein A. (1923): Zur allgemeinen Relativitätstheorie. Sitzungsber. Preuss. Akad. Wiss. (Berlin), 32–38
6. Einstein A. (1923): Zur affinen Feldtheorie. Sitzungsber. Preuss. Akad. Wiss. (Berlin), 137–140
7. Schrödinger E. (1943): The general unitary theory of the physical fields. Proc. R. Irish Acad. A **49**, 43–58
8. Schrödinger E. (1946): The general affine field laws. Proc. R. Irish Acad. A **51**, 41–50
9. Schrödinger E. (1947): The final affine field laws, I. Proc. R. Irish Acad. A **52**, 163–171
10. Schrödinger E. (1948a): The final affine field laws, II. Proc. R. Irish Acad. A **51**, 205–216
11. Schrödinger E. (1948): The final affine field laws, III. Proc. R. Irish Acad. A **52**, 1–9
12. Kijowski J. (1978): On a new variational principle in general relativity and the energy of the gravitational field. Gen. Rel. Grav. **9**, 857–877
13. Ferraris M., Kijowski J. (1981): General relativity is a Gauge–type theory. Lett. Math. Phys. **5**, 127–135
14. Ferraris M., Kijowski J. (1982): Unified geometric theory of electromagnetic and gravitational interactions. Gen. Rel. Grav. **14** (1), 37–47
15. Ferraris M. (1984): Affine Unified Theories of Gravitation and Electromagnetism, in *Proc. Journées Relativistes 1983 (Torino, 1983)*, ed. by S. Benetti, M. Ferraris, M. Francaviglia, Tecnoprint, Bologna, pp. 125–153
16. Raiteri M., Ferraris M., Francaviglia M. (1996): General Relativity as a Gauge Theory of Orthogonal Groups in Three Dimensions, in *Gravity, Particles and Space-Time*, ed. by Pronin P. and Sardanashvily G., World Scientific, Singapore, pp. 81–98
17. Ferraris M., Francaviglia M., Raiteri M.: Dual Lagrangian Formulation of Gauge Theories, in preparation
18. Urbantke H. (1984): J. Math. Phys. **25**, 2321
19. Lunev F.A. (1992): Phys. Lett. B **295**, 99
20. Bengtsson I. (1995): Class. Quantum Grav. **12**, 1581
21. Ganor O., Sonnenschein J. (1995): hep-th/9507036
22. Capovilla R., Jacobson T., Dell J. (1989): Phys. Rev. Lett. **63**, 2325
23. Capovilla R., Jacobson T., Dell J., Mason L. (1991): Class. Quantum Grav. **8**, 41
24. Capovilla R., Jacobson T., Dell J. (1991): Class. Quantum Grav. **8**, 59
25. Capovilla R., Jacobson T. (1992): Modern Phys. Lett. A **7** (21), 1871
26. Ashtekar A. (1988): New Perspectives in Canonical Gravity. Bibliopolis, Napoli
27. Fatibene L., Ferraris M., Francaviglia M., Raiteri M. (1999): Phys. Rev. D **60**, 124013

The Boundary State Description of D-Branes

A. Lerda

Abstract. We review the boundary state description of the Dirichlet branes and show how to obtain the Dirac–Born–Infeld action and the classical supergravity D-brane solutions using the closed string operator formalism.

1 Introduction

Classical solutions of various low-energy string actions carrying a non-vanishing charge with respect to some $(p+1)$-form field have been studied intensively in the last few years. The simplest of them are discussed in detail in [1] where one can also find references to the original papers. The classical solutions having a non-vanishing electric or magnetic charge under the Neveu-Schwarz–Neveu-Schwarz (NS-NS) 2-form correspond, respectively, to the fundamental string and the solitonic 5-brane, or, in the dual formulation, to the solitonic string and the fundamental 5-brane. On the contrary, the classical solutions with a non-vanishing charge under the various $(p+1)$-forms of the Ramond-Ramond (R-R) sector do not have any relation to the perturbative closed string or its solitons. In fact, as has been recognized by Polchinski [2], these solutions, which are required by various string dualities, correspond to membranes on which open strings can end with Dirichlet boundary conditions in the transverse directions and the usual Neumann boundary conditions in the longitudinal directions. For this reason they are called Dirichlet branes (or D-branes for short).

The boundary state, originally introduced for factorizing the planar and non-planar open string one-loop diagrams in the closed string channel [3], has been very useful recently for describing the properties and interactions of D-branes in the framework of the operator formalism of string theory (see, for example, [4–6] and references therein). In particular in [4] it has been shown that the boundary state encodes all relevant properties of classical D-branes since it correctly reproduces the couplings of the Dirac–Born–Infeld action as well as the large distance behavior of the classical Dp-brane supergravity solutions. In this contribution we briefly review these results.

2 The boundary state with an external field

In the closed string operator formalism the supersymmetric Dp-branes of Type II theories are described by means of boundary states $|B\rangle$ [7, 8]. These are closed string states which insert a boundary on the world-sheet and enforce on it the appropriate boundary conditions. Both in the NS-NS and in the R-R sectors, there are two possible implementations for the boundary conditions of a Dp-brane which correspond to two boundary states $|B, \eta\rangle$, with $\eta = \pm 1$. However, only the combinations

$$|B\rangle_{\rm NS} = \frac{1}{2}\Big[|B, +\rangle_{\rm NS} - |B, -\rangle_{\rm NS}\Big] \tag{2.1}$$

and

$$|B\rangle_{\rm R} = \frac{1}{2}\Big[|B, +\rangle_{\rm R} + |B, -\rangle_{\rm R}\Big] \tag{2.2}$$

are selected by the GSO projection in the NS-NS and in the R-R sectors respectively. As discussed in [5], the boundary state $|B, \eta\rangle$ is the product of a matter element and a ghost element

$$|B, \eta\rangle = \frac{T_p}{2}\,|B_{\rm mat}, \eta\rangle |B_{\rm g}, \eta\rangle, \tag{2.3}$$

where

$$|B_{\rm mat}, \eta\rangle = |B_X\rangle |B_\psi, \eta\rangle, \quad |B_{\rm g}, \eta\rangle = |B_{\rm gh}\rangle |B_{\rm sgh}, \eta\rangle. \tag{2.4}$$

The overall normalization T_p can be fixed unambiguously from the factorization of amplitudes of closed strings emitted from a disk [9, 4] and is the brane tension [10]

$$T_p = \sqrt{\pi}\left(2\pi\sqrt{\alpha'}\right)^{3-p}. \tag{2.5}$$

Explicit expressions for the various components of $|B\rangle$ have been given in [5] in the simplest case of a static D-brane. However, the operator structure of the boundary state does not change even when more general configurations are considered and is always of the form

$$|B_X\rangle = \exp\left[-\sum_{n=1}^{\infty}\frac{1}{n}\,\alpha_{-n}\cdot S\cdot\tilde{\alpha}_{-n}\right]|B_X\rangle^{(0)}, \tag{2.6}$$

and

$$|B_\psi, \eta\rangle_{\rm NS} = -\mathrm{i}\,\exp\left[\mathrm{i}\,\eta\sum_{m=1/2}^{\infty}\psi_{-m}\cdot S\cdot\tilde{\psi}_{-m}\right]|0\rangle \tag{2.7}$$

for the NS-NS sector, and

$$|B_\psi, \eta\rangle_{\rm R} = -\exp\left[\mathrm{i}\,\eta\sum_{m=1}^{\infty}\psi_{-m}\cdot S\cdot\tilde{\psi}_{-m}\right]|B, \eta\rangle_{\rm R}^{(0)} \tag{2.8}$$

for the R-R sector. The matrix S and the zero-mode contributions $|B_X\rangle^{(0)}$ and $|B, \eta\rangle_{\rm R}^{(0)}$ encode all information about the overlap equations that the string coordinates have to satisfy, which in turn depend on the boundary conditions of the open strings ending on the Dp-brane. Since the ghost and superghost fields are not affected by the type of boundary conditions that are imposed, the ghost part of the boundary state is always the same. Its explicit expression can be found in [5] but we do not give it again since it will not play any significant role for our present purposes. However, we would like to recall that the boundary state must be written in the $(-1,-1)$ superghost picture in the NS-NS sector, and in the asymmetric $(-1/2,-3/2)$ picture in the R-R sector in order to saturate the superghost number anomaly of the disk [11, 5].

When a constant gauge field F is present on the D-branes world-volume, the overlap conditions that the boundary state must satisfy are [7]

$$\begin{aligned}\Big\{(\mathbb{I}+\hat{F})^{\alpha}{}_{\beta}\,\alpha_n^{\beta}+(\mathbb{I}-\hat{F})^{\alpha}{}_{\beta}\,\tilde{\alpha}_{-n}^{\beta}\Big\}\,|B_X\rangle=0\\ \Big\{\alpha_n^i-\tilde{\alpha}_{-n}^i\Big\}\,|B_X\rangle=0\end{aligned}\tag{2.9}$$

for the bosonic part, and

$$\begin{aligned}\Big\{(\mathbb{I}+\hat{F})^{\alpha}{}_{\beta}\,\psi_m^{\beta}-\mathrm{i}\,\eta\,(\mathbb{I}-\hat{F})^{\alpha}{}_{\beta}\,\tilde{\psi}_{-m}^{\beta}\Big\}\,|B_\psi,\eta\rangle=0\\ \Big\{\psi_m^i+\mathrm{i}\,\eta\,\tilde{\psi}_{-m}^i\Big\}\,|B_\psi,\eta\rangle=0\end{aligned}\tag{2.10}$$

for the fermionic part. In these equations, Greek indices $\alpha,\beta,\ldots$ label the world-volume directions $0,1,\ldots,p$ along which the Dp-brane extends, while Latin indices $i,j,\ldots$ label the transverse directions $p+1,\ldots,9$; moreover $\hat{F}=2\pi\alpha F$. These equations are solved by the "coherent states" (2.6)–(2.8) with a matrix S given by

$$S_{\mu\nu}=\Big(\big[(\eta-\hat{F})(\eta+\hat{F})^{-1}\big]_{\alpha\beta}\,;\,-\delta_{ij}\Big)\tag{2.11}$$

and with the zero-mode parts given by

$$|B_X\rangle^{(0)}=\sqrt{-\det(\eta+\hat{F})}\;\delta^{(9-p)}(q^i-y^i)\prod_{\mu=0}^{9}|k^{\mu}=0\rangle\tag{2.12}$$

for the bosonic sector, and by

$$|B_\psi,\eta\rangle_{\mathbb{R}}^{(0)}=\left(C\Gamma^0\Gamma^1\ldots\Gamma^p\,\frac{1+\mathrm{i}\eta\Gamma_{11}}{1+\mathrm{i}\eta}\,U\right)_{AB}|A\rangle\,|\tilde{B}\rangle\tag{2.13}$$

for the R sector. In writing these formulas we have denoted by y^i the position of the D-brane, by C the charge conjugation matrix and by U the following matrix

$$U=\frac{1}{\sqrt{-\det(\eta+\hat{F})}}\;;\exp\Big(-\frac{1}{2}\,\hat{F}_{\alpha\beta}\Gamma^{\alpha}\Gamma^{\beta}\Big);\tag{2.14}$$

where the symbol ; ; means that one has to expand the exponential and then antisymmetrize the indices of the Γ-matrices. Finally, $|A\rangle\,|\tilde{B}\rangle$ stands for the spinor vacuum of the R-R sector[1]. We note that the overlap equations (2.9) and (2.10) do not permit the detremination of the overall normalization of the boundary state, nor even that of the Born–Infeld prefactor of Eq. (2.12). The latter can be introduced by hand as in [7] but can also be derived by boosting the boundary state and then performing a T-duality as explicitly shown in [12].

We end this section with a few comments. If F is an external *magnetic* field, the corresponding boundary state describes a stable BPS bound state formed by a Dp-brane with other lower dimensional D-branes (as, for example, the Dp-D$(p-2)$ bound state). This case was considered explicitly in [4] where the long distance behavior of the massless fields of these configurations was determined by use of the boundary state approach. On the contrary, if F is an external *electric* field, then the boundary state describes a stable bound state between a fundamental string and a Dp-brane that preserves one half of the space-time supersymmetries. This kind of bound state denoted by (F,Dp) is a generalization of the dyonic string configurations of Schwarz [13] which has recently been studied from the supergravity point of view in [14] and from the operator formalism point of view in [15].

3 D-branes effective action

We now show how the low-energy effective action of a D-branes is related to the boundary state we have just constructed. As we have mentioned before, the boundary state is the exact conformal description of a D-branes and therefore it contains complete information about the interactions between a D-branes and the closed strings that propagate in the bulk. In particular it encodes the couplings with the bulk massless fields which can be simply obtained by saturating the boundary state $|B\rangle$ with the massless states of the closed string spectrum. In order to find a non-vanishing result, it is necessary to soak up the superghost number anomaly of the disk and thus, as a consequence of the superghost charge of the boundary state, we have to use closed string states in the $(-1,-1)$ picture in the NS-NS sector and states in the asymmetric $(-1/2,-3/2)$ picture in the R-R sector.

In the NS-NS sector, the states that represent the graviton $h_{\mu\nu}$, the dilaton ϕ and the Kalb–Ramond antisymmetric tensor $A_{\mu\nu}$ are of the form

$$\epsilon_{\mu\nu}\,\tilde{\psi}^{\mu}_{-\frac{1}{2}}\,\psi^{\nu}_{-\frac{1}{2}}\,|k/2\rangle_{-1}\,|\widetilde{k/2}\rangle_{-1} \tag{3.1}$$

with

$$\epsilon_{\mu\nu}=h_{\mu\nu},\quad h_{\mu\nu}=h_{\nu\mu},\quad k^{\mu}h_{\mu\nu}=\eta^{\mu\nu}h_{\mu\nu}=0 \tag{3.2}$$

for the graviton,

$$\epsilon_{\mu\nu}=\frac{\phi}{2\sqrt{2}}\,(\eta_{\mu\nu}-k_{\mu}\ell_{\nu}-k_{\nu}\ell_{\mu}),\quad \ell^2=0,\quad k\cdot\ell=1 \tag{3.3}$$

[1] For our conventions on Γ-matrices, spinors, etc., see, for example [4, 5].

for the dilaton, and

$$\epsilon_{\mu\nu} = \frac{1}{\sqrt{2}} A_{\mu\nu}, \quad A_{\mu\nu} = -A_{\nu\mu}, \quad k^\mu A_{\mu\nu} = 0 \tag{3.4}$$

for the Kalb–Ramond field. In order to obtain their couplings with the boundary state it is useful to first compute the quantity

$$J^{\mu\nu} \equiv {}_{-1}\langle \widetilde{k/2}| \, {}_{-1}\langle k/2| \, \psi^\nu_{\frac{1}{2}} \, \tilde{\psi}^\mu_{\frac{1}{2}} \, |B\rangle_{\rm NS} = -\frac{T_p}{2} \, V_{p+1} \sqrt{-\det(\eta + \hat{F})} \, S^{\nu\mu} \tag{3.5}$$

where V_{p+1} is the (infinite) world-volume of the brane, and then to project it on the various independent fields using their explicit polarizations. We thus obtain: for the graviton,

$$J_h \equiv J^{\mu\nu} h_{\mu\nu} = -T_p \, V_{p+1} \sqrt{-\det(\eta + \hat{F})} \left[(\eta + \hat{F})^{-1} \right]^{\alpha\beta} h_{\beta\alpha} \tag{3.6}$$

where we have used the tracelessness of $h_{\mu\nu}$; for the dilaton,

$$\begin{aligned} J_\phi &\equiv \frac{1}{2\sqrt{2}} J^{\mu\nu} \left(\eta_{\mu\nu} - k_\mu \ell_\nu - k_\nu \ell_\mu \right) \phi \\ &= \frac{T_p}{2\sqrt{2}} V_{p+1} \sqrt{-\det(\eta + \hat{F})} \left[3 - p + \mathrm{Tr}\left(\hat{F}(\eta + \hat{F})^{-1} \right) \right] \phi \ ; \end{aligned} \tag{3.7}$$

and, finally, for the Kalb–Ramond field,

$$\begin{aligned} J_A &\equiv \frac{1}{\sqrt{2}} J^{\mu\nu} A_{\mu\nu} \\ &= -\frac{T_p}{2\sqrt{2}} V_{p+1} \sqrt{-\det(\eta + \hat{F})} \left[(\eta - \hat{F})(\eta + \hat{F})^{-1} \right]^{\alpha\beta} A_{\beta\alpha} \\ &= -\frac{T_p}{\sqrt{2}} V_{p+1} \sqrt{-\det(\eta + \hat{F})} \left[(\eta + \hat{F})^{-1} \right]^{\alpha\beta} A_{\beta\alpha} \end{aligned} \tag{3.8}$$

where in the last line we have used the antisymmetry of $A_{\mu\nu}$.

We now show that the couplings J_h, J_ϕ and J_A are precisely those which are produced by the Dirac–Born–Infeld action which governs the low-energy dynamics of the D-brane. In the string frame, this action reads as follows:

$$S_{\rm DBI} = -\frac{T_p}{\kappa} \int_{V_{p+1}} d^{p+1}\xi \; \mathrm{e}^{-\varphi} \sqrt{-\det\left[G + \mathcal{A} + \hat{F} \right]} \tag{3.9}$$

where $2\kappa^2 = (2\pi)^7 (\alpha')^4 g_s^2$ is Newton's constant (g_s being the string coupling), and $G_{\alpha\beta}$ and $\mathcal{A}_{\alpha\beta}$ are respectively the pullbacks of the space-time metric and of the NS-NS antisymmetric tensor on the D-branes world-volume.

In order to compare the couplings described by this action with those obtained from the boundary state, it is first necessary to rewrite $S_{\rm DBI}$ in the Einstein frame. In fact, like any string amplitude computed with the operator formalism, the couplings J_h, J_ϕ and J_A also are written in the Einstein frame. Furthermore, it is also convenient to introduce canonically normalized fields. These two goals can be realized by means of the following field redefinitions

$$G_{\mu\nu} = \mathrm{e}^{\varphi/2}\, g_{\mu\nu}, \quad \varphi = \sqrt{2}\kappa\,\phi, \quad \mathcal{A}_{\mu\nu} = \sqrt{2}\kappa\,\mathrm{e}^{\varphi/2}\, A_{\mu\nu}. \tag{3.10}$$

Using the new fields in Eq. (3.9), we easily obtain

$$S_{\mathrm{DBI}} = -\frac{T_p}{\kappa}\int_{V_{p+1}} d^{p+1}\xi\; \mathrm{e}^{-\frac{\kappa(3-p)}{2\sqrt{2}}\phi}\sqrt{-\det\left[g + \sqrt{2}\kappa\, A + \hat{F}\,\mathrm{e}^{-\frac{\kappa}{\sqrt{2}}\phi}\right]}. \tag{3.11}$$

By expanding the metric around the flat background

$$g_{\mu\nu} = \eta_{\mu\nu} + 2\kappa\, h_{\mu\nu}, \tag{3.12}$$

and keeping only the terms which are linear in h, ϕ and A, the action (3.11) reduces to the following expression

$$S_{\mathrm{DBI}} \simeq -\,T_p\int_{V_{p+1}} d^{p+1}\xi\sqrt{-\det\left[\eta + \hat{F}\right]}\left\{\left[(\eta + \hat{F})^{-1}\right]^{\alpha\beta} h_{\beta\alpha} \right. \tag{3.13}$$
$$\left. -\frac{1}{2\sqrt{2}}\left[3 - p + \mathrm{Tr}\left(\hat{F}(\eta + \hat{F})^{-1}\right)\right]\phi + \frac{1}{\sqrt{2}}\left[(\eta + \hat{F})^{-1}\right]^{\alpha\beta} A_{\beta\alpha}\right\}.$$

It is now easy to see that the couplings with the graviton, the dilaton and the Kalb–Ramond field that can be obtained from this action are exactly the same as those obtained from the boundary state and given in Eqs. (3.6), (3.7) and (3.8) respectively.

Let us now turn to the R-R sector. As we mentioned above, in this sector we have to use states in the asymmetric $(-1/2, -3/2)$ picture in order to soak up the superghost number anomaly of the disk. In the more familiar symmetric $(-1/2, -1/2)$ picture the massless states are associated to the field strengths of the R-R potentials. On the contrary, in the $(-1/2, -3/2)$ picture, the massless states are associated directly to the R-R potentials which, in the notation of forms, we denote by

$$C_{(n)} = \frac{1}{n!}\, C_{\mu_1\ldots\mu_n}\, dx^{\mu_1} \wedge \ldots \wedge dx^{\mu_n} \tag{3.14}$$

with $n = 1, 3, 5, 7, 9$ in the Type IIA theory and $n = 0, 2, 4, 6, 8, 10$ in the Type IIB theory. The string states $|C_{(n)}\rangle$ representing these potentials have a rather non-trivial structure. In fact, as shown in [5], the natural expression

$$|C_{(n)}\rangle \simeq \frac{1}{n!}\, C_{\mu_1\ldots\mu_n}\left(C\Gamma^{\mu_1\ldots\mu_n}\frac{1+\Gamma_{11}}{2}\right)_{AB} |A; k/2\rangle_{-1/2}\, |\widetilde{B}; \widetilde{k/2}\rangle_{-3/2} \tag{3.15}$$

is BRST invariant only if the potential is pure gauge. To avoid this restriction, in general it is necessary to add to Eq. (3.15) a whole series of terms with the same structure but with different contents of superghost zero-modes. However, in the present situation there is a short-cut that considerably simplifies the analysis. In fact, one can use the incomplete states (3.15) and ignore the superghosts, whose contribution can then be recovered simply by changing at the end the overall normalizations of the amplitudes[2]. With this in mind, the couplings between the R-R potentials (3.14) and the Dp-brane can therefore be obtained by computing the overlap between the states (3.15) and the R-R component of the boundary state, namely

$$J_{C_{(n)}} \equiv \langle C_{(n)}|B\rangle_{\mathrm{R}}. \tag{3.16}$$

The evaluation of $J_{C_{(n)}}$ is straightforward, even if a bit lengthy (for details see [15]); and the final result is

$$J_{C_{(n)}} = -\frac{T_p}{16\sqrt{2}\,n!}\, V_{p+1}\, C_{\mu_1\ldots\mu_n}\, \mathrm{Tr}\left(\Gamma^{\mu_n\ldots\mu_1}\Gamma^0\ldots\Gamma^p\,; \mathrm{e}^{-\frac{1}{2}\hat{F}_{\alpha\beta}\Gamma^\alpha\Gamma^\beta} ;\right).$$

It is easy to see that the trace in this equation is non-vanishing only if $n = p + 1 - 2\ell$, where ℓ denotes the power of $\hat{F}$ which is produced by expanding the exponential term. Due to the antisymmetrization ; ; prescription, the integer ℓ takes only a finite number of values up to a maximum $\ell_{\max}$ which is $p/2$ for the Type IIA string and $(p+1)/2$ for the Type IIB string. The simplest term to compute, corresponding to $\ell = 0$, describes the coupling of the boundary state with a $(p+1)$-form potential of the R-R sector and is given by

$$J_{C_{(p+1)}} = \frac{\sqrt{2}\,T_p}{(p+1)!}\, V_{p+1}\, C_{\alpha_0\ldots\alpha_p}\, \varepsilon^{\alpha_0\ldots\alpha_p} \tag{3.17}$$

where $\varepsilon^{\alpha_0\ldots\alpha_p}$ is the completely antisymmetric tensor on the D-branes world-volume[3]. From Eq. (3.17) we can immediately deduce that the charge μ_p of a Dp-brane with respect to the R-R potential $C_{(p+1)}$ is

$$\mu_p = \sqrt{2}\,T_p, \tag{3.18}$$

in agreement with Polchinski's original calculation [2].

The next term in the expansion of the exponential of Eq. (3) corresponds to $\ell = 1$ and yields the coupling of the Dp-brane with a $(p-1)$-form potential which is given by

$$J_{C_{(p-1)}} = \frac{\mu_p}{2(p-1)!}\, V_{p+1}\, C_{\alpha_0\ldots\alpha_{p-2}}\, \hat{F}_{\alpha_{p-1}\alpha_p}\, \varepsilon^{\alpha_0\ldots\alpha_p}. \tag{3.19}$$

By proceeding in the same way, one can easily evaluate the higher order terms generated by the exponential which describe the interactions of the D-branes with

---

[2] Note that this procedure is not allowed when the odd-spin structure contributes; see [5].

[3] Our convention is that $\varepsilon^{0\ldots p} = -\varepsilon_{0\ldots p} = 1$.

potential forms of lower degree. All these couplings can be encoded in the following Wess–Zumino-like term

$$S_{\mathrm{WZ}} = \mu_p \int_{V_{p+1}} \left[\sum_{\ell=0}^{\ell_{\max}} C_{(p+1-2\ell)} \wedge \mathrm{e}^{\hat{F}} \right]_{p+1} \tag{3.20}$$

where $\hat{F} = \frac{1}{2} \hat{F}_{\alpha\beta}\, d\xi^\alpha \wedge d\xi^\beta$, and $C_{(n)}$ is the pullback of the n-form potential (3.14) on the D-branes world-volume. The square bracket in Eq. (3.20) means that in expanding the exponential form one picks up only the terms of total degree $(p+1)$, which are then integrated over the $(p+1)$-dimensional world-volume.

In conclusion we have shown explicitly that, by projecting the boundary state $|B\rangle$ with an external field onto the massless states of the closed string spectrum, one can reconstruct the linear part of the low-energy effective action of a Dp-brane. This is the sum of the Dirac–Born–Infeld part (3.13) and the (anomalous) Wess-Zumino term (3.20) which are produced respectively by the NS-NS and the R-R components of the boundary state.

4 The classical solution from the boundary state

In this section we show that the boundary state is also a very efficient tool to obtain the classical solution corresponding to a Dp-brane at long distances.

For simplicity, from now on we will consider only the case of a pure Dp-brane, with no external gauge field on its world-volume, which is described by a boundary state like the one given in Sect. 2 with $F = 0$ and a diagonal S matrix given by

$$S = (\eta_{\alpha\beta}; -\delta_{ij}) \; ; \tag{4.1}$$

the procedure can however be applied to more general cases, as shown in [15].

To obtain the long distance behavior of the fields emitted by a Dp-brane, one simply adds a closed string propagator D to the boundary state B and then projects the resulting expression onto the various massless states of the closed string spectrum. According to this procedure, the long-distance fluctuation of a field Ψ is then given by

$$\delta\Psi \equiv \langle P^{(\Psi)}|D|B\rangle \tag{4.2}$$

where $\langle P^{(\Psi)}|$ denotes the projector associated to Ψ, i.e., the operator which, when applied to an arbitrary massless state of the closed string, selects the Ψ component contained in that state.

Before giving the details of this calculation, we observe that, since we are not using explicitly the ghost and superghost degrees of freedom, we must take into account their contribution by shifting appropriately the zero-point energy and use for the closed string propagator the following expression:

$$D = \frac{\alpha'}{4\pi} \int_{|z|\leq 1} \frac{d^2 z}{|z|^2}\, z^{L_0 - a}\, \bar{z}^{\tilde{L}_0 - a} \tag{4.3}$$

where the operators L_0 and $\tilde{L}_0$ depend only on the orbital oscillators and the intercept is $a = 1/2$ in the NS-NS sector and $a = 0$ in the R-R sector. We now begin our analysis by studying the projection (4.2) in the NS-NS sector. The projector operators onto the states of the NS-NS sector can be easily obtained from Eqs. (3.1)–(3.4) and are

$$\langle P^{(\phi)}| = {}_{-1}\langle \widetilde{k/2}|\; {}_{-1}\langle k/2|\; \psi^{\nu}_{\frac{1}{2}} \tilde{\psi}^{\mu}_{\frac{1}{2}} \frac{1}{\sqrt{8}} \left(\eta_{\mu\nu} - k_\mu \ell_\nu - k_\nu \ell_\mu\right), \tag{4.4}$$

$$\begin{aligned} \langle P^{(h)\mu\nu}| &= {}_{-1}\langle \widetilde{k/2}|\; {}_{-1}\langle k/2|\; \frac{1}{2}\left(\psi^{\nu}_{\frac{1}{2}} \tilde{\psi}^{\mu}_{\frac{1}{2}} + \psi^{\mu}_{\frac{1}{2}} \tilde{\psi}^{\nu}_{\frac{1}{2}}\right) \\ &\quad - \langle P^{(\phi)}| \frac{1}{\sqrt{8}} \left(\eta^{\mu\nu} - k^\mu \ell^\nu - k^\nu \ell^\mu\right), \end{aligned} \tag{4.5}$$

$$\langle P^{(A)\mu\nu}| = {}_{-1}\langle \widetilde{k/2}|\; {}_{-1}\langle k/2|\; \frac{1}{\sqrt{2}}\left(\psi^{\nu}_{\frac{1}{2}} \tilde{\psi}^{\mu}_{\frac{1}{2}} - \psi^{\mu}_{\frac{1}{2}} \tilde{\psi}^{\nu}_{\frac{1}{2}}\right). \tag{4.6}$$

Since they all contain the following structure

$${}_{-1}\langle \widetilde{k/2}|\; {}_{-1}\langle k/2|\; \psi^{\nu}_{\frac{1}{2}}\; \tilde{\psi}^{\mu}_{\frac{1}{2}}, \tag{4.7}$$

it is first convenient to compute the matrix element

$$T^{\mu\nu} \equiv {}_{-1}\langle \widetilde{k/2}|\; {}_{-1}\langle k/2|\; \psi^{\nu}_{\frac{1}{2}}\; \tilde{\psi}^{\mu}_{\frac{1}{2}} |\, D\, |B\rangle_{\rm NS} = -\frac{T_p}{2} \frac{V_{p+1}}{k_\perp^2}\; S^{\nu\mu} \tag{4.8}$$

where $k_\perp$ is the momentum in the transverse directions which is emitted by the brane. Notice that the matrix $T^{\mu\nu}$ differs from the matrix $J^{\mu\nu}$ defined in Eq. (3.5) (computed for vanishing external field F) simply by the factor of $1/k_\perp^2$ coming from the insertion of the propagator.

Using this result and the explicit form of the dilaton projector (4.4), after some straighforward algebra, we find that the long-distance behavior of the dilaton emitted by the Dp-brane is given by

$$\delta\phi \equiv \langle P^{(\phi)}|D|B\rangle_{\rm NS} = \frac{1}{2\sqrt{2}} \left(\eta^{\mu\nu} - k^\mu \ell^\nu - k^\nu \ell^\mu\right) T_{\mu\nu}. \tag{4.9}$$

Using the explicit expression for the matrix $T_{\mu\nu}$, we get

$$\delta\phi = \mu_p \frac{V_{p+1}}{k_\perp^2} \frac{3-p}{4} \tag{4.10}$$

where μ_p is the unit of R-R charge of a Dp-brane defined in Eq. (3.18). Similarly, using the projector (4.6) for the antisymmetric Kalb–Ramond field, we find that

$$\delta A_{\mu\nu} \equiv \langle P^{(A)}_{\mu\nu}|D|B\rangle_{\rm NS} = \frac{1}{\sqrt{2}}\left(T_{\mu\nu} - T_{\nu\mu}\right). \tag{4.11}$$

Since in our case the matrix $T_{\mu\nu}$ is symmetric, we immediately conclude that the Kalb–Ramond field emitted by the pure Dp-brane identically vanishes. Finally, using Eq. (4.5) we find that the components of the metric tensor are

$$\delta h_{\mu\nu} \equiv \langle P^{(h)}_{\mu\nu}|D|B\rangle_{\rm NS} = \frac{1}{2}\Big(T_{\mu\nu} + T_{\nu\mu}\Big) - \frac{\delta\phi}{2\sqrt{2}}\,\eta_{\mu\nu} \tag{4.12}$$

which explicitly read

$$\begin{aligned} \delta h_{00} &= -\delta h_{11} = \ldots = -\delta h_{pp} = \mu_p\,\frac{V_{p+1}}{k_\perp^2}\,\frac{7-p}{8\sqrt{2}}, \\ \delta h_{p+1,p+1} &= \ldots = \delta h_{99} = \mu_p\,\frac{V_{p+1}}{k_\perp^2}\,\frac{p+1}{8\sqrt{2}}. \end{aligned}$$

Let us now turn to the R-R sector. In this case, after the insertion of the closed string propagator, we have to saturate the R-R boundary state (2.2) with the projector $\langle P^{(C)}|$ on the R-R massless field. Using Eq. (3.15) one can show that

$$\langle P^{(C)}_{\mu_1\ldots\mu_n}| = {}_{-1/2}\langle \widetilde{B}, \widetilde{k/2}|\; {}_{-3/2}\langle A, k/2|\left(C\Gamma_{\mu_1\ldots\mu_n}\frac{1-\Gamma_{11}}{2}\right)_{AB}\frac{(-1)^n}{2\sqrt{2}}. \tag{4.13}$$

This calculation is completely analogous to the one described in the previous section for obtaining the couplings of a Dp-brane with the R-R potentials; the only new feature is the presence of the factor of $1/k_\perp^2$ produced by the closed string propagator. Due to the structure of the R-R component of the boundary state, it is not difficult to see that the only projector of the form (4.13) that can give a non-vanishing result is the one corresponding to a $(p+1)$-form with all indices along the world-volume directions, and to find that its long distance behavior is given by

$$\delta C_{01\cdots p} \equiv \langle P^{(C)}_{01\cdots p}|\, D\,|B\rangle_{\rm R} = -\mu_p\,\frac{V_{p+1}}{k_\perp^2}. \tag{4.14}$$

We can now rewrite the long distance behavior of the massless fields produced by the brane in a more suggestive way. First of all, we perform a Fourier transformation to work in configuration space. This is readily computed by observing that, for $p<7$, one has

$$\int d^{(p+1)}x\, d^{(9-p)}y\,\frac{e^{ik_\perp\cdot y}}{(7-p)\,r^{7-p}\,\Omega_{8-p}} = \frac{V_{p+1}}{k_\perp^2} \tag{4.15}$$

where $\Omega_q = 2\pi^{(q+1)/2}/\Gamma((q+1)/2)$ is the surface area of a unit q-dimensional sphere and $r^2 = y_i y^i$ measures the distance from the branes. For later convenience, we also introduce the quantity

$$Q_p = \mu_p\,\frac{\sqrt{2}\,\kappa}{(7-p)\,\Omega_{8-p}}. \tag{4.16}$$

Then, using Eq. (4.10) and assuming that the dilaton has vanishing vacuum expectation value, after some elementary steps, we obtain that the long-distance behavior of the dilaton is

$$\varphi = \sqrt{2}\kappa\,\phi \simeq -\frac{p-3}{4}\,\frac{Q_p}{r^{7-p}}. \tag{4.17}$$

Since we are going to compare our results with the standard supergravity description of D-brane, we have reintroduced the field φ which differs from the canonically normalized dilaton ϕ by a factor of $\sqrt{2}\kappa$ (see also Eq. (3.10)). Similarly, recalling that $g_{\mu\nu} = \eta_{\mu\nu} + 2\kappa\, h_{\mu\nu}$, from Eq. (4.13) we find that

$$\begin{aligned} g_{00} &= -g_{11} = \dots = -g_{pp} =\simeq -1 - \frac{(p-7)}{8}\,\frac{Q_p}{r^{7-p}}, \\ g_{p+1,p+1} &= \dots = g_{99} \simeq 1 + \frac{p+1}{8}\,\frac{Q_p}{r^{7-p}}. \end{aligned} \tag{4.18}$$

Finally, rescaling the R-R potential by a factor of $\sqrt{2}\kappa$ to obtain the standard supergravity normalization we easily obtain

$$\widehat{C}_{(p+1)} = \sqrt{2}\kappa\, C_{(p+1)} \simeq -\frac{Q_p}{r^{7-p}}\, dx^0 \wedge \ldots \wedge dx^p. \tag{4.19}$$

Equations (4.17)–(4.19) represent the leading long-distance behavior of the massless fields emitted by the Dp-brane. It is reasonable to expect that, by inserting more boundaries on the closed string world-sheet, i.e., by introducing more boundary states, one can perturbatively reconstruct the exact brane-solution of the supergravity field equations. In fact, this has been verified recently in [16], even if with a different formalism. We then assume that this is indeed what happens so that the exact solution can be written in terms of powers of the harmonic function

$$H(r) = 1 + \frac{Q_p}{r^{7-p}}. \tag{4.20}$$

Under this assumption, from Eqs. (4.17)–(4.19) we can infer that, in the exact solution corresponding to a Dp-brane, the dilaton is

$$e^{\varphi} = H^{(3-p)/4}, \tag{4.21}$$

the metric is

$$\begin{aligned} ds^2 &= H^{(p-7)/8}\left[-\left(dx^0\right)^2 + \left(dx^1\right)^2 + \cdots + \left(dx^p\right)^2\right] \\ &+ H^{(p+1)/8}\left[\left(dx^{p+1}\right)^2 + \cdots + \left(dx^9\right)^2\right], \end{aligned} \tag{4.22}$$

and, finally, the R-R potential is

$$\widehat{C}_{(p+1)} = \left(H^{-1} - 1\right) dx^0 \wedge \cdots \wedge dx^p. \tag{4.23}$$

In writing this solution we have assumed that all fields except the metric have vanishing asymptotic values. This explains why we have subtracted the 1 in the last three equations. Our solution agrees exactly with that derived in [1] from the supergravity point of view.

We can therefore conclude that the boundary state provides the complete conformal description of the D-branes of string theory; in fact it generates the correct D-Brane effective action and reduces to the classical D-Brane solution when it is projected onto the massless states of the closed string spectrum.

References

1. Duff M.J., Khuri R.R., Lu J.X. (1995): Phys. Rep. **259**, 213, hep-th/ 9412184
2. Polchinski J. (1995): Phys. Rev. Lett. **75**, 4724, hep-th/9510017
3. Cremmer E., Scherk J. (1972): Nucl. Phys. B **50**, 222;
   Clavelli L., Shapiro J. (1973): Nucl. Phys. B **57**, 490
4. Di Vecchia P., Frau M., Pesando I., Sciuto S., Lerda A., Russo R. (1997): Nucl. Phys. B **507**, 259, hep-th/9707068
5. Billó M., Di Vecchia P., Frau M., Lerda A., Pesando I., Russo R., Sciuto S. (1998): Nucl. Phys. B **526**, 199, hep-th/9802088
6. Billó M., Di Vecchia P., Frau M., Lerda A., Russo R., Sciuto S. (1998): Mod. Phys. Lett. A **13**, 2977, hep-th/9805091
7. Callan C.G., Lovelace C., Nappi C.R., Yost S.A. (1987): Nucl. Phys. B **288**, 525;
   Callan C.G., Lovelace C., Nappi C.R., Yost S.A. (1987): Nucl. Phys. B **293**, 83;
   Callan C.G., Lovelace C., Nappi C.R., Yost S.A. (1988): Nucl. Phys. B **308**, 221
8. Polchinski J., Cai Y. (1988): Nucl. Phys. B **296**, 91
9. Frau M., Pesando I., Sciuto S., Lerda A., Russo R. (1997): Phys. Lett. B **400**, 52, hep-th/9702037
10. Polchinski J. (1996): TASI Lectures on D-branes, hep-th/9611050
11. Friedan D., Martinec E., Shenker S. (1986): Nucl. Phys. B **271**, 93
12. Billó M., Cangemi D., Di Vecchia P. (1997): Phys. Lett. B **400**, 63, hep-th/9701190
13. Schwarz J.H. (1995): Phys. Lett. B **360** 13; Erratum Phys. Lett. B **364**, 252, hep-th/9508143
14. Lu J.X., Roy S. (1999): (m, n) String-like Dp-brane Bound States, hep-th/ 9904112; Non-threshold (F, Dp) Bound States, hep-th/9904129
15. Di Vecchia P., Frau M., Lerda A., Liccardo A. (1999): Boundary State for (F, Dp) Bound States, NORDITA preprint, NORDITA-1999/34 HE and Torino preprint, DFTT 33/99 (June 1999)
16. Berenstein D., Leigh R. (1999): Superstring Perturbation Theory and Ramond–Ramond Background, hep-th/9904104

The N-Body Problem in Tetrad Gravity: A First Step towards a Unified Description of the Four Interactions

L. Lusanna

Abstract. After a review of the canonical reduction to the rest-frame Wigner-covariant instant form of standard theories in Minkowski spacetime, a new formulation of tetrad gravity is introduced. Its canonical reduction, in the presence of N scalar particles, is carried out. A modification of the ADM formulation to solve the deparametrization problem of general relativity (how to recover the rest-frame instant form for $G = 0$) is presented.

1 Electromagnetic, weak and strong interactions in Minkowski spacetime

The attempt to arrive at a unified description of the four interactions [with the matter being either Grassmann-valued Dirac fields or relativistic particles] based on the Dirac–Bergmann theory of constraints, which is needed for the Hamiltonian formulation of both gauge theories and general relativity, motivated the study of the canonical reduction of a new formulation of tetrad gravity [1–4]. This requires the examination of general relativity from the canonical point of view and the generalization to it of all the results already obtained in the canonical study of gauge theories, since neither a complete reduction of gravity with an identification of the physical canonical degrees of freedom of the gravitational field nor a detailed study of its Hamiltonian group of gauge transformations (whose infinitesimal generators are the first class constraints) has ever been carried through to the end in an explicit way.

A research program aiming to express the special relativistic strong, weak and electromagnetic interactions in terms of Dirac's observables [5] is in an advanced stage of development [6]. This program is based on the Shanmugadhasan canonical transformations [7]: if a system has first class constraints at the Hamiltonian level, then, at least locally, one can find a canonical basis with as many new momenta as first class constraints (Abelianization of first class constraints), with their conjugate canonical variables as Abelianized gauge variables and with the remaining pairs of canonical variables as pairs of canonically conjugate Dirac observables (canonical basis of physical variables adapted to the chosen Abelianization; they give a trivialization of the BRST construction of observables). Setting the Abelianized gauge variables equal to zero, one defines a local gauge of the model. If a system with constraints admits one (or more) global Shanmugadhasan canonical transformations, one obtains one (or more) privileged global gauges in which the physical Dirac observables are globally defined and globally separated from the gauge degrees of freedom [for systems with a compact configuration space this is impossible]. These privileged gauges (when they exist) can be called generalized Coulomb gauges. Second class constraints, when present, are also taken into account by the Shanmugadhasan canonical transformation [7].

Firstly, inspired by [8], the canonical reduction to noncovariant generalized Coulomb gauges, with the determination of the physical Hamiltonian as a function of a canonical basis of Dirac observables, has been achieved for the following isolated systems (for them one asks that the 10 conserved generators of the Poincaré algebra are finite so as to be able to use group theory; theories with external fields can only be recovered as limits in some parameter of a subsystem of the isolated system):

a) Yang–Mills theory with Grassmann-valued fermion fields [9] in the case of a trivial principal bundle over a fixed-x^o R^3 slice of Minkowski spacetime with suitable Hamiltonian-oriented boundary conditions; this excludes monopole solutions and, since R^3 is not compactified, one only has a winding number and no instanton number. After a discussion of the Hamiltonian formulation of Yang–Mills theory, of its group of gauge transformations and of the Gribov ambiguity, the theory has been studied in suitable weighted Sobolev spaces where the Gribov ambiguity is absent. The global Dirac observables are the transverse quantities $\boldsymbol{A}_{a\perp}(\boldsymbol{x}, x^o)$, $\boldsymbol{E}_{a\perp}(\boldsymbol{x}, x^o)$ and fermion fields dressed with Yang–Mills (gluonic) clouds. The nonlocal and nonpolynomial (due to the presence of classical Wilson lines along flat geodesics) physical Hamiltonian has been obtained: it is nonlocal but without any kind of singularities, it has the correct Abelian limit if the structure constants are turned off, and it contains the explicit realization of the abstract Mitter–Viallet metric.
b) The Abelian and non-Abelian SU(2) Higgs models with fermion fields [10], where the symplectic decoupling is a refinement of the concept of unitary gauge. There is an ambiguity in the solutions of the Gauss law constraints, which reflects the existence of disjoint sectors of solutions of the Euler–Lagrange equations of Higgs models. The physical Hamiltonian and Lagrangian of the Higgs phase have been found; the self-energy turns out to be local and contains a local four-fermion interaction.
c) The standard $SU(3) \times SU(2) \times U(1)$ model of elementary particles [11] with Grassmann-valued fermion fields. The final reduced Hamiltonian contains nonlocal self-energies for the electromagnetic and color interactions, but "local ones" for the weak interactions implying the non-perturbative emergence of 4-fermion interactions.

The next problem is how to make covariant these results valid in Minkowski spacetime with Cartesian coordinates. Again the starting point was given by Dirac [5] with his reformulation of classical field theory on spacelike hypersurfaces foliating Minkowski spacetime M^4 [the foliation is defined by an embedding $R \times \Sigma \to M^4$, $(\tau, \boldsymbol{\sigma}) \mapsto z^{(\mu)}(\tau, \boldsymbol{\sigma}) \in \Sigma_\tau$, with Σ an abstract 3-surface diffeomorphic to R^3, with Σ_τ its copy embedded in M^4 labelled by the value τ (the Minkowski flat indices are (μ); the scalar parameter τ labels the leaves of the foliation, $\boldsymbol{\sigma}$ gives the curvilinear coordinates on Σ_τ and $(\tau, \boldsymbol{\sigma})$ gives the Σ_τ-adapted holonomic coordinates for M^4); this is the classical basis of Tomonaga–Schwinger quantum field theory]. In this way one gets a parametrized field theory with a covariant 3+1 splitting of Minkowski spacetime and already in a form suited to the coupling to general relativity in its ADM canonical formulation (see also [12], where a theoretical study of this problem is carried out in curved spacetimes). The price is that one has to add as new configuration variables the points $z^{(\mu)}(\tau, \boldsymbol{\sigma})$ of the spacelike hypersurface Σ_τ [the only ones carrying Lorentz indices] and then to define the fields on Σ_τ so that they know the hypersurface Σ_τ of τ-simultaneity [for a Klein–Gordon field $\phi(x)$, this new field is $\tilde{\phi}(\tau, \boldsymbol{\sigma}) = \phi(z(\tau, \boldsymbol{\sigma}))$: it contains the nonlocal information about the embedding]. Besides a Lorentz-scalar form of the constraints of the given system, from the Lagrangian rewritten on the hypersurface [a function of $z^{(\mu)}$ through the induced metric $g_{AB} = z_A^{(\mu)} \eta_{(\mu)(\nu)} z_B^{(\nu)}$, $z_A^{(\mu)} = \partial z^{(\mu)}/\partial \sigma^A$, $\sigma^A = (\tau, \sigma^r)$] one gets four further first class constraints $\mathcal{H}_{(\mu)}(\tau, \boldsymbol{\sigma}) = \rho_{(\mu)}(\tau, \boldsymbol{\sigma}) - l_{(\mu)}(\tau, \boldsymbol{\sigma}) T_{\text{sys}}^{\tau\tau}(\tau, \boldsymbol{\sigma}) - z_{r(\mu)}(\tau, \boldsymbol{\sigma}) T_{\text{sys}}^{\tau r}(\tau, \boldsymbol{\sigma}) \approx 0$ [where $T_{\text{sys}}^{\tau\tau}(\tau, \boldsymbol{\sigma})$, $T_{\text{sys}}^{\tau r}(\tau, \boldsymbol{\sigma})$, are the components of the energy-momentum tensor of the system in the holonomic coordinate system corresponding to the energy- and momentum-densities of the isolated system; one has $\{\mathcal{H}_{(\mu)}(\tau, \boldsymbol{\sigma}), \mathcal{H}_{(\nu)}(\tau, \boldsymbol{\sigma}')\} = 0$] which implies the independence of the description from the choice of the foliation with spacelike hypersurfaces. The evolution vector is given by $z_\tau^{(\mu)} = N_{[z](\text{flat})} l^{(\mu)} + N^r_{[z](\text{flat})} z_r^{(\mu)}$, where $l^{(\mu)}(\tau, \boldsymbol{\sigma})$ is the normal to Σ_τ in $z^{(\mu)}(\tau, \boldsymbol{\sigma})$ and $N_{[z](\text{flat})}(\tau, \boldsymbol{\sigma}) = \sqrt{{}^4g_{\tau\tau} + {}^3g^{rs}\, {}^4g_{\tau r}\, {}^4g_{\tau s}} = \sqrt{{}^4g/{}^3\gamma}$, $N_{[z](\text{flat})r}(\tau, \boldsymbol{\sigma}) = {}^3g_{rs}(\tau, \boldsymbol{\sigma}) N^s_{[z](\text{flat})}(\tau, \boldsymbol{\sigma}) = {}^4g_{\tau r}$, are the flat lapse and shift functions defined through the metric as in general relativity; however, they are not independent variables but functionals of $z^{(\mu)}(\tau, \boldsymbol{\sigma})$ in Minkowski spacetime.

The original Dirac Hamiltonian contains an element given by $\int d^3\sigma \lambda^{(\mu)}(\tau, \boldsymbol{\sigma}) \mathcal{H}_{(\mu)}(\tau, \boldsymbol{\sigma})$ with $\lambda^{(\mu)}(\tau, \boldsymbol{\sigma})$ being Dirac multipliers. By using

$$ {}^4\eta^{(\mu)(\nu)} = [l^{(\mu)} l^{(\nu)} - z_r^{(\mu)}\, {}^3g^{rs} z_s^{(\nu)}](\tau, \boldsymbol{\sigma}), $$

we can write

$$ \begin{aligned} \lambda_{(\mu)}(\tau, \boldsymbol{\sigma}) \mathcal{H}^{(\mu)}(\tau, \boldsymbol{\sigma}) &= [(\lambda_{(\mu)} l^{(\mu)})(l_{(\nu)} \mathcal{H}^{(\nu)}) - (\lambda_{(\mu)} z_r^{(\mu)})({}^3g^{rs} z_{s(\nu)} \mathcal{H}^{(\nu)})](\tau, \boldsymbol{\sigma}) \\ &\stackrel{\text{def}}{=} N_{(\text{flat})}(\tau, \boldsymbol{\sigma})(l_{(\mu)} \mathcal{H}^{(\mu)})(\tau, \boldsymbol{\sigma}) \\ &\quad - N_{(\text{flat})r}(\tau, \boldsymbol{\sigma})({}^3g^{rs} z_{s(\nu)} \mathcal{H}^{(\nu)})(\tau, \boldsymbol{\sigma}) \end{aligned} $$

with the (non-holonomic form of the) constraints

$$ (l_{(\mu)} \mathcal{H}^{(\mu)})(\tau, \boldsymbol{\sigma}) \approx 0, \qquad ({}^3g^{rs} z_{s(\mu)} \mathcal{H}^{(\mu)})(\tau, \boldsymbol{\sigma}) \approx 0, $$

satisfying the universal Dirac algebra. In this way we have defined new flat lapse and shift functions

$$ \begin{aligned} N_{(\text{flat})}(\tau, \boldsymbol{\sigma}) &= \lambda_{(\mu)}(\tau, \boldsymbol{\sigma}) l^{(\mu)}(\tau, \boldsymbol{\sigma}), \\ N_{(\text{flat})r}(\tau, \boldsymbol{\sigma}) &= \lambda_{(\mu)}(\tau, \boldsymbol{\sigma}) z_r^{(\mu)}(\tau, \boldsymbol{\sigma}), \end{aligned} \tag{1} $$

which have the same content as the arbitrary Dirac multipliers $\lambda_{(\mu)}(\tau, \boldsymbol{\sigma})$, namely, they multiply primary first class constraints satisfying the Dirac algebra. In Minkowski spacetime they are quite distinct from the previous lapse and shift functions $N_{[z](\mathrm{flat})}$, $N_{[z](\mathrm{flat})r}$, defined starting from the metric. Instead in general relativity the lapse and shift functions defined starting from the 4-metric are also the coefficients (in the canonical part of the Hamiltonian) of secondary first class constraints satisfying the Dirac algebra.

In special relativity, it is convenient to restrict ourselves to arbitrary spacelike hyperplanes $z^{(\mu)}(\tau, \boldsymbol{\sigma}) = x_s^{(\mu)}(\tau) + b_r^{(\mu)}(\tau)\sigma^r$. Since they are described by only 10 variables [an origin $x_s^{(\mu)}(\tau)$ and, on it, three orthogonal spacelike unit vectors generating the fixed constant timelike unit normal to the hyperplane], we still only have 10 first class constraints determining the 10 variables conjugate to the hyperplane [they are a 4-momentum $p_s^{(\mu)}$ and the six independent degrees of freedom hidden in a spin tensor $S_s^{(\mu)(\nu)}$; with these 20 canonical variables it is possible to build 10 Poincaré generators $p_s^{(\mu)}$, $J_s^{(\mu)(\nu)} = x_s^{(\mu)} p_s^{(\nu)} - x_s^{(\nu)} p_s^{(\mu)} + S_s^{(\mu)(\nu)}$] in terms of the variables of the system: $\tilde{\mathcal{H}}^{(\mu)}(\tau) = p_s^{(\mu)} - p_{(\mathrm{sys})}^{(\mu)} \approx 0$, $\tilde{\mathcal{H}}^{(\mu)(\nu)}(\tau) = S_s^{(\mu)(\nu)} - S_{(\mathrm{sys})}^{(\mu)(\nu)} \approx 0$.

After the restriction to spacelike hyperplanes this element of the Dirac Hamiltonian is reduced to $\tilde{\lambda}^{(\mu)}(\tau)\tilde{\mathcal{H}}_{(\mu)}(\tau) - \frac{1}{2}\tilde{\lambda}^{(\mu)(\nu)}(\tau)\tilde{\mathcal{H}}_{(\mu)(\nu)}(\tau)$. Since at this stage we have $z_r^{(\mu)}(\tau, \boldsymbol{\sigma}) \approx b_r^{(\mu)}(\tau)$ so that

$$
\begin{aligned}
z_\tau^{(\mu)}(\tau, \boldsymbol{\sigma}) &\approx N_{[z](\mathrm{flat})}(\tau, \boldsymbol{\sigma}) l^{(\mu)}(\tau, \boldsymbol{\sigma}) + N^r_{[z](\mathrm{flat})}(\tau, \boldsymbol{\sigma}) \\
b_r^{(\mu)}(\tau, \boldsymbol{\sigma}) &\approx \dot{x}_s^{(\mu)}(\tau) + \dot{b}_r^{(\mu)}(\tau)\sigma^r = -\tilde{\lambda}^{(\mu)}(\tau) - \tilde{\lambda}^{(\mu)(\nu)}(\tau) b_{r(\nu)}(\tau)\sigma^r,
\end{aligned}
$$

it is only now that we obtain the coincidence of the two definitions of flat lapse and shift functions:

$$
\begin{aligned}
N_{[z](\mathrm{flat})}(\tau, \boldsymbol{\sigma}) &\approx N_{(\mathrm{flat})}(\tau, \boldsymbol{\sigma}) = -\tilde{\lambda}_{(\mu)}(\tau) l^{(\mu)} - l^{(\mu)} \tilde{\lambda}_{(\mu)(\nu)}(\tau) b_s^{(\nu)}(\tau)\sigma^s, \\
N_{[z](\mathrm{flat})r}(\tau, \boldsymbol{\sigma}) &\approx N_{(\mathrm{flat})}(\tau, \boldsymbol{\sigma}) \\
&= -\tilde{\lambda}_{(\mu)}(\tau) b_r^{(\mu)}(\tau) - b_r^{(\mu)}(\tau)\tilde{\lambda}_{(\mu)(\nu)}(\tau) b_s^{(\nu)}(\tau)\sigma^s.
\end{aligned}
\tag{2}
$$

The 20 variables for the phase space description of a hyperplane are:

i) $x_s^{(\mu)}(\tau)$, $p_s^{(\mu)}$, parametrizing the origin of the family of spacelike hyperplanes. The four constraints $\mathcal{H}^{(\mu)}(\tau, \boldsymbol{\sigma}) \approx 0$ express the facts that:
   a) $p_s^2 \approx M_{(\mathrm{sys})}^2$ [$M_{(\mathrm{sys})}$ is the invariant mass of the system];
   b) $u_s^{(\mu)}(p_s) = p_s^{(\mu)}/p_s^2 \approx$ [the orientation of the four-momentum of the isolated system with respect to an arbitrary given external observer].

   The origin $x_s^{(\mu)}(\tau)$ plays role of a kinematical center of mass for the isolated system and may be interpreted as a decoupled observer with his parametrized clock.

ii) The other six independent pairs of degrees of freedom are contained in $b_A^{(\mu)}(\tau)$ (with $b_\tau^{(\mu)} = l^{(\mu)}$ τ-independent and normal to the hyperplanes) and $S_s^{(\mu)(\nu)} = -S_s^{(\nu)(\mu)}$ with the orthonormality constraints $b_A^{(\mu)}\, {}^4\eta_{(\mu)(\nu)} b_B^{(\nu)} = {}^4\eta_{AB}$.

However, for each configuration of an isolated system, there is a privileged family of hyperplanes (the Wigner hyperplanes orthogonal to $p_s^{(\mu)}$, which exist when it is timelike) corresponding to the intrinsic rest-frame of the isolated system. To get this result, we must boost at rest all the variables with Lorentz indices by using the standard Wigner boost $L^{(\mu)}{}_{(\nu)}(p_s, \overset{\circ}{p}_s)$ for timelike Poincaré orbits, and then add the gauge-fixings $b_r^{(\mu)}(\tau) - L^{(\mu)}{}_r(p_s, \overset{\circ}{p}_s) \approx 0$. Since these gauge-fixings depend on $p_s^{(\mu)}$, the final canonical variables, apart from $p_s^{(\mu)}$ itself, are of 3 types: i) there is a non-covariant center-of-mass variable $\tilde{x}^{(\mu)}(\tau)$ [it is only covariant under the little group of timelike Poincaré orbits like the Newton–Wigner position operator]; ii) all the 3-vector variables become Wigner spin 1 3-vectors [boosts in M^4 induce Wigner rotations on them];
iii) all the other variables are Lorentz scalars.

One obtains in this way a new kind of instant form of the dynamics (see [13]), the "Wigner-covariant 1-time rest-frame instant form" [16] with a universal breaking of Lorentz covariance. It is the special relativistic generalization of the nonrelativistic separation of the center of mass from the relative motion [$H = \frac{\boldsymbol{P}^2}{2M} + H_{rel}$]. The role of the center of mass is taken by the Wigner hyperplane, identified by the point $\tilde{x}^{(\mu)}(\tau)$ and by its normal $p_s^{(\mu)}$.

The only surviving four first class constraints can be put into the following form:

i) the vanishing of the total (Wigner spin 1) 3-momentum of the system $\boldsymbol{p}_{\text{sys}} \approx 0$ [with $\boldsymbol{p}_{\text{sys}}$ = [the intrinsic three-momentum of the isolated system inside the Wigner hyperplane]]: instead of putting a restriction on $u_s^{(\mu)}(p_s)$ they say that the Wigner hyperplane $\Sigma_{W\,\tau}$ is the intrinsic rest-frame [instead, $\boldsymbol{p}_s$ is left arbitrary, since p_s^{μ} depends upon the orientation of the Wigner hyperplane with respect to arbitrary reference frames in M^4];
ii) $\pm\sqrt{p_s^2} - M_{(\text{sys})} \approx 0$, expressing the fact that the invariant mass $M_{(\text{sys})}$ of the system replaces the nonrelativistic Hamiltonian H_{rel} for the relative degrees of freedom, after the addition of the gauge-fixing $T_s - \tau \approx 0$ [identifying the time parameter τ, labelling the leaves of the foliation, with the Lorentz scalar time of the center of mass in the rest-frame, $T_s = p_s \cdot \tilde{x}_s / M_{(\text{sys})}$; $M_{(\text{sys})}$ generates the evolution in this time].

In this special gauge, 3 degrees of freedom of the isolated system [a 3-center-of-mass (Wigner spin 1) variable $\boldsymbol{x}_{(\text{sys})}$ defined inside the Wigner hyperplane and conjugate to $\boldsymbol{p}_{(\text{sys})}$] become gauge variables [the natural gauge-fixing for $\boldsymbol{p}_{(\text{sys})} \approx 0$ is $\boldsymbol{x}_{(\text{sys})} \approx 0$, so that it coincides with the origin $\boldsymbol{\sigma} = 0$ of the Wigner hyperplane], while $\tilde{x}^{(\mu)}$ describes a physical decoupled observer. For N free particles [16] one has $\boldsymbol{x}_{(\text{sys})} = \boldsymbol{\eta}_+(\tau) = \sum_{i=1}^N \boldsymbol{\eta}_i(\tau)$. After the gauge-fixing $\boldsymbol{\eta}_+(\tau) \approx 0$ we are left only with Newtonian-like degrees of freedom with rotational covariance: i) a 3-coordinate (not Lorentz covariant) $\boldsymbol{z}_s = \sqrt{p_s^2}(\tilde{\boldsymbol{x}}_s - \frac{\boldsymbol{p}_s}{p_s^o}\tilde{x}^o)$ and its conjugate momentum $\boldsymbol{k}_s = \boldsymbol{p}_s/\sqrt{p_s^2}$ for the decoupled center of mass in M^4; ii) a set of relative conjugate pairs of variables with Wigner covariance inside the Wigner hyperplane. When fields are present, one needs to find a rest-frame canonical basis of center-of-mass and relative variables for fields (in analogy with particles) to identify $\boldsymbol{x}_{(sys)}$. Such a basis has already been found for a real Klein–Gordon field [14] and it is under reformulation on spacelike hypersurfaces [15].

The isolated systems analyzed to date for their rest-frame Wigner-covariant generalized Coulomb gauges [i.e., the subset of global Shanmugadhasan canonical bases, which, for each Poincaré stratum, are also adapted to the geometry of the corresponding Poincaré orbits with their little groups; these special bases can be named Poincaré–Shanmugadhasan bases for the given Poincaré stratum of the presymplectic constraint manifold (every stratum requires an independent canonical reduction); at present only the main stratum with P^2 timelike and $W^2 \neq 0$ has been investigated] are:

a) The system of N scalar particles with Grassmann electric charges plus the electromagnetic field [16]. The starting configuration variables are a 3-vector $\boldsymbol{\eta}_i(\tau)$ for each particle $[x_i^{(\mu)}(\tau) = z^{(\mu)}(\tau, \boldsymbol{\eta}_i(\tau))]$ and the electromagnetic gauge potentials $A_A(\tau, \boldsymbol{\sigma}) = \frac{\partial z^{(\mu)}(\tau, \boldsymbol{\sigma})}{\partial \sigma^A} A_{(\mu)}(z(\tau, \boldsymbol{\sigma}))$, which know the embedding of Σ_τ into M^4. One has to choose the sign of the energy of each particle, because there are no mass-shell constraints (like $p_i^2 - m_i^2 \approx 0$) among the constraints of this formulation, due to the fact that one has only three degrees of freedom for particles, determining the intersection of a timelike trajectory and of the spacelike hypersurface Σ_τ. For each choice of sign of the energy of the N particles, one describes only one of the branches of the mass spectrum of the manifestly covariant approach based on the coordinates $x_i^{(\mu)}(\tau)$, $p_i^{(\mu)}(\tau)$, $i = 1, \cdot, N$, and on the constraints $p_i^2 - m_i^2 \approx 0$ (in the free case). In this way, one gets a description of relativistic particles with a given sign of the energy with consistent couplings to fields and valid independently of the quantum effect of pair production [in the manifestly covariant approach, containing all possible branches of the particle mass spectrum, the classical counterpart of pair production is the intersection of different branches deformed by the presence of fields]. The final Dirac observables are:

i) the transverse radiation field variables $\boldsymbol{A}_\perp$, $\boldsymbol{E}_\perp$;
ii) the particle canonical variables $\boldsymbol{\eta}_i(\tau)$, $\boldsymbol{\kappa}_i(\tau)$, dressed with a Coulomb cloud.

The physical Hamiltonian contains the Coulomb potentials extracted from field theory and there is a regularization of the Coulomb self-energies due to the Grassmann character of the electric charges Q_i $[Q_i^2 = 0]$. The no-radiation conditions $\boldsymbol{A}_\perp = \boldsymbol{E}_\perp = 0$ identify an approximate (delay is neglected) canonical subspace containing only physical charged particles with mutual instantaneous Coulomb potentials. In [17] a study is made of the Lienard–Wiechert potentials and of the Abraham–Lorentz–Dirac equations in this rest-frame Coulomb gauge; scalar electrodynamics also is reformulated in it. In addition rest-frame 1-time relativistic statistical mechanics has been developed [16].

b) The system of N scalar particles with Grassmann-valued color charges plus the color SU(3) Yang–Mills field [18]. This gives the pseudoclassical description of the relativistic scalar-quark model, deduced from the classical QCD Lagrangian and with the color field present. The physical invariant mass of the system is given in terms of the Dirac observables. From the reduced Hamilton equations the second order equations of motion for both the reduced transverse color field and the particles, are extracted. Then, the $N = 2$ (meson) case is studied. A special form of the requirement of having only color singlets, suited for a field-independent quark model, produces a "pseudoclassical asymptotic freedom" and a regularization of the quark self-energy. With these results one can make covariant the bosonic part of the standard model given in [11].

c) The system of N spinning particles of definite energy [$(\frac{1}{2}, 0)$ or $(0, \frac{1}{2})$ representation of $SL(2, C)$] with Grassmann electric charges plus the electromagnetic field [19] and that of a Grassmann-valued Dirac field plus the electromagnetic field (the pseudoclassical basis of QED) [20]. In both cases there are geometrical complications connected with the spacetime description of the path of electric currents and not only of their spin structure, which suggests a reinterpretation of the supersymmetric scalar multiplet as a spin fibration with the Dirac field in the fiber and the Klein–Gordon field in the base; a new canonical decomposition of the Klein–Gordon field into center-of-mass and relative variables [14, 15] would be helpful in clarifying these problems. With their solution and with a description of Grassmann-valued chiral fields [this will require the transcription of the front form of the dynamics in the instant form for the Poincaré strata with $P^2 = 0$] the rest-frame form of the full standard $SU(3) \times SU(2) \times U(1)$ model can be achieved.

All these new pieces of information will allow one, after quantization of this new consistent relativistic mechanics without the classical problems connected with pair production, to find the asymptotic states of the covariant Tomonaga–Schwinger formulation of quantum field theory on spacelike hypersurfaces: these states are needed for the theory of quantum bound states [since Fock states do not constitute a Cauchy problem for the field equations, because an in (or out) particle can be in the absolute future of another one due to the tensor product nature of these asymptotic states, bound state equations like those of Bethe–Salpeter have spurious solutions which are excitations in relative energies, the variables being conjugate to relative times (which are gauge variables [16])]. Moreover, it will be possible to include bound states among the asymptotic states.

As stated in [17, 18], the quantization of these rest-frame models has to overcome two problems. On the particle side, the complication is the quantization of the square roots associated with the relativistic kinetic energy terms: in the free case this has been done in [21] [see [22] for the complications induced by the Coulomb potential]. On the field side (all physical Hamiltonians are nonlocal and, with the exception of the Abelian case, nonpolynomial, but quadratic in the momenta), the obstacle is the absence (notwithstanding the fact that there is no no-go theorem) of a complete regularization and renormalization procedure of electrodynamics (to start with) in the Coulomb gauge: see Chap. 9 of [23] (and its bibliography) for existing results for QED.

However, as shown in [16, 9], the rest-frame instant form of dynamics automatically gives a physical ultraviolet cutoff in the spirit of Dirac and Yukawa: it is the Møller radius [24] $\rho = \sqrt{-W^2 c/P^2} = |S|c/\sqrt{P^2}$ ($W^2 = -P^2 S^2$ is the Pauli-

Lubanski Casimir when $P^2 > 0$), namely, the classical intrinsic radius of the worldtube, around the covariant noncanonical Fokker-Pryce center of inertia Y^μ, inside which the noncovariance of the canonical center of mass $\tilde{x}^\mu$ is concentrated. At the quantum level ρ becomes the Compton wavelength of the isolated system multiplied by its spin eigenvalue $\sqrt{s(s+1)}$, $\rho \mapsto \hat{\rho} = \sqrt{s(s+1)}\hbar/M = \sqrt{s(s+1)}\lambda_M$ with $M = \sqrt{P^2}$ the invariant mass and $\lambda_M = \hbar/M$ its Compton wavelength. Therefore, the criticism of classical relativistic physics, based on quantum pair production, concerns the testing of distances where, due to the Lorentz signature of spacetime, one has intrinsic classical covariance problems: it is impossible to localize the canonical center of mass $\tilde{x}^\mu$ adapted to the first class constraints of the system (also called the Pryce center of mass and having the same covariance of the Newton–Wigner position operator) in a frame independent way.

Recall [16] that ρ is also a remnant in flat Minkowski spacetime of the energy conditions of general relativity: since the Møller noncanonical, noncovariant center of energy has its noncovariance localized inside the same worldtube with radius ρ (it was discovered in this way) [24], it turns out that, for an extended relativistic system with the material radius smaller than its intrinsic radius ρ, one has: i) its peripheral rotation velocity can exceed the velocity of light; ii) its classical energy density cannot be positive definite everywhere in every frame.

Now, the really relevant point is that this ultraviolet cutoff determined by ρ also exists in Einstein's general relativity (which is not power counting renormalizable) in the case of asymptotically flat spacetimes, taking into account the Poincaré Casimirs of its asymptotic ADM Poincaré charges (when supertranslations are eliminated with suitable boundary conditions; note that Einstein and Wheeler use closed universes because they do not want to introduce boundary conditions, but in this way they loose Poincaré charges and also the possibility of making contact with particle physics and of defining spin). The generalization of the worldtube of radius ρ to asymptotically flat general relativity with matter, could be connected with the unproved cosmic censorship hypothesis.

Moreover, the extended Heisenberg relations of string theory [25], i.e., $\Delta x = \frac{\hbar}{\Delta p} + \frac{\Delta p}{T_{cs}} = \frac{\hbar}{\Delta p} + \frac{\hbar \Delta p}{L_{cs}^2}$ implying the lower bound $\Delta x > L_{cs} = \sqrt{\hbar/T_{cs}}$ due to the $y + 1/y$ structure, have a counterpart in the quantization of the Møller radius [16]: if we ask that, at the quantum level too, one cannot test the inside of the worldtube, we must ask that$\Delta x > \hat{\rho}$ which is the lower bound implied by the modified uncertainty relation $\Delta x = \frac{\hbar}{\Delta p} + \frac{\hbar \Delta p}{\hat{\rho}^2}$. This could imply that the center-of-mass canonical noncovariant 3-coordinate $z = \sqrt{P^2}(\tilde{\boldsymbol{x}} - \frac{\boldsymbol{P}}{P^o}\tilde{x}^o)$ [16] cannot become a self-adjoint operator. See Hegerfeldt's theorems (quoted in [9, 16]) and his interpretation suggesting the impossibility of a good localization of relativistic particles (experimentally one determines only a worldtube in spacetime emerging from the interaction region). Since the eigenfunctions of the canonical center-of-mass operator are playing the role of the wave function of the universe, one could also say that the center-of-mass variable is not to be quantized, because it lies on the classical macroscopic side of the Copenhagen interpretation and, moreover, because, in the spirit of Mach's principle that only relative motions can be observed, no one can observe it (it is only used to define a decoupled "point particle clock"). On the other hand, if one rejects the canonical noncovariant center of mass in favor of the covariant noncanonical Fokker–Pryce center of inertia Y^μ, $\{Y^\mu, Y^\nu\} \neq 0$, one could invoke the philosophy of quantum groups to quantize Y^μ to obtain a kind of quantum plane for the center-of-mass description. Note that the quantization of the square root Hamiltonian done in [21] is consistent with this discussion.

In conclusion, the best set of canonical coordinates adapted to the constraints and to the geometry of Poincaré orbits in Minkowski spacetime and naturally predisposed to the coupling to canonical tetrad gravity, is emerging for the electromagnetic, weak and strong interactions with matter described either by fermion fields or by relativistic particles with a definite sign of the energy.

2 Canonical reduction of tetrad gravity, the deparametrization problem and the N-body problem

Tetrad gravity is the formulation of general relativity natural for the coupling to fermion fields of the standard model. However, we need a formulation of it which allows us to solve its constraints for doing the canonical reduction and to solve the deparametrization problem of general relativity (how to recover the rest-frame instant form when the Newton constant is put equal to zero, G = 0).

To implement this program we shall restrict ourselves to the simplest class of spacetimes [time-oriented pseudo-Riemannian or Lorentzian 4-manifold $(M^4, {}^4g)$ with signature $\epsilon\,(+---)$ $(\epsilon = \pm 1)$ and with a choice of time orientation], assumed to be:

i) Globally hyperbolic 4-manifolds, i.e., topologically they are $M^4 = R \times \Sigma$, and so a well-having posed Cauchy problem [with Σ the abstract model of the Cauchy surface] at least until no singularity develops in M^4 [see the singularity theorems]. It follows that, these spacetimes admit regular foliations with orientable, complete, non-intersecting spacelike 3-manifolds Σ_τ [$\tau : M^4 \to R$, $z^\mu \mapsto \tau(z^\mu)$, is a global timelike future-oriented function labelling the leaves (surfaces of simultaneity)]. In this way, one obtains 3+1 splittings of M^4 and the possibility of a Hamiltonian formulation.
ii) Asymptotically flat at spatial infinity, and so giving the possibility of defining asymptotic Poincaré charges [26–31]: they allow the definition of a Møller radius in general relativity as well and are a bridge towards a future welding together with the theory of elementary particles in Minkowski spacetime defined as irreducible representations of its kinematical, globally implemented Poincaré group according to Wigner.
iii) Admitting a spinor (or spin) structure [32] for the coupling to fermion fields. Since we consider noncompact space- and time-orientable spacetimes, spinors can be defined if and only if they are "parallelizable" [33], as in our case. This implies that the orthonormal frame principal SO(3)-bundle over Σ_τ (whose connections are the spin connections determined by the cotriads) is trivial.

iv) Having its noncompact parallelizable simultaneity 3-manifolds (the Cauchy surfaces) Σ_τ topologically trivial, geodesically complete and, finally, diffeomorphic to R^3. These 3-manifolds have the same manifold structure as Euclidean spaces: a) the geodesic exponential map $Exp_p : T_p\Sigma_\tau \rightarrow \Sigma_\tau$ is a diffeomorphism ; b) the sectional curvature is less than or equal to zero everywhere; c) they have no "conjugate locus" [i.e., there are no pairs of conjugate Jacobi points (intersection points of distinct geodesics through them) on any geodesic] and no "cut locus" [i.e., no closed geodesics through any point].
v) Having, as in the Yang–Mills case [9], the 3-spin-connection on the orthogonal frame SO(3)-bundle (and therefore cotriads) restricted to suitable weighted Sobolev spaces to avoid Gribov ambiguities [9, 34]. In turn, this implies the absence of isometries of the noncompact Riemannian 3-manifold $(\Sigma_\tau, {}^3g)$ [see, for instance, the review paper in [35]].

Diffeomorphisms on Σ_τ (Diff Σ_τ) are interpreted in the passive sense, following [36], in accord with the Hamiltonian point of view that infinitesimal diffeomorphisms are generated by taking the Poisson bracket with the first class supermomentum constraints [passive diffeomorphisms are also named 'pseudodiffeomorphisms'].

The new formulation of tetrad gravity [see [37–47] for the existing versions of the theory] utilizes the ADM action of metric gravity with the 4-metric expressed in terms of arbitrary cotetrads. Note that, in both the ADM metric and the tetrad formulation, one has to introduce an extra ingredient of the 3+1 splittings of M^4 with foliations whose leaves Σ_τ are spacelike 3-hypersurfaces. However, their points $z^\mu(\tau, \boldsymbol{\sigma})$ [$(\tau, \boldsymbol{\sigma})$ are Σ_τ-adapted holonomic coordinates of M^4] are not configurational variables of these theories in contrast to what happens in Minkowski parametrized theories [$\frac{\partial z^\mu}{\partial \sigma^A}$ are not tetrads when M^4 is not Minkowski spacetime with Cartesian coordinates, because ${}^4g^{AB} \frac{\partial z^\mu}{\partial \sigma^A} \frac{\partial z^\nu}{\partial \sigma^B} = {}^4g^{\mu\nu} \neq {}^4\eta^{(\mu)(\nu)}$].

By using Σ_τ-adapted holonomic coordinates for M^4, one has found a new parametrization of arbitrary tetrads and cotetrads on M^4 in terms of cotriads on Σ_τ [${}^3e_{(a)r}(\tau, \boldsymbol{\sigma})$], of lapse [$N(\tau, \boldsymbol{\sigma})$] and shift [$N_{(a)}(\tau, \boldsymbol{\sigma}) = \{{}^3e_{(a)r} N^r\}(\tau, \boldsymbol{\sigma})$] functions and of 3 parameters [$\varphi_{(a)}(\tau, \boldsymbol{\sigma})$] parametrizing point-dependent Wigner boosts for timelike Poincaré orbits. Putting these variables in the ADM action for metric gravity [26] (with the 3-metric on Σ_τ expressed in terms of cotriads: ${}^3g_{rs} = {}^3e_{(a)r}\, {}^3e_{(a)s}$ with positive signature), one gets a new action depending only on lapse, shifts and cotriads, but not on the boost parameters (therefore, there is no need to use Schwinger's time gauge). There are 10 primary and 4 secondary first class constraints and a vanishing canonical Hamiltonian. Besides the 3 constraints associated with vanishing Lorentz boost momenta, there are 4 constraints stating that the momenta associated with lapse and shifts vanish, 3 constraints describing rotations, 3 constraints generating space-diffeomorphisms on the cotriads induced by those (Diff Σ_τ) on Σ_τ (a linear combination of supermomentum constraints and of the rotation constraints; a different combination of these constraints generates SO(3) Gauss law constraints for the momenta ${}^3\tilde{\pi}^r_{(a)}$ conjugated to cotriads with the covariant derivative constructed with the spin connection) and one superhamiltonian constraint. It turns out that, with the techniques developed for Yang–Mills theory, one can Abelianize the 3 rotation constraints and then the space-diffeomorphism constraints as well. In the Abelianization of the rotation constraints one needs the Green function of the 3-dimensional covariant derivative containing the spin connection, well-defined only if there is no Gribov ambiguity in the SO(3)-frame bundle and no isometry of the Riemannian 3-manifold $(\Sigma_\tau, {}^3g)$. The Green function is similar to the Yang–Mills function for a principal SO(3)-bundle [9], but, instead of the Dirac distribution for the Green function of the flat divergence, it contains the DeWitt function or bitensor [48] defining the tangent in one endpoint of the geodesic arc connecting two points (which reduces to the Dirac distribution only locally in normal coordinates). Moreover, the definition of the Green function now requires the geodesic exponential map.

In the resulting quasi-Shanmugadhasan canonical basis, the original cotriad can be expressed in closed form in terms of 3 rotation angles, 3 diffeomorphism-parameters and a reduced cotriad depending only on 3 independent variables (they are Dirac's observables with respect to 13 of the 14 first class constraints) and with their conjugate momenta, still subject to the reduced form of the superhamiltonian constrain: this is the phase space over the superspace of 3-geometries [49].

So far no coordinate condition [50] has been imposed. It turns out that these conditions are hidden in the choice of how to parametrize the reduced cotriads in terms of three independent functions. The simplest parametrization (the only one studied so far) corresponds to choosing a system of global 3-orthogonal coordinates on Σ_τ, in which the 3-metric is diagonal. With a further canonical transformation on the reduced cotriads and conjugate momenta, one arrives at a canonical basis containing the conformal factor $\phi(\tau, \boldsymbol{\sigma}) = e^{q(\tau,\boldsymbol{\sigma})/2}$ of the 3-geometry and its conjugate momentum $\rho(\tau, \boldsymbol{\sigma})$ plus two other pairs of conjugate canonical variables $r_{\bar{a}}(\tau, \boldsymbol{\sigma})$, $\pi_{\bar{a}}(\tau, \boldsymbol{\sigma})$, $\bar{a} = 1, 2$. The reduced superhamiltonian constraint, expressed in terms of these variables, turns out to be an integral equation for the conformal factor (reduced Lichnerowicz equation) whose conjugate momentum is, therefore, the last gauge variable. If we replace the gauge fixing of the Lichnerowicz [51] and York [52–54] approach [namely, the vanishing of the trace of the extrinsic curvature of Σ_τ, ${}^3K(\tau, \boldsymbol{\sigma}) \approx 0$, also called the internal extrinsic York time [55]] with the natural one $\rho(\tau, \boldsymbol{\sigma}) \approx 0$ and we pass to Dirac brackets, we find that $r_{\bar{a}}(\tau, \boldsymbol{\sigma})$, $\pi_{\bar{a}}(\tau, \boldsymbol{\sigma})$ give the canonical basis for the physical degrees of freedom or Dirac's observables of the gravitational field.

Since we then have a physical canonical basis, it is possible to define "void spacetimes" as the equivalence class of spacetimes "without gravitational field", whose members in the 3-orthogonal gauge are obtained by adding by hand the second class constraints $r_{\bar{a}}(\tau, \boldsymbol{\sigma}) \approx 0, \pi_{\bar{a}}(\tau, \boldsymbol{\sigma}) \approx 0$ [one gets $\phi(\tau, \boldsymbol{\sigma}) = 1$ as the relevant solution of the reduced Lichnerowicz equation]. The members of this equivalence class (the extension to general relativity of the Galilean non-inertial coordinate systems with their Newtonian inertial forces) are in general "not flat" [their Σ_τ's are 3-conformally flat] but gauge equivalent to Minkowski spacetime with Cartesian coordinates (this holds in the absence of matter).

The next step is to find their physical Hamiltonian and to solve the deparametrization problem. If we wish to arrive at the welding togething of tetrad gravity with matter and the parametrized Minkowski formulation for the same matter, we must require that the lapse and shift functions of tetrad gravity [which must grow linearly in $\boldsymbol{\sigma}$, in suitable asymptotic Minkowski coordinates, according to the existing literature on asymptotic Poincaré charges at spatial infinity [27, 28]] agree asymptotically with the flat lapse and shift functions, which, however, are unambigously defined only on Minkowski spacelike hyperplanes.

In metric ADM gravity the canonical Hamiltonian is $H_{(c)\mathrm{ADM}} = \int d^3\sigma[N\tilde{\mathcal{H}} + N_r\tilde{\mathcal{H}}^r](\tau,\boldsymbol{\sigma}) \approx 0$, where $\tilde{\mathcal{H}}(\tau,\boldsymbol{\sigma}) \approx 0$ and $\tilde{\mathcal{H}}^r(\tau,\boldsymbol{\sigma}) \approx 0$ are the superhamiltonian and supermomentum constraints. It is differentiable and finite only for suitable $N(\tau,\boldsymbol{\sigma}) = n(\tau,\boldsymbol{\sigma}) \rightarrow_{|\boldsymbol{\sigma}|\rightarrow\infty} 0$, $N_r(\tau,\boldsymbol{\sigma}) = n_r(\tau,\boldsymbol{\sigma}) \rightarrow_{|\boldsymbol{\sigma}|\rightarrow\infty} 0$ defined in [28] in suitable asymptotic coordinate systems. For more general lapse and shift functions one must add a surface term [49] to $H_{(c)\mathrm{ADM}}$, which contains the "strong" Poincaré charges [26] P^A_{ADM}, J^{AB}_{ADM} [they are conserved and gauge invariant surface integrals]. To have well defined asymptotic Poincaré charges at spatial infinity [26–29] one needs: i) the selection of a class of coordinate systems for Σ_τ asymptotic to flat coordinates; ii) the choice of a class of Hamiltonian boundary conditions in these coordinate systems [all the fields must belong to some functional space of the same type as that of the weighted Sobolev spaces]; iii) a definition of the Hamiltonian group $\mathcal{G}$ of gauge transformations (and, in particular, of proper gauge transformations) with a well-defined limit at spatial infinity so as to respect i) and ii). The scheme is the same needed to define the non-Abelian charges in Yang–Mills theory [9]. The delicate point is to be able to exclude supertranslations [32], because the presence of these extra asymptotic charges leads to the replacement of the asymptotic Poincaré group by the infinite-dimensional spi group [31], which does not allow the definition of the Poincaré spin due to the absence of the Pauli-Lubanski Casimir. This can be done with suitable boundary conditions (in particular all the fields and gauge transformations must have direction independent limits at spatial infinity) respecting the "parity conditions" of [28] [see also [29]].

We remark that in [56] and in the book in [5] (see also [27]), Dirac introduced asymptotic Minkowski rectangular coordinates $z^{(\mu)}_{(\infty)}(\tau,\boldsymbol{\sigma}) = x^{(\mu)}_{(\infty)}(\tau) + b^{(\mu)}_{(\infty)\,r}(\tau)\sigma^r$ in M^4 at spatial infinity $S_\infty = \cup_\tau S^2_{\tau,\infty}$. For each value of τ, the coordinates $x^{(\mu)}_{(\infty)}(\tau)$ label a point near spatial infinity, chosen as origin. On it there is a flat tetrad $b^{(\mu)}_{(\infty)\,A}(\tau) = \{l^{(\mu)}_{(\infty)} = b^{(\mu)}_{(\infty)\,\tau} = \epsilon^{(\mu)}{}_{(\alpha)(\beta)(\gamma)} b^{(\alpha)}_{(\infty)\,1}(\tau) b^{(\beta)}_{(\infty)\,2}(\tau) b^{(\gamma)}_{(\infty)\,3}(\tau);\ b^{(\mu)}_{(\infty)\,r}(\tau)\}$, with $l^{(\mu)}_{(\infty)}$ τ-independent, satisfying $b^{(\mu)}_{(\infty)\,A}\, {}^4\eta_{(\mu)(\nu)}\, b^{(\nu)}_{(\infty)\,B} = {}^4\eta_{AB}$ for every τ [at this level we do not assume that $l^{(\mu)}_{(\infty)}$ is tangent to S_∞, as the normal l^μ to Σ_τ]. There will be transformation coefficients $b^\mu_A(\tau,\boldsymbol{\sigma})$ from the holonomic adapted coordinates $\sigma^A = (\tau,\sigma^r)$ to coordinates $x^\mu = z^\mu(\sigma^A)$ in an atlas of M^4 such that, in a chart at spatial infinity, one has $z^\mu(\tau,\boldsymbol{\sigma}) = \delta^\mu_{(\mu)} z^{(\mu)}(\tau,\boldsymbol{\sigma})$ and $b^\mu_A(\tau,\boldsymbol{\sigma}) = \delta^\mu_{(\mu)} b^{(\mu)}_{(\infty)A}(\tau)$ [for $r \rightarrow \infty$ one has ${}^4g_{\mu\nu} \rightarrow \delta^{(\mu)}_\mu \delta^{(\nu)}_\nu {}^4\eta_{(\mu)(\nu)}$ and ${}^4g_{AB} = b^\mu_A\, {}^4g_{\mu\nu} b^\nu_B \rightarrow b^{(\mu)}_{(\infty)A}\, {}^4\eta_{(\mu)(\nu)} b^{(\nu)}_{(\infty)B} = {}^4\eta_{AB}$].

Dirac [56] and, then, Regge and Teitelboim [27] proposed that the asymptotic Minkowski rectangular coordinates $z^{(\mu)}_{(\infty)}(\tau,\boldsymbol{\sigma}) = x^{(\mu)}_{(\infty)}(\tau) + b^{(\mu)}_{(\infty)r}(\tau)\sigma^r$ should define 10 new independent degrees of freedom at the spatial boundary S_∞, as happens for Minkowski parametrized theories [16] when restricted to spacelike hyperplanes [defined by $z^{(\mu)}(\tau,\boldsymbol{\sigma}) \approx x^{(\mu)}_s(\tau) + b^{(\mu)}_r(\tau)\sigma^r$]; then, 10 conjugate momenta should exist. These 20 extra variables of the Dirac proposal can be put in the form: $x^{(\mu)}_{(\infty)}(\tau)$, $p^{(\mu)}_{(\infty)}$, $b^{(\mu)}_{(\infty)A}(\tau)$ [with $b^{(\mu)}_{(\infty)\tau} = l^{(\mu)}_{(\infty)}$ τ-independent], $S^{(\mu)(\nu)}_{(\infty)}$, with Dirac brackets implying the orthonormality constraints $b^{(\mu)}_{(\infty)A}\,{}^4\eta_{(\mu)(\nu)}b^{(\nu)}_{(\infty)B} = {}^4\eta_{AB}$ [so that $p^{(\mu)}_{(\infty)}$ and $J^{(\mu)(\nu)}_{(\infty)} = x^{(\mu)}_{(\infty)}p^{(\nu)}_{(\infty)} - x^{(\nu)}_{(\infty)}p^{(\mu)}_{(\infty)} + S^{(\mu)(\nu)}_{(\infty)}$ satisfy a Poincaré algebra]. In analogy with Minkowski parametrized theories restricted to spacelike hyperplanes, one expects to have 10 extra first class constraints of the type $p^{(\mu)}_{(\infty)} - P^{(\mu)}_{\rm ADM} \approx 0$, $S^{(\mu)(\nu)}_{(\infty)} - S^{(\mu)(\nu)}_{\rm ADM} \approx 0$ with $P^{(\mu)}_{\rm ADM}$, $S^{(\mu)(\nu)}_{\rm ADM}$ related to the ADM Poincaré charges $P^A_{\rm ADM}$, $J^{AB}_{\rm ADM}$. The origin $x^{(\mu)}_{(\infty)}$ will play the role of a decoupled observer with his parametrized clock.

Note that, if we replace $p^{(\mu)}_{(\infty)}$ and $S^{(\mu)(\nu)}_{(\infty)}$, whose Poisson algebra is the direct sum of an Abelian algebra of translations and a Lorentz algebra, with the new variables (with holonomic indices with respect to Σ_τ) $p^A_{(\infty)} = b^A_{(\infty)(\mu)}p^{(\mu)}_{(\infty)}$, $J^{AB}_{(\infty)} = b^A_{(\infty)(\mu)}b^B_{(\infty)(\nu)}S^{(\mu)(\nu)}_{(\infty)}$ [$\neq b^A_{(\infty)(\mu)}b^B_{(\infty)(\nu)}J^{(\mu)(\nu)}_{(\infty)}$], the Poisson brackets for $p^{(\mu)}_{(\infty)}$, $b^{(\mu)}_{(\infty)A}$, $S^{(\mu)(\nu)}_{(\infty)}$ imply that $p^A_{(\infty)}$, $J^{AB}_{(\infty)}$ satisfy a Poincaré algebra. This implies that the Poincaré generators $P^A_{\rm ADM}$, $J^{AB}_{\rm ADM}$ define in the asymptotic Dirac rectangular coordinates a momentum $P^{(\mu)}_{\rm ADM}$ and only an ADM spin tensor $S^{(\mu)(\nu)}_{\rm ADM}$ [to define an angular momentum tensor $J^{(\mu)(\nu)}_{\rm ADM}$ one should find a "center of mass of the gravitational field" $X^{(\mu)}_{\rm ADM}[{}^3g, {}^3\tilde{\Pi}]$ (see [14] for the Klein–Gordon case) conjugate to $P^{(\mu)}_{\rm ADM}$, so that $J^{(\mu)(\nu)}_{\rm ADM} = X^{(\mu)}_{\rm ADM}P^{(\nu)}_{\rm ADM} - X^{(\nu)}_{\rm ADM}P^{(\mu)}_{\rm ADM} + S^{(\mu)(\nu)}_{\rm ADM}$].

The following splitting of the lapse and shift functions and the following set of boundary conditions [consistent with those of [57]] fulfill all the previous requirements [welding together with the lapse and shift functions on Minkowski hyperplanes; absence of supertranslations (strictly speaking, one gets $P^r_{\rm ADM} = 0$ as in [57]; how to weaken the condition on ${}^3\tilde{\Pi}^{rs}$ without reintroducing supertranslations is an open problem); $r = |\boldsymbol{\sigma}|$]

$$
\begin{aligned}
&{}^3g_{rs}(\tau,\boldsymbol{\sigma}) \rightarrow_{r\rightarrow\infty} \left(1+\frac{M}{r}\right)\delta_{rs} + {}^3h_{rs}(\tau,\boldsymbol{\sigma}) = \left(1+\frac{M}{r}\right)\delta_{rs} + o_4(r^{-3/2}),\\
&{}^3\tilde{\Pi}^{rs}(\tau,\boldsymbol{\sigma}) \rightarrow_{r\rightarrow\infty} {}^3k^{rs}(\tau,\boldsymbol{\sigma}) = o_3(r^{-5/2}),
\end{aligned}
$$

$$
\begin{aligned}
N(\tau,\boldsymbol{\sigma}) &= N_{(as)}(\tau,\boldsymbol{\sigma}) + n(\tau,\boldsymbol{\sigma}), &\quad n(\tau,\boldsymbol{\sigma}) &= O(r^{-(3+\epsilon)}),\\
N_r(\tau,\boldsymbol{\sigma}) &= N_{(as)r}(\tau,\boldsymbol{\sigma}) + n_r(\tau,\boldsymbol{\sigma}), &\quad n_r(\tau,\boldsymbol{\sigma}) &= O(r^{-\epsilon}),
\end{aligned}
$$

$$
\begin{aligned}
N_{(as)}(\tau,\boldsymbol{\sigma}) &= -\tilde{\lambda}_{(\mu)}(\tau)l^{(\mu)}_{(\infty)} - l^{(\mu)}_{(\infty)}\tilde{\lambda}_{(\mu)(\nu)}(\tau)b^{(\nu)}_{(\infty)s}(\tau)\sigma^s =\\
&= -\tilde{\lambda}_\tau(\tau) - \tilde{\lambda}_{\tau s}(\tau)\sigma^s,
\end{aligned}
$$

$$N_{(as)r}(\tau,\boldsymbol{\sigma}) = -b^{(\mu)}_{(\infty)r}(\tau)\tilde{\lambda}_{(\mu)}(\tau) - b^{(\mu)}_{(\infty)r}(\tau)\tilde{\lambda}_{(\mu)(\nu)}(\tau)b^{(\nu)}_{(\infty)s}(\tau)\sigma^s =$$
$$= -\tilde{\lambda}_r(\tau) - \tilde{\lambda}_{rs}(\tau)\sigma^s,$$
$$\Rightarrow N_{(as)A}(\tau,\boldsymbol{\sigma}) \stackrel{\text{def}}{=} (N_{(as)}\,;\; N_{(as)r})(\tau,\boldsymbol{\sigma}) = -\tilde{\lambda}_A(\tau) - \frac{1}{2}\tilde{\lambda}_{As}(\tau)\sigma^s, \tag{3}$$

with ${}^3h_{rs}(\tau,-\boldsymbol{\sigma}) = {}^3h_{rs}(\tau,\boldsymbol{\sigma})$, ${}^3k^{rs}(\tau,-\boldsymbol{\sigma}) = -{}^3k^{rs}(\tau,\boldsymbol{\sigma})$; here ${}^3\tilde{\Pi}^{rs}(\tau,\boldsymbol{\sigma})$ is the momentum conjugate to the 3-metric ${}^3g_{rs}(\tau,\boldsymbol{\sigma})$ in ADM metric gravity.

After the addition of the surface term, the resulting canonical and Dirac Hamiltonians of ADM metric gravity are

$$\begin{aligned}
H_{(c)\mathrm{ADM}} &= \int d^3\sigma[(N_{(as)}+n)\tilde{\mathcal{H}} + (N_{(as)r}+n_r)\,{}^3\tilde{\mathcal{H}}^r](\tau,\boldsymbol{\sigma}) \mapsto \\
&\mapsto H'_{(c)\mathrm{ADM}} = \int d^3\sigma[(N_{(as)}+n)\tilde{\mathcal{H}} + (N_{(as)r}+n_r)\,{}^3\tilde{\mathcal{H}}^r](\tau,\boldsymbol{\sigma}) + \\
&\quad + \tilde{\lambda}_A(\tau)P^A_{\mathrm{ADM}} + \tilde{\lambda}_{AB}(\tau)J^{AB}_{\mathrm{ADM}} = \\
&= \int d^3\sigma[n\tilde{\mathcal{H}} + n_r\,{}^3\tilde{\mathcal{H}}^r](\tau,\boldsymbol{\sigma}) + \tilde{\lambda}_A(\tau)\hat{P}^A_{\mathrm{ADM}} + \tilde{\lambda}_{AB}(\tau)\hat{J}^{AB}_{\mathrm{ADM}} \approx \\
&\approx \tilde{\lambda}_A(\tau)\hat{P}^A_{\mathrm{ADM}} + \tilde{\lambda}_{AB}(\tau)\hat{J}^{AB}_{\mathrm{ADM}}, \\
H'_{(D)\mathrm{ADM}} &= H'_{(c)ADM} + \int d^3\sigma[\lambda_n\tilde{\pi}^n + \lambda^{\boldsymbol{n}}_r\tilde{\pi}^r_{\boldsymbol{n}}](\tau,\boldsymbol{\sigma}) + \\
&\quad + \zeta_A(\tau)\tilde{\pi}^A(\tau) + \zeta_{AB}(\tau)\tilde{\pi}^{AB}(\tau),
\end{aligned} \tag{4}$$

with the "weak conserved improper charges" $\hat{P}^A_{\mathrm{ADM}}$, $\hat{J}^{AB}_{\mathrm{ADM}}$ [they are volume integrals differing from the weak charges by terms proportional to integrals of the constraints]. The previous splitting implies that the variables $N(\tau,\boldsymbol{\sigma})$, $N_r(\tau,\boldsymbol{\sigma})$ be replaced by the variables $\tilde{\lambda}_A(\tau)$, $\tilde{\lambda}_{AB}(\tau) = -\tilde{\lambda}_{BA}(\tau)$, $n(\tau,\boldsymbol{\sigma})$, $n_r(\tau,\boldsymbol{\sigma})$ [with conjugate momenta $\tilde{\pi}^A(\tau)$, $\tilde{\pi}^{AB}(\tau) = -\tilde{\pi}^{BA}(\tau)$, $\tilde{\pi}^n(\tau,\boldsymbol{\sigma})$, $\tilde{\pi}^r_{\boldsymbol{n}}(\tau,\boldsymbol{\sigma})$] in the ADM theory.

With these assumptions one has the following form of the line element (also its form in tetrad gravity is given)

$$\begin{aligned}
ds^2 &= \epsilon([N_{(as)}+n]^2 - [N_{(as)r}+n_r]^3g^{rs}[N_{(as)s}+n_s])(d\tau)^2 - \\
&\quad - 2\epsilon[N_{(as)r}+n_r]d\tau d\sigma^r - \epsilon\,{}^3g_{rs}d\sigma^r d\sigma^s = \\
&= \epsilon([N_{(as)}+n]^2 - [N_{(as)r}+n_r]^3e^r_{(a)}\,{}^3e^s_{(a)}[N_{(as)s}+n_s])(d\tau)^2 - \\
&\quad - 2\epsilon[N_{(as)r}+n_r]d\tau d\sigma^r - \epsilon\,{}^3e_{(a)r}\,{}^3e_{(a)s}d\sigma^r d\sigma^s.
\end{aligned} \tag{5}$$

The final suggestion of Dirac is to modify ADM metric gravity in the following way:

i) add the 10 new primary constraints $p^A_{(\infty)} - \hat{P}^A_{\mathrm{ADM}} \approx 0$, $J^{AB}_{(\infty)} - \hat{J}^{AB}_{\mathrm{ADM}} \approx 0$, where $p^A_{(\infty)} = b^A_{(\infty)(\mu)}p^{(\mu)}_{(\infty)}$, $J^{AB}_{(\infty)} = b^A_{(\infty)(\mu)}b^B_{(\infty)(\nu)}S^{(\mu)(\nu)}_{(\infty)}$ [remember that $p^A_{(\infty)}$ and $J^{AB}_{(\infty)}$ satisfy a Poincaré algebra];
ii) consider $\tilde{\lambda}_A(\tau)$, $\tilde{\lambda}_{AB}(\tau)$, as Dirac multipliers for these 10 new primary constraints, and not as configurational (arbitrary gauge) variables coming from the lapse and

shift functions [so that there are no conjugate momenta $\tilde{\pi}^A(\tau)$, $\tilde{\pi}^{AB}(\tau)$ and no associated Dirac multipliers $\zeta_A(\tau)$, $\zeta_{AB}(\tau)$], in the assumed Dirac Hamiltonian [it is finite and differentiable]

$$H_{(D)\mathrm{ADM}} = \int d^3\sigma [n\tilde{\mathcal{H}} + n_r \tilde{\mathcal{H}}^r + \lambda_n \tilde{\pi}^n + \lambda^{\boldsymbol{n}}_r \tilde{\pi}^r_{\boldsymbol{n}}](\tau, \boldsymbol{\sigma}) - \\ - \tilde{\lambda}_A(\tau)[p^A_{(\infty)} - \hat{P}^A_{\mathrm{ADM}}] - \tilde{\lambda}_{AB}(\tau)[J^{AB}_{(\infty)} - \hat{J}^{AB}_{\mathrm{ADM}}] \approx 0, \tag{6}$$

The reduced phase space is still the ADM phase space: on the ADM variables there are only the secondary first class constraints $\tilde{\mathcal{H}}(\tau, \boldsymbol{\sigma}) \approx 0$, $\tilde{\mathcal{H}}^r(\tau, \boldsymbol{\sigma}) \approx 0$ [generators of proper gauge transformations], because the other first class constraints $p^A_{(\infty)} - \hat{P}^A_{\mathrm{ADM}} \approx 0$, $J^{AB}_{(\infty)} - \hat{J}^{AB}_{\mathrm{ADM}} \approx 0$ do not generate improper gauge transformations but eliminate 10 of the extra 20 variables.

In this modified ADM metric gravity, one has restricted the 3+1 splittings of M^4 to foliations whose leaves Σ_τ tend to Minkowski spacelike hyperplanes asymptotically at spatial infinity in a direction-independent way. Therefore, these Σ'_τ should be determined by the 10 degrees of freedom $x^{(\mu)}_{(\infty)}(\tau)$, $b^{(\mu)}_{(\infty)A}(\tau)$, as happens for flat spacelike hyperplanes: this means that it must be possible to define a "parallel transport" of the asymptotic tetrads $b^{(\mu)}_{(\infty)A}(\tau)$ to get well-defined tetrads in each point of Σ'_τ. While it is not yet clear whether this can be done for $\tilde{\lambda}_{AB}(\tau) \neq 0$ [maybe Nester's teleparallelism [58] can be used], there is a solution for $\tilde{\lambda}_{AB}(\tau) = 0$. This case corresponds to passing to the Wigner-like hypersurfaces [the analogue of the Minkowski Wigner hyperplanes with the asymptotic normal $l^{(\mu)}_{(\infty)} = l^{(\mu)}_{(\infty)\Sigma}$ parallel to $\hat{P}^{(\mu)}_{\mathrm{ADM}}$]. Following the same procedure defined for Minkowski spacetime, one gets $\bar{S}^{rs}_{(\infty)} \equiv \hat{J}^{rs}_{\mathrm{ADM}}$ [see [16] for the definition of $\bar{S}^{AB}_{(\infty)}$], $\tilde{\lambda}_{AB}(\tau) = 0$ and $[\epsilon_{(\infty)} = \sqrt{\epsilon^{p^2}_{(\infty)}}]$ $-\tilde{\lambda}_A(\tau)[p^a_{(\infty)} - \hat{P}^A_{\mathrm{ADM}}] = -\tilde{\lambda}_\tau(\tau)[\epsilon_{(\infty)} - \hat{P}^\tau_{\mathrm{ADM}}] + \tilde{\lambda}_r(\tau)\hat{P}^r_{\mathrm{ADM}}$, so that the final form of these four surviving constraints is $\epsilon_{(\infty)} - \hat{P}^\tau_{\mathrm{ADM}} \approx 0$, $\hat{P}^r_{\mathrm{ADM}} \approx 0$.

On this subclass of foliations [whose leaves $\Sigma^{(\mathrm{WSW})}_\tau$ will be called Wigner–Sen–Witten hypersurfaces] one can introduce a parallel transport by using the interpretation of [59] of the Witten spinorial method of demonstrating the positivity of the ADM energy [60]. Let us consider the Sen–Witten connection [61, 60] restricted to $\Sigma^{(\mathrm{WSW})}_\tau$ (it depends on the trace of the extrinsic curvature of $\Sigma^{(\mathrm{WSW})}_\tau$) and the spinorial Sen–Witten equation associated with it. As shown in [62], this spinorial equation can be rephrased as an equation whose solution determines (in a surface-dependent dynamical way) a tetrad in each point of $\Sigma^{(\mathrm{WSW})}_\tau$ once it is given at spatial infinity (again this requires a direction-independent limit).

On the Wigner–Sen–Witten hypersurfaces (the intrinsic asymptotic rest-frame of the gravitational field), the remaining four extra constraints are: $\hat{P}^r_{\mathrm{ADM}} \approx 0$ (this is automatically implemented with the boundary conditions of [57]) and $\epsilon_{(\infty)} = \sqrt{\epsilon^{p^2}_{(\infty)}} \approx \hat{P}^\tau_{\mathrm{ADM}} \approx M_{\mathrm{ADM}} = \sqrt{\epsilon^{\hat{P}^2}_{AM}}$. Now the spatial indices have become spin-1 Wigner indices [they transform with Wigner rotations under asymptotic Lorentz transformations]. As stated for parametrized theories in Minkowski spacetime, in this special gauge, 3 degrees of freedom of the gravitational field [a 3-center-of-mass variable $X_{\text{ADM}}[{}^3g, {}^3\tilde{\Pi}]$ inside the Wigner–Sen–Witten hypersurface] become gauge variables, while $\tilde{x}^{(\mu)}_{(\infty)}$ [the canonical non-covariant variable replacing $x^{(\mu)}_{(\infty)}$] becomes a decoupled observer with his "point particle clock" [63, 64] near spatial infinity.

Since the positivity theorems for the ADM energy imply that one has only timelike or lightlike orbits of the asymptotic Poincaré group, the restriction to universes with timelike ADM 4-momentum allows us to define the Møller radius $\rho_{\text{AMD}} = \sqrt{-\epsilon \hat{W}^2_{\text{ADM}}/\hat{P}^2_{\text{ADM}}c}$ from the asymptotic Poincaré Casimirs $\hat{P}^2_{\text{ADM}}$, $\hat{W}^2_{\text{ADM}}$ [this is an intrinsic classical unit of length as in parametrized Minkowski theories, to be used as an ultraviolet cutoff in a future attempt at quantization].

By going from $\tilde{x}^{(\mu)}_{(\infty)}$, $p^{(\mu)}_{(\infty)}$, to the canonical basis

$$
\begin{aligned}
&T_{(\infty)} = p_{(\infty)(\mu)}\tilde{x}^{(\mu)}_{(\infty)}/\epsilon_{(\infty)} = p_{(\infty)(\mu)}x^{(\mu)}_{(\infty)}/\epsilon_{(\infty)},\\
&\epsilon_{(\infty)},\\
&z^{(i)}_{(\infty)} = \epsilon_{(\infty)}(\tilde{x}^{(i)}_{(\infty)} - p^{(i)}_{(\infty)}\tilde{x}^{(o)}_{(\infty)}/p^{(o)}_{(\infty)}),\\
&k^{(i)}_{(\infty)} = p^{(i)}_{(\infty)}/\epsilon_{(\infty)} = u^{(i)}(p^{(\rho)}_{(\infty)}),
\end{aligned}
$$

as in the flat case one finds that the final reduction requires the gauge-fixings $T_{(\infty)} - \tau \approx 0$ and $X^r_{\text{ADM}} \approx 0$, where $\sigma^r = X^r_{\text{ADM}}$ is a variable representing the "center of mass" of the 3-metric of the slice Σ_τ of the asymptotically flat spacetime M^4. Since $\{T_{(\infty)}, \epsilon_{(\infty)}\} = -\epsilon$, with the gauge fixing $T_{(\infty)} - \tau \approx 0$ one gets $\tilde{\lambda}_\tau(\tau) \approx \epsilon$, and the final Dirac Hamiltonian is $H_D = M_{\text{ADM}} + \tilde{\lambda}_r(\tau)\hat{P}^r_{\text{ADM}}$ with M_{ADM} [the ADM mass of the universe] the natural physical Hamiltonian to reintroduce an evolution in the "mathematical" $T_{(\infty)} \equiv \tau$: namely, in the rest-frame time identified with the parameter τ labelling the leaves $\Sigma^{(\text{WSW})}_\tau$ of the foliation of M^4. Physical times (atomic clocks, ephemeridis time, ...) must be put in a local 1-1 correspondence with this "mathematical" time.

The asymptotic rest-frame instant form realization of the Poincaré generators becomes (no more reference to the boosts $\hat{J}^{\tau r}_{\text{ADM}}$)

$$
\begin{aligned}
&\epsilon_{(\infty)} = M_{\text{ADM}},\\
&p^{(i)}_{(\infty)},\\
&J^{(i)(j)}_{(\infty)} = \tilde{x}^{(i)}_{(\infty)}p^{(j)}_{(\infty)} - \tilde{x}^{(j)}_{(\infty)}p^{(i)}_{(\infty)} + \delta^{(i)r}\delta^{(j)s}\hat{J}^{rs}_{\text{ADM}},\\
&J^{(o)(i)}_{(\infty)} = p^{(i)}_{(\infty)}\tilde{x}^{(o)}_{(\infty)} - \sqrt{M^2_{\text{ADM}} + \boldsymbol{p}^2_{(\infty)}}\,\tilde{x}^{(i)}_{(\infty)} - \frac{\delta^{(i)r}\hat{J}^{rs}_{\text{ADM}}\delta^{(s(j)}p^{(j)}_{(\infty)}}{M_{\text{ADM}} + \sqrt{M^2_{\text{ADM}} + \boldsymbol{p}^2_{(\infty)}}}.
\end{aligned}
\tag{7}
$$

This whole construction also holds in our formulation of tetrad gravity (since it uses the ADM action) and in its canonically reduced form in the 3-orthogonal gauge. In particular the Poincaré charges of void spacetimes vanish.

Therefore, the final physical Hamiltonian for the physical gravitational field is the reduced volume form of the ADM energy $\hat{P}^{\tau}_{\mathrm{ADM}}[r_{\bar{a}}.\pi_{\bar{a}}, \phi(r_{\bar{a}}, \pi_{\bar{a}})]$ with ϕ a solution of the reduced Lichnerowicz equation in the 3-orthogonal gauge with $\rho(\tau, \boldsymbol{\sigma}) \approx 0$.

We next compare the standard general covariant formulation of gravity based on the Hilbert action with its invariance under Diff M^4 with the ADM Hamiltonian formulation.

As regards the 10 Einstein equations of the standard approach, the Bianchi identities imply that four equations are linearly dependent on the other six and their gradients. Moreover, the four combinations of Einstein's equations projectable to phase space (where they become the secondary first class superhamitonian and supermomentum constraints of canonical metric gravity) are independent of the accelerations being restrictions on the Cauchy data. As a consequence the Einstein equations have solutions, in which the ten components ${}^4g_{\mu\nu}$ of the 4-metric depend on only two truly dynamical degrees of freedom (defining the physical gravitational field) and on eight undetermined degrees of freedom. This transition from the ten components ${}^4g_{\mu\nu}$ of the tensor ${}^4g$ in some atlas of M^4 to the 2 (deterministic)+8 (undetermined) degrees of freedom breaks general covariance because these quantities are neither tensors nor invariants under diffeomorphisms (their functional form is atlas-dependent).

Since the Hilbert action is invariant under Diff M^4, one usually says that a "gravitational field" is a 4-geometry over M^4, namely, an equivalence class of spacetimes $(M^4, {}^4g)$, a solution of Einstein's equations, modulo Diff M^4. See, however, the interpretational problems concerning what is observable in general relativity, for instance, in [65, 66]; in particular the facts that

i) scalars under Diff M^4, like ${}^4R$, are not Dirac observables but gauge dependent quantities;
ii) the functional form of ${}^4g_{\mu\nu}$ in terms of the physical gravitational field and, therefore, the angle and distance properties of material bodies and the standard procedures for defining measures of length and time based on the line element ds^2, are gauge-dependent.

Instead, in the ADM formalism with the extra notion of 3+1 splittings of M^4, the (tetrad) metric ADM action (differing from the Hilbert action by a surface term) is quasi-invariant under the (14) 8 types of gauge transformations which are the pull-back of the Hamiltonian group $\mathcal{G}$ of gauge transformations, whose generators are the first class constraints of the theory. The Hamiltonian group $\mathcal{G}$ has a subgroup (whose generators are the supermomentum and superhamiltonian constraints) formed by the diffeomorphisms of M^4 adapted to its 3+1 splittings, Diff M^{3+1} [which is different from Diff M^4]. Moreover, the Poisson algebra of the supermomentum and superhamiltonian constraints reflects the embeddability in M^4 of the foliation associated with the 3+1 splitting [67]. Now a "gravitational field" is an equivalence class of spacetimes modulo the Hamiltonian group $\mathcal{G}$. [These equivalence classes are connected with the conformal 3-geometries of the Lichnerowicz–York approach].

The interpretation of the 14 gauge transformations and of their gauge fixings in tetrad gravity (it is independent from the presence of matter) is the following [a tetrad in a point of Σ_τ is a local observer]:

i) the gauge fixings of the gauge boost parameters associated with the 3 boost constraints are equivalent to the choice of local observer either at rest or Lorentz-boosted;
ii) the gauge fixings of the gauge angles associated with the 3 rotation constraints are equivalent to the fixing of the standard of non-rotation of the local observer;
iii) the gauge fixings of the 3 gauge parameters associated with passive space diffeomorphisms [Diff Σ_τ; change of coordinates charts] are equivalent to the fixing of 3 standards of length by means of choice of coordinate system on Σ_τ [the measuring apparatus (the "rods") should be defined in terms of Dirac's observables for some kind of matter, after its introduction into the theory];
iv) according to constraint theory the choice of 3-coordinates on Σ_τ induces the gauge fixings of the 3 shift functions [i.e., of ${}^4g_{oi}$], whose gauge nature is connected with the "conventionality of simultaneity" [68] [therefore, the gauge fixings are equivalent to a choice of simultaneity and, as a consequence, to a statement about the isotropy or anisotropy of the velocity of light in that gauge];
v) the gauge fixing on the the momentum $\rho(\tau,\sigma)$ conjugate to the conformal factor of the 3-metric is a nonlocal statement about the extrinsic curvature of the leaves Σ_τ of the given 3+1 splitting of M^4; since the superhamiltonian constraint produces normal deformations of Σ_τ [67] and, therefore, transforms a 3+1 splitting of M^4 into another one (the ADM formulation is independent of the choice of 3+1 splitting), this gauge fixing is equivalent to the choice of a particular 3+1 splitting;
vi) the previous gauge fixing induces the gauge fixing of the lapse function (which determines the packing of the leaves Σ_τ in the chosen 3+1 splitting) and, therefore, is equivalent to the fixing of a standard of proper time [again "clocks" should be built with the Dirac observables of some kind of matter].

The 3-orthogonal gauge of tetrad gravity is the equivalent of the Coulomb gauge in classical electrodynamics (like the harmonic gauge is the equivalent of the Lorentz gauge). Only after a complete gauge fixing the 4-tensors and the scalars under Diff M^4 become measurable quantities (like the electromagnetic vector potential in the Coulomb gauge). At this stage one may accept the proposal of Bergmann [36] of identifying the points of a spacetime $(M^4, {}^4g)$, a solution of the Einstein's equations, in the absence of matter, in a way which isinvariant under spacetime diffeomorphisms extended to 4-tensors, by using four invariants bilinear in the Weyl tensors called "individuating fields" (also see [65, 66]).

Our approach breaks the general covariance of general relativity completely by passing to the special 3-orthogonal gauge. But this is done in a way naturally associated with theories with first class constraints (like all formulations of general relativity and the standard model of elementary particles with or without supersymmetry): the global Shanmugadhasan canonical transformations (when they exist) correspond to privileged Darboux charts for presymplectic manifolds defined by the first class constraints. Therefore, the gauges identified by these canonical transformations should have a special (as yet unexplored) role in generally covariant theories as well, in which traditionally one looks for observables invariant under diffeomorphisms and not for Dirac observables not general covariant.

Recall that Bergmann [36] made the following critique of general covariance: it would be desirable to restrict the group of coordinate transformations (spacetime diffeomorphisms) in such a way that it contain an invariant subgroup describing the coordinate transformations that change the frame of reference of an outside observer (these transformations could be called Lorentz transformations; see also the comments in [69] on the asymptotic behaviour of coordinate transformations); the remaining coordinate transformations would be like the gauge transformations of electromagnetism. This is what we have done. In this way "preferred" coordinate systems will give rise to WSW hypersurfaces), which, as stated by Bergmann, are not "flat": while the inertial coordinates are determined experimentally by the observation of trajectories of force-free bodies, these intrinsic coordinates can be determined only by much more elaborate experiments, since they depend, at the least, on the inhomogeneities of the ambient gravitational fields. See [70] for other criticisms of general covariance: very often, to get physical results, one uses preferred coordinates not merely for calculational convenience, but also for understanding. In [71] this fact has been formalized as the "principle of restricted covariance". In our case the choice of the gauge fixings has been dictated by the Shanmugadhasan canonical transformations, which produce generalized Coulomb gauges, in which one can put into normal form the Hamilton equations for the canonical variables of the gravitational field [and, therefore, they also produce a normal form of the two associated combinations of the Einstein equations which depend on the accelerations].

If we add to the tetrad ADM action the action for N scalar particles with positive energy in the form of [16] [where it was given on arbitrary Minkowski spacelike hypersurfaces], the only constraints which are modified are the superhamiltonian constraint, which becomes dependent on the matter energy density $\mathcal{M}(\tau, \boldsymbol{\sigma})$, and the 3 space diffeomorphism constraints, which become dependent on the matter momentum density $\mathcal{M}_r(\tau, \boldsymbol{\sigma})$. The canonical reduction and the determination of the Dirac observables can be done as in absence of matter. However, the reduced Lichnerowicz equation for the conformal factor of the 3-metric in the 3-orthogonal gauge and with $\rho(\tau, \boldsymbol{\sigma}) \approx 0$ now acquires an extra dependence on $\mathcal{M}(\tau, \boldsymbol{\sigma})$ and $\mathcal{M}_r(\tau, \boldsymbol{\sigma})$.

Since, as a preliminary result, we are interested in identifying explicitly the action-at-a-distance (Newton-like and gravitomagnetic) potentials among particles hidden in tetrad gravity (just us the Coulomb potential is hidden in the electromagnetic gauge potential), we shall restrict ourselves to spacetimes without gravitational field by adding the second class constraints $r_{\bar{a}}(\tau, \boldsymbol{\sigma}) \approx 0$, $\pi_{\bar{a}}(\tau, \boldsymbol{\sigma}) \approx 0$. Moreover, since, in the presence of matter, the equations of motion do not imply $r_{\bar{a}} = \pi_{\bar{a}} = 0$, void spacetimes have now to be understood as a strong approximation identifying the instantaneous action-at-a-distance interaction among the particles contained in tetrad gravity. If we develop the conformal factor $\phi(\tau, \sigma)$ in a formal series in the Newton constant G $[\phi = 1 + \sum_{n=1}^{\infty} G^n \phi_n]$, one can find a solution $\phi = 1 + G\phi_1$ at order G (post-Minkowskian approximation) of the reduced Lichnerowicz equation. However, due to a self-energy divergence in ϕ evaluated at the positions $\boldsymbol{\eta}_i(\tau)$ of the particles, one needs to rescale the bare masses to physical ones, $m_i \mapsto \phi^{-2}(\tau, \boldsymbol{\eta}_i(\tau)) m_i^{(phys)}$, and to make a regularization of the type defined in [72]. Then, the regularized solution for ϕ can be put into the reduced form of the ADM energy, which becomes [$\boldsymbol{\kappa}_i(\tau)$ are the particle momenta conjugate to $\boldsymbol{\eta}_i(\tau)$; $\boldsymbol{n}_{ij} = [\boldsymbol{\eta}_i - \boldsymbol{\eta}_j]/|\boldsymbol{\eta}_i - \boldsymbol{\eta}_j|$]

$$\begin{aligned}\hat{P}^{\tau}_{ADM} &= \sum_{i=1}^{N} c\sqrt{m_i^{(phys)2} c^2 + \boldsymbol{\kappa}_i^2(\tau)} - \\ &- \frac{G}{c^2} \sum_{i \neq j} \frac{\sqrt{m_i^{(phys)2} c^2 + \boldsymbol{\kappa}_i^2(\tau)} \sqrt{m_j^{(phys)2} c^2 + \boldsymbol{\kappa}_j^2(\tau)}}{|\boldsymbol{\eta}_i(\tau) - \boldsymbol{\eta}_j(\tau)|} - \\ &- \frac{G}{8c^2} \sum_{i \neq j} \frac{3\boldsymbol{\kappa}_i(\tau) \cdot \boldsymbol{\kappa}_j(\tau) - 5\boldsymbol{\kappa}_i(\tau) \cdot \boldsymbol{n}_{ij}(\tau) \boldsymbol{\kappa}_j(\tau) \cdot \boldsymbol{n}_{ij}(\tau)}{|\boldsymbol{\eta}_i(\tau) - \boldsymbol{\eta}_j(\tau)|} + O(G^2). \quad (8)\end{aligned}$$

One sees the Newton-like and the gravitomagnetic (in the sense of York) potentials (both of them need regularization) at the post-Minkowskian level (order G but exact in c) emerging from the tetrad ADM version of Einstein general relativity. For G=0 we recover N free scalar particles on the Wigner hyperplane in Minkowski spacetime, as required by deparametrization. As $c \to \infty$, we get the post-Newtonian Hamiltonian

$$\begin{aligned}H_{PN} &= \sum_{i=1}^{N} \frac{\boldsymbol{\kappa}_i^2(\tau)}{2m_i^{(phys)}} \left(1 - \frac{2G}{c^2} \sum_{j \neq i} \frac{m_j^{(phys)}}{|\boldsymbol{\eta}_i(\tau) - \boldsymbol{\eta}_j(\tau)|}\right) - \frac{G}{2} \sum_{i \neq j} \frac{m_i^{(phys)} m_j^{(phys)}}{|\boldsymbol{\eta}_i(\tau) - \boldsymbol{\eta}_j(\tau)|} \\ &- \frac{G}{8c^2} \sum_{i \neq j} \frac{3\boldsymbol{\kappa}_i(\tau) \cdot \boldsymbol{\kappa}_j(\tau) - 5\boldsymbol{\kappa}_i(\tau) \cdot \boldsymbol{n}_{ij}(\tau) \boldsymbol{\kappa}_j(\tau) \cdot \boldsymbol{n}_{ij}(\tau)}{|\boldsymbol{\eta}_i(\tau) - \boldsymbol{\eta}_j(\tau)|} + O(G^2),\end{aligned} \quad (9)$$

which is of the same type as those implied by the results of [72, 73] [the differences are probably connected with the use of different coordinate systems and with the fact that one has essential singularities on the particle worldlines and with the need of regularization].

Our future research program will concentrate on the following subjects:

1) the post-Minkowskian 2-body problem in these spacetimes, to see the relevance of exact relativistic recoil effects in the motion of binaries;
2) the replacement of scalar particles with spinning ones to identify the precessional effects (like that of Lense–Thirring) of gravitomagnetism;
3) the linearization of the theory in the 3-orthogonal gauge in the presence of matter: besides finding the Coulomb gauge description of gravitational waves, one expects to find a consistent (post-Minkowskin) coupling of the linearized gravitational

field with matter, since the Bianchi identities have been solved, and to go beyond the strong approximation of void spacetimes;
4) perfect fluids and, then, extended relativistic bodies;
5) the coupling of tetrad gravity to the electromagnetic field, to fermion fields and then to the standard model, trying to achieve reduction to Dirac observables in all these cases and to study their post-Minkowskian approximations;
6) quantization of tetrad gravity in the 3-orthogonal gauge with $\rho(\tau, \boldsymbol{\sigma}) \approx 0$: for each perturbative (in G) solution of the reduced Lichnerowicz equation one defines a Schroedinger equation in τ for a wave functional $\Psi[\tau; r_{\bar{a}}]$ with the associated quantized ADM energy $\hat{P}^{\tau}_{\text{ADM}}[r_{\bar{a}}, i\frac{\delta}{\delta r_{\bar{a}}}]$ as Hamiltonian; no problem of physical scalar product is present, but only ordering problems in the Hamiltonian; moreover, one has the Møller radius as an ultraviolet cut-off.

For a more recent review with improvement as the interpretational level see: L. Lusanna, "Towards a unified description of the four interactions in terms of Dirac–Bergmann Observables", invited contribution to the book of the Indian National Science Academy for the International Mathematics Year 2000 AD (HEP-TH/9907081).

References

1. Lusanna L., Russo S. (1998): Tetrad gravity I): A new formulation. Firenze Univ., preprint, gr-qc/9807073
2. Lusanna L., Russo S. (1998): Tetrad gravity II): Dirac's observables. Firenze Univ., preprint gr-qc/9807074
3. De Pietri R., Lusanna, L. (1999): Tetrad gravity III): Asymptotic Poincaré charges, the physical Hamiltonian and void spacetimes. Firenze Univ., preprint gr-qc/9909025
4. De Pietri R., Lusanna L., Vallisneri M.: Tetrad gravity IV): The N-body problem, in preparation
5. Dirac P.A.M. (1950): Can. J. Math. **2**, 129; (1964): Lectures on Quantum Mechanics. Belfer Graduate School of Science, Monographs Series. Yeshiva University, New York
6. Lusanna L. (1997): Solving Gauss' Laws and Searching Dirac Observables for the Four Interactions, in *Second Conf. on Constrained Dynamics and Quantum Gravity, S. Margherita Ligure 1996*, ed. by V. De Alfaro, J.E. Nelson, G. Bandelloni, A. Blasi, M. Cavaglià, A.T. Filippov, Nucl. Phys. (Proc. Suppl.) B **57**, 13, hep-th/9702114; Unified Description and Canonical Reduction to Dirac's Observables of the Four Interactions, in *Int. Workshop, New Non Perturbative Methods and Quantization on the Light Cone, Les Houches School, 1997*, ed. by P. Grangé, H.C. Pauli, A. Neveu, S. Pinsky, A. Werner, Springer, Berlin–Heidelberg–New York, hep-th/9705154; (1998): The Pseudoclassical Relativistic Quark Model in the Rest-Frame Wigner-Covariant Gauge, in *Euroconference QCD97, Montpellier 1997*, ed. by S. Narison, Nucl. Phys. (Proc. Suppl.) B **64**, 306
7. Shanmugadhasan S. (1973): J. Math. Phys. **14**, 677;
Lusanna L. (1993): Int. J. Mod. Phys. A **8**, 4193;
Chaichian M., Martinez D.L., Lusanna L. (1994): Ann. Phys. (N.Y.) **232**, 40;
Lusanna L. (1990): Phys. Rep. **185**, 1; (1991): Riv. Nuovo Cimento **14** 3, 1; (1990): J. Math. Phys. **31**, 2126; (1990): J. Math. Phys. **31**, 428
8. Dirac P.A.M. (1955): Can. J. Phys. **33**, 650
9. Lusanna L. (1995): Int. J. Mod. Phys. A **10**, 3531, 3675

10. Lusanna L., Valtancoli P. (1997): Int. J. Mod. Phys. A **12**, 4769, hep-th/9606078; (1997): Int. J. Mod. Phys. A **12**, 4797, hep-th/9606079
11. Lusanna L., Valtancoli P. (1998): Int. J. Mod. Phys. A **13**, 4605, hep-th/9707072
12. Kuchar K. (1976): J. Math. Phys. **17**, 777, 792, 801; (1976): **18**, 1589
13. Dirac P.A.M. (1949): Rev. Mod. Phys. **21**, 392
14. Longhi G., Materassi M. (1999): J. Math. Phys. **40**, 480, hep-th/9803128; (1998): Collective and relative variables for a classical Klein–Gordon field. Firenze University preprint, hep-th/9890024, and Int. J. Mod. Phys. A, to appear
15. Lusanna L., Materassi M.: The canonical decomposition in center-of-mass and relative variables of a Klein–Gordon field in the rest-frame Wigner-covariant instant form. Firenze University preprint, hep-th/9904202
16. Lusanna L. (1997): Int. J. Mod. Phys. A **12**, 645
17. Alba D., Lusanna L. (1998): Int. J. Mod. Phys. A **13**, 2791, hep-th/9705155
18. Alba D., Lusanna L. (1998): Int. J. Mod. Phys. A **13**, 3275, hep-th/9705156
19. Bigazzi F., Lusanna L. (1999): Int. J. Mod. Phys. A **14**, 1429, hep-th 9807052
20. Bigazzi F., Lusanna L. (1999): Int. J. Mod. Phys. A **14**, 1877, hep-th/9807054
21. Lämmerzahl C. (1993): J. Math. Phys. **34**, 3918
22. Herbst, I. (1977): Commun. Math. Phys. **53**, 285; (1997): **55**, 316;
Durand B., Durand L. (1983): Phys. Rev. D **28**, 396; (1994): erratum, Phys. Rev. D **50**, 6642;
Basdevant J.J., Boukraa S. (1985): Z. Phys. C **28**, 413;
Martin A., Roy S.M. (1989): Phys. Lett. B **233**, 407;
Le Yaouanc A.,Oliver L., Raynal J.C. (1995): Ann. Phys. (N.Y.) **239**, 243;
Lucha W., Schöberl F.F. (1994): Phys. Rev. D **50**, 5443
23. Leibbrandt G. (1994): Non-Covariant Gauges. World Scientific, Singapore
24. Møller C. (1949): Ann. Inst. H. Poincaré **11**, 251; (1957): The Theory of Relativity. Oxford University Press, Oxford
25. Veneziano, G. (1990): Quantum Strings and the Constants of Nature, in *The Challenging Questions*. The Subnuclear Series 27, ed. by A. Zichichi, Plenum Press, New York
26. Arnowitt R., Deser S., Misner C.W. (1960): Phys. Rev. **117**, 1595; (1962): The Dynamics of General Relativity, in *Gravitation: an Introduction to Current Research*, ed. by L. Witten, Wiley, New York, p. 227
27. Regge T., Teitelboim C. (1974), Ann. Phys. (N.Y.) **88**, 286
28. Beig R., Murchadha, Ó. (1987): Ann. Phys. (N.Y.) **174**, 463
29. Andersson L. (1987): J. Geom. Phys. **4**, 289
30. Thiemann T. (1995): Class. Quantum Grav. **12**, 181
31. A. Ashtekar (1980): Asymptotic Structure of the Gravitational Field at Spatial Infinity, in *General Relativity and Gravitation*, ed. by A. Held, Vol. 2, Plenum, New York;
Ashtekar A., Hansen R.O. (1978): J. Math. Phys. **19**, 1542;
Ashtekar A., Magnon A. (1984): J. Math. Phys. **25**, 2682;
Ashtekar A., Romano J.D. (1992): Class. Quantum Grav. **9**, 1069
32. Wald R.M. (1984): General Relativity, Chicago University Press, Chicago
33. Geroch R. (1968): J. Math. Phys. **9**, 1739; (1970): J. Math. Phys. **11**, 343
34. Moncrief V. (1975): J. Math. Phys. **16**, 1556
35. Choquet-Bruhat Y., York jr. J.W. (1980): The Cauchy Proble, in *General Relativity and Gravitation*, ed. by A. Held Plenum, Vol. 1, New York, p. 99
36. Bergmann P.G. (1961): Rev. Mod. Phys. **33**, 510
37. Weyl H. (1929): Z. Physik **56**, 330
38. Dirac P.A.M. (1962): in *Recent Developments in General Relativity*, Pergamon Press, Oxford, and PWN-Polish Scientific Publishers, Warsaw

39. Schwinger J. (1963): Phys. Rev. **130**, 1253
40. Kibble T.W.B. (1963): J. Math. Phys. **4**, 1433
41. Deser S., Isham, C.J. (1976): Phys. Rev. D **14**, 2505;
Nelson J.E., Teitelboim C. (1978): Ann. Phys. (N.Y.) **116**, 86;
Pilati M. (1978): Nucl. Phys. B **132**, 138;
Castellani L., van Nieuwenhuizen P., Pilati M. (1982): Phys. Rev. D **26**, 352;
Nelson J.E., Regge T. (1986): Ann. Phys. (N.Y.) **166**, 234; (1989): Int. J. Mod. Phys. A **4**, 2021
42. Charap J.M., Nelson J.E. (1983): J. Phys. A **16**, 1661, 3355; (1986): Class. Quantum Grav. **3**, 1061;
Charap J.M. (1987): The Constraints in Vierbein General Relativity, in *Constraint's Theory and Relativistic Dynamics*, ed. by G. Longhi, L. Lusanna, World Scientific, Singapore, p. 84
43. Maluf J.W. (1991): Class. Quantum Grav. **8**, 287
44. Henneaux M. (1978): Gen. Rel. Grav. **9**, 1031;
Geheniau J., Henneaux M. (1977): Gen. Rel. Grav. **8**, 611
45. Henneaux M. (1983): Phys. Rev. D **27**, 986
46. Charap J.M., Henneaux M., Nelson J.E. (1988): Class. Quantum Grav. **5**, 1405
47. Henneaux M., Nelson J.E., Schonblond C. (1989): Phys. Rev. D **39**, 434
48. De Witt B.S. (1967): Phys. Rev. **162**, 1195; (1967): The Dynamical Theory of Groups and Fields, Gordon and Breach, New York; (1964): Dynamical Theory of Groups and Fields, in *Relativity, Groups and Topology, Les Houches, 1963*, ed. by C. De Witt, B.S. De Witt, Gordon and Breach, London, p. 587; (1984): The Spacetime Approach to Quantum Field Theory, in *Relativity, Groups and Topology II, Les Houches 1983*, ed. by B.S. DeWitt, R. Stora, North-Holland, Amsterdam, p. 381
49. De Witt B.S. (1967): Phys. Rev. **160**, 1113
50. Isham C.J., Kuchar K. (1984): Ann. Phys. (N.Y.) **164**, 288 and 316;
Kuchar K. (1986): Found. Phys. **16**, 193
51. Lichnerowicz A. (1944): J. Math. Pure Appl. **23**, 37;
Choquet-Bruhat Y. (1948): C. R. Acad. Sci. Paris **226**, 1071; (1956): J. Rat. Mech. Anal. **5**, 951; (1962): The Cauchy Problem, in *Gravitation: An Introduction to Current Research*, ed. by L. Witten, Wiley, New York, p. 130
52. York jr J.W. (1971): Phys. Rev. Lett. **26**, 1656; (1972): **28**, 1082; (1972): J. Math. Phys. **13**, 125; (1972): **14**, 456; (1974): Ann. Ins. H. Poincaré **21**, 318;
O'Murchadha N., York Jr. J.W. (1972): J. Math. Phys. **14**, 1551; (1974): Phys. Rev. D **10**, 428
53. York Jr. J.W. (1979): Kinematics and Dynamics of General Relativity, in *Sources of Gravitational Radiation, Battelle-Seattle Workshop, 1978*, ed. by L.L. Smarr, Cambridge University Press, Cambridge, p. 83
54. Ciufolini I., Wheeler J.A. (1995): Gravitation and Inertia, Princeton University Press, Princeton
55. Qadir A., Wheeler J.A. (1985): York's Cosmic Time Versus Proper Time, in *From SU(3) to Gravity, Y. Ne'eman's Festschrift*, ed. by E. Gotsma, G. Tauber, Cambridge University Press, Cambridge, p. 128
56. Dirac P.A.M. (1951): Canad. J. Math. **3**, 1
57. Christodoulou D., Klainerman S. (1993): The Global Nonlinear Stability of the Minkowski Space, Princeton University Press, Princeton
58. Nester J.M. (1988): Class.Quantum Grav. **5**, 1003;
Cheng W.H., Chern D.C., Nester J.M. (1988): Phys. Rev. D **38**, 2656;

Nester J.M. (1980): J. Math. Phys. **30**, 624; (1992): **33**, 910;
J.M. Nester (1989): Int. J. Mod. Phys. A **4**, 1755;
Nester J.M. (1991): Class. Quantum Grav. **8**, L19;
Dimakis A., Müller-Hoissen F. (1989): Phys. Lett. A **142**, 73

59. Ashtekar A., Horowitz G.T. (1984): J. Math. Phys. **25**, 1473
60. Witten E. (1981): Commun. Math. Phys. **80**, 381
61. Sen A. (1981): J. Math. Phys. **22**, 1781; (1982): Phys. Lett. B 119, 89
62. Frauendiener J. (1991): Class. Quantum Grav. **8**, 1881
63. Isham C.J. (1993): Canonical Quantum Gravity and the Problem of Time, in *Integrable Systems, Quantum Groups and Quantum Field Theories, Salamanca, 1993*, ed. by L.A. Ibort, M.A. Rodriguez, Kluwer, London, p. 21; (1991): Conceptual and Geometrical Problems in Quantum Gravity, in *Recent Aspects of Quantum Fields, Schladming, 1991*, ed. by H. Mitter, H. Gausterer, Springer, Berlin, p. 58; (1994): Prima Facie Questions in Quantum Gravity; Canonical Quantum Gravity and the Question of Time, in *Canonical Gravity: From Classical to Quantum*, ed. by J. Ehlers, H. Friedrich, Springer, Berlin, p. 1
64. Kuchar K. (1992): Time and Interpretations of Quantum Gravity, in *Proc. 4th Canadian Conf. on General Relativity and Relativistic Astrophysics*, ed. by G. Kunstatter, D. Vincent, J. Williams, World Scientific, Singapore, p. 211
65. Stachel J. (1987): The Meaning of General Covariance, in *General Relativity and Gravitation, GR11, Stockholm, 1986*, ed. by M.A.H. Mac Callum, Cambridge University Press, Cambridge; (1993): The Meaning of General Covariance, in *Philosophical Problems of the Internal and External Worlds, Essays in the Philosophy of A. Grünbaum*, ed. by J. Earman, A.I. Janis, G.J. Massey, N. Rescher, Pittsburgh University Press, Pittsburgh
66. Rovelli C. (1991): Class. Quantum Grav. **8**, 297, 317
67. Teitelboim C. (1980): The Hamiltonian Structure of Space-Time, in *General Relativity and Gravitation*, ed. by A. Held, Vol. I, Plenum, New York;
Hojman, A.S., Kuchar K., Teitelboim C. (1971): Ann. Phys. (N.Y.) **96**, 88
68. Havas P. (1987): Gen. Rel. Grav. **19**, 435;
Anderson R., Vetharaniam I., Stedman G.E. (1998): Phys. Rep. **295**, 93
69. Landau L., Lifschitz E. (1951): The Classical Theory of Fields, Addison-Wesley, Cambridge
70. Ellis G.F.R., Matravers D.R. (1995): Gen. Rel. Grav. **27**, 777
71. Zalaletdinov R., Tavakol R., Ellis G.F.R. (1996): Gen. Rel. Grav. **28**, 1251
72. Einstein A., Hoffman B., Infeld L. (1938): Ann. Math. **39**, 66;
Einstein A., Infeld L. (1940): Ann. Math. **41**, 797; (1949): Canad. J. Math. **1**, 209;
Infeld L. (1957): Rev. Mod. Phys. **29**, 398
73. Lorentz H.A., Droste L. (1917): Amst. Akad. Versl. **26**, 392;
Eddington A., Clarke G.L. (1938): Proc. Roy. Soc. London A **166**, 465;
Fock V. (1939): J. Phys. (U.S.S.R.) **1**, 81; Papapetrou A. (1951): Proc. Phys. Soc.(London) **64**, 57

Applications of Calculus of Variations to General Relativity

A. Masiello

Abstract. We present some global results on Lorentzian geometry obtained by using global variational methods. In particular some results on the geodesic connectedness of Lorentzian manifolds and on the multiplicity of lightlike geodesics joining a point with a timelike curve are presented. Such results allow to give a mathematical description of the *gravitational lens effect*.

1 An historical introduction to the calculus of variations

The modern story of the calculus of variations (for the ancient one see, for instance, Dido's problem in Virgil's *Aeneid*) began in 1696 with the challenge of Johann Bernoulli of finding the *brachystocrone*, the curve joining two points in a vertical plane along which a frictionless bead will descend in the least possible time with null initial speed. Put $A = (0, 0)$ and $B = (x_1, y_1)$, with $x_1 > 0$, $y_1 < 0$, then the brachystocrone between A and B is the curve that minimizes the function

$$T(y) = \int_0^{x_1} \frac{\sqrt{1 + y'(x)^2}}{\sqrt{2gy(x)}} \mathrm{d}x \tag{1.1}$$

among all C^1-curves $y(x)\colon [0, x_1] \longrightarrow \mathbf{R}$ such that $y(0) = 0$, $y(x_1) = y_1$. The problem fascinated many scientists of the time and the solution, an arc of a *cycloid*, was found first by Johann Bernoulli and also by Newton and by Leibniz. The proof of Johann Bernoulli is based on the analogy of the brachystocrone problem with *Fermat's Principle in optics* formulated by Fermat in 1662 for the propagation of light in an isotropic medium. Consider a region Ω of space occupied by an isotropic medium. Let $v(x) > 0$, the light speed in the medium; then the Fermat principle claims that the light rays joining two points p, q in Ω are the trajectories that minimizes the functional

$$T(x) = \int_0^1 \frac{\sqrt{\langle \dot{x}(s), \dot{x}(s) \rangle}}{v(x(s))} \mathrm{d}s$$

among all C^1-curves $x(s)\colon [0, 1] \longrightarrow \Omega$ such that $x(0) = p$, $x(1) = q$. Note that, if the medium is *homogeneous*, the speed of light in Ω is a constant and the trajectory is a line segment. In general the trajectories of light rays are *geodesics* for the conformal metric

$$\mathrm{d}\sigma^2 = \frac{\mathrm{d}s^2}{v^2(x)}.$$

In the 18th century the calculus of variations had an enormous growth trough the work of Euler, Lagrange and Maupertuis. Euler first introduced the classical functional

$$I(x) = \int_a^b \mathcal{L}(t, x(t), \dot{x}(t))\mathrm{d}t \tag{1.2}$$

and looked for a function $x(t)$ that minimizes (1.2). He found the so-called *Euler–Lagrange* equations

$$\frac{d}{dt}\frac{\partial L}{\partial \dot{x}} = \frac{\partial L}{\partial x} \tag{1.3}$$

representing necessary conditions satisfied by a minimum point of (1.2). Euler's method was considerably simplified by Lagrange who proved (1.3) using the well-known method of taking variations of (1.2) with respect to a parameter. The work of Lagrange provided the basis of *Analytical Mechanics*.

In the 19th century Hamilton stated the *least action principle*. We present a simplified version of it. Consider the dynamical system in $\mathbf{R}^3$ (or in any Euclidean space)

$$\begin{cases} \ddot{x} = \nabla V(x) \\ x(0) = p;\, x(1) = q. \end{cases} \tag{1.4}$$

Then the solutions of (1.4) are the curves that minimize the *action functional*

$$I(x) = \frac{1}{2}\int_0^1 |\dot{x}(t)|^2 \mathrm{d}t + \int_0^1 V(x(t))\mathrm{d}t \tag{1.5}$$

on the space $\Omega(\mathbf{R}^3; p, q)$ of the C^1-curves $y\colon [0, 1] \longrightarrow \mathbf{R}^3$ such that $y(0) = p$, $y(1) = q$.

On the other hand there are examples of systems of differential equations like (1.4), whose solutions are not minima, *saddle points* for the action integral. Hamilton's principle should be known as the *stationary action principle* [1].

A system all of whose solutions are saddle points for the action integral can be easily constructed and we will find a basic example in general relativity. Here we present an example from differential geometry, which was developing with the work of Gauss, Lobachevski and Riemann in the same period as Hamilton's principle. Consider a surface $\mathcal{M}$ contained in $\mathbf{R}^3$ (or, in general, a submanifold of Euclidean space $\mathbf{R}^n$ or a Riemannian manifold). A smooth curve $\gamma : [0, 1] \longrightarrow \mathcal{M}$ is a *geodesic* for the surface $\mathcal{M}$ if, for any $s \in [0, 1]$,

$$P_{T_{\gamma(s)}}(\ddot{\gamma}(s)) = 0\,, \tag{1.6}$$

where $P_{T_{\gamma(s)}}$ denotes the orthogonal projection of $\mathbf{R}^3$ on the tangent space $T_{\gamma(s)}\mathcal{M}$ to $\mathcal{M}$ at $\gamma(s)$. From a physical point of view a geodesic represents the trajectory of a massive point, constrained on the surface $\mathcal{M}$, on which no force acts. Geodesics satisfy a variational principle. Indeed, fix two points p and q on $\mathcal{M}$; then (1.6) gives the Eulero–Lagrange equations for two functionals, the *length*,

$$L(x) = \int_0^1 \sqrt{\langle \dot{x}, \dot{x} \rangle} \mathrm{d}s$$

and the *action*,

$$E(x) = \int_0^1 \langle \dot{x}, \dot{x} \rangle \mathrm{d}s\ , \tag{1.7}$$

on the space $\Omega(p, q; \mathcal{M})$ of the C^1-curves $x(s)\colon [0, 1] \longrightarrow \mathcal{M}$ such that $x(0) = p$, $x(1) = q$. An example of geodesic which is not a minimum for E (and for L) is the following. Consider the 2-sphere S^2; it is well–known that the geodesics on S^2 are the great circles. Let p, q be two nonantipodal points of S^2. Then a great circle joining p and q can be divided into two arcs, the first of mimimal length (and action), which is the absolute minimum for L and E, the second which is a saddle point for L and E (indeed it contains a conjugate point).

Another basic variational principle stated in the first part of the 19th century and playing a key role in mathematical physics (in particular in potential theory) is the *Dirichlet Principle*. Let Ω be a bounded open region of the space $\mathbf{R}^3$ with smooth boundary $\partial\Omega$ and let $f\colon \partial\Omega \longrightarrow \mathbf{R}$ be a smooth function. Consider the *Dirichlet problem* for f,

$$\begin{cases} \Delta u = 0 \\ u_{|\partial\Omega} = f \end{cases} \tag{1.8}$$

where Δ is the Laplace operator. The *Dirichlet principle* claims that a solution of (1.8) is a function which minimizes the *Dirichlet integral*

$$E(v) = \iiint_\Omega |\nabla v|^2 \mathrm{d}^3 x$$

among all functions $v \in C^2(\Omega)C^1(\overline{\Omega})$ such that $v_{|\partial\Omega} = f$.

Dirichlet and Riemann were fully convinced of the existence of a solution of (1.8). Their argument were very simple. Since $E(v)$ is nonnegative, it necessarily has a minimum point which solves (1.8) by the Dirichlet principle. In particular Riemann applied the existence of harmonic functions on open regions of the plane to get very deep results in complex function theory.

In 1870 a counterexample due to Weierstrass showed that the argument of Dirichlet and Riemann was wrong. Consider the functional

$$I(x) = \int_0^1 t^2 \dot{x}^2 \mathrm{d}t$$

on the space of C^1-curves $x(t)\colon [-1, 1] \longrightarrow \mathbf{R}$ such that $x(-1) = -1$, $x(1) = 1$. $I(x)$ is nonnegative and hence, according to Dirichlet and Riemann, it has a minimum point. On the other hand it is easy to see that for any $\epsilon > 0$ there exists a function x_ϵ such that $I(x_\epsilon) < \epsilon$. Then the minimum of I is 0. Therefore, if x is the minimum point of I, then $I(x) = 0$ and necessarily

$$x(t) = \begin{cases} -1, & t \in [-1, 0[\\ 1, & t \in]0, 1] \end{cases}$$

Hence the minimum is not a continuous function and we can not evaluate $I(x)$. Another point of view of the counterexample of Weierstrass is the following: the minimum exists but it is *irregular*.

After the work of Weierstrass, the problem of the existence and of the regularity of minima for variational problems played a central role in mathematics. In 1900, Hilbert announced his famous 23 problems for mathematics in 20th century at the International Mathematical Congress in Paris . Hilbert's problem XX was the existence and regularity of minima for variational problems. In 1901 Hilbert rigorously proved the Dirichlet principle and the existence and the regularity for the solution of (1.8). Modern calculus of variations of minima begins with the work of Hilbert and such other mathematicians as Lebesgue, Caratheodory and Frechet and is still a very active field of research. The contribution of the Italian school of calculus of variations of minima is enormous. We could cite for example, Tonelli, Caccioppoli and De Giorgi.

At the end of the 19th century a new point of view for variational problems was introduced by Poincaré in his study of dynamical systems and periodic orbits. For instance, consider again a surface $\mathcal{M}$ in $\mathbf{R}^3$ (or a Riemannian manifold); then a geodesic $\gamma : [0, 1] \longrightarrow \mathcal{M}$ is said to be *closed* if $\gamma(0) = \gamma(1)$, $\dot{\gamma}(0) = \dot{\gamma}(1)$. Closed geodesics are the stationary points of the action integral (1.7) on the space $\Lambda(\mathcal{M})$ of C^1-curves $y(t)$ on $\mathcal{M}$ such that $y(0) = y(1)$. But they are not the absolute minimum points for E, since the minimum points of E on $\Lambda(\mathcal{M})$ are the trivial curves with image reduced to a point. Thus if one looks for closed geodesics, on a Riemannian manifold, one has to find critical points which are not absolute minima. The first result in this direction was obtained by Birkhoff who proved the existence of a closed geodesics on a compact surface of genus 0 (without *holes*, for instance, the sphere). This can be considered the first result in what is now called *Calculus of Variations in the Large* or *Critical Point Theory*, which studies not only the minimum points of a function, but all of its critical points. By the second decade of the 20th century, with the seminal works of Morse in the U.S.A. [2] and Lusternik–Schnirelmann in the U.S.S.R. [3], critical point theory for functionals has had an enormous impact and has had its applications in analysis, geometry, mathematical physics, ...

Let $\mathcal{M}$ be a smooth (for instance C^∞) compact manifold and $f : \mathcal{M} \longrightarrow \mathbf{R}$ a smooth function. A point $p \in \mathcal{M}$ is said to be a *critical point* for f if $f'(p) = 0$, where $f'(p)$ denotes the differential of f at p. A number $c \in \mathbf{R}$ is said to be a *critical value* for f if there exists a critical point p of f such that $f(p) = c$, otherwise c is called a *regular value*. The key idea of critical point theory is the following (Morse, Lusternik–Schnirelmann):

The topological structure of the manifold $\mathcal{M}$ has a deep influence on the (differential) structure of the set of critical points of f for any smooth function f.

The influence of the topology of $\mathcal{M}$ on the critical points of f is exploited by the basic *Deformation Lemma* due to Morse (cf. [4] for the proof of a more general version of the lemma). Let $f: X \longrightarrow \mathbf{R}$ be a function defined on a set X. For any $c \in \mathbf{R}$ the sublevel f^c of f is defined as

$$f^c = \{x \in X : f(x) \leq c\} . \tag{1.9}$$

Moreover, for any pair of real numbers $a < b$, the strip f_a^b is defined as

$$f_a^b = \{x \in X : a \leq f(x) \leq b\} .$$

Lemma 1.1 (Deformation Lemma). *Let $f: \mathcal{M} \longrightarrow \mathbf{R}$ be a smooth function defined on the compact manifold $\mathcal{M}$. Let $a < b$ and assume that the strip f_a^b does not contain critical points of f.*

Then the sublevel f^b can be continuously deformed by a homotopy on the sublevel f^a. This means that there exists a continuous function $H: [0,1] \times f^b \longrightarrow f^b$ such that:

(a) $H(0, x) = x$ *for any* $x \in f^b$;
(b) $H(1, x) \in f^a$ *for any* $x \in f^b$;
(c) $H(t, y) = y$ *for any* $y \in f^a$.

In other words, if there are no critical values between the numbers a and b, the sublevel f^b can be continuously deformed onto the sublevel f^a and f^a remains fixed during the deformation.

The proof of the Deformation Lemma for C^2-functions is obtained by constructing the homotopy $H(t, x)$ by using the flow lines of the vector field ∇f, where ∇f is the *gradient* of f with respect to some Riemannian metric on $\mathcal{M}$ (any metric works since $\mathcal{M}$ is compact). If the function f is only of class C^1, the gradient vector field is replaced by the *pseudogradient vector field* introduced in [5].

The Deformation Lemma immediately gives some consequences on the properties of the critical points of a function.

1. *If there are no critical points in the strip f_a^b, the homotopy–invariant topological properties of sublevels f^b and f^a (homotopy groups, homology and cohomology groups) are the same.*
2. *If a sublevel f^b can not be deformed onto a lower sublevel f^a, there exists a critical point of f in the strip f_a^b.*

The Deformation Lemma is the starting point for all the results relating the critical points of a function to the topology of the manifold on which the function is defined. We start with the results of Lusternik–Schnirelmann, which give a lower bound on the number of critical points of a function, measured by a homotopical invariant of the manifold, the *category*.

Definition 1.2. *Let X be a topological space and A a subset of X. A is contractible in X if there exist a point $q \in X$ and a continuous map $H\colon [0,1] \times A \longrightarrow X$ such that*

$$H(0,x) = x \quad \text{for any } x \in A;$$
$$H(1,x) = p \quad \text{for any } x \in A.$$

Definition 1.3. *Let X be a topological space; the category of X is the minimal number* $\mathrm{cat}X$ *of closed and contractible subsets in X covering the whole space X. If such a minimal number does not exist, we set* $\mathrm{cat}X = +\infty$.

Theorem 1.4 (Lusternik–Schnirelmann). *Let $f\colon \mathbf{M} \longrightarrow \mathbf{R}$ be a C^1 function defined on a compact manifold $\mathcal{M}$. Then f has at least* $\mathrm{cat}\mathcal{M}$ *critical points.*

The Lusternik–Schnirelmann Theorem claims that the category is a lower bound on the number of critical points of a function on a compact manifold. For instance any function on the unit sphere S^2 in $\mathbf{R}^3$, $S^2 = \{(x,y,z) \in \mathbf{R}^3 : x^2 + y^2 + z^2 = 1\}$, has at least 2 critical points (the absolute minimum and maximum). Any smooth function on the torus $T = S^1 \times S^1$ has at least 3 critical points since $\mathrm{cat}T = 3$. In general the category of a compact surface of genus g is equal to $2g + 1$.

Lusternik and Schnirelmann applied their result to obtain some beautiful results on critical points for functions. For instance, they extended results on the existence of N real eigenvalues for a symmetric $N \times N$ matrix to nonlinear functions. Let $f\colon \mathbf{R}^N \longrightarrow \mathbf{R}$ be a smooth even function ($f(-x) = f(x)$). The critical points of f on the sphere S^{N-1} are the solutions of the nonlinear eigenvalue problem

$$\nabla f(x) = \lambda x. \tag{1.10}$$

Lusternik and Schnirelmann proved the existence of *at least* N pairs $\{x, -x\}$ of solutions of (1.10). The proof is obtained showing that $\mathrm{cat}\mathbf{P}^{N-1} = N$, where $\mathbf{P}^{N-1}$ is projective N-space. Lusternik and Schnirelmann also applied their ideas on critical point theory to infinite dimensional variational problems, in particular, to closed geodesics on compact surfaces. For instance, they proved the existence of three distinct prime geodesics on any compact surface of genus [6].

At the same times as the work of Lusternik–Schnirelmann, Morse developed his critical point theory for functions on compact manifolds. Morse theory gives more accurate results on the relations between the critical points of a function $f\colon \mathcal{M} \longrightarrow \mathbf{R}$, $\mathcal{M}$ compact, and the topology of $\mathcal{M}$. But one has to pay for this by a loss of generality in the choice of f. Indeed Morse theory applies only to functions f whose critical points x are *nondegenerate*, which means that the Hessian matrix $H_f(x)$ at any critical point has nonnull eigenvalues.

Let $f\colon \mathcal{M} \longrightarrow \mathbf{R}$ be a *Morse function*, that is, any critical point of f is nondegenerate; let x be a critical point of f; then the *Morse index* of f at x is the number $m(x,f)$ of negative eigenvalues of the Hessian matrix $H_f(x)$. In other words $m(x,f)$ denotes the number of directions where f decreases in a neighborhood of the critical point x.

The Morse polynomial of f is defined as

$$\mathcal{M}_f(t) = \sum_{x \in K(f)} t^{m(x,f)}, \tag{1.11}$$

where $K(f)$ is the set of critical points of f. Any critical point x of f contributes to the Morse polynomial with a monomial of power $m(x, f)$. In particular, setting $t = 1$, we see that $\mathcal{M}_f(1)$ is equal to the number $\mathrm{card}K(f)$ of critical points of f.

Now let $\mathcal{K}$ be an algebraic field; the *Poincaré polynomial* $\mathcal{P}(\mathcal{M}, \mathcal{K})$ of the manifold $\mathcal{M}$ (with coefficients in $\mathcal{K}$) is defined as

$$\mathcal{P}(\mathcal{M}, \mathcal{K})(t) = \sum_{q=0}^{\infty} \beta_q(\mathcal{M}, \mathcal{K}) t^q, \tag{1.12}$$

where $\beta_q(\mathcal{M}, \mathcal{K})$ is the dimension of the q-th homology group $H_q(\mathcal{M}, \mathcal{K})$ (the so-called *Betti numbers*).

We now state the results of Morse in a modern fashion [7]. It shows how the set of critical points of a Morse function, depending on the differential structure, is related to the topological structure of the manifold which is exploited by the Poincaré polynomial with respect to some field.

Theorem 1.5. *Let $f : \mathcal{M} \longrightarrow \mathbf{R}$ be a Morse function defined on the compact manifold $\mathcal{M}$. Let $\mathcal{K}$ be a field. Then there exists a polynomial $Q(t)$ with nonnegative integer coefficients such that*

$$\mathcal{M}_f(t) = \mathcal{P}(\mathcal{M}, \mathcal{K})(t) + (1 + t)Q(t). \tag{1.13}$$

In other words, (1.13) says that the critical points of f are of two types:

- the first type comes from the topology of the manifold and the critical points are counted by the Betti numbers;
- the second comes from the form of the function and the critical points are counted by the polynomial $Q(t)$.

Putting $t = 1$ in (1.13), we obtain an evaluation of the number of critical points of f:

$$\mathrm{card}K(f) = \mathcal{B} + 2Q \tag{1.14}$$

where $\mathcal{B}$ is the sum of all the Betti numbers and Q is a nonnegative integer. In particular the nuber of *topological* critical points is equal to the *total Betti number* $\mathcal{B} = \sum_q \beta_q(\mathcal{M}, \mathcal{K})$, while the number of all the other critical points is *even* (this fact plays a basic role in the proof that the number of images in the gravitational lens effect is *odd*).

Whenever the polynomial $Q(t)$ in Theorem 1.5 is null, f is said to be a *perfect function*: it has only topological critical points. For instance, consider the function $f(x, y, z) = z$ on the unit sphere S^2. f has only two critical points: the north pole $N = (0, 0, 1)$, which is the absolute maximum point, and the south pole $S = (0, 0, -1)$, which is the absolute minimum point. The function f is perfect. Indeed S and N have Morse index respectively equal to 0 and 2. Here the Morse polynomial is equal to $1 + t^2$. On the other hand the Betti numbers of S^2 with respect to any field $\mathcal{K}$ are the following: $\beta_0(S^2, \mathcal{K}) = 1$ (S^2 is connected), $\beta_1(S^2, \mathcal{K}) = 0$ (S^2 is simply connected), $\beta_2(S^2, \mathcal{K}) = 1$ (S^2 is a compact oriented surface). The higher order Betti numbers are null. Then the Poincaré polynomial is $1 + t^2$, and $Q(t) \equiv 0$.

Morse theory gives more precise results than Lusternik–Schnirelmann theory. For instance, the function $f(x, y, z) = z$ on the torus $T = S^1 \times S^1$ has, according to Lusternik–Schnirelmann, at least 3 critical points. On the other hand the total Betti number of the torus with respect to the field $\mathbf{R}$ is 4, so f has at least 4 critical points. The number of critical points is exactly 4, so f is a perfect function. Indeed, $\beta_0(T, \mathbf{R}) = 1$, $\beta_1(T, \mathbf{R}) = 2$ (by the Künneth formula; see [8]), $\beta_2(T, \mathbf{R}) = 1$. Hence $\mathcal{B} = 4$. Moreover it follows from (1.14) that f has 2 saddle points.

Morse was interested in his study of critical points for functionals in variational problems for geodesics on Riemannian manifolds. Let $(\mathcal{M}, \langle\cdot, \cdot\rangle)$ be a complete Riemannian manifold; a smooth curve $\gamma : I \longrightarrow \mathcal{M}$ is said to be *geodesic* if $D_s \dot{\gamma} = 0$, where D_s is the covariant derivative along γ. The intrinsic notion of geodesic on a Riemannian manifold (and, in general, on a semi-Riemannian one) extends the notion of geodesic on a surface.

Fix $p, q \in \mathcal{M}$ and let $\Omega(p, q, \mathcal{M})$ be the set of continuous and piecewise smooth curves $x(s) : [0, 1] \longrightarrow \mathcal{M}$ such that $x(0) = p, x(1) = q$. On the space $\Omega(p, q, \mathcal{M})$ we define the *action integral*

$$E(x) = \int_0^1 \langle \dot{x}(s), \dot{x}(s) \rangle \, \mathrm{d}s. \tag{1.15}$$

Unfortunately the results of Theorem 1.5 can not be applied to the functional E, since $\Omega(p, q, \mathcal{M})$ is not a compact manifold. Indeed it is modeled on a function space which is *infinite dimensional*.

To be more precise, $\Omega(p, q, \mathcal{M})$ is not equipped with any structure of infinite differentiable manifold. However it is possible to define a vector space of admissible variations (in other words, a *tangent space*) to a curve $x \in \Omega(p, q, \mathcal{M})$. It consists of continuous and piecewise smooth vector fields along x having null boundary conditions [9]. Moreover the derivatives of E at a curve x along the admissible directions can be evaluated and the following result holds.

Theorem 1.6. *A curve $x \in \Omega(p, q, \mathcal{M})$ is a stationary point for the action integral E if and only if x is a geodesic joining p and q.*

Now, let $c \in \mathbf{R}$ be a regular value of E, i.e., there are no geodesics x in $\Omega(p, q, \mathcal{M})$ with $E(x) = c$ and consider the sublevel $E^c = \{x \in \Omega(p, q, \mathcal{M}) : E(x) \le c\}$. Even if $\Omega(p, q, \mathcal{M})$ is not a compact manifold, Morse proved that the set E^c is *homotopically equivalent* to a finite product of compact manifolds. The proof is obtained essentially by approximating any curve x of E^c with *broken geodesics*; the number of breaks is finite and depends only on c. Transferring the problem to the homotopic manifolds (depending on c), a Morse theory for the geodesics joining p and q can be developed. The Hessian form $H_E(x)$ and the index $m(x, E)$ for a geodesic x joining p and q can be defined. Namely, the Hessian $H_E(x)\colon T_x\Omega(p, q, \mathcal{M}) \times T_x\Omega(p, q, \mathcal{M}) \longrightarrow \mathbf{R}$ is defined for any "tangent vector" ξ along x by

$$H_E(x)[\xi, \xi] = \int_0^1 (\langle D_s\xi, D_s\xi\rangle - \langle R(\dot{x}, \xi)\dot{x}, \xi\rangle)\, ds,$$

where $R(\cdot, \cdot)[\cdot]$ is the Riemann tensor for the Riemannian metric $\langle\cdot, \cdot\rangle$. Moreover the index $m(x, E)$ is defined as the maximal number of linearly independent tangent directions ξ such that $H_E(x)[\xi, \xi] < 0$.

One of the most important results of Morse is the *Index Theorem* which relates the analytical index $m(x, E)$ to the geometrical index $\mu(x)$, the number of conjugate points along x ([2, 9] and see [10] for a proof based on arguments of functional analysis).

Theorem 1.7. *The Morse index $m(x, E)$ is equal to the geometric index $\mu(x)$ of x, defined as the number of conjugate points $x(s)$, $s \in]0, 1[$, counted with their multiplicities.*

We recall that a point $x(s)$, $s \in]0, 1]$ is said to be *conjugate* to $x(0)$ along x if there exists a smooth vector field ξ along $x_{|[0,s]}$ such that

$$\begin{cases} D^2\xi + R(\dot{x}, \xi)\dot{x} = 0 \\ \xi(0) = 0, \xi(s) = 0. \end{cases} \tag{1.16}$$

The multiplicity of $x(s)$ is the number of linearly independent solutions of (1.16). It is always finite.

The results obtained by Morse for Riemannian geodesics were the starting point for the study of the relations between the geometry and the topology of a Riemannian manifold.

After the Second World War results of functional analysis on infinite dimensional spaces (Banach spaces, Hilbert spaces) were extended to differentiable manifolds modeled on infinite dimensional spaces (Hilbert manifolds, Finsler manifolds). In particular in a number of papers [11–13]) Palais and Smale extended the results of Lusternik and Schnirelmann and of Morse theory to Banach and Hilbert manifolds. Palais, in the seminal paper [11], extended Morse theory to Hilbert manifolds, and using some ideas introduced by Sobolev in the study of partial differential equations, applied his abstract results to prove, in a completely new way, all the results of Morse theory for Riemannian geodesics. From 1970 critical point theory exploded with many new abstract results and applications to partial differential equations of elliptic and hyperbolic type, Lagrangian and Hamiltonian systems, differential geometry and mathematical physics, and it is still a very active field of research. In particular many results have been obtained for critical points which are not minima for functionals on Banach spaces such as the *Mountain Pass Theorem*, the *Saddle Point Theorem* and the *Linking Theorem*. We refer [4, 14, 15] for the main results and applications in critical point theory for functionals on infinite dimensional manifolds.

The basic result of critical point theory for functionals defined on Hilbert manifolds, for instance, a Hilbert space or the unit sphere of a Hilbert space (for results on Banach–Finsler manifolds see [5]) is the extension of the Deformation Lemma to noncompact manifolds (moreover, in the case of finite dimensional manifolds, the result is false; it is not satisfied by the function e^x). Let X be a Hilbert manifold and $f: X \longrightarrow \mathbf{R}$ a smooth functional. Let a and b be two regular values for f, $a < b$. The main difficulty in proving the Deformation Lemma is that the stripe f_a^b is not a compact subset, so the vector field ∇f can not be complete and the associated flow can escape to infinity in finite time.

In order to avoid this situation a compactness property of the functional is introduced. This property is known as the *Palais–Smale* condition.

Definition 1.8. *Let $f: X \longrightarrow \mathbf{R}$ be a smooth functional defined on the Hilbert manifold X. We say that f satisfies the Palais–Smale (PS) condition if any sequence $(x_m)_{m\in\mathbf{N}}$ of points of X such that*

the sequence $\{f(x_m)\}_{m\in\mathbf{N}}$ is bounded;
the sequence $\{f'(x_m)\}_{m\in\mathbf{N}}$ tends to 0,

contains a convergent subsequence. Here $f'(x)$ denotes the differential of f at x (an element of the cotangent bundle of X).

The Palais–Smale condition permits us to to prove the Deformation Lemma for a functional defined on a not necessarily compact manifold (note that it is always satisfied on a compact manifold). Lusternik–Schnirelmann and Morse theories can be developed for smooth functionals defined on a Hilbert manifold, bounded from below and satisfying the Palais–Smale condition [15].

2 Geodesics on Riemannian manifolds

The first application of critical point theory for infinite dimensional spaces was to Riemannian geodesics. Let $\mathcal{M}$ be a smooth manifold; in [11] is constructed the *Sobolev manifold* $H^{1,2}([0,1],\mathcal{M})$ of the curves $x(s)\colon [0,1] \longrightarrow \mathcal{M}$ such that, for any local chart (U,φ) of the manifold satisfying $Ux([0,1]) \neq \emptyset$, the Euclidean curve $\varphi\circ x\colon x^{-1}(U) \longrightarrow \mathbf{R}$ belongs to the *Sobolev space* $H^{1,2}(x^{-1}(U),\mathbf{R}^n)$, $n = \dim\mathcal{M}$. In [11] it is proved that $H^{1,2}([0,1],\mathcal{M})$ is equipped with a structure of *infinite dimensional Hilbert manifold* modeled on the Sobolev space $H^{1,2}([0,1],\mathbf{R}^n)$ [16] and the tangent space at a curve x consists of the vector fields of class $H^{1,2}$ along x. Moreover, let $p, q \in \mathcal{M}$; then the set $\Omega^{1,2}(p,q,\mathcal{M})$ of the curves $x \in H^{1,2}([0,1],\mathcal{M})$ such that $x(0) = p$ and $x(1) = q$ is a submanifold of $H^{1,2}([0,1],\mathcal{M})$. The tangent space to a curve $x \in \Omega^{1,2}(p,q,\mathcal{M})$ consists of the vector fields $\xi(s)$ along x of class $H^{1,2}$ such that $\xi(0) = 0$ and $\xi(1) = 0$.

Now let $(\mathcal{M}, \langle \cdot, \cdot \rangle)$ be a complete Riemannian manifold and let p, q be two points on $\mathcal{M}$. Consider the action integral $E: \Omega^{1,2}(p, q, \mathcal{M}) \longrightarrow \mathbf{R}$,

$$E(x) = \int_0^1 \langle \dot{x}, \dot{x} \rangle \, \mathrm{d}s. \tag{2.1}$$

The following results are proved in [11].

Theorem 2.1. *The action integral is a smooth functional on the manifold $\Omega^{1,2}(p, q, \mathcal{M})$. Moreover, a curve $x \in \Omega^{1,2}(p, q, \mathcal{M})$ is a critical point of E on $\Omega^{1,2}(p, q, \mathcal{M})$ if and only if x is a geodesic joining p and q. Finally, under the assumption of completeness of $(\mathcal{M}, \langle \cdot, \cdot \rangle)$, the functional E satisfies the Palais–Smale condition on E.*

The results of Theorem 2.1 permit us to develop Lusternik–Schnirelmann and Morse theories for Riemannian geodesics. For the Lusternik–Schnirelmann category, the following basic result is due to Fadell and Husseini [17].

Theorem 2.2. *Let $\mathcal{M}$ be a manifold and $p, q \in \mathcal{M}$. Then:*

- *if $\mathcal{M}$ is contractible into itself (cf. Definition 1.2), then* $\mathrm{cat}\Omega^{1,2}(p, q, \mathcal{M}) = 1$*;*
- *if $\mathcal{M}$ is noncontractible into itself, then* $\mathrm{cat}\Omega^{1,2}(p, q, \mathcal{M}) = +\infty$.

From the results above we obtain the following result on the existence and multiplicity of Riemannian geodesics.

Theorem 2.3. *Let $(\mathcal{M}, \langle \cdot, \cdot \rangle)$ be a complete Riemannian manifold and p, q two points on $\mathcal{M}$. Then there exists at least one geodesic joining p and q. Moreover, if $\mathcal{M}$ is noncontractible, there exist infinitely many geodesics joining p and q. Moreover, it can be proved that the action integral assumes arbitratrily large values on the geodesics.*

The first part of Theorem 2.3 can also be proved geometrically (Hopf–Rinow Theorem) or by a minimum argument (Weierstrass Theorem). The case of a noncontractible manifold can be proved in general only with a *global analytical* argument. It extends a previous result of Serre obtained for compact simply connected manifolds, using deep results of algebraic topology.

Theorem 2.2 also allows us to develop an infinite dimensional Morse theory for Riemannian geodesics [11]. Since $\Omega^{1,2}(p, q, \mathcal{M})$ is noncompact, it could possess infinitely many nonnull Betti numbers (and this occurs if $\mathcal{M}$ is noncontractible! see [17]). Moreover the action integral could have (and it really does if $\mathcal{M}$ is noncontractible; see the previous theorem) infinitely many critical points. So the notions of Morse polynomial and Poincaré polynomial have to be replaced by the notions of *Morse series* and *Poincaré series*. Moreover, as shown by Morse, the index of a geodesic x as a critical point of E is finite and is equal to its *geometric index* $\mu(x)$, the number of conjugate points along x counted with their multiplicities. Collecting together all the results we obtain the Morse relations for two nonconjugate points on a Riemannian manifold.

Theorem 2.4. *Let $(\mathcal{M}, \langle\cdot,\cdot\rangle)$ be a complete Riemannian manifold and let p, q be two points on $\mathcal{M}$. Assume that p and q are nonconjugate, that is, q is not conjugate to p along any geodesic joining them. Let $\mathcal{K}$ be a field. Then there exists a series $Q(t)$ with coefficients in $\mathbf{N} \cup \{+\infty\}$ such that, with $\mathcal{G}_{p,q}$ the set of the geodesics joining p and q,*

$$\sum_{x \in \mathcal{G}_{p,q}} t^{\mu(x)} = \mathcal{P}(\Omega^{1,2}(p,q,\mathcal{M}),\mathcal{K})(t) + (1+t)Q(t). \tag{2.2}$$

Note that the assumption on nonconjugacy of p and q implies that the set of geodesics joining p and q is discrete; hence the sum in the left side of (2.2) makes sense. Moreover, setting $t = 1$ in (2.2), we get

$$\operatorname{card} \mathcal{G}_{p,q} = \mathcal{B}_{\mathcal{K}} + 2Q_{\mathcal{K}}, \tag{2.3}$$

where $\mathcal{B}_{\mathcal{K}}$ is the total Betti number of $\Omega^{1,2}(p,q,\mathcal{M})$ with respect to the field $\mathcal{K}$ and $Q_{\mathcal{K}}$ is a number in $\mathbf{N} \cup \{+\infty\}$, the sum of all the coefficients of Q. In particular, if $\mathcal{M}$ is contractible and $Q_{\mathcal{K}}$ is finite, the set of geodesics joining p and q is *odd*.

3 Geodesics on Lorentzian manifolds

In the 1980s Benci and Fortunato first applied the results of critical point theory to Lorentzian geometry, and in particular to geodesics [18]. After [18], many results have been obtained in the study of geodesics on space–times by variational methods. Here we present the main results concerning two problems: the existence and multiplicity of geodesics joining two points on a Lorentzian manifold and, in the next section, applications of critical point theory to the light rays joining a point with a timelike curve.

The only global result in Lorentzian geometry before the paper [18] was the Seifert theorem on the existence of causal geodesics in *globally hyperbolic space–times* (we refer to the books [19–21] for the basic notions on Lorentzian geometry).

Theorem 3.1. *Let $(\mathcal{M}, \langle\cdot,\cdot\rangle)$ be a globally hyperbolic Lorentzian manifold and let p and q be two causally related points of $\mathcal{M}$. Then there exists a causal geodesic joining p and q.*

We recall that p and q are causally related if there exists a causal curve joining them. Moreover a geodesic is said to be causal if it is nonspacelike. The proof of the Avez–Seifert theorem [21] is based on the maximization of the *length functional*

$$L(z) \int_0^1 \sqrt{-\langle \dot z, \dot z\rangle}\, \mathrm{d}s \tag{3.1}$$

on the space of causal curves joining p and q. The global hyperbolicity of the metric guarantees *good compactness properties* of the space of such causal curves and the proof follows via the Weierstrass Theorem.

If one is interested in finding a geodesic joining two arbitrary points on a Lorentzian manifold, the situation is more delicate. First of all, the length functional is not well-defined because the metric is not definite and good intrinsic notions of *distance* and *completeness* cannot be introduced on a Lorentzian manifold. The Theorem of Hopf–Rinow [19] claims that a Riemannian manifold is complete if and only if it is *geodesically complete*, that is, any maximal geodesic for the metric is defined on the whole real line **R**. Moreover any geodesically complete (or, equivalently, metrically complete) Riemannian manifold is *geodesically connected*, that is, any pair of points can be joined by a geodesic.

Unfortunately in Lorentzian geometry there are simple counterexamples which show that geodesical completeness does *not* imply geodesic connectedness. For instance, *de Sitter* and *Anti–de Sitter* space–times are not geodesically connected [21]. Note that the de Sitter space–time is globally hyperbolic; hence such a condition also does not guarantee the existence of spacelike geodesics. Geodesical connectedeness also does not hold in physically pathological situations. For instance, there exist *compact* Lorentzian manifolds which are not geodesically connected [22].

We now present the main difficulties (from a variational point of view) in establishing the existence of geodesics. Let $(\mathcal{M}, \langle\cdot,\cdot\rangle)$ be a Lorentzian manifold and let p and q be two points of $\mathcal{M}$; then the geodesics joining p and q are the critical points, as in the Riemannian case, of the action integral $E\colon \Omega^{1,2}(p,q;\mathcal{M}) \longrightarrow \mathbf{R}$,

$$E(z) = \int_0^1 \langle \dot{z}(s), \dot{z}(s)\rangle \, ds. \tag{3.2}$$

The proof is the same as in the Riemannian case since only the properties of the Levi–Civita connection and the nondegeneracy of the metric tensor are applied. But now, different from the Riemannian case, E is an unbounded functional both from below and from above. Moreover it can be proved that the Morse index of the critical points of E is equal to $+\infty$, since the Hessian of E at a geodesic z has infinitely many positive and negative eigenvalues (a situation similar to Hamiltonian systems and wave equations). We recall that the Hessian form $H_E(z)\colon T_z\Omega^{1,2}(p,q;\mathcal{M}) \times T_z\Omega^{1,2}(p,q;\mathcal{M}) \longrightarrow \mathbf{R}$ is given by

$$H_E(z)[\zeta,\zeta'] = \int_0^1 \big(\langle D_s\zeta, D_s\zeta'\rangle - \langle R(\dot{z},\zeta)\dot{z},\zeta'\rangle\big)\, ds, \tag{3.3}$$

where $R(\cdot,\cdot)[\cdot]$ denotes the Riemann tensor for the metric. This spectral property of H_E makes it very difficult to apply critical point theory to the functional. This fact is strictly related to the contractibility of the unit sphere of an infinite dimensional Hilbert space. Another problem comes from the fact that there are no good completeness or compactness properties for a Lorentzian manifold: compactness or global hyperbolicity do not guarantee the existence of a critical point for E.

On the other hand, the functional E has a good *geometry*, since it naturally has the geometry of the Rabinowitz *Saddle Point Theorem* [14].

Theorem 3.2. *Let $\mathcal{M} = \mathcal{M}_0 \times H_n$ be a product manifold such that $\mathcal{M}_0$ is a Riemannian manifold (possibly infinite dimensional) and H_n is a linear manifold of* *finite dimension n embedded in a Hilbert space H. Let* $f : \mathcal{M} \longrightarrow \mathbf{R}$ *be a smooth functional and assume that* f *satisfies the Palais–Smale condition and there exist two points* $(x_0, y_0) \in \mathcal{M}$ *and* $\bar{x} \in \mathcal{M}$ *such that*

$$f(x, y_0) \to +\infty \text{ as } \mathrm{d}(x, \bar{x}) \to +\infty;$$
$$f(x_0, y) \to -\infty \text{ as } \|y\| \to +\infty.$$

Then f *admits a critical point.*

An example of a functional as in the previous theorem is the classical *saddle function* $f(x, y) = x^2 - y^2$.

In fact, a purely geometric theorem which guarantees the geodesic connectedness of a Lorentzian manifold is not known. However some results have been obtained for suitable classes of Lorentzian manifolds. Such classes are sufficiently large to contain some of the most physically interesting classes of space–times in general relativity.

We first consider the class of *stationary Lorentzian manifolds*. A Lorentzian manifold $(\mathcal{M}, \langle\cdot,\cdot\rangle)$ is said to be *stationary* if it admits a timelike, Killing vector field [21]. A stationary Lorentzian manifold is said to be *regular* if the quotient space obtained from $\mathcal{M}$ by identifying the points on the same integral curve of Y is a manifold $\mathcal{M}_0$. Note that, in this case, $\mathcal{M}_0$ is a spacelike hypersurface of $\mathcal{M}$. Indeed the restriction of the metric to the fiber bundle of $\mathcal{M}_0$ is positive definite. If the field Y is complete, $\mathcal{M}$ is diffeomorphic to the product manifold $\mathcal{M}_0 \times \mathbf{R}$ and the metric $\langle\cdot,\cdot\rangle$ can be written in the following way. Let $z = (x, t) \in \mathcal{M} \equiv \mathcal{M}_0 \times \mathbf{R}$ and let $\zeta = (\xi, \tau) \in T_z\mathcal{M} \equiv T_x\mathcal{M}_0 \times \mathbf{R}$. Moreover, denote by $\langle\cdot,\cdot\rangle_0$ the (Riemannian) restriction of $\langle\cdot,\cdot\rangle$ to the tangent space at $\mathcal{M}_0 \times \{0\}$. Then

$$\langle\zeta, \zeta\rangle = \langle\xi, \xi\rangle_0 + 2\langle\delta(x), \xi\rangle_0\tau - \beta(x)\tau^2,$$

where $\delta(x)$ and $\beta(x)$ are, respectively, a smooth vector field and a smooth positive scalar field on $\mathcal{M}_0$.

Some examples of regular stationary manifolds are given by *Minkowski space–time*, *Schwarzschild space–time* outside the event horizon, *Reissner–Nördstrom space–time* outside the event horizons and *Kerr space–time* outside the *stationary limit surface*.

Let $z_0 = (x_0, t_0)$ and $z_1 = (x_1, t_1)$ be two points of a regular stationary space–time; then the geodesics joining them are the critical points of the action integral (3.2) which is, in this case, $E : \Omega^{1,2}(z_0, z_1; \mathcal{M}) = \Omega^{1,2}(x_0, x_1; \mathcal{M}_0) \times \Omega^{1,2}(t_0, t_1; \mathbf{R}) \longrightarrow \mathbf{R}$,

$$E(z) = E(x, t) = \int_0^1 \left[\langle\dot{x}, \dot{x}\rangle_0 + 2\langle\delta(x), \dot{x}\rangle_0\dot{t} - \beta(x)\dot{t}^2\right] \mathrm{d}s. \tag{3.4}$$

From (3.4) the saddle structure of the functional E is immediately seen. Unfortunately, E has no critical points even if the Riemannian metric $\langle\xi, \xi\rangle_0$ is complete. A classical counterexample is *anti de Sitter space–time*. In the 2-dimensional version

[23], $\mathcal{M} =]-\frac{\pi}{2}, \frac{\pi}{2}[\times \mathbf{R}$, $\beta(x) = 1/\cos^2(x)$, $\delta(x) = 0$ (the metric is *static*) and the Riemannian metric $\langle \cdot, \cdot \rangle_0$ is conformal to the Euclidean by the factor $1/\cos^2(x)$. Note that such a metric is complete, but such a Lorentzian manifold is *not* geodesically connected [23].

Some asymptotical properties on the coefficients of the metric are needed in rder to obtain geodesic connectedness of a regular stationary Lorentzian manifold. The following result is proved in [24] and it extends to regular stationary Lorentzian manifolds the existence and multiplicity results for Riemannian geodesics.

Theorem 3.3. *Let $(\mathcal{M}\langle \cdot, \cdot \rangle)$ be a regular stationary Lorentzian manifold and assume that:*

(a) *the Riemannian metric $(\mathcal{M}, \langle \cdot, \cdot \rangle_0)$ is complete;*
(b) $\sup_{x \in \mathcal{M}_0} \langle \delta(x), \delta(x) \rangle_0 < +\infty$;
(c) $0 < inf_{x \in \mathcal{M}_0} \beta(x) \leq \sup_{x \in \mathcal{M}_0} \beta(x) < +\infty$.

Then $(\mathcal{M}\langle \cdot, \cdot \rangle)$ is geodesically connected. Moreover, if $\mathcal{M}$ is noncontractible into itself, then for any pair of points of $\mathcal{M}$, there exist infinitely many geodesics joining them on which the action integral E tends to $+\infty$.

In [24] the case of manifolds with boundary is also considered, since many of the physically interesting space–times have boundaries. In particular the results of Theorem 3.3 are proved for some open subsets of Schwarzschild, Reissner–Nördstrom and Kerr space–times (cf. also [10]). The results of Theorem 3.3 also have been proved for the Schwarzschild space–time outside the event horizon, but the proof is rather delicate [25]. An extension to nonregular stationary Lorentzian space–time has been proved in [26].

The proof of Theorem 3.3 does not need the saddle point theorem and can be obtained with arguments similar to those for Riemannian geodesics. Indeed, using a global version of the implicit function theorem, we can see that the search for critical points of the functional E is equivalent to the search for the critical points of a functional defined only on the Riemannian coordinate x. Assumptions (a)–(c) allow us to show that such a new functional is bounded from below and satisfies the Palais–Smale condition. So the abstract results of critical point theory can be applied.

We consider now the class of *splitting* Lorentzian manifolds. A Lorentzian manifold $(\mathcal{M}, \langle \cdot, \cdot \rangle)$ is said to be splitting if $\mathcal{M} = \mathcal{M}_0 \times \mathbf{R}$, where $\mathcal{M}_0$ is a smooth manifold, and the metric $\langle \cdot, \cdot \rangle$ has the following form: for any $z = (x, t) \in \mathcal{M}$ and for any $\zeta = (\xi, \tau) \in T_z\mathcal{M} = T_x\mathcal{M}_0 \times \mathbf{R}$,

$$\langle \zeta, \zeta \rangle = \langle \alpha(x, t)\xi, \xi \rangle_0 + 2\langle \delta(x, t), \xi \rangle_0 \tau - \beta(x, t)\tau^2, \tag{3.5}$$

where $\langle \cdot, \cdot \rangle_0$ is a Riemannian metric on $\mathcal{M}_0$, $\alpha(x, t)$ is a positive definite operator on $T_x\mathcal{M}_0$ smoothly depending on (x, t), $\delta(x, t)$ is a tangent vector to $\mathcal{M}_0$ smoothly depending on z and $\beta\colon \mathcal{M} \longrightarrow \mathbf{R}$ a smooth positive scalar field. Whenever the vector field $\delta(x, t) \equiv 0$, $(\mathcal{M}, \langle \cdot, \cdot \rangle)$ is said to be *orthogonal splitting*.

The class of orthogonal splitting Lorentzian manifolds contains many physically relevant space–times. A deep result of Geroch [27] shows that any *globally hyperbolic* Lorentzian manifold is isometric to an orthogonal splitting Lorentzian manifold.

Let $z_0 = (x_0, t_0)$ and $z_1 = (x_1, t_1)$ be two points of a splitting Lorentzian manifold $(\mathcal{M}, \langle\cdot,\cdot\rangle)$; then the geodesics joining z_0 and z_1 are the critical points of the action integral (3.2). In this case $E: \Omega^{1,2}(z_0, z_1; \mathcal{M}) = \Omega^{1,2}(x_0, x_1; \mathcal{M}_0) \times \Omega^{1,2}(t_0, t_1; \mathbf{R}) \longrightarrow \mathbf{R}$ is given by

$$E(z) = E(x,t) = \int_0^1 \left[\langle\alpha(z)\dot{x}, \dot{x}\rangle_0 + 2\langle\delta(z), \dot{x}\rangle_0 \dot{t} - \beta(z)\dot{t}^2\right] \mathrm{d}s. \tag{3.6}$$

A splitting manifold is not geodesically connected in general. Some asymptotical assumptions on the coefficients α, β and δ are needed. A counterexample is *de Sitter space–time*, which has an orthogonal splitting structure with $\mathcal{M}_0 = S^n$, the standard n-dimensional unit sphere, $\langle\cdot,\cdot\rangle_0$ the standard Riemannian metric on S^n, $\alpha(x,t) = \cosh^2 t I$, where I is the identity operator and $\beta(z) = 1$.

We present now a result on the existence and multiplicity of geodesics on a orthogonal splitting Lorentzian manifold.

Theorem 3.4. *Let $(\mathcal{M}, \langle\cdot,\cdot\rangle)$ be an orthogonal splitting Lorentzian manifold and assume that:*

(a) *$(\mathcal{M}, \langle\cdot,\cdot\rangle_0)$ is a complete Riemannian manifold;*
(b) *there exists $\lambda > 0$ such that, for any $z = (x,t) \in \mathcal{M}$ and $\xi \in T_x\mathcal{M}_0$,*

$$\langle\alpha(x,t)\xi, \xi\rangle_0 \geq \lambda\langle\xi,\xi\rangle_0;$$

(c) *there exist two constants $0 < \nu \leq M$ such that for any $z \in \mathcal{M}$,*

$$\nu \leq \beta(z) \leq M;$$

(d) *there exists a constant $L > 0$ such that, for any $z \in \mathcal{M}$,*

$$\|\alpha_t(z)\| \leq L, \qquad |\beta_t(z)| \leq L;$$

(e) *the following relations hold uniformly with respect to $x \in \mathcal{M}_0$ and $\xi \in T_x\mathcal{M}_0$:*

$$\begin{cases} \limsup_{t\to+\infty} \dfrac{\langle\alpha_t(x,t)\xi,\xi\rangle_0}{\langle\xi,\xi\rangle_0} \leq 0 \\ \liminf_{t\to-\infty} \dfrac{\langle\alpha_t(x,t)\xi,\xi\rangle_0}{\langle\xi,\xi\rangle_0} \geq 0. \end{cases}$$

Then $(\mathcal{M}, \langle\cdot,\cdot\rangle)$ is geodesically connected. Moreover, if $\mathcal{M}$ is noncontractible to itself, for any pair of points z_0 and z_1, there exist infinitely many geodesics joining them on which the action integral takes arbitrarily large values.

The result on geodesic connectedness in Theorem 3.4 is proved in [28]. The multiplicity result is proved in [29], where geodesics on nonorthogonal splitting manifolds are considered. The coefficients α, β and δ depend on the *time coordinate* t. For this reason it is not possible to have a reduction to a purely spatial problem as in the stationary case. On the other hand, the functional (3.6) has a saddle structure under suitable assumptions on the coefficients α, β and δ. A critical point of E can be found by using Theorem 3.1 modulo some hard analysis. Indeed it can be proved that E does *not* satisfy the Palais–Smale condition. The functional E has to be approximated by a family E_δ of functionals, with $E_0 = E$, satisfying the Palais–Smale condition and still having the saddle geometry. A priori estimates on the critical points of the approximating functionals E_δ allow us to pass to the limit as $\delta \to 0$ to find a critical point of E. Moreover, a finite dimensional reduction is also needed since, in Theorem 3.2, the linear manifold H_n is finite dimensional, while the functionals E_δ goes to $-\infty$ along infinitely many directions.

The existence of infinitely many geodesics requires a more subtle topological invariant, the *relative category*, a generalization of the category of a topological space that works for unbounded functionals ([30]. Fadell and Husseini showed [31] that, if $\mathcal{M}$ is noncontractible, there are subsets having arbitrarily large relative category ([10, 29] for more details).

The assumptions of Theorem 3.4 can be weakened. For instance, one can assume that the constants λ, ν, M and L are functions depending only on the t-variable, $\lambda(t)$, $\nu(t)$, $M(t)$ and $L(t)$. Assumption (e) is replaced by the assumption that $\mathcal{M}$ is covered by a sequence of *vertical* convex subsets (see [32]). In this case de Sitter space–time fails to satisfy only the last condition. Moreover, manifolds with boundary have also been studied. In [33] it is proved that the second region of Reissner–Nördstrom space–time contained between the two event horizons is geodesically connected.

4 The Fermat principle in general relativity and the multiple image effect

In this section we present some results on lightlike geodesics (*light rays*) joining a point with a timelike curve in a Lorentzian manifold. Let $(\mathcal{M}, \langle\cdot,\cdot\rangle)$ be a Lorentzian manifold, consider an event $p \in \mathcal{M}$ and a timelike curve $\gamma\colon \mathbf{R} \longrightarrow \mathcal{M}$. We are interested in smooth curves $z\colon [0,1] \longrightarrow \mathcal{M}$ which are light rays joining p and γ. Then $D_s\dot{z} \equiv 0$, $\langle \dot{z}, \dot{z}\rangle \equiv 0$, $z(0) = p$, $z(1) \in \gamma(\mathbf{R})$. Whenever there exist two or more light rays joining p and γ we say that there is a *multiple image effect*. Such a multiple image effect gives a mathematical model of the *gravitational lensing effect* in astrophysics. Indeed, assume that γ is the world line of a pointwise light source and p an event in the space–time. Then the past pointing lightlike geodesics joining p and γ represent the images of the source seen by an observer at the event p. Such a multiple image effect is due to the bending of light near to a gravitational source. Altough a multiple image effect was predicted by some astronomers at the beginning of general relativity, the first candidate for a gravitational lens was detected in 1977. We refer to [34] for a detailed description of gravitational lensing.

The topological and geometrical structure of a space–time has a big influence on the multiple image effect. This fact can be exploited in developing a variational theory for the light rays joining p and γ. Such a theory can be considered as an extension of Fermat's principle of classical optics to general relativity. The role of a variational theory for light rays in the context of general relativity was noted by Weyl in [35], in which a Fermat principle in *static space–times* is proved. Indeed, in [35] it is shown that light rays joining a point with a timelike line are the critical points of a functional which, in modern language, is called *the arrival time* (see below). This functional be considered as an extension of the travel time of classical optics. Such Fermat principle was extended to stationary metrics by Levi–Civita [36]. In [37] a Fermat principle is proved for splitting Lorentzian manifolds satisfying a suitable growth condition. In [38] a Fermat principle different from that of Levi–Civita is proved for conformally stationary space–times. In [39] a Fermat principle in orthogonal splitting space–times (in particular, in globally hyperbolic space–times) is obtained. Such a variational principle is extended to the more general class of *stably causal* Lorentzian manifolds in [40]. Finally a Fermat principle in an arbitrary Lorentzian manifold is proved in [41] (cf. [42] also).

Let $(\mathcal{M}, \langle\cdot,\cdot\rangle)$ be a Lorentzian manifold and assume that it is time–oriented by a timelike vector field Y. Let p be a point and $\gamma : \mathbf{R} \longrightarrow \mathcal{M}$ be an injective timelike curve. A *future pointing Fermat principle* is a pair $(\mathcal{L}^+, \tau^+)$, where $\mathcal{L}^+$ is a smooth manifold consisting of *future pointing lightlike curves* joining p and γ and $\tau^+ : \mathcal{L}^+ \longrightarrow \mathbf{R}$ is a smooth functional on $\mathcal{L}^+$ such that the *critical points of* τ^+ *are all future pointing light rays joining* p *and* γ.

Analogously a *past pointing Fermat principle* can be defined as a pair $(\mathcal{L}^-, \tau^-)$, characterizing the *past pointing* light rays joining p and γ. Since there is no difference between the past and the future pointing cases, we shall focus our attention on the future pointing case.

The natural candidate for a Fermat principle is the pair $(\mathcal{L}^+_{p,\gamma}, \tau_{p,\gamma})$, where

$$\mathcal{L}^+_{p,\gamma} = \{z : [0,1] \longrightarrow \mathcal{M} : z \in H^{1,2}(p,\gamma;\mathcal{M}), \\ \langle\dot{z},\dot{z}\rangle = 0 \ a.e., \ \langle Y(z), \dot{z}\rangle \geq 0 \ a.e. \ \}$$

and $\tau_{p,\gamma} : \mathcal{L}^+_{p,\gamma} \longrightarrow \mathbf{R}$ is the *arrival time* defined as

$$\tau_{p,\gamma}(z) = \gamma^{-1}(z(1)) \ .$$

Note that $H^{1,2}(p,\gamma;\mathcal{M}) = \{z \in H^{1,2}([0,1];\mathcal{M}) : z(0) = p, z(1) \in \gamma(\mathbf{R})\}$.

In other words the arrival time is the value of the parameter of the point of γ reached by z. It is not difficult to show that $\mathcal{L}^+_{p,\gamma}$ is *not* equipped with the structure of a differentiable manifold (cf. [39, 43]), so a variational calculus cannot be developed for $(\mathcal{L}^+_{p,\gamma}, \tau_{p,\gamma})$. In [42] it is proved that *directional derivatives* can be defined for $\tau_{p,\gamma}$ on the set $\mathcal{L}^+_{p,\gamma}$ and that light rays are the stationary points of $\tau_{p,\gamma}$. In [41] a space equipped with the structure of infinite differentiable manifold is introduced and it is proved that the critical points of τ on such a space are the light rays joining p and γ.

Unfortunately the functional $\tau_{p,\gamma}$ does not seem to be useful in developing a critical point theory for light rays. Indeed $\tau_{p,\gamma}$ is invariant with respect to reparameterizations of curves. By virtue of this invariance $\tau_{p,\gamma}$ does not satisfy the Palais–Smale condition on any *reasonable manifold* and some alternative Fermat principles for light rays have to be proved. Moreover a good approximation for the *nonsmooth manifold* $\mathcal{L}^+_{p,\gamma}$ is needed. We refer to [43] for a survey of the evolution of Morse relations for light rays.

We present a very general result on Morse relations for light rays. Note that, even if $\mathcal{L}^+_{p,\gamma}$ nonsmooth, it carries the right topology to develop a Morse theory.

Let $(\mathcal{M}, \langle\cdot,\cdot\rangle)$ be a *strongly causal* [21] Lorentzian manifold, p a point of $\mathcal{M}$, $\gamma\colon \mathbf{R} \longrightarrow \mathcal{M}$ a smooth timelike curve. Consider the pair $(\mathcal{L}^+_{p,\gamma}, \tau_{p,\gamma})$ introduced above. We give some definitions.

Definition 4.1. *The pair* $(\mathcal{L}^+_{p,\gamma}, \tau_{p,\gamma})$ *is called* precompact *if, for any* $c \in \mathbf{R}$, *there exists a compact subset* $K(c)$ *of* $\mathcal{M}$ *such that, for any* $z \in \mathcal{L}^+_{p,\gamma}$ *with* $\tau_{p,\gamma}(z) \le c$, $z([0,1]) \subset K(c)$.

Definition 4.2. p *and* γ *are said to be nonconjugate if, for any future pointing lightlike geodesic* $z\colon [0,1] \longrightarrow \mathcal{M}$ *with* $z(0) = p$ *and* $z(1) \in \gamma(\mathbf{R})$, $z(1)$ *is nonconjugate to* p *along* z.

Theorem 4.3 (Morse relations for light rays). *Let* $(\mathcal{M}, \langle\cdot,\cdot\rangle)$ *be a time oriented, strongly causal Lorentzian manifold,* p *a point of* $\mathcal{M}$ *and* $\gamma\colon \mathbf{R} \longrightarrow \mathcal{M}$ *a smooth timelike curve. Assume that:*

(a) $\mathcal{L}^+_{p,\gamma} \neq \emptyset$;
(b) p *and* γ *are nonconjugate;*
(c) $(\mathcal{L}^+_{p,\gamma}, \tau_{p,\gamma})$ *is precompact.*

Then, for any field $\mathcal{K}$, *there exists a formal series* $Q(t)$ *with coefficients in* $\mathbf{N} \cup \{+\infty\}$ *such that*

$$\sum_{z \in \mathcal{G}^+_{p,\gamma}} t^{\mu(z)} = \mathcal{P}(\mathcal{L}^+_{p,\gamma}, \mathcal{K})(t) + (1+t)Q(t) \tag{4.1}$$

where $\mathcal{G}^+_{p,\gamma}$ *denotes the set of future pointing light rays joining* p *and* γ, $\mu(z)$ *is the geometric index of the light ray* z *(the number of conjugate points to* p *along* z *counted with their multiplicities) and* $\mathcal{P}(\mathcal{L}^+_{p,\gamma}, \mathcal{K})(t)$ *is the Poincaré series of* $\mathcal{L}^+_{p,\gamma}$ *with coefficients in* $\mathcal{K}$.

Morse relations also holds for past pointing light rays joining p and γ. Note that the past pointing case is the most interesting for the gravitational lens effect.

Assumption (a) in Theorem 4.3 is necessary. Indeed, there are simple examples in Minkowski space–time for which $\mathcal{L}^+_{p,\gamma}$ is the empty set. Assumption (b) introduces a nondegeneration of lightlike geodesics similar to that for Riemannian geodesics. Finally assumption (c) is a compactness assumption for the functionals used to prove Theorem 4.3.

The results of Theorem 4.3 are proved in [37] for orthogonal splitting Lorentzian manifolds satisfying a suitable growth condition, in [38] for conformally stationary Lorentzian manifolds (also having a boundary, with applications to Schwarzschild,

Reissner–Nördstrom and Kerr space–times; cf. [10] also), in [44] for globally hyperbolic space–times and in [43] for stably causal Lorentzian manifolds having a boundary eventually see also [45]. In all the papers [37, 38, 44, 43] a Fermat principle alternative to $(\mathcal{L}^+_{p,\gamma}, \tau_{p,\gamma})$ is proved for light rays and the abstract results of critical point theory are applied to functionals different from the arrival time. Moreover, in [39, 43], the space $\mathcal{L}^+_{p,\gamma}$ is approximated by a suitable family of manifolds consisting of timelike curves. A priori estimates on timelike geodesics permit us to pass to the limit and prove Theorem 4.3. In the case of conformally stationary metrics a direct approach is possible [38]. However the good topology of Morse theory for light rays is carried by the nonsmooth space $\mathcal{L}^+_{p,\gamma}$.

Theorem 4.3 is proved in the context of *strongly causal* and *time oriented* Lorentzian manifolds, and by working directly on the arrival time functional, in the paper [46].

Putting $t = 1$ in (4.1) we get the relations for the number of light rays

$$\operatorname{card} \mathcal{G}^+_{p,\gamma} = \mathcal{B}(\mathcal{L}^+_{p,\gamma}, \mathcal{K}) + 2Q\,, \tag{4.2}$$

where $\mathcal{B}(\mathcal{L}^+_{p,\gamma}, \mathcal{K})$ is the total Betti number of the space $\mathcal{L}^+_{p,\gamma}$ with coefficients in $\mathcal{K}$.

We now present some corollaries to Theorem 4.3.

Theorem 4.4. *Under the assumptions of Theorem 4.3, we have:*

(i) *if $\mathcal{L}^+_{p,\gamma}$ is not contractible into itself, there exists at least two light rays joining p and γ;*
(ii) *if $\mathcal{L}^+_{p,\gamma}$ is contractible into itself, the number of light rays joining p and γ is infinite or odd.*

Theorem 4.4 relates the number of light rays joining p and γ to the topological properties of the function space $\mathcal{L}^+_{p,\gamma}$. We can relate the number of such light rays to the topology of the manifold $\mathcal{M}$ if we assume that

(d) $\mathcal{L}^+_{p,\gamma}$ is *homotopically equivalent* to the based loop space $\Omega(\mathcal{M})$.

We recall that $\Omega(\mathcal{M})$ is the set of continuous paths on $\mathcal{M}$ with initial and final point given by a fixed point. The topology of this set is well-known.

Assumption (d) is certainly satisfied if $(\mathcal{M}, \langle\cdot,\cdot\rangle)$ is a conformally stationary Lorentzian manifold or a splitting Lorentzian manifold satisfying the growth condition introduced in [37, 43].

Under assumption (d), (4.2) becomes

$$\operatorname{card} \mathcal{G}^+_{p,\gamma} = \mathcal{B}(\Omega(\mathcal{M}), \mathcal{K}) + 2Q\,. \tag{4.3}$$

By (4.3) and the topological properties of $\Omega(\mathcal{M})$ [17], we get the following result.

Theorem 4.5. *Under assumptions (a)–(c) of Theorem 4.3 and assumption (d), we have:*

(i) *if $\mathcal{M}$ is not contractible, there exist infinitely many light rays joining p and γ;*
(ii) *if $\mathcal{M}$ is contractible, the number of light rays joining p and γ is infinite or odd.*

Using Morse theory we have shown that the number of images in the multiple image effect is odd whenever this number is finite. It is interesting to observe that, in fact, the number of images observed in any gravitational lens effect is *even*.

In case (ii) of Theorem 4.5, the number of images can be infinite even if the topology of the space–time is trivial. In order to prove the existence of only a finite number of images, geometric restrictions on the Lorentzian metric have to be imposed. A finiteness result of the images is contained in [47].

Morse relations have been proved under a nondegeneration assumption on the images. On the other hand, degenerate situations can occur physically, for instance, extended images as *arcs*. In the degenerate situation, a Lusternik–Schnirelmann theory can be developed and it is proved that, under assumptions (a) and (c) of Theorem 4.3, there are at least $\text{cat}\mathcal{L}^{+}_{p,\gamma}$ light rays joining p and γ. An analogous result holds for past pointing light rays.

Finally we would like to mention that a variational principle for timelike geodesics has been proved in [48]. Here it is proved that the critical points in the arrival on suitable manifolds of timelike curves are timelike geodesics. Such a principle is used in [49] to develop a Morse theory for timelike geodesics and to obtain Theorem 4.3 as a limit of such Morse theories for timelike geodesics.

5 Conclusions

Some application of variational methods to general relativity have been presented in this paper. We think that the results obtained up to now are only the tip of the iceberg of all the possible applications of the calculus of variations to general relativity.

References

1. Arnold V.I. (1989): Mathematical Methods of Classical Mechanics. Springer, Berlin Heidelberg New York
2. Morse M. (1934): The Calculus of Variations in the Large. Coll. Lect. Am. Math. Soc. **18**
3. Lusternik L.A., Schnirelmann L. (1934): Methodes Topologiques dans les Problèmes Variationelles. Gautier–Villars, Paris
4. Struwe M. (1996): Variational Methods. Springer, Berlin Heidelberg New York
5. Palais R. (1963): Lusternik–Schnirelmann theory on Banach manifolds. Topology **5**, 115–132
6. Klingenberg W. (1978): Lecture on Closed Geodesics. Springer, Berlin Heidelberg New York
7. Bott R. (1982): Lectures on Morse Theory Old and New. Bull. Am. Math. Soc. **7**, 331–358
8. Spanier E.H. (1966): Algebraic Topology. McGraw Hill, New York
9. Milnor J. (1963): Morse Theory. Ann. of Math. Studies **51**, Princeton University Press, Princeton
10. Masiello A. (1994): Variational Methods in Lorentzian Geometry. Pitman Research Notes in Mathematics **309**, Longman, London
11. Palais R. (1963): Morse theory on Hilbert manifolds. Topology **2**, 299–340
12. Palais R. (1970): Critical Point Theory and the Min–Max Principle. Proc. Symp. Pure Appl. Math. **15**, 185–212

13. Palais R., Smale S. (1964): A Generalized Morse theory. Bull. Am. Math. Soc. **70**, 165–171
14. Rabinowitz P.H. (1984): Min–Max Methods in Critical Point Theory with Applications to Differential Equations. CBMS Reg. Conf. Soc. in Math. **65** Amer. Math. Soc., Providence
15. Mawhin J.,Willem M. (1989): Critical Point Theory and Hamiltonian Systems. Springer, Berlin Heidelberg New York
16. Adams R. (1975): Sobolev spaces. Acad. Press, New York
17. Fadell E., Susseini S. (1991): Category of Loop Spaces of Open Subsets in Euclidean Space. Nonlinear Analysis T.M.A. **17**, 1153–1161
18. Benci V., Fortunato D. (1990): Existence of Geodesics for the Lorentz Metric of a Stationary Gravitational Field. Ann. H. Poincaré Analyse Non Linéaire **7**, 27–35
19. Beem J.K., Ehrlich P.E., Easley K. (1996): Global Lorentzian Geometry. Marcel Dekker, New York
20. Hawking S.W., Ellis G.F. (1973): The Large Scale Structure of Space–Time. Cambridge University Press, Cambridge
21. O'NeilL B. (1983): Semi-riemannian Geometry with Applications to Relativity. Academic Press, New York
22. Beem J.K., Parker R.E. (1989): Pseudoconvexity and Geodesic Connectedness. Ann. Mat. Pura Appl. **155** (4), 137–142
23. Penrose R. (1972): Techniques of Differential Topology in Relativity. Conf. Board Math. Sci **7**. SIAM, Philadelphia
24. Giannoni F., Masiello A. (1991): On the Existence of Geodesics on Stationary Lorentz Manifolds with Convex Boundary. J. Funct. Anal. **101**, 340–369
25. Benci V., Fortunato D., Giannoni F. (1992): On the existence of geodesics in static Lorentz Manifolds with Singular Boundary. Ann. Sc. Norm. Sup. Pisa **19** (4), 255–289
26. Giannoni F., Piccione P. (1999): An Intrinsic Approach to the Geodesical Connectedness of Stationary Lorentzian Manifolds. Comm. Anal. Geom. **7**, 157–197
27. Geroch R. (1970): Domains of Dependence. J. Math. Phys. **11**, 437–449
28. Benci V., Fortunato D., Masiello A. (1994): On the geodesic connectedness of Lorentzian manifolds. Math. Z. **217**, 74–94
29. Giannoni F., Masiello A. (1995): Geodesics on Product Lorentzian Manifolds. Ann. Inst. H. Poincaré Analyse non linéaire **12**, 27–60
30. Fadell E. (1985): Lectures in Cohomological Index Theories of G-Spaces with Applications to Critical Point Theory. Sem. Dip. Mat. Università della Calabria, Cosenza
31. Fadell E., Husseini S. (1994): Relative category and coproducts. Rend. Sem. Mat. Fis. Milano **64**, 99–117
32. Masiello A. (1994): Convex regions of Lorentzian manifolds. Ann. Mat. Pura Appl. **167** (4), 299-322
33. Giannoni F., Masiello A. (1993): Geodesics on Lorentzian Manifolds with Quasi-Convex Boundary. Man. Math. **78**, 381–396
34. Schneider P., Ehlers J., Falco E. (1992): Gravitational Lensing. Springer, Berlin Heidelberg New York
35. Weyl H. (1917): Zur Gravitationstheorie. Ann. Phys. **54**, 117–145
36. Levi–Civita T. (1928): Fondamenti di Meccanica Relativistica. Zanichelli, Bologna
37. Uhlenbeck K. (1975): A Morse Theory for Geodesics on a Lorentz Manifold. Topology **14**, 69–90
38. Fortunato D., Giannoni F., Masiello A. (1995): A Fermat Principle for Stationary Space–Times with Applications to Light Rays. J. Geom. Phys. **15**, 159–188
39. Giannoni F., Masiello A. (1998): On a Fermat Principle in General Relativity. A Lusternik–Schnirelmann Theory of Light Rays. Ann. Mat. Pura Appl. **174** (4), 161–207

40. Antonacci F., Piccione P. (1996): A Fermat principle on Lorentzian manifolds and applications. Appl. Math. Lett. **9**, 91–96
41. Perlick V. (1995): Infinite dimensional Morse theory and Fermat's principle in general relativity. I. J. Math. Phys. **36**, 6915–6928
42. Perlick V. (1990): On Fermat's principle in general relativity: I. The general case. Class. Quant. Grav. **7**, 1319–1331
43. Giannoni F., Masiello A., Piccione P. (1998): A Morse Theory for Light Rays on Stably Causal Lorentzian Manifolds. Ann. Inst. H. Poincaré Physique Theorique **69**, 359–412
44. Giannoni F., Masiello A. (1996): On a Fermat principle in General Relativity. A Morse Theory for light rays. Gen. Rel. Grav. **28**, 855–897
45. Giannoni F., Masiello A., Piccione P. (1997): A Variational Theory for Light Rays on Stably Causal Lorentzian Manifolds. Existence, Regularity and Multiplicity Results, Commun. Math. Phys **187**, 375–415
46. Giannoni F., Masiello A., Piccione P.: The Fermat Principle in General Relativity and Applications, in preparation
47. Giannoni F., Masiello A., Piccione P. (1999): Convexity and the Finiteness of the Number of Geodesics. Applications to the Multiple-Image Effect. Class. Quant. Grav. **16**, 731–748
48. Giannoni F., Masiello A., Piccione P. (1998): A Timelike Extension of Fermat's Principle in General Relativity and Applications. Calc. Var. P.D.E. **6**, 263–283
49. Giannoni F., Masiello A., Piccione P.: A Morse Theory for Massive Particles and Photons. J. Geom. Phys. in press

The Search for Gravitational Waves with Resonant Detectors

G. Pizzella

Abstract. An overview of experiments in the search for gravitational waves by means of resonant detectors is given. Since 1990 cryogenic resonant antennas have been in operation and data have been recorded by EXPLORER, ALLEGRO, NIOBE, NAUTILUS and AURIGA. The sensitivity for pulse detection with SNR=1 is now $h \sim 4 \cdot 10^{-19}$ (corresponding to a total energy of less than 0.001 solar masses for a source in the Galactic Center). The sensitivity for monochromatic waves is $h \sim 2 \cdot 10^{-25}$ for one year of integration and $\tilde{h} \sim 7 \cdot 10^{-22} \frac{1}{\sqrt{\mathrm{Hz}}}$ for stochastic background detection.
Comparison of data from Explorer and Nautilus has given 19 coincidences with an expectation of 11 accidental coincidences during 29.2 days of non-continuous operation. Cross-correlation analysis between Explorer and Nautilus has given an upper limit of $\Omega = 60$ for the ratio of gravitational wave stochastic background energy density to the energy density needed for a closed Universe, at a frequency of 900 Hz.
Plans for the construction of large spherical detectors that should reach, at 1 kHz, a spectral sensitivity on the order of $\tilde{h} \sim 7 \cdot 10^{-24} \frac{1}{\sqrt{\mathrm{Hz}}}$ are reported.

1 Introduction

Experiments in the search for gravitational waves (GW) started in the early sixties with Joseph Weber at the University of Maryland, who brought the problem of GW into the domain of observable physical phenomena.

During the last thirty years detection techniques have been improved enormously, and further remarkable steps can be foreseen in the near and distant future. The sensitivity reached by today's resonant detectors is $h \simeq 4 \cdot 10^{-19}$, differing by slightly more than two orders of magnitude in amplitude from that estimated to be necessary for the detection of GW signals originating from various astrophysical sources expected to have satisfactory statistical occurrence (supernovae in the Virgo cluster, coalescence binaries, pulsars and stochastic background).

Improvements in the techniques have been accompanied by a better understanding of resonant detector sensitivity in terms of the technical characteristics of the apparatus, making more visible the road to follow for reaching the final goal of a sensitivity near to the quantum limit. In the present day scenario we associate to the basic components of the antenna, bar and transducer, two important features of the detecting apparatus. The bar must be associated with the sensitivity of a stochastic background, that is, to the spectral sensitivity, the transducer must be associated which the detection bandwidth.

2 Astrophysical sources of gravitational waves

Because of the small value of the gravitational constant G, the irradiated power of gravitational waves is so feeble that, at present, it is not possible to detect GW that could be generated in a laboratory on the earth. It is necessary that sources consist of enormous masses moving with extreme accelerations. Astrophysical sources are the best candidates for GW: e.g., supernovae, coalescence of binary systems, pulsars. Some estimation of the strength of GW signals (in terms of the perturbation h of the metric tensor) and their occurrence rate, as reported in the review by Thorne [1], are given in Table 1.

Table 1. Astrophysical sources of gravitational waves

source	distance	rate per year	h
pulsar	1 kpc	several sources	continuous 10^{-26}
binary coalescence	100 Mpc	1	10^{-21}
galactic supernovae	10 kpc	0.1	$3 \cdot 10^{-18}$

For calculating value of h corresponding to supernovae we have taken the optimistic estimate that 1 per cent of the collapsing mass goes into GW. The strength varies with the inverse of the distance. Thus, supernovae occurring in the Virgo cluster at a distance of 10 Mpc have strength $h \sim 3 \cdot 10^{-21}$.

An interesting feature of the source due to the coalescence of a binary system is that it allows us, from the detected amplitude and its variation with frequency, to determine the distance of the source and the mass of the two orbiting collapsed objects. Knowing the distance one can estimate the Hubble constant with an accuracy of a few per cent, in this way setting an accurate scale for measuring cosmic distances.

Among the possible sources of GW we also have to consider a possible stochastic cosmological background. Nucleosynthesis considerations put an upper limit on the ratio $\Omega \leq 10^{-5}$ of GW energy density to the critical density needed for a closed universe. The observation of stochastic GW would give important information on the extremely early period of the Universe, close to the Planck time, about 10^{-43} s.

3 Sensitivity of a resonant detector for gravitational waves

The sensitivity of a cylindrical resonant detector for gravitational waves can be expressed [2] at a level of signal-to-noise ratio (SNR) equal to unity by means of the formula

$$S_h(f_o) = \tilde{h}^2(f_o) = \frac{\pi}{2} \frac{kT_e}{MQv^2} \frac{1}{f_o} \tag{1}$$

with the following meanings for the symbols:

f_o resonance frequency of the detector;
$S_h(f_o)$ the power spectrum at the resonance frequency of the gravitational potential h that can be detected with SNR=1 (spectral sensitivity);
T_e the thermodynamic temperature of the detector plus a term due to the transducer back-action. This is negligible when a SQUID amplifier is used;
M mass of the bar;
Q quality factor of the detector;
v sound velocity in the bar material.

We note that Eq.1 does not depend on transducer characteristics, provided that the back-action is neglected. The frequency bandwidth of the detector is given by

$$\delta f = \frac{f_o}{Q}\frac{4T_e}{T_{\text{eff}}} \tag{2}$$

where T_{eff} is the noise temperature for burst detection, that is, the average value of the noise after applying to the GW data a filter matched to delta-like signals. It can be shown [2] that the frequency bandwidth depends on the transducer and associated electronics.

It turns out that δf can approach 100 Hz if one is able to operate the detector near to the quantum limit, that is, with $T_{\text{eff}} \sim 10^{-7} K$.

From Eqs. 1 and 2 one can derive the antenna sensitivity for various types of GW.

For delta-like bursts with duration τ_g the sensitivity is given by

$$h \simeq \frac{1}{\tau_g}\sqrt{\frac{2S_h}{\pi\delta f}}. \tag{3}$$

Another useful formula which can be derived from Eqs. 1 and 2 is

$$h \simeq \frac{L}{\tau_g v^2}\sqrt{\frac{kT_{\text{eff}}}{M}} \tag{4}$$

where L is the length of the bar.

For monochromatic GW, integrating over a continuous time t_m, the minimum wave amplitude that can be detected with SNR=1 is

$$h = \sqrt{\frac{2S_h}{t_m}}. \tag{5}$$

Finally Eq. 1 gives immediately that the sensitivity to a GW stochastic background measures only an upper limit , since it is practically impossible to subtract from the measured power spectrum the contribution due to noise.

In order to measure the stochastic background one needs to cross-correlate the output of two antennas [2], obtaining the measurement of the cross-spectrum

$$S_h(f) = \frac{\sqrt{S_{1h}S_{2h}}}{\sqrt{t_m\delta f}} \tag{6}$$

where t_m is the total time of cross-correlation and δf is the frequency bandwidth in common between the two detectors.

The standard deviation of this measurement is equal [3] to the same value S_h as given by Eq. 6. From the measured S_h we can calculate [2] the value of Ω, the ratio between GW energy density and the energy density needed for a closed universe, by using the following formula

$$\Omega = \frac{4\pi^2}{3} \frac{f^3}{H^2} S_h(f) \tag{7}$$

where H is the Hubble constant. We note that, while it is important to have a large frequency bandwidth (attainable with a good transducer followed by a very low noise electronic amplifier) for burst detection, this is less important for the stochastic measurement.

For monochromatic waves the sensitivity is independent of the bandwidth, but a larger bandwidth permits the exploration of a larger frequency region.

There are at present five cryogenic bars in operation [4–8] (Allegro, Auriga, Explorer, Nautilus and Niobe). They have roughly the same experimental sensitivity, as given in Table 2.

Table 2. Sensitivity of the resonant detectors in operation

resonance frequency [Hz]	$\tilde{h} = \sqrt{S_h}$ at resonance $[\frac{1}{\sqrt{\text{Hz}}}]$	frequency bandwidth δf [Hz]	minimum h for 1 ms bursts	minimum h for continuos waves	minimum Ω
900–700	$7 \cdot 10^{-22}$	0.5–1	$4 \cdot 10^{-19}$	$2 \cdot 10^{-25}$	0.1

Niobe, made of niobium, has a resonance frequency of 700 Hz, the others, of aluminum, have resonance frequencies near 900 Hz. The above minimum values for monochromatic waves and for the quantity Ω have been estimated by considering one year of integration time (for Ω we assume that cross-correlation of two identical antennas has been done).

The burst sensitivity for all bars can be increased by improving the transducer and associated electronics. It has been estimated [9] that a factor of 50 is within technical possibilities.

In addition to increasing the bandwidth, Auriga and Nautilus can improve their spectral sensitivities (see Table 3) by making full use of their capability to go down in temperature to $T = 0.10$ K. At present the major difficulty is due to excess noise, sometimes of unknown origin, and work is in progress for eliminating this noise.

The search for signals due to GW bursts is made after the raw data have been filtered with optimum filter algorithms [10–12].

These algorithms may have various expressions but they all have in common an optimum integration time that is roughly the inverse of the detector bandwidth and

Table 3. Target sensitivity for Auriga and Nautilus

$\tilde{h} = \sqrt{S_h}$ at resonance $[\frac{1}{\sqrt{\text{Hz}}}]$	frequency bandwidth δf [Hz]	minimum h for 1 ms bursts	minimum h for continuous waves	minimum Ω
$6 \cdot 10^{-23}$	50	$3 \cdot 10^{-21}$	$2 \cdot 10^{-26}$	10^{-4}

all give approximately the same value of T_{eff} (all algorithms being optimal filters for short bursts).

4 Major problems encountered with resonant detectors

The design and construction of resonant bar cryogenic detectors started in 1970. Reaching an operational phase has taken much longer than expected. This happened because the research program was rather optimistic and the work started when many necessary techniques had not yet been developed.

Finally, in 1990, Explorer, installed at CERN, was the first cryogenic antenna to enter into steady operation, joined soon after, at various times, by the other four antennas.

Here we present, and briefly discuss, the major problems faced during almost thirty years of work.

a) Unexpected noise. This is perhaps the most difficult problem. It has been found experimentally that the noise distribution does not follow the behavior calculated under the assumption that the only noise present is thermal and electronic noise. Additional noise of a mechanical and electromagnetic nature enters into the apparatus from the laboratory environment. Therefore it is important to improve the mechanical as well the electrical shields. The design of mechanical filters is today much simpler than in the past, due to the availability of powerful computer utilities. Also electromagnetic shields can take advantage of special materials and diagnostic instrumentation.

b) The quality factor Q. It is important to have detectors with high quality factors. The original decision to cool the antenna was due to the need to reduce thermal noise. But an unexpected bonus came with it. It was found that, at low temperature, the quality factor of various materials increases by orders of magnitude [**?**]. However it was soon also found that, in practice, Q is obtained after the electromechanical transducer is mounted on the bar. This Q also includes electrical losses due to the transducer, which can be very large [14]. For all experiments in operation today the loaded Q is less than 10 million. In some cases it is as low as a few hundred thousands. We believe that a detector with an operating Q on the order of 10 million is already a good achievement and so we shall consider in our calculations the value $Q = 10^7$ for future resonant cryogenic detectors.

c) The matching of the electromechanical transducer to the bar. GW interact with the antenna leaving only a very tiny amount of energy in it. Thus it is important that as much of this energy as possible be extracted for measurement. Then the requirements for the transducer are: its mechanical part must be well matched to the antenna, and its electrical part well matched to the electronic amplifier. In the past this has posed severe problems and it is worth investing more effort in this area.

5 Recent results obtained with resonant detectors

The data recorded by the various detectors are now being analyzed, seeking, in particular, for coincidences above the background. The coincidence technique is a powerful method for reducing noise. As an example we give in Table 4 some resulst obtained very recently [15] when searching for coincidences between data from Explorer and Nautilus during the years 1995 and 1996.

Table 4. Results of a coincidence search on data from Explorer and Nautilus. See text for explanation

number of days	number of Explorer events	number of Nautilus events	$< n >$	n_c	p_{poisson} per cent	p_{exp} per cent
29.2	8527	5679	11.0	19	1.5	1.44

In the first column we give the number of days when both antennas were operating. The small number of useful days shows that it is difficult to keep a GW antenna in operation continuously with good behavior. In future it should be possible to increase the useful time to 70% of the total time, considering that time is always lost for cryogenic maintenance. For the coincidence search reported Nautilus had just started to operate and various interruptions occurred for instrumental improvements.

In the second and third column we show the number of *candidate events*. The candidate events are obtained by introducing a proper threshold on the data filtered with an optimum filter for short burst detection. We notice the large number of candidate events that make it practically impossible to search for a particular signal due to GW. Much improvement is obtained by comparing data from the two detectors. Then we only have to worry about the number of accidental coincidences.

In the fourth column we give the expected number of accidental coincidences measured by means of 10,000 shifts of the event times of one detector with respect to the other and by using a coincidence window of $w = \pm 0.29$ s (one sampling time). This number of accidentals is small enough to start considering the possibility of searching for a coincidence excess (though, according to astrophysical expectations, this excess should be much smaller than the observed accidentals). In this case the number of coincidences n_c, reported in column five, turns out to be slightly larger than the expected number of accidentals.

Finally, in columns six and seven, we report the probability calculated with Poisson's formula and the *experimental probability*, obtained by counting how many times a number of accidental coincidences was equal to or larger than n_c and dividing this number by 10,000 the number of trials. The agreement between theoretical values and experimental values is good, indicating a Poissonian statistical behavior of the data.

A new and interesting result has been obtained recently by cross-correlating the data recorded with Explorer and Nautilus, as reported at the Eighth MG Conference [16]. The measured spectral amplitudes of Explorer and Nautilus were correlated. The result of such cross-correlation at a resonance of 907.2 Hz (the same for the two detectors) has been the determination of an upper limit for the parameter Ω that measures the closure of the universe. It has been found that $\Omega \geq 60$, where a factor of 6 has been also included from taking into consideration the fact that Explorer and Nautilus are separated by about 600 km.

6 Sensitivity of a large spherical detector

New resonant detectors of different shape and much larger mass [17–19] have been proposed for obtaining greater sensitivity.

The best geometry for a resonant detector is spherical, because a sphere has the largest possible mass for a given occupied space and because a spherical detector can be instrumented with transducers installed in various locations on its surface, allowing the best detection of GW with any direction and polarization.

Among various proposals an aluminum sphere with a diameter of 3 m, having mass $M = 38\,\text{ton}$ and operating at $T = 20\,\text{mK}$ has been considered.

The estimated sensitivity for this detector is shown in Table 5. We have assumed that the detector operates near the quantum limit, that is, with $T_{\text{eff}} \sim 10^{-7}$ K.

Furthermore we must consider that the sphere is sensitive to GW with any incoming direction and degree of polarization. Among other proposals a sphere made from a different material with a mass greater than 100 tons has also been considered [20].

Table 5. Target sensitivity for a 38 ton resonant spherical detector

$\tilde{h} = \sqrt{S_h}$ at resonance $[\frac{1}{\sqrt{\text{Hz}}}]$	frequency bandwidth δf [Hz]	minimum h for 1 ms bursts	minimum h for continuous waves	minimum Ω
$6 \cdot 10^{-24}$	50	$3 \cdot 10^{-22}$	$2 \cdot 10^{-27}$	10^{-6}

7 Conclusions

For reaching the target sensitivity, further developments of the various techniques are needed. Since it is not possible to guess when a new technique will be implemented, it remains very difficult to predict the time when detectors will be sufficiently sensitive to observe GW as foreseen by the theorists.

However, it is important to realize that there are now sensitive resonant antennas in operation, with a duty cycle that is on the order of 50 per cent.

New upper limits have been obtained. Also some intriguing results on coincidences between antennas have been found. It is very important to continue to take data, even with detectors which have not yet reached their target sensitivity, to advance the exploration of a still unknown field of physics.

References

1. Thorne K.S. (1987): in *300 Years of Gravitation*, ed. by S.W. Hawking, W. Israel, Cambridge Univ. Press, Cambridge,
2. Astone P., Pallottino G.V., Pizzella G. (1997): Class. Quantum Grav. **14**, 2019–2030
3. Bendat J.S., Piersol A.G. (1966): Measurement and Analysis of Random Data. Wiley, New York
4. Mauceli E., Geng Z.K., Hamilton W.O. (1996): Phys. Rev. D **54**, 1264
5. Cerdonio M., Franceschini L, Fontana G. (1995): Status of the AURIGA gravitational wave antenna and perspectives for the gravitational wave search with ultracryogenic resonant detectors, in *First Edoardo Amaldi Conference on Gravitational wave Experiments, Frascati, 14–17 June 1994*, ed. by E. Coccia, G. Pizzella, F. Ronga, World Scientific, Singapore
6. Astone P., Bassan M., Bonifazi V. (1993): Phys. Rev. D **47**, 362
7. Astone P. et al. (1997): Astroparticle Physics **7**, 231–243
8. Heng I.S. et al. (1996): Physics Letters A **218**, 190–196
9. Bassan M., Pizzella G. (1995): Sensitivity of a capacitive transducer for resonant gravitational wave antennas. Internal Report, LNF-95/064, Frascati
10. Astone P., Bonifazi P., Palottino G.V., Pizzella G. (1994): Il Nuovo Cimento **17**, 713
11. Astone P. et al. (1997): Il Nuovo Cimento **20**, 9
12. Astone P., Pallottino G.V., Pizzella G. (1998): Journal of General Relativity and Gravitation **30**, 105–114
13. Carelli P., Foco A., Giovanardi V. (1975): Cryogenics **15**, 406–408
14. Boughn S.P., Fairbank W.M., Giffard R.P. (1977): Phys. Rev. Lett. **38**, 454
15. Astone P., Bassan M., Blair D.G. (1999): Astroparticle Physics **10**, 83–92
16. Fafone V., Astone P., Bassan M.: Operation of the gravitational wave detector NAUTILUS, in *Proc. of the Eight Marcel Grossmann Meeting on General Relativity, 22–27 June 1997, Jerusalem*, ed. by T. Piran, World Scientific, Singapore, p. 1039
17. Forward R. (1971): Journal of General Relativity and Gravitation **2**, 199
18. Johnson W.W., Merkowitz S.M. (1993): Phys. Rev. Letters **70**, 2367
19. Coccia E., Lobo J.A., Ortega J.A. (1995): Phys. Rev. D **52**, 3735
20. Frossati G. (1997): A 100 ton 10 mK Gravitational Wave Antenna, in *Proceedings of the First Intern. Workshop OMNI 1, 26–31 May 1996, Sao Jose dos Campos (Brazil)*. World Scientific, Singapore, p. 252

Cauchy-Perturbative Matching and Outer Boundary Conditions

L. Rezzolla

Abstract. We present a new method of extracting gravitational radiation from three-dimensional numerical relativity codes and providing outer boundary conditions. Our approach matches the solution of a Cauchy evolution of Einstein's equations to a set of one-dimensional linear wave equations on a curved background. We illustrate the mathematical properties of our approach and discuss a numerical module we have constructed for this purpose. This module implements the perturbative matching approach in connection with a generic three-dimensional numerical relativity simulation. Tests of its accuracy are presented.

1 Introduction

In the past few years considerable effort has been devoted to the solution of Einstein's equations in numerical simulations of strong-field, highly dynamical sources of gravitational radiation. This effort is partly motivated by the development and construction of gravitational wave detectors such as LIGO, VIRGO, GEO and TAMA. Knowing the theoretical waveform produced by the most likely astrophysical sources of gravitational radiation will not only increase the probability of a successful detection but, most importantly, will allow for the extraction of astrophysically significant information from the observations.

The Binary Black Hole "Grand Challenge" *Alliance* [1], is a major example of this effort, in which a multi-institutional collaboration in the United States was created in order to study the inspiral and coalescence of a binary black hole system, one of the most significant source of signals for the interferometric gravity wave detectors. Present three-dimensional (3D) numerical relativity simulations face a number of fundamental and in some cases unsolved problems, including: coordinate choice, most suitable form of Einstein's equations, singularity avoiding techniques, gravitational wave extraction and outer boundary conditions. While a robust solution to the generic problem is still awaited, some interesting results have already been obtained, for instance, in the evolution of a generic 3D black hole [2], in the translation of a 3D black hole across a numerical grid [3], and in the extraction of gravitational wave information and imposition of outer boundary conditions [4].

Determining the asymptotic form of the gravitational waves produced in a dynamical evolution of Einstein's equations is an important goal of many numerical relativity simulations. This goal, however, necessarily requires accurate techniques which compute waveforms from numerical relativity simulations on 3D spacelike hypersurfaces with finite extents. In an ideal situation with unlimited numerical resources, the computational domain can extend into the distant wave zone [5], where the geometric optics approximation is valid and the gravitational waves approach their asymptotic form. With present-day computational limitations, however, the outer boundary of typical numerical relativity simulations lies rather close to the highly dynamical and strong-field region where backscatter of waves off curvature can be significant. As a result, additional techniques need to be implemented in order to "extract" such information from the strong-field region and "evolve" it out to a large distance.

In a series of recent papers [4, 6, 7], we have presented a new method for extracting gravitational wave data from a 3D numerical relativity simulation and evolving it out to an arbitrary distant zone. Our method has been developed in order to match a generic full 3D Cauchy solution of nonlinear Einstein equations on spacelike hypersurfaces with a linear solution in a region where the waveforms can be treated as perturbations on a spherically symmetric curved background. This "perturbative module" is used not only to extract gravitational wave data from the Cauchy evolution but, at the same time, to impose outer boundary conditions (a parallel development aims to match interior Cauchy solutions to exterior solutions on characteristic hypersurfaces [8]). While the problem of radiation extraction is important for computing observable waveforms from numerical simulations, imposition of correct outer boundary conditions is essential for maintaining the integrity of the simulations themselves, as incorrect outer boundary conditions are often a likely source of numerical instabilities. One of the most important requirements for any radiation-extraction and outer-boundary module is that it provides for stable evolution of the interior equations and minimizes the spurious (numerical) reflection of radiation at the boundary. This requirement is particularly important for the "Grand Challenge" investigation in which the computations need to be performed on a gravitational wave emission timescale, which is much longer than the orbital one.

In this review we briefly recall the mathematical basis of our approach and present its application to the evolution of 3D linear and mildly nonlinear gravitational waves. A more detailed discussion of the features of the Cauchy-perturbative matching method can be found in [6, 7].

2 The Cauchy-perturbative matching method

The Cauchy-perturbative matching method involves replacing, at least in parts of the 3D numerical domain, the solution of the full nonlinear Einstein equations with the solution of a set of simpler linear equations that can be integrated to high accuracy with minimal computational cost.

In order to do this, it is necessary to determine the region of spacetime where a perturbative approach can be applied. In general, the 3D numerical grid (indicated as $\mathtt{N}$ in Fig. 1) will comprise an isolated region of spacetime where the gravitational fields are strong and highly dynamical. In this region, indicated as $\mathcal{S}$ in Fig. 1, the full nonlinear Einstein equations must be solved. Outside of $\mathcal{S}$, however, in what we will refer to as the perturbative region $\mathcal{P}$, a perturbative approach is not only possible but highly advantageous. Anywhere in the portion of $\mathcal{P}$ covered by $\mathtt{N}$ we can place a two-dimensional (2D) surface which will serve as the surface joining numerically the highly dynamical strong-field region $\mathcal{S}$ and the perturbative region $\mathcal{P}$. Here, we have chosen this surface to be a 2-sphere of radius r_E, indicated as **E** in Fig. 1. It is important to emphasize that the 2-sphere **E** *need not* be in a region of spacetime where the gravitational fields are *weak* or the curvature is *small*. In contrast to previous investigations which matched Einstein's equations onto a Minkowski background [9] the matching is here made on a Schwarzschild background. As a result, the only requirement is that the spacetime outside of $\mathcal{S}$ approaches a Schwarzschild spacetime. Even in the case of a binary black hole merger, it will be possible to find a region of spacetime, sufficiently distant from binary black holes, where this requirement is met [10].

In a practical implementation of the Cauchy-perturbative method, a numerical code provides the solution to the full nonlinear Einstein equations everywhere in the 3D grid N except at its outer boundary surface **B**. At the extraction 2-sphere **E**, a perturbative module "extracts" the gravitational wave information and transforms it into a set of multipole amplitudes which we have here chosen to depend only on the radial and time coordinates of the background Schwarzschild metric[1].

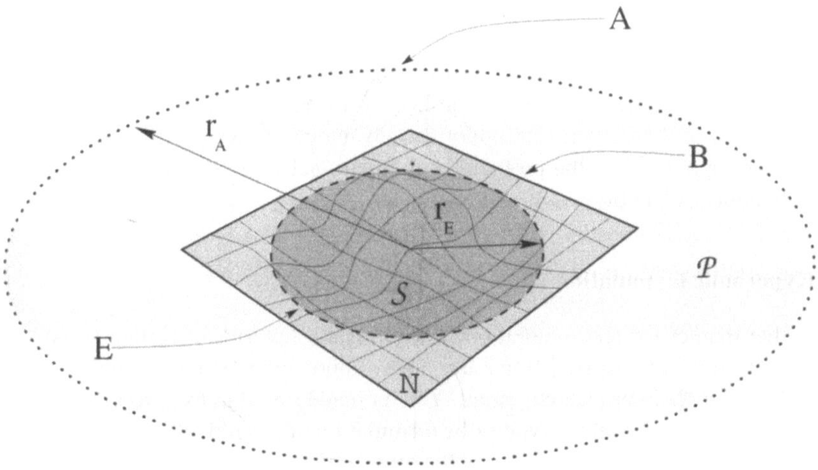

Fig. 1. Schematic picture of the Cauchy-perturbative matching procedure for a spacelike slice of spacetime (one dimension has been suppressed). N is the 3D numerical grid in which the full Einstein equations are solved and **B** its 2D outer boundary. The interior (dark shaded) region $\mathcal{S}$ shows the strong-field highly dynamical region of spacetime fully covered by N. $\mathcal{P}$ is the region of spacetime where a perturbative solution can be performed and extends from the 2-sphere **E** (of radius r_E) to the 2-sphere **A** (of radius r_A) located in the asymptotically flat region of spacetime. $\mathcal{P}$ is covered entirely by a 1D grid L (not shown) and partially by the 3D grid N

[1] Note that, although highly convenient, the suppression of the angular part of the multipoles is not strictly necessary. Indeed, different linear perturbation equations can be derived in which the angular dependence is explicitly contained in the evolution equations.

In this way, two of the three spatial dimensions of the problem are suppressed and the propagation of gravitational waves on a curved background is reduced to a one-dimensional (1D) problem. During each timestep, information about the gravitational field read-off at **E** is propagated by the perturbative module out to the 2-sphere **A** in the asymptotic flat region of spacetime. This is done by solving a set of coupled 1D linear differential equations (one for each of the multipoles extracted at **E**) on the 1D grid $\mathtt{L}$ (not shown in Fig. 1) covering the perturbative region $\mathcal{P}$ and ranging between r_E and $r_A \gg r_E$. From a computational point of view, this represents an enormous advantage: with a few straightforward transformations, the computationally expensive 3D evolution of the gravitational waves via the nonlinear Einstein equations is replaced with a set of 1D linear equations that can be integrated to high accuracy with minimal computational cost. Although linear, these equations account for all of the effects of wave propagation in a curved spacetime and, in particular, automatically incorporate the effects of backscatter off the curvature (only the wave-wave effects are omitted).

As a result of our construction (and as shown in Fig. 1), the perturbative region $\mathcal{P}$ is entirely covered by a 1D grid $\mathtt{L}$ and only partially by a 3D grid in the complement to $\mathcal{S}$ in $\mathtt{N}$. The overlap between these two grids is essential. In fact, the knowledge of the solution on $\mathcal{P}$ allows the perturbative module to provide boundary conditions at the outer boundary surface **B** and, if useful, Dirichlet data at every gridpoint of $\mathtt{N}$ outside the strong region $\mathcal{S}$. As we will further discuss in Sect. 6, this freedom to specify boundary data on a 2-surface of arbitrary shape as well as on a whole 3D region of $\mathtt{N}$ represents an important advantage of the perturbative module over similar approaches to the problem of gravitational wave extraction and imposition of boundary conditions.

Hyperbolic formulation

Rather than characterize radiation asymptotically in terms of certain variables constructed from the metric [9], we use a new approach which characterizes radiation in terms of the extrinsic curvature. This is made possible by a recently developed spatially gauge-covariant hyperbolic formulation of general relativity. This system is constructed from first derivatives of the spacetime Ricci tensor [11] and may therefore appropriately be called the "Einstein–Ricci" system.

The Einstein–Ricci equations are obtained from the "3 + 1" form of the metric,

$$ds^2 = -N^2 dt^2 + g_{ij}(dx^i + \beta^i dt)(dx^j + \beta^j dt), \tag{1}$$

where N is the lapse function, β^i is the shift vector, and g_{ij} is the spatial metric in the slice Σ. An appropriate time derivative operator that evolves spatial quantities along the normal to the slice Σ is

$$\hat{\partial}_0 = \partial_t - \mathcal{L}_\beta, \tag{2}$$

where $\mathcal{L}_\beta$ is the Lie derivative along the shift vector in Σ.

The extrinsic curvature K_{ij} of Σ can be defined by

$$\hat{\partial}_0 g_{ij} = -2N K_{ij}, \tag{3}$$

which serves also as the evolution equation for the spatial metric. By working out the expression $\Omega_{ij} \equiv \hat{\partial}_0 R_{ij} - 2\overline{\nabla}_{(i} R_{j)0}$ in 3+1 form, where R_{ij} and R_{j0} are components of the spacetime Ricci tensor and $\overline{\nabla}_i$ denotes the spatial covariant derivative, we find a wave-like equation which governs the evolution of K_{ij}:

$$-N\hat{\Box}K_{ij} = J_{ij} + S_{ij} - \Omega_{ij}, \tag{4}$$

where the physical wave operator for arbitrary shift is $\hat{\Box} \equiv -N^{-1}\hat{\partial}_0 N^{-1}\hat{\partial}_0 + \overline{\nabla}_k\overline{\nabla}^k$. Equation (4) is an identity until we substitute the Einstein equations $R_{\alpha\beta} = 8\pi(T_{\alpha\beta} - \frac{1}{2}T^{\lambda}_{\lambda} g_{\alpha\beta})$ into Ω_{ij} ($G = c = 1$).

The detailed form of the right-hand side of (4) can be found in [11]; the present conventions are those in [11]. Here we simply point out that Ω_{ij} has become a matter source that is zero here, J_{ij} is the nonlinear self-interaction term in $3+1$ form, and S_{ij} is a slicing-dependent term that must involve fewer than second derivatives of K_{ij} to render (4) a true (hyperbolic) wave equation. A simple way to satisfy the restriction on S_{ij} is to invoke the harmonic slicing condition

$$\hat{\partial}_0 N + N^2 H = 0, \tag{5}$$

where H is the trace of K_{ij}, and from which it follows that $S_{ij} = 0$.

For appropriate choice of initial data [11], Eqs. (3), (4), and (5) represent the dynamical part of Einstein's equations. Combining them we obtain a quasi-diagonal hyperbolic equation for g_{ij}, with principal (highest-order) part $\hat{\Box}\hat{\partial}_0$. Hence (3), (4), and (5) may be said to give the "third-order" form of the Einstein–Ricci system.

We note that the third-order Einstein–Ricci system can also be cast into a first-order symmetric hyperbolic form [11]. It also possesses a higher order form (the "fourth-order Einstein–Ricci system"), essentially a wave equation for $(\hat{\partial}_0 K_{ij})$, obtained from $\hat{\partial}_0\Omega_{ij} + \overline{\nabla}_i\overline{\nabla}_j R_{00}$ [11]. This system has a well-posed Cauchy problem and complete freedom in choosing both β^i and N: it has no analog of a slicing term like S_{ij}.

Perturbative expansion

The first step in obtaining radial wave equations is to linearize the hyperbolic Einstein-Ricci equations around a static Schwarzschild background. We separate the gravitational quantities of interest into background (denoted by a tilde) and perturbed parts: the 3-metric $g_{ij} = \widetilde{g}_{ij} + h_{ij}$, the extrinsic curvature $K_{ij} = \widetilde{K}_{ij} + \kappa_{ij}$, the lapse $N = \widetilde{N} + \alpha$ and the shift vector $\beta^i = \widetilde{\beta}^i + v^i$. In Schwarzschild coordinates (t, r, θ, ϕ), the background quantities are given by

$$\widetilde{N} = \left(1 - \frac{2M}{r}\right)^{1/2}, \tag{6}$$

$$\begin{aligned} \widetilde{g}_{ij}dx^i dx^j &= \widetilde{N}^{-2}dr^2 + r^2(d\theta^2 + \sin^2\theta d\phi^2), & (7)\\ \widetilde{\beta}^i &= 0, & (8)\\ \widetilde{K}_{ij} &= 0, & (9) \end{aligned}$$

while the perturbed quantities have arbitrary angular dependence. The background quantities satisfy the dynamical equations $\partial_t \widetilde{g}_{ij} = 0$, $\partial_t \widetilde{N} = 0$, and thus remain constant for all time. The perturbed quantities, on the other hand, obey the following evolution equations

$$\begin{aligned} \partial_t h_{ij} &= -2\widetilde{N}\kappa_{ij} + 2\widetilde{\nabla}_{(i}v_{j)}, & (10)\\ \partial_t \alpha &= v^i \widetilde{\nabla}_i \widetilde{N} - \widetilde{N}^2 \kappa, & (11) \end{aligned}$$

$$\begin{aligned} \widetilde{N}^{-1}\partial_t^2 \kappa_{ij} - \widetilde{N}\widetilde{\nabla}^k\widetilde{\nabla}_k\kappa_{ij} &= -4\widetilde{\nabla}_{(i}\kappa^k_{\ j)}\widetilde{\nabla}_k\widetilde{N} + \widetilde{N}^{-1}\kappa_{ij}\widetilde{\nabla}^k\widetilde{N}\widetilde{\nabla}_k\widetilde{N} + 3\widetilde{\nabla}^k\widetilde{N}\widetilde{\nabla}_k\kappa_{ij}\\ &+ \kappa_{ij}\widetilde{\nabla}^k\widetilde{\nabla}_k\widetilde{N} - 2\kappa^k_{\ (i}\widetilde{\nabla}_{j)}\widetilde{\nabla}_k\widetilde{N} - 2\widetilde{N}^{-1}\kappa^k_{\ (i}\widetilde{\nabla}_{j)}\widetilde{N}\widetilde{\nabla}_k\widetilde{N} + 2\kappa\widetilde{\nabla}_i\widetilde{\nabla}_j\widetilde{N}\\ &+ 4\partial_{(i}\kappa\partial_{j)}\widetilde{N} + 2\widetilde{N}^{-1}\kappa\widetilde{\nabla}_i\widetilde{N}\widetilde{\nabla}_j\widetilde{N} - 2\widetilde{N}\widetilde{R}_{k(i}\kappa^k_{\ j)} - 2\widetilde{N}\widetilde{R}_{kijm}\kappa^{km}, \end{aligned} \quad (12)$$

where $\kappa \equiv \kappa^i_{\ i}$ and the tilde denotes a spatial quantity defined in terms of the background metric, $\widetilde{g}_{ij}$. Note that the wave equation (12) for κ_{ij} involves only the background lapse and curvature.

Angular decomposition

We can further simplify the evolution equation (12) by separating out the angular dependence, thus reducing it to a set of 1D equations. We accomplish this by expanding the extrinsic curvature in Regge–Wheeler tensor spherical harmonics [12] and substituting this expansion into (12). Using the notation of Moncrief [13] we express the expansion as

$$\begin{aligned} \kappa_{ij} &= a_\times(t,r)(\hat{e}_1)_{ij} + rb_\times(t,r)(\hat{e}_2)_{ij} + \widetilde{N}^{-2}a_+(t,r)(\hat{f}_2)_{ij} + rb_+(t,r)(\hat{f}_1)_{ij}\\ &+ r^2c_+(t,r)(\hat{f}_3)_{ij} + r^2d_+(t,r)(\hat{f}_4)_{ij}, \end{aligned} \quad (13)$$

where $(\hat{e}_1)_{ij}, \cdots, (\hat{f}_4)_{ij}$ are the Regge–Wheeler harmonics, which are functions of (θ, ϕ) and have suppressed angular indices (ℓ, m) for each mode. The odd-parity multipoles ($a_\times$ and $b_\times$) and the even-parity multipoles (a_+, b_+, c_+ and d_+) also have suppressed indices for each angular mode and there is an implicit sum over all modes in (13). The six multipole amplitudes correspond to the six components of κ_{ij}. However, using the linearized momentum constraints

$$\widetilde{\nabla}_j(\kappa^j_{\ i} - \delta^j_{\ i}\kappa) = 0, \quad (14)$$

we reduce the number of independent components of κ_{ij} to three. An important relation is also obtained through the wave equation for κ, whose multipole expansion is simply given by $\kappa = h(t,r)Y_{\ell m}$ where $Y_{\ell m}(\theta,\phi)$ is the standard scalar spherical harmonic and again there is an implicit sum over suppressed indices (ℓ, m). Using this expansion, in conjunction with the momentum constraints (14), we derive a set of radial constraint equations which relate the dependent amplitudes $(b_\times)_{\ell m}$, $(b_+)_{\ell m}$, $(c_+)_{\ell m}$ and $(d_+)_{\ell m}$ to the three independent amplitudes $(a_\times)_{\ell m}$, $(a_+)_{\ell m}$, $(h)_{\ell m}$:

$$
\begin{aligned}
(b_\times)_{\ell m} &= -\frac{1}{(\ell+2)(\ell-1)}[(1+3\widetilde{N}^2) + 2\widetilde{N}^2 r\,\partial_r]\,(a_\times)_{\ell m}, && (15)\\
(b_+)_{\ell m} &= \frac{1}{\ell(\ell+1)}[(3+r\partial_r)\,(a_+)_{\ell m} - (1+r\partial_r)\,(h)_{\ell m}], && (16)\\
(c_+)_{\ell m} &= \frac{1}{2(\ell+2)(\ell-1)}\{2(1-\ell-\ell^2)\,(a_+)_{\ell m}\\
&\quad -2(h)_{\ell m} + \ell(\ell+1)[(1+5\widetilde{N}^2) + 2\widetilde{N}^2 r\,\partial_r]\,(b_+)_{\ell m}\}, && (17)\\
(d_+)_{\ell m} &= \frac{1}{\ell(\ell+1)}[(a_+)_{\ell m} + 2(c_+)_{\ell m} - (h)_{\ell m}], && (18)
\end{aligned}
$$

for each (ℓ, m) mode.

Substituting (13) into (12) and using the constraint equations (15)–(18), we obtain a set of linearized radial wave equations for each independent amplitude. For each (ℓ, m) mode we have one odd-parity equation

$$
\left\{\partial_t^2 - \widetilde{N}^4\partial_r^2 - \frac{2}{r}\widetilde{N}^2\partial_r - \frac{2M}{r^3}\left(1-\frac{3M}{2r}\right) + \widetilde{N}^2\left[\frac{\ell(\ell+1)}{r^2} - \frac{6M}{r^3}\right]\right\}(a_\times)_{\ell m} = 0, \qquad (19)
$$

and two coupled even-parity equations,

$$
\begin{aligned}
&\left[\partial_t^2 - \widetilde{N}^4\partial_r^2 - \frac{6}{r}\widetilde{N}^4\partial_r + \widetilde{N}^2\frac{\ell(\ell+1)}{r^2} - \frac{6}{r^2} + \frac{14M}{r^3} - \frac{3M^2}{r^4}\right](a_+)_{\ell m}\\
&\quad + \left[\frac{4}{r}\widetilde{N}^2\left(1-\frac{3M}{r}\right)\partial_r + \frac{2}{r^2}\left(1-\frac{M}{r}-\frac{3M^2}{r^2}\right)\right](h)_{\ell m} = 0, \qquad (20)
\end{aligned}
$$

$$
\begin{aligned}
&\left[\partial_t^2 - \widetilde{N}^4\partial_r^2 - \frac{2}{r}\widetilde{N}^2\partial_r + \widetilde{N}^2\frac{\ell(\ell+1)}{r^2} + \frac{2M}{r^3} - \frac{7M^2}{r^4}\right](h)_{\ell m}\\
&\quad - \frac{2M}{r^3}\left(3-\frac{7M}{r}\right)(a_+)_{\ell m} = 0. \qquad (21)
\end{aligned}
$$

These equations are related to the standard Regge–Wheeler and Zerilli equations [12], which can be derived in a more complete analysis of gauge-invariant hyperbolic formulations.

The radial wave equations (19)–(21) for each (ℓ, m) mode of the independent multipole amplitudes $(a_\times)_{\ell m}$, $(a_+)_{\ell m}$, $(h)_{\ell m}$ form the basis for our approach. In the perturbative region, they replace the nonlinear Einstein equations and determine the evolution of K_{ij}. They can be used to evolve, with minimal computational cost, gravitational wave data to arbitrarily large distances from the highly dynamical strong field region. The evolution equations for h_{ij} (10) and α (11) can also be integrated using the data for K_{ij} computed in this region. Note that, because h_{ij} and α evolve along the coordinate time axis, these equations need only be integrated in the region in which their values are desired, not over the whole region L (these quantities have characteristic speed zero).

3 Numerical implementation

This section is a general guide for the numerical implementation of the Cauchy-perturbative matching method for radiation extraction and outer boundary conditions described so far.

Consider a 3D numerical relativity code which solves the Cauchy problem of Einstein's equations in either the standard ADM form [3] or the hyperbolic form [14]. During each timestep the procedure followed by our module for extracting radiation and imposing outer boundary conditions can be summarized in three successive steps: *(1) extraction* of the independent multipole amplitudes on **E**, *(2) evolution* of the radial wave equations (19)–(21) on L out to the distant wave zone, *(3) reconstruction* of K_{ij} and $\partial_t K_{ij}$ at specified gridpoints at the outer boundary of N. We now discuss each of these steps in detail.

Extraction – As mentioned in Sect. 2, the extraction 2-sphere **E** acts as the joining surface between the evolution of the highly dynamical, strong field region (dark shaded area of Fig. 1) and the perturbative regions (light shaded areas). At each timestep, K_{ij} and $\partial_t K_{ij}$ are computed on N as a solution to Einstein's equations. In the test cases presented here, N uses topologically Cartesian coordinates, although there are no restrictions on the choice of the coordinate system. The Cartesian components of these tensors are then transformed into their equivalents in a spherical coordinate basis and their traces are computed using the inverse background metric, i.e., $H = \tilde{g}^{ij} K_{ij}$, $\partial_t H = \tilde{g}^{ij}\partial_t K_{ij}$. From the spherical components of K_{ij} and $\partial_t K_{ij}$, the independent multipole amplitudes for each (ℓ, m) mode are then derived by an integration over the 2-sphere:

$$(a_\times)_{\ell m} = \frac{1}{\ell(\ell+1)} \int \frac{1}{\sin\theta} \left[K_{r\phi}\,\partial_\theta - K_{r\theta}\,\partial_\phi\right] Y^*_{\ell m}\, d\Omega, \tag{22}$$

$$(a_+)_{\ell m} = \int \tilde{N}^2\, K_{rr}\, Y^*_{\ell m}\, d\Omega, \tag{23}$$

$$(h)_{\ell m} = \int H\, Y^*_{\ell m} d\Omega. \tag{24}$$

Their time derivatives are computed similarly. Rather than performing the integrations (22)–(24) using spherical polar coordinates, it is useful to cover **E** with two stereographic coordinate "patches". These are uniformly spaced 2D grids onto which the values of K_{ij} and $\partial_t K_{ij}$ are interpolated using either a three-linear or a three-cubic polynomial interpolation scheme. Each point on the 2-sphere, denoted by spherical coordinate values (θ, ϕ), corresponds to a point (q, p) on a stereographic grid whose coordinates can be combined into a single complex number ζ:

$$\zeta_N \equiv q_N + ip_N = \tan\left(\frac{\theta}{2}\right) e^{i\phi}, \tag{25}$$

$$\zeta_S \equiv q_S + ip_S = \frac{1}{\zeta_N}, \tag{26}$$

where N and S denote the northern $(0 \leq \theta \leq \pi/2)$ and southern $(\pi/2 \leq \theta \leq \pi)$ hemispheres, respectively. As a result of this transformation, the integrals over the 2-sphere in (3) are computed over the stereographic patches, which naturally avoid polar singularities.

Evolution – Once the multipole amplitudes, $(a_\times)_{\ell m}$, $(a_+)_{\ell m}$, $(h)_{\ell m}$ and their time derivatives are computed on **E** in the timeslice $t = t_0$, they are imposed as inner boundary conditions on the 1D grid. Using a second-order integration scheme, our module then evolves the radial wave equations (19)–(21) for each (ℓ, m) mode forward to the next timeslice at $t = t_1$. The outer boundary of the 1D grid is always placed at a distance large enough so that background field and near-zone effects are unimportant, and a radial Sommerfeld condition for the wave equations (19)–(21) can be imposed there. Of course, the initial data on L must be consistent with the initial data on N. This can either be imposed analytically or determined by applying the aforementioned extraction procedure to the initial data set at each gridpoint of L in the region of overlap with N. In the latter case, initial data outside the overlap region can be set by considering the asymptotic fall-off of each variable.

Reconstruction and matching – From the perturbative data evolved to time t_1, outer boundary values for N can now be computed. The procedure for doing this differs depending on whether a hyperbolic or an ADM formulation of Einstein's equations is used by the 3D "interior code". For a hyperbolic code (cf. [14]), it is necessary to provide boundary data for K_{ij} and $\partial_t K_{ij}$. For an ADM code (cf. [3]), on the other hand, outer boundary data are necessary only for K_{ij}, since the interior code can calculate g_{ij} at the outer boundary by integrating in time the boundary values for K_{ij}. In either case, if outer boundary values for the lapse N are needed (e.g., for integrating harmonic slicing condition (5)), these can be computed by the perturbative module or by integration of H at the boundary.

In order to compute K_{ij} at an outer boundary point of N (or at any other point in the overlap between N and $\mathcal{P}$), it is necessary to reconstruct K_{ij} from the multipole amplitudes and tensor spherical harmonics. The Schwarzschild coordinate values (r, θ, ϕ) of the relevant gridpoint are first determined. Next, $(a_\times)_{\ell m}$, $(a_+)_{\ell m}$ and $(h_{\ell m})$ for each (ℓ, m) mode are interpolated to the radial coordinate value of that point. The dependent multipole amplitudes $(b_\times)_{\ell m}$, $(b_+)_{\ell m}$, $(c_+)_{\ell m}$ and $(d_+)_{\ell m}$ are then computed using the constraint equations (15)–(18). Finally, the Regge–Wheeler tensor spherical harmonics $(\hat{e}_1)_{ij}$–$(\hat{f}_4)_{ij}$ are computed for the angular coordinates (θ, ϕ) for each (ℓ, m) mode and the sum in equation (13) is performed. This leads to the reconstructed component of κ_{ij} (and therefore K_{ij}); a completely analogous algorithm is used to reconstruct $\partial_t K_{ij}$.

It is important to emphasize that this procedure allows one to compute K_{ij} at any point of N which is covered by the perturbative region. As a result, the numerical module can reconstruct the values of K_{ij} and $\partial_t K_{ij}$ on a 2-surface of arbitrary shape, or for any collection of points outside of $\mathbf{E}$.

Numerical implementation of this method is rather straightforward. Very few modifications to a standard 3D numerical relativity code are necessary in order to allow for the simultaneous evolution of the highly dynamical region and of the perturbative one. Because of the use of numerically inexpensive integration of 1D wave equations, implementation of this module provides gravitational wave extraction *and* stable outer boundary conditions with only minimal additional computational cost.

Finally, it should be noted that, in practice, we may not know *a priori* if the Schwarzschild-perturbative approximation is valid near the outer boundary of a given numerical relativity simulation. Through experimentation, however, it is possible to test the validity of the approximation. This can be done, for instance, by extracting data at different radii and comparing the waveforms computed at the outer sphere with those evolved from the inner sphere. This makes it possible to determine if the neglected terms in the approximation have a significant effect. At any point in the overlap region between N and $\mathcal{P}$, it is possible to reconstruct gravitational wave data and compare these values with those computed by the full nonlinear evolution.

4 Numerical setup

We report numerical results for the propagation of $\ell = 2$, $m = 0$ (unless otherwise stated) even-parity linear waves, initially modulated by a Gaussian envelope with amplitude $A = 10^{-6}$ and width parameter $b = 1$. Being time-symmetric at the initial time, these waves have ingoing and outgoing parts. At each time level, the extrinsic curvature and 3-metric are computed using a 3D code solving the full Einstein equations with a geodesic slicing condition (i.e., $N = 1$, $\beta^i = 0$) on a 3D vertex-centered grid, with extents $(x, y, z) \in [-4, 4]$. We have also used a number of different grid resolutions ranging from $(17)^3$ to $(129)^3$ grid points and comparable resolutions have been used on the extraction 2-sphere. In the following sections we will discuss the results of our computations in detail and concentrate on two different but interrelated aspects, namely, the "short term" and the "long term" behaviours. In the first we will consider gravitational wave extraction and imposition of boundary conditions on timescales comparable with the crossing timescale of the numerical grid (i.e., $t \sim 8$). In the second section we consider the opposite regime and investigate the effects of a Cauchy-perturbative matching on timescales much larger than the crossing timescale (i.e., $t \gg 8$) when most of the radiation has left the numerical grid and the stability properties of the module are put to a test.

5 The short term behaviour

In the case of a flat background spacetime and for weak waves on Schwarzschild-like backgrounds, the position of the extraction 2-sphere is arbitrary. This gives us the important possibility of analyzing the influence of the position of the extraction 2-sphere on the accuracy of the gravitational information read-off, and of seeing how this then affects the accuracy of the boundary conditions which are provided.

In Fig. 2 we show the time series of the multipole amplitude $(a_+)_{20}$ extracted at **E** (in the case of an initial traceless $\ell = 2$, $m = 0$ wave packet, this is the only analytically non-zero multipole). Other multipoles of the same mode (e.g. $(a_+)_{21}$, $(a_+)_{2-1}$, $(a_+)_{22}$, $(a_+)_{2-2}$) as well as other parity amplitudes (i.e., $(a_\times)_{\ell m}$, $(h)_{\ell m}$) are also extracted, but their amplitudes are generally several orders of magnitude smaller. The six different diagrams refer to the six different positions at which we have placed the extraction 2-sphere (i.e., $r_E = 1.0,\ 1.5,\ 2.0,\ 2.5,\ 3.0,\ 3.5$). Each diagram also shows the same quantity computed at three different resolutions (namely, with $(129)^3$, $(65)^3$ and $(33)^3$ grid points) and we scale the amplitude by r^3 to compensate for the leading-order radial fall-off of $(a_+)_{20}$.

It is clear from Fig. 2 that there is an increasing relative error between the analytic solution and the extracted data as the extraction 2-sphere is placed at larger radii while the resolution is held constant (e.g., compare results at $r_E = 1.0$ and $r_E = 3.5$). Since the results shown in Fig. 2 do not vary if the resolution on the two-sphere (i.e., the number of grid points used to cover the extraction 2-sphere) is increased or decreased, the origin of this behaviour has to be found in the intrinsic numerical error which is introduced by the solution of Einstein's equation by the interior 3D code and which becomes larger as the waves propagate outwards. In fact, a more careful investigation of the behaviour of the multipole amplitudes other than $(a_+)_{20}$ shows that the initially traceless linear waves develop a non-zero trace of K_{ij} as the evolution proceeds (the presence of non-zero trace becomes apparent by looking at the amplitudes of the extracted $(h)_{\ell m}$). A non-zero trace of K_{ij} is due to truncation error and it rapidly converges to zero as the resolution is increased, but it has a subtle effect on the accuracy of the extracted data. While the multipole amplitudes fall approximately as $\sim r^{-3}$, the non-zero trace of K_{ij} remains constant during the time evolution.

As a result, for increasing extraction radii, the difference in the amplitudes of say, $(a_+)_{20}$ and $(h)_{20}$, becomes smaller and smaller. For $r_E \gtrsim 3$ the two multipoles are comparable and this error becomes more severe as a coarser resolution is used. However, it is also clear that all of these pathologies can be cured by simply increasing the resolution and Fig. 2 shows the time series rapidly converging to the analytic solution as the resolution is increased, even in the most extreme case of an extraction radius $r_E = 3.5$. In view of this, we can summarize the properties of the *perturbative radiation extraction* as follows: for any extraction 2-sphere location, it is always possible to find a resolution for the interior grid which will provide gravitational wave information with the required accuracy.

Having shown the ability of the module to extract convergent gravitational wave data from a fully nonlinear 3D numerical relativity code, we next turn to examine the corresponding ability to "reconstruct" the extrinsic curvature K_{ij} from the information extracted at the 2-sphere (Fig. 2).

For the computations discussed in this section, we have chosen to impose the simplest type of boundary conditions that can be implemented within a Cauchy-

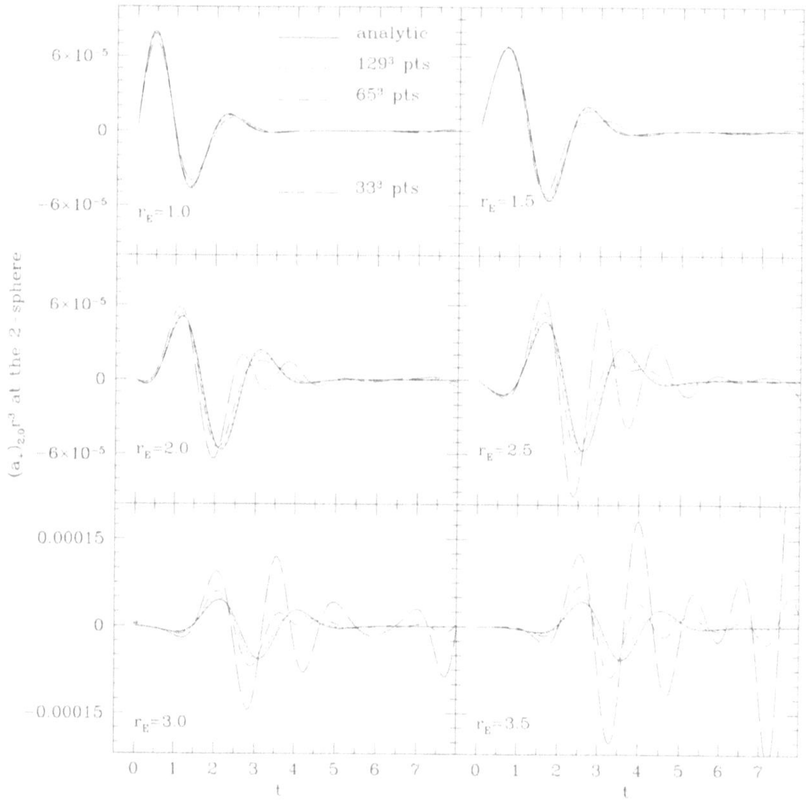

Fig. 2. Time series of the multipole amplitude $(a_+)_{20}$ extracted at a 2-sphere of radius $r_E = 1.0,\ 1.5,\ 2.0,\ 2.5,\ 3.0,\ 3.5$. Different grid resolutions are indicated with different line types, with a dotted line referring to $(129)^3$ grid points, a short dashed line referring to $(65)^3$ grid points and a long dashed line referring to $(33)^3$ grid points. The analytic solution is indicated by a continuous line. Note that we have scaled the amplitude by r^3 to compensate for the radial fall-off. We have here used a leapfrog integration scheme

perturbative approach. From the solution at the new time level of the evolution equations (19)–(21) for each (ℓ, m) mode we calculate the value of the extrinsic curvature at all the grid points on **B**. We then impose these values as the outer boundary conditions and will refer to this implementation as the *Dirichlet injection* to distinguish it from other types of boundary conditions which will be discussed in the next section.

Figure 3 shows a time series of the reconstructed values of the K_{zz} component of the extrinsic curvature for different positions of the extraction 2-sphere and different

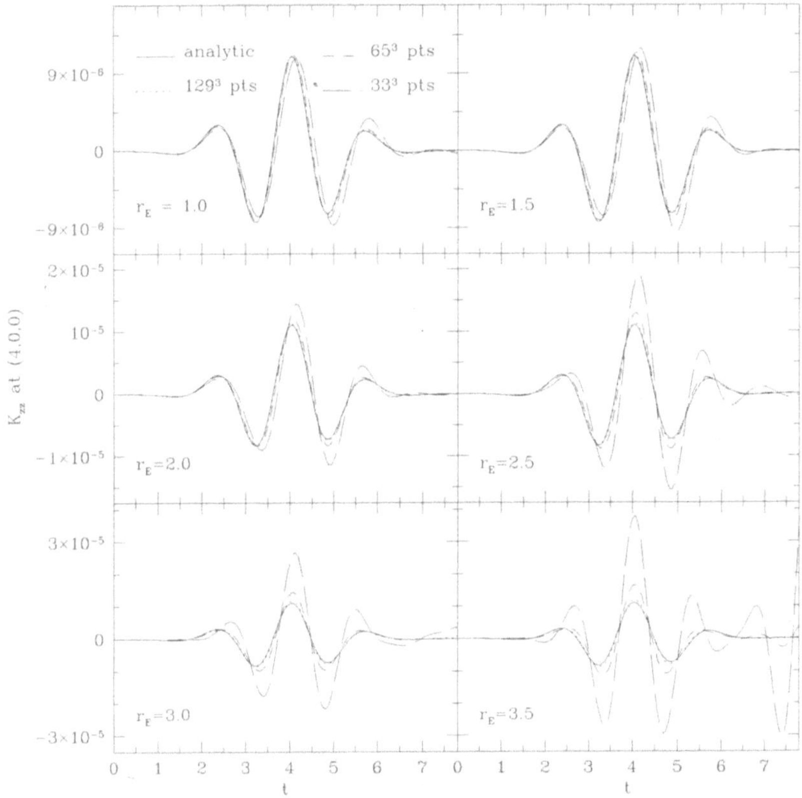

Fig. 3. Time series of the reconstructed values of K_{zz}. The six different diagrams refer to the six different positions of the extraction 2-sphere, and, for each diagram, results obtained with different grid resolutions are indicated with different line types. The time series is computed on a gridpoint at the outer boundary aligned on the x−axis (i.e., at a coordinate location (4, 0, 0)) and a leapfrog evolution scheme has been used

grid resolutions. The time series is computed on a gridpoint at the outer boundary aligned on the x-axis (i.e., at a coordinate location (4, 0, 0).

It is not surprising that the behaviour of the time series for K_{ij} in Fig. 3 mimics that seen in Fig. 2 for the extracted multipole amplitudes. In fact, as we will discuss further in the following section, the extraction 2-sphere and the outer boundary are closely coupled. The accuracy of the boundary conditions imposed is clearly dependent on the accuracy of the extracted gravitational wave information. A large relative error between the extracted and analytic data will translate into a proportionally large relative error between the injected values for K_{ij} and the analytic values for the same quantities.

It is also important to emphasize that the imposition of poor boundary conditions does, in turn, produce spurious reflection of radiation at the outer boundary. This reflected gravitational wave information will contaminate the radiation signal read off at the extraction 2-sphere leading to increasingly larger differences from the purely outgoing analytic solution. In a loose sense, the extraction 2-sphere and the outer boundary behave as a coupled system of microphones and loud-speakers with the 2-sphere playing the role of the microphones. It is clear that such a coupling can be extremely delicate and might be the cause for exponentially growing instabilities as we will discuss in the following section, where we also indicate a number of prescriptions that make this coupling less important.

6 The long term behaviour

Providing an accurate and numerically convergent approximation to the gravitational waveform in the wave zone surrounding an isolated source represents a very important feature of any radiation-extraction and outer-boundary module. However, even more important is the fact that the module provides a stable evolution of the interior equations, minimizing the numerical reflection of radiation at the boundary. Particularly for systems of evolution equations in which radiation and background dominated metric/extrinsic curvature variables are not easily defined, matching techniques may be the only way to achieve a stable evolution. In this section we discuss a number of approaches which were applied to this problem. A more detailed discussion can be found in [7].

Perturbative Sommerfeld boundary conditions

The results presented in the previous section were obtained by imposing as outer boundary values for K_{ij}, those reconstructed from the values, at the new time level, of the multipole amplitudes. Although straightforward to implement and very accurate, a "Dirichlet injection" of outer boundary data leads to a rapid error growth when the time evolution is carried on for sufficiently long periods of time. A careful analysis reveals that these boundary conditions seem to produce a large amount of reflection as the gravitational waves leave the numerical grid. This is basically the result of the slight mismatch between the wave phases and amplitudes imposed at the outer boundary and those in the interior of the 3D numerical grid.

It is clear that a finite discretization will always produce a certain amount of reflection. It is thus important to study and develop new techniques that tend to suppress this reflection and allow, as much as possible, the outgoing radiation to escape freely to infinity. Interesting results in this direction have been obtained by implementing a boundary condition which we have named *perturbative Sommerfeld* [4]. While a simple Sommerfeld outgoing wave condition applied to a tensor quantity such as the extrinsic curvature or the three-metric is (strictly speaking) incorrect, it is often satisfactory, especially if applied in the distant wave zone (cf. [15] for the case of linear waves). In view of this, we have related, at the outer boundary **B**, the null derivatives of the extrinsic curvature obtained from the interior grid (i.e., K_{ij}) with that obtained from the perturbative module (κ_{ij}, since background extrinsic curvature is assumed to be zero):

$$\frac{\partial}{\partial t}\left(K_{ij}-\kappa_{ij}\right)+\frac{\partial}{\partial r}\left(K_{ij}-\kappa_{ij}\right)+\frac{q}{r}\left(K_{ij}-\kappa_{ij}\right)=0, \tag{27}$$

where q is a positive integer[2]. In other words, we "correct" the Sommerfeld outgoing wave condition $\partial_t K_{ij}+\partial_r K_{ij}+(2/r)K_{ij}=0$, with a right-hand side which is usually taken to be zero but which is not (in general) zero. As a result, this prescription resembles a Sommerfeld condition but is effectively much more general since (i) it can be used in regions where the radiation is not dominated by the asymptotic outgoing behavior, and (ii) it takes into account arbitrary angular dependence, as well as the effects of a Schwarzschild black hole background. Since the perturbative correction can be very small and is the result of a near cancellation of several terms involving space and time derivatives, it is important to implement Eq. (7) so that the same numerical differential operator acts on both K_{ij} and κ_{ij}.

Implementation of this method has shown that the perturbative Sommerfeld boundary conditions represent a very promising implementation of the Cauchy-perturbative matching and provide both high accuracy and very small reflection at the outer boundary. Despite these appealing features, they do not provide for very long term stability of the interior code and they eventually suffer from exponential growth. Of course, in the computations reported here, the boundary conditions provided by the Cauchy-perturbative matching are totally adequate on the timescale necessary for the gravitational waves to leave the numerical grid. Indeed, the instabilities induced by the outer boundary become relevant only long after the crossing timescale, when the grid basically contains numerical noise. However, providing accurate boundary conditions on a dynamical timescale is usually not sufficient and the achievement of unconditionally stable codes is not only of academic interest. In order to successfully model the problem of binary black hole coalescence, the numerical code will have to be able to solve Einstein's equations stably on a timescale (the gravitational wave emission timescale) which is much longer than the dynamical timescale. It is therefore imperative to devise a technique which provides unconditionally stable boundary conditions. A first successful step in this direction has been made with the implementation of the "perturbative-blending" technique we will discuss next.

Perturbative boundary conditions with "blending"

The perturbative Sommerfeld boundary conditions are much more effective in providing accuracy and reducing reflection at the outer boundary than is the simple Dirichlet injection. One of the major differences between the two types of boundary condition is in their numerical implementation. This involves only the outermost boundary grid points (i.e., those on **B** of Fig. 1) in the case of a Dirichlet injection

[2] Results do not depend sensitively on the value of q, which we have chosen to be 2 in these tests in order to reproduce the leading order term in the asymptotic radial fall-off.

but also the closest interior neighbouring gridpoints in the case of a perturbative Sommerfeld condition (necessary for taking a finite difference form of the spatial derivatives in (27)). Perturbative blending can then be considered as the extension of the perturbative Sommerfeld condition to a larger set of gridpoints. The basic idea is simple and based on attempting to modify the propagation characteristics in the vicinity of the outer boundary with the goal of acting distinctively on the outgoing and ingoing parts of the gravitational waves [7]. A detailed description of the basic properties of "*sponge filters*" in conjunction with absorbing boundary conditions in one-dimensional wave propagation can be found in [16].

Figure 4 illustrates the radical changes in the long time behaviour introduced by the use of a perturbative blending match.

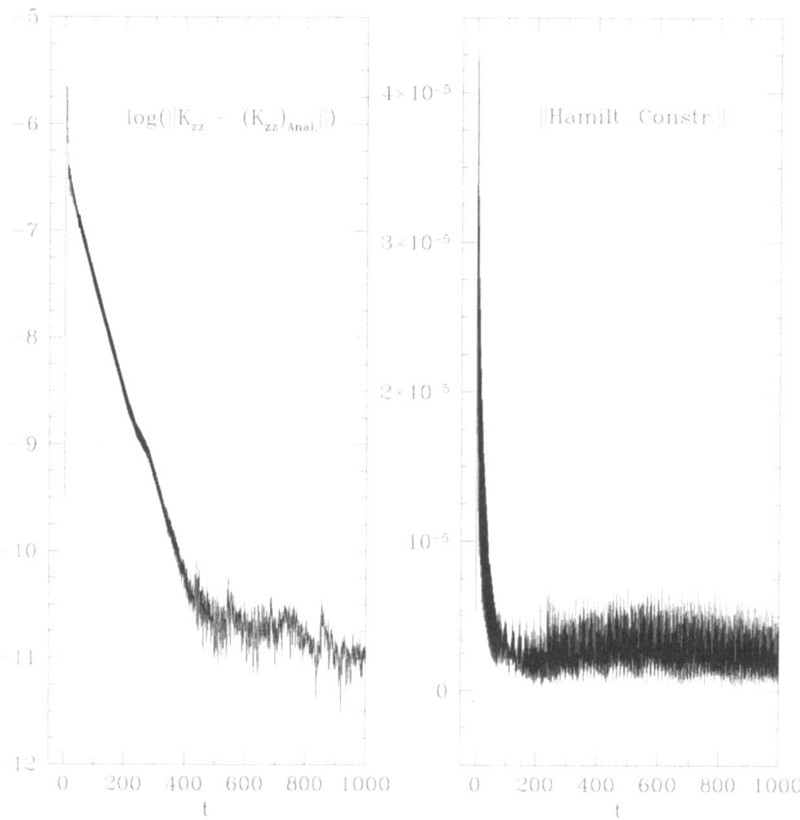

Fig. 4. Time series of the errors in the K_{zz} component of the extrinsic curvature and of the violation of the Hamiltonian constraint when a perturbative blending match is used. The resolution is that of $(33)^3$ gridpoints and the extraction radius is $r_E = 1$ and the blending region is covered with 10 gridpoints. The integration scheme used is Crank-Nicholson

The long evolution times reached and the quiescent behaviour of the evolved variables reduced to their roundoff error values, clearly show that the use of a perturbative blending match does provide the desired long term stability. This is evident from the gradual decay of the norms and the stationary behaviour of the violation of the Hamiltonian constraint which does not show any sign of instability on a timescale more than 125 times the crossing timescale. We have also verified that the perturbative-blending boundary conditions provide *accurate* short term extraction of the waveform, comparable to the results in Figs. 2–3.

The use of perturbative blending boundary conditions has provided the requisite unconditional long-term stability we were requiring to our radiation-extraction and outer-boundary module. Given the versatility of the perturbative matching, this approach could represent a very powerful tool in other numerical relativity applications as well. Further work is necessary in this direction and experimentation with more complex physical configurations.

7 Conclusions

We have presented a method of extracting gravitational radiation from three-dimensional numerical relativity codes and providing outer boundary conditions. Our approach matches the solution of a Cauchy evolution of Einstein's equations to a set of one-dimensional linear wave equations on a curved background. We have also presented results of the evolution of linear and mildly nonlinear waves. In particular, we have shown the ability of the perturbative module to extract convergent gravitational wave information at different locations within the 3D numerical grid by solving the nonlinear form of Einstein's equation. Our results indicate that, given an extraction 2-sphere radius, a resolution can be found which provides extraction and reinjection with the required accuracy.

A number of different approaches to the problem of imposing outer boundary conditions have been discussed. Relying on the important advantage of being able to provide information on the whole portion of the 3D numerical grid outside the extraction 2-sphere, we have investigated Dirichlet boundary conditions, perturbative Sommerfeld boundary conditions and "perturbative blended" boundary conditions. Each of these approaches has been shown to provide convergent boundary conditions, but only the latter provides stable evolutions.

These first successes of the Cauchy perturbative matching method motivate further work in this direction, both in the application of our numerical module to fully nonlinear spacetimes and in the extension of the mathematical apparatus to more general background spacetimes.

Acknowledgements

This method and technique was developed in collaboration with Andrew M. Abrahams, Richard A. Matzner, Mark E. Rupright, and Stuart L. Shapiro. It is also a pleasure to thank Roberto Gómez and Jeff Winicour for helpful discussions and suggestions.

References

1. Information about the Binary Black Hole Grand Challenge Alliance, its goals and the present status of its research can be found at the web page: `http://www.npac.syr.edu/projects/bh/`.
2. Gómez R., Lehner L., Marsa R.L., Winicour J. et al. (1998): Phys. Rev. Lett. **80**, 3915
3. Cook G.B., Huq M.F., Klasky S.A., Scheel M.A.. et al. (1998): Phys. Rev. Lett. **80**, 2512
4. Abrahams A.M., Rezzolla L., Rupright M.E. et al. (1998): Phys. Rev. Lett. **80**, 1812
5. Thorne K.S. (1980): Rev. Mod. Phys. **52**, 299
6. Rupright M.E., Abrahams A.M., Rezzolla L. (1998): Phys. Rev. D **58**, 044005
7. Rezzolla, L.,Abrahams A.M., Matzner R.A., Rupright M.E., Shapiro S.L. (1999): Phys. Rev. D **59**, 064001
8. Bishop N.T. et al. (1996): Phys. Rev. Lett. **76**, 4303;
   Bishop N.T., Gómez R., Lehner L., Winicour J. (1996): Phys. Rev. D **54**, 6153
9. Abrahams A.M., Evans C.R. (1988): Phys. Rev. D **37**, 317; (1990): Phys. Rev. D **42**, 2585
10. Price R.H., Pullin J. (1994): Phys. Rev. Lett. **72**, 3297;
    Abrahams A.M., Cook G. (1994): Phys. Rev. D **50**, 2364;
    Abrahams A.M., Shapiro S.L., Teukolsky S.A. (1995): Phys. Rev. D **51**, 4295;
    Abrahams A.M., Price R.H. (1996): Phys. Rev. D **53**, 1963 and 1972
11. Choquet-Bruhat Y., York Jr. J.W. (1995): C. R. Acad. Sci. Série I, Paris **321**, 1089;
    Abrahams A.M., Anderson A., Choquet-Bruhat Y., York Jr., J.W. (1995): Phys. Rev. Lett. **75**, 3377
12. Regge T., Wheeler J.A. (1957): Phys. Rev. **108**, 1063;
    Zerilli F. (1970): Phys. Rev. Lett. **24** 737
13. Moncrief V. (1974): Ann. Phys. **88**, 323
14. Scheel M.A. et al. (1997): Phys. Rev. D **56**, 6320
15. Rezzolla L. et al. (1998): Phys. Rev. D **57**, 1084
16. Israeli M., Orszag S.A. (1981): Journ. Comp. Phys. **41**, 115;
    Marsa R.L., Choptuik M.W. (1996): Phys. Rev. D **54**, 4929
17. Nakamura T., Oohara K., Kojima Y. (1987): Prog. Theor. Phys. Suppl. **90**, 76;
    Shibata M., Nakamura T. (1995): Phys. Rev. D **52**, 5428

Branes in Anti-de Sitter Space-Time

M. Trigiante

Abstract. An intense study of the relationship between certain quantum theories of gravity realized on curved backgrounds and suitable gauge theories, has been initiated by a remarkable conjecture put forward by Maldacena almost one year ago. Among the possible curved vacua of superstring or M-theory, spaces having the form of an anti-de Sitter space-time times a compact Einstein manifold, have been playing a special role in this correspondence, since the quantum theory realized on them, in the original formulation of the conjecture, was identified with the effective superconformal theory on the world volume of parallel p-branes set on the boundary of such a space (holography). An important step in verifying such a conjecture and eventually generalizing it, consists in a precise definition of the objects entering both sides of the holographic correspondence. Indeed in the most general case it turns out that important features of the field theory on the boundary of the curved background, identified with the quantum theory of gravity in the bulk, are encoded in the dynamics of the coinciding parallel p-branes set on the boundary of the same space. The study of p-brane dynamics in curved space-times which are vacua of superstring or M-theory, turns out therefore to be a relevant issue in verifying the existence of the holographic correspondence. In the present paper, besides providing a hopefully elementary introduction to Maldacena's duality, I shall deal in a tentatively self-contained way with a particular aspect of the problem of p-brane dynamics in anti-de Sitter space-time, discussing some recent results towards the definition of a method for retrieving important features of the field theory on the boundary from the quantization for small oscillations of the world-volume theory of a single p-brane around the same boundary.

1 Introduction

Superstring theory has been regarded for a long time as the most promising quantum theory of gravity [1, 2]. Indeed it naturally includes the graviton among its massless states and it is finite. Nevertheless it has intrinsic limits deriving from its own definition. As is well-known, this theory describes one-dimensional extended objects (closed or open strings) vibrating in a ten-dimensional Minkowski space-time, whose spectrum of vibrational modes are expected to correspond to the observed fields in nature. From a mathematical point of view the theory is a σ-model defined on the $1+1$ world-sheet of a string propagating in a 10-dimensional target space. A dimensionful parameter characterizing the string and defining the scale of its vibrational levels is the string length ℓ_s, whose square will be denoted by α'.

Consistency of the theory requires its local invariance with respect to diffeomorphism and conformal transformations on the world-sheet coordinates and metric. Among the massless fields in the spectrum of the theory there is a scalar ϕ (the dilaton) and a spin–2 field $G_{\mu\nu}$ which is identified with the graviton field. One of the vacuum solutions of the theory is the one on which the symmetric tensor gives the Minkowski metric on the target space-time ($\langle G_{\mu\nu}\rangle = \eta_{\mu\nu}$) and $\langle\phi\rangle = \phi_0$. Superstring theory is quantized on this vacuum, perturbatively with respect to an effective coupling constant $g = \exp(\phi_0)$. Versions of the theory on certain other vacua of the form $M_d \times K$ representing compactifications of the ten-dimensional target space to a lower dimensional Minkowski space M_d ($d \leq 10$) times a suitable compact Ricci-flat manifold K have been constructed as well. The low energy limit of these theories is described by a *d-dimensional supergravity*. On the other hand a second quantized formulation of superstring theory on a curved background characterized by a non-Ricci-flat solution of the low–energy effective supergravity theory, is not yet known (except for very special cases) and represents in general a complicated problem [3]. As we shall see, one of the potential achievements of the duality recently conjectured by Maldacena (see [4],[5] and references therein), is the shedding of light on the quantum spectrum of superstring theory realized on a particular kind of non-Ricci-flat vacuum, namely of the form $AdS \times K$, where AdS stands for a d-dimensional anti-de Sitter space-time and K is a suitable compact $(10 - d)$-dimensional Einstein space. These vacua correspond to the geometry of some regions of space-time in the presence of extended objects which are believed to belong to the non-perturbative spectrum of superstring theories and therefore may be regarded as non-perturbative vacua of these theories. In the remaining part of the present introduction I shall try to formalize these concepts further in a pedagogical fashion, preparing a physical scenario in which to set Maldacena's conjecture.

One of the great achievements of the last ten years is the concept of *duality*: correspondences have been conjectured and, in part, verified between regimes of various superstring theories realized on different backgrounds of the form $M_d \times K$ (see [6] and references therein). The existence of such dualities would allow us to consider the known superstring theories as local effective formulations on different backgrounds of a unique, yet unknown, quantum theory living in a higher dimensional space-time. Among the candidates for this larger quantum theory there is the so-called M-theory in $(10 + 1)$-dimensions, whose low energy effective field theory is the well-known 11-dimensional supergravity, and F-theory in $(10 + 2)$-dimensions. Duality, being a mapping between the spectra of two different superstring theories, is realized at the massless level by means of suitable discrete transformations on the background fields, which form a group $G(Z)$. The largest duality which has been conjectured is called U-duality and the corresponding action on the massless fields $G(Z)$ is believed to be a suitable discrete version of the largest global symmetry group G of the field equations and the Bianchi identities of the underlying low energy supergravity [7]. The study of dualities, and in particular of the non-perturbative dualities, sheds some light on the non-perturbative side of superstring theories. For instance, it became clear that the full spectrum of these theories should contain not only 0 and 1-dimensional objects (particles and strings), but also p-extended objects (p-branes) which were not present in the perturbative spectrum, but nonetheless corresponded to solutions of the low energy supergravity. Among them, those coupled with a certain kind of background field, namely the Ramond–Ramond fields, are of particular interest. The main feature of the Ramond–Ramond (R–R) fields is that they do not couple directly to the fundamental closed string. On the other hand they are mixed with the other background fields which do couple directly to the closed string (Neveu-Schwarz–Neveu-Schwarz fields) through the U-duality group $G(Z)$. Therefore, if the U-duality is meant to be an exact symmetry of a larger quantum theory, the distinction between R–R and other fields is to be considered an artifact of perturbative superstring theory and therefore one expects to find in the non-perturbative spectrum of superstring theory objects coupled to R–R fields and thus carrying an R–R charge. Since the R–R fields are in general $(p+1)$-forms, they can minimally couple to p-extended objects. A particular kind of these extended objects (namely, those saturating the BPS bound on the mass) have been described in the context of perturbative superstring theory on flat space as Dirichlet surfaces [8], that is (roughly speaking), as hypersurfaces on which open strings start and end. The dynamics of these so-called Dp-branes is determined by the oscillations of the open strings attached to them. In ten dimensions the quantum world may therefore be represented figuratively as containing closed strings free to move in the bulk and to interact among themselves and with dynamical hypersurfaces on which open strings are attached. In eleven dimensions, on the other hand, the conjectured M-theory is expected to contain p-extended objects as well (Mp-branes), which are solutions of low energy supergravity.

Going back to the suggestive 10-dimensional picture of the quantum world previously portrayed, one may ask what the limits of such a representation are, i.e., in what physical regime does the framework of superstring theory on a Minkowski background break down? A D-brane is a massive object, which means that it interacts with the gravitational field and thus deforms the surrounding space-time. From the superstring point of view, being the quantum fluctuation of the background metric associated with a vibrational 0-mode of the closed string, this interaction is described through the emission of a closed string from the brane in the surrounding flat space-time and the amplitude of such a process, computed in the framework of perturbative string theory by means, for instance, of boundary state methods, yields the deformation of the Minkowski metric due to the presence of the extended object. For a single D-brane this deformation is localized around the source in a region R_1 whose thickness is of the order of the string length ℓ_s. This is an *a posteriori* consistency check that the interaction between the D-brane and the closed strings can be correctly computed in the framework of string theory on Minkowski space, since the closed strings taking part in this interaction "live" much longer in the flat region R_2, a complement of R_1 in the whole space, than in R_1. The situation changes when we consider N coinciding Dp-branes. An important feature of these objects is that parallel D-branes do not exert any force on each other and therefore may be set to coincide without expense of energy. For this system, the thickness of the curved region R_1 around the brane would be proportional to a certain (positive) power of N, and for N large enough, it would contain all the physical information on the interaction between the brane and the closed strings. In the limit $N \to \infty$, R_1 would fill the whole space and the interaction of the strings with the branes could be described only in the framework of a superstring theory realized on a new curved vacuum defined by the near horizon geometry of the branes. We will look at this scenario from the low energy supergravity point of view.

The system consisting of a large number N of coinciding D branes, is well described by a solution of the low energy supergravity theory since the curvature in the region R_1 is small with respect to $1/\ell_s^2$. In general, p-brane solutions of supergravity are associated with a $(p+1)$-form potential $A^{(p+1)}$ to which they couple and whose integral on their world-volume gives their *electric* charge. (A simple example is that of an elementary particle (0-brane) minimally coupled to a vector potential. Its *electric* charge is given by the integral of the vector potential along the particle world-line.) If $A^{(p+1)}$ is an elementary field of the theory, then the solution is an *elementary* p-brane, characterized by a singular p-hypersurface hidden by an event horizon. To an elementary p-brane there corresponds a *solitonic* $(D-p-4)$-brane coupled to the $(D-p-3)$-form potential *dual* to $A^{(p+1)}$. This *bona fide* solution of the theory is non-singular. Elementary and solitonic p-branes in D-dimensions can be found as solutions of a truncated supergravity model consisting only of the metric, the dilaton and the $(p+1)$-form potential [9, 5, 10] which are described by the following metrics:

Elementary solution:

$$ds^2 \left(1+\frac{k}{r^{\tilde{d}}}\right)^{-\frac{4\tilde{d}}{\Delta(D-2)}} dx^\mu \otimes dx^\nu \eta_{\mu\nu} - \left(1+\frac{k}{r^{\tilde{d}}}\right)^{\frac{4d}{\Delta(D-2)}} dz^p \otimes dz^q \delta_{pq}.$$

Solitonic solution:

$$ds^2 =, \left(1+\frac{k}{r^{d}}\right)^{-\frac{4d}{\Delta(D-2)}} dx^\mu \otimes dx^\nu \eta_{\mu\nu} - \left(1+\frac{k}{r^{d}}\right)^{\frac{4\tilde{d}}{\Delta(D-2)}} dz^p \otimes dz^q \delta_{pq}. \tag{1.1}$$

The matrix $\eta_{\mu\nu}$ is the metric on a $(p+1)$-dimensional Minkowski space-time and has the following signature: $\eta_{\mu\nu} = \mathrm{diag}(+, -, \dots, -)$. These elementary (solitonic) solutions have a flat world-volume parametrized by the $d = p+1$ ($\tilde{d} = D-p-3$) coordinates x^μ, while z^q denote the coordinates along the directions orthogonal to the world-volume ($q = 1, \dots, d-p-1$). The parameter Δ is a characteristic quantity which will have the value 4 in all the cases we shall deal with. Finally k represents the charge of the solution with respect to the corresponding potential. The $r=0$ hypersurface represents the *event horizon* of the solution and is a coordinate singularity.

In the 11-dimensional supergravity (low energy limit of the conjectured M-theory) there is an $M2$ (elementary) and an $M5$ (solitonic) brane solution. In type IIB supergravity ($D = 10$) of particular interest, is the self–dual (since it is coupled to the R–R self-dual 4-form $A^{(4)}$) 3-brane solution. Their metrics are immediately computed from Eqs. (1.1):

$D = 11$ $M2$-brane:

$$ds^2 = \left(1+\frac{k_2}{r^6}\right)^{-\frac{2}{3}} dx^\mu \otimes dx^\nu \eta_{\mu\nu} - \left(1+\frac{k_2}{r^6}\right)^{\frac{1}{3}} \left(d^2r + r^2 d^2\Omega_7\right).$$

$D = 11$ $M5$-brane:

$$ds^2 = \left(1.+\frac{k_5}{r^3}\right)^{-\frac{1}{3}} dx^\mu \otimes dx^\nu \eta_{\mu\nu} - \left(1+\frac{k_5}{r^3}\right)^{\frac{2}{3}} \left(d^2 r + r^2 d^2 \Omega_4\right).$$

$D = 10$ 3-brane:

$$ds^2 = \left(1+\frac{k_3}{r^4}\right)^{-\frac{1}{2}} dx^\mu \otimes dx^\nu \eta_{\mu\nu} - \left(1+\frac{k_3}{r^4}\right)^{\frac{1}{2}} \left(d^2 r + r^2 d^2 \Omega_5\right). \quad (1.2)$$

The linear extension of the near horizon region R_1 may be characterized by the condition $r \ll R$, R being $k_2^{1/6}$ for the $M2$-brane, $k_5^{1/3}$ for the $M5$-brane and $k_3^{1/4}$ for the 3-brane. In this region the metrics in Eqs. (1.2) may be rewritten in the following form:

$$\begin{aligned} ds^2 &= \rho^2 dx^\mu \otimes dx^\nu \eta_{\mu\nu} - (Rw)^2 \frac{d\rho^2}{\rho^2} - R^2 d\Omega_{(D-p-2)} \\ w &= \frac{2(D-2)}{\tilde{d}}\,;\; \rho = \left(\frac{r}{R}\right)^{\frac{1}{w}}. \end{aligned} \quad (1.3)$$

The above metric describes a space-time of the form $AdS_{p+2} \times S^{D-p-2}$ (with abuse of language we shall denote by anti-de Sitter space a particular compactification of anti-de Sitter space characterized by the condition $\rho > 0$). The parameters k_2, k_5, k_3 are related to the charges these solutions have with respect to the potentials they couple to. Indeed, in suitable units, they represent the flux of the corresponding field strengths $F^{(4)}$, $F^{(7)}$, $F^{(5)}$ through the spheres S^7, S^4, S^5 respectively. One may interpret the solutions in Eqs. (1.2) from a microscopic point of view as bound states describing N coinciding $M2$-, $M5$-, $D3$-branes, respectively, each of them carrying a unit of the charge associated with the potential they couple to. In this picture the parameters k_i ($i = 2, 5, 3$) are proportional to N.

The solutions (1.2) describe a space-time continuum which starts from a flat Minkowski geometry at infinite distance from the extended object, then acquires a non-vanishing curvature at finite r until it forms a "throat" of width R near the horizon at $r = 0$, with an anti-de Sitter geometry.

We should say a bit more about the self–dual 3-brane solution in type IIB supergravity. If interpreted from the string theory point of view as a bound state of N coinciding $D3$-branes, each carrying, unit of R–R charge, its charge with respect to $A^{(4)}$ will be proportional to $k_3 = gN\ell_s^4$ (g being the string coupling constant). According to the $D3$-brane interpretation of this solution, $R \approx \ell_s (gN)^{1/4}$ and therefore in the limit $N \to \infty$ the whole space-time has the form $AdS_5 \times S^5$. It follows that, in this regime, a microscopic description of the $D3$-branes would be related to superstring theory realized on this new curved background. Indeed spaces of the form $AdS_{p+2} \times M^{D-p-2}$, where M^{D-p-2} is a suitable Einstein manifold, have been shown to be exact solutions of superstring or M-theory ($D = 10, 11$ respectively) [11] and thus possible vacua for these theories (the space $AdS_5 \times S^5$ could be interpreted as a non-perturbative vacuum of superstring theory exhibiting a condensation of $D3$-branes). Nevertheless, as previously emphasized, finding the spectrum of physical excitations of a string on a curved background (such as the non– perturbative vacua described above) is a very complicated issue and in general represents an unsolved problem (different from the case of a conformal σ-model on a Minkowski background, in a general curved target space; indeed, it is no longer possible to construct the Hilbert space of quantum states as a Fock space built through the action on a vacuum state of creation operators corresponding to free oscillators). A hint as to the physical content of some of these theories is provided by a powerful duality conjectured by Maldacena almost one year ago.

In Sect. 2 I shall give a pedagogical and tentatively self-consistent review of this conjecture and its formulation as a duality between a *singleton* super-conformal field theory on the boundary of anti-de Sitter space and superstring or M-theory in the bulk. In Sect. 3 I shall focus on the particular case of the $M2$-brane and discuss some recent results in the analysis of the dynamics of branes in anti-de Sitter space.

2 AdS/CFT duality

We focus for the moment on a 10-dimensional space-time in which the physics is described by type IIB superstring theory in the presence of $D3$- branes. The main idea is that, as the number N of coinciding $D3$-branes increases ($N \gg 1$), the physics is described by a new non-perturbative vacuum which is no longer the flat Minkowski vacuum on which the superstring theory is perturbatively defined but has the geometry of $(AdS_5 \times K^5)_N$ in which the N $D3$-branes are suitably embedded (the subscript "N" reminds us that the background space-time is the near horizon geometry of the N branes and therefore its "radius" is proportional to $N^{1/4}$) . It is reasonable to think that the effective low energy theory around this vacuum has $D3$-branes as fundamental objects instead of strings and describes fluctuations of these extended objects around their static configuration described by Eq. (1.2), which have energy much low er than the string scale $(1/\alpha')$. This theory is the effective field theory on the world-volume on the N coinciding branes in the limit $\alpha' \to 0$ (in this limit the interaction of the branes with the bulk through emission of closed strings is suppressed). Intuitively, in much the same way as a field theory of elementary particles and of solitons can be viewed as effective descriptions of the same quantum theory on two different vacua, the theory of superstrings and that of branes may be considered as "dual" to each other, i.e., the theory defined on the world-volume of the N coinciding branes (in which the branes are the fundamental objects) embedded in their near horizon geometry, for small fluctuations, can be viewed as the effective low energy realization of the quantum theory of strings on an $(AdS_5 \times K^5)_N$ vacuum. This is the basic idea which has been formalized by Maldacena in his powerful duality conjecture:

Type IIB superstring theory realized on $(AdS_5 \times S^5)_N$ and the low energy effective field theory on the world-volume of the N coinciding D3-branes are the same quantum theory.

The same kind of duality was originally conjectured also between M-theory (whose physical content is not as yet known) on $(AdS_{p+2} \times S^{9-p})_N$, describing the near horizon geometry of N Mp-branes for $p = 2, 5$, and the low energy effective field theory on the world-volume of these branes. In this case the low energy condition on the world-volume theory is expressed in terms of the only length scale of 11-dimensional supergravity theory, namely, the Plank scale ℓ_p: $\ell_p \to 0$. This duality is also referred to as *holographic* correspondence since it states that the quantum dynamics of fields in a $(p+2)$-dimensional space-time ($AdS_{(p+2)}$) is encoded in a theory defined on a $(p+1)$-dimensional subspace.

In the original formulation of the conjecture the inner compact space was a $(D-p-2)$-dimensional sphere $S^{(D-p-2)}$. Eventually it was suggested that the duality could be formulated on more general spaces in which the compact Einstein space $K^{(D-p-2)}$ is a homogeneous space of the form G/H [12]. These spaces are maximally symmetric solutions of supergravity theory and near horizon geometry of certain p-brane solutions of the same theory. In contrast to the case in which K = Sphere, solutions of the form $AdS_{(p+2)} \times (G/H)$ are in general not maximally supersymmetric, i.e., they preserve less supersymmetry than the original theory. Maldacena's conjecture has been further extended recently to space-time geometries of the form $AdS \times K$ in which there is an even more general Einstein manifold related to the so-called Sasaki manifolds [13].

A special role in this duality is played by the anti-de Sitter space-time representing the non–compact factor of this vacuum. Superstring theory (supergravity) realized on $AdS_{(p+2)} \times K^{(D-p-2)}$ is *locally* invariant under the general coordinate transformations generated by the space-time isometry group:

$$\begin{aligned} \mathcal{G} &= \mathcal{I}\text{som}\left(AdS_{(p+2)} \times K^{(D-p-2)}\right) \\ &= \mathrm{SO}(2, p+1) \otimes \mathcal{I}\text{som}\left(K^{(D-p-2)}\right). \end{aligned} \tag{2.1}$$

Since we are dealing with a locally supersymmetric theory, we may include supersymmetry transformations in the previous statement and say that the quantum theory in the bulk is locally invariant with respect to the superextension $\mathcal{SG}$ of the isometry group in (2.1), whose structure will be described in more detail in the next section. However it is useful to write the explicit form of $\mathcal{SG}$ in the following way:

$$\begin{aligned} \mathcal{I}\text{som}\left(K^{(D-p-2)}\right) &\to \mathrm{SO}\,(N) \oplus \mathcal{K}' \\ \mathcal{SG} &= \mathcal{SC} \oplus \mathcal{K}' \\ \mathcal{SC} &= \begin{cases} OSp(4/\mathcal{N}) & M2\text{-brane} \\ OSp(6,2/\mathcal{N}) & M5\text{-brane} \\ \mathrm{SU}\,(2,2/\mathcal{N}) & D3\text{- brane} \end{cases} \end{aligned} \tag{2.2}$$

where $\mathcal{N}\times$(dimension of the spinors in $(p+2)$ dimensions) gives the number of supercharges preserved by the background solution ($\mathcal{N}$ is the number of *Killing spinors* of the supergravity solution and depends on the compact manifold K). The group $\mathcal{SC}$ is the supersymmetric extension of the group SO $(2, p+1)$, that is, it consists of the SO $(2, p+1)$ bosonic generators plus a number $\mathcal{N}$ of fermionic generators. An important property of the groups SO $(2, p+1)$ and $\mathcal{SC}$ is that they act respectively as the *conformal* and *superconformal* groups in $p+1$ dimensions.

One of the features of an anti-de Sitter space is having a boundary. This boundary $\partial AdS_{(p+2)}$ is a $(p+1)$-dimensional locus of points having the property of being stable with respect to the action of the $AdS_{(p+2)}$ isometry group SO $(2, p+1)$ (or the superisometry group $\mathcal{SC}$ of the anti-de Sitter superspace, in the framework of a supersymmetric theory).

Parametrizing $AdS_{(p+2)}$ by means of the coordinates (ρ, x^{μ}) in terms of which the metric is written in the form (1.3), the boundary may be characterized as the $(p+1)$-dimensional Minkowski space $M_{(p+1)}$ spanned by the coordinates x^{μ} in the limit $\rho \to 0$ plus a point at infinity $\rho \to \infty$. Therefore the Minkowski part of $\partial AdS_{(p+2)}$ $(M_{(p+1)})$ coincides with the world-volume of the N overlapping p-branes, which is thus stable under the action of the SO $(2, p+1)$ isometry group.

The effective low energy theory on the world-volume $M_{(p+1)}$ of the N p-branes is obtained by expanding the Born–Infeld action S_{BI} on $M_{(p+1)}$ for small oscillations of the branes around the static position at the boundary of $AdS_{(p+2)}$. The action S_{BI} describes a σ-model on $M_{(p+1)}$ in which the local fields are the coordinates $X^M(\xi) = (\rho(\xi), x^{\mu}(\xi); y^m(\xi))$ of the target space-time $AdS_{(p+2)} \times K^{(D-p-2)}$, locally depending on the $p+1$ world-volume coordinates ξ^a. The main feature of S_{BI} is that it is expressed as the integral over $M_{(p+1)}$ of the invariant volume element and thus the isometries of the target space are *global* symmetries of the theory. In particular, since the anti-de Sitter isometry group SO $(2, p+1)$ acts on $M_{(p+1)}$ as a conformal group, the theory on the world-volume of the p-branes on the boundary is a *conformal field theory* (CFT). If supersymmetries are taken into account, the theory is expected to be globally invariant under the whole superisometry group $\mathcal{SG}$, which contains the superconformal group $\mathcal{SC}$, and thus to be a *superconformal* field theory (SCFT).

The conjectured duality is therefore between superstring (M) theory on $AdS_{(p+2)} \times K^{(D-p-2)}$ and an SCFT on the boundary $\partial AdS_{(p+2)}$.

This conjecture has been subsequently made more precise by Witten [14] who defined a precise correspondence between quantum states $\Phi(x)$ of the theory in the bulk and superconformal operators $\mathcal{O}(\xi)$ on the boundary. In particular, according to this picture, the operators $\mathcal{O}(\xi)$ are interpreted as vertex operators defining the emission of the corresponding state $\Phi(x)$ from the membrane in the bulk. At the heart of this correspondence is the consideration that either the states $\Phi(x)$ or the operators $\mathcal{O}(\xi)$ must transform in representations of the same group $\mathcal{G}$ defined in Eq. (2.1) or, more generally, of its superextension $\mathcal{SG}$. Witten also found a relation between the masses of the states in the bulk and the conformal dimensions of the operators on the boundary.

As I previously emphasized, the whole spectrum of the quantum theory in the bulk is not known, nevertheless the low energy part of it consists of the Kaluza–Klein states of D-dimensional supergravity compactified on $AdS_{(p+2)} \times K^{(D-p-2)}$, which were studied extensively in the eighties [15]. These states are in one-to-one correspondence with the harmonic functions on the compact space $K^{(D-p-2)}$ and their mass is of order $m_{KK} \approx 1/R$. They sit in *short* supermultiplets of the supergroup $\mathcal{SG}$ since their masses are related to quantities characteristic of the isometry algebra of $K^{(D-p-2)}$ and this property protects them, for sufficiently large supersymmetry, from quantum corrections. In the previously outlined correspondence, the Kaluza–Klein states naturally correspond to conformal operators sitting in short representations of the superconformal group $\mathcal{SC}$ acting on the boundary. They are the *chiral* operators. Besides the Kaluza–Klein states there are the string or M-theory states which sit on *long* supermultiplets of $\mathcal{SG}$. Their mass is of order $m \approx 1/\ell$ (ℓ being ℓ_s or ℓ_p in the superstring or M-theory case respectively) and are not protected from quantum corrections. These states would naturally correspond to *non-chiral* operators in the SCFT on the boundary.

Recall that the "radius" of the anti-de Sitter space R depends on both ℓ and N in the following way:

$$R \approx \ell N^{\beta} \tag{2.3}$$

where β is 1/4, 1/6, 1/3 for the $D3$, $M2$, $M6$-brane respectively. If, together with $\ell \to 0$ (low energy limit on world volume theory), we send N to infinity in such a way that R is kept fixed, the string or M-theory modes will decouple since their mass will grow as N^{β}. In this limit the duality can be restated as a mapping between an SCFT on the boundary and tree–level Kaluza–Klein supergravity in the bulk. Since a good deal is known about the latter, Maldacena's duality has so far received several checks from the study of the large N limit of certain SCFTs (especially in the case of the low energy SCFT on the world-volume of N coinciding $D3$-branes which is known to be a superconformal SU (N) Yang–Mills theory. In this framework a stringent check of the duality was carried out in [16]), for instance.

In the correspondence described above between states of the theory in the bulk and conformal operators on the boundary belonging to the same representation of the superconformal group, there is a piece missing. Indeed, among the unitary irreducible representations, besides the long and short supermultiplet, the superconformal group $\mathcal{SC}$ admits *ultrashort* representations (*super-singletons*) (in a non-supersymmetric framework, the ultrashort representations of the AdS isometry group SO $(2, p+1)$ are called *singletons* and were first found by Dirac [17]). Supersingletons may be characterized as those states which saturate the lowest bound on the energy in order for the representation to be unitary. States in these multiplets do not have a representation in terms of local fields in the bulk since they can be gauged away everywhere, except on the boundary of the anti-de Sitter space, where they admit a field representation. Super-singletons are therefore the natural candidates for describing the elementary fields of the SCFT on the boundary. They do not have a counterpart in the theory defined on the bulk. Superconformal field theories of singletons on ∂AdS were studied extensively in the eighties [18]. An important property of these ultrashort representations is that all the other unitary irreducible representations (UIR) of $\mathcal{SC}$ (in which, for instance, all the fields in the bulk transform) may be built from tensor products of two or more singleton states (composite states). This led, during the early stages of research on singleton field theories, to the hypothesis that all the excitations of the quantum theory in the bulk could be expressed as composite states of elementary super-singletons living in the boundary (see [5] and references therein) which was a first kind of *holographic* correspondence.

Already in these early years, the super–singleton SCFT was suggested (Nicolai et al.) to describe the dynamics of physical objects embedded on the boundary of $AdS_{(p+2)}$, namely, p-branes. However this correspondence was never proven rigorously and there are two main differences between the super-singleton SCFT studied in the eighties and those which are expected to describe one side of Maldacena duality. The former were defined on a boundary of $AdS_{(p+2)}$ with topology $S^p \times S^1$, they were (for sufficiently enough supersymmetry) free field theories with a mass term and were related to the dynamics of a single *spherical* p-brane embedded in $\partial AdS_{(p+2)}$. On the other hand SCFTs relevant to the Maldacena conjecture (as previously pointed out) are defined on a boundary $\partial AdS_{(p+2)}$ which is given the topology of a $(p+1)$-dimensional Minkowski space (plus a point at infinity), are interacting (since they have a non-abelian $\mathrm{SU}(N)$ gauge invariance) massless field theories and are expected to describe the dynamics of N coinciding p-branes set on $\partial AdS_{(p+2)}$.

Nevertheless, the characterization of the SCFT entering Maldacena duality as a super-singleton theory allowed for a recent group theoretical analysis of the conjecture [19].

There is, however, an important feature of the super-singleton theory which is, in general, not fixed by its invariance with respect to $\mathcal{SC}$. Indeed we recall that this theory is globally invariant with respect to the whole superisometry group $\mathcal{SG}$ in Eq. (2.2) which contains, besides $\mathcal{SC}$, a bosonic subgroup $\mathcal{K}'$. This group acts as a *flavour* group on the singleton theory and the flavour representation of the singletons with respect to $\mathcal{K}'$ is independent of their superconformal transformation properties and can in general be retrieved only from the p-brane dynamics on the boundary. In the maximally supersymmetric case $K^{(D-p-2)} = S^{(D-p-2)}$, the whole group $\mathcal{I}som\left(S^{(D-p-2)}\right) = \mathrm{SO}(D-p-1)$ enters the superconformal group $\mathcal{SC}$ in Eqs. (2.2) and the group $\mathcal{K}'$ is trivial; therefore all the transformation properties of the singletons with respect to $\mathcal{SG}$ are characterized by the $\mathcal{SC}$ quantum numbers. Verifying the Maldacena conjecture in the most general case requires the knowledge of the flavour representation of the singletons with respect to $\mathcal{K}'$. Indeed since all the conformal operators of the theory are constructed as composite structures of the singleton fields, their $\mathcal{K}'$ quantum numbers will depend on the singleton flavour representation. Verifying the matching between the $\mathcal{K}'$ quantum numbers of the boundary operators with those of the bulk states would provide a new check of the Maldacena conjecture. To this end it is crucial to develop a procedure for retrieving the supersingleton action from the quantization of N coinciding p-branes set on the boundary of $AdS_{(p+2)}$. For $N > 1$ a classical action on the world-volume of the overlapping branes would be a non-abelian Born–Infeld action, which is not known at present. However, if our aim is just to determine the $\mathcal{K}'$ flavour representation of the singletons, this information could be inferred from the study of a single p-brane on the boundary. Indeed one may consider a system of $N+1$ parallel p-branes, of which the first $N \gg 1$ are coinciding and set on the boundary of $AdS_{(p+2)} \times K^{(D-p-2)}$ while the remaining one (*probe* brane) is set at a distance r from them. In principle, one could then quantize (for small fluctuations) the world-volume Born–Infeld action on the probe brane in the limit in which the latter is set to coincide with the other N ($r \to 0$) (superconformal limit) and retrieve the superconformal action of free (abelian) massless singletons, together with their $\mathcal{K}'$ representation.

Quantizing for small oscillations the B–I action on the world-volume of a brane embedded in a curved space is a difficult problem because of technical difficulties related to the gauge fixing of the local invariances of the theory (in particular the fermionic κ symmetry needed in order for the world-volume theory to be supersymmetric). A general method for retrieving the supersingleton action from the world-volume theory of a probe brane on the boundary of an anti-de Sitter space was defined in [20], and applied to the case of an $M2$- brane on the boundary of $AdS_4 \times S^7$. This method relied on a particular parametrization of the anti-de Sitter superspace defined by using a *supersolvable* subalgebra of the superisometry group. The supersolvable parametrization is the superextension of the *solvable* parametrization of the anti-de Sitter space defined in [12], which is related to the coordinates (ρ, x^{μ}) in terms of which the metric has the expression (1.3). The main goal of the supersolvable parametrization is to allow us to write the B–I action which is κ-symmetry fixed from the very beginning, that is, which depends only on the physical fermionic fields. This κ-gauge fixing turns out to be equivalent to fixing the *Killing spinor* gauge defined in [21].

Solvable and supersolvable parametrizations, defined in [12] and [20], proved to be powerful algebraic tools for describing brane dynamics in anti-de Sitter spaces. They have been applied as well to the $D3$-brane in $AdS_5 \times S^5$ [22].

The final goal, however, is to apply this method to a background of the form $AdS_{(p+2)} \times K^{(D-p-2)}$ where K is a homogeneous manifold of the form G/H or related to a more general Sasaki manifold. In these cases indeed, as previously emphasized, the AdS/CFT correspondence from a group-theoretical point of view becomes non-trivial because the $\mathcal{K}'$ flavour representation of the singletons enters the game on the SCFT side, as an input independent of the superconformal invariance of the theory. This analysis is still work in progress.

3 Supermembrane in anti-de Sitter space

In the present section I shall recall the main facts about anti-de Sitter spaces and review some recent results in the study of the $M2$-brane dynamics in $AdS_4 \times S^7$.

An n-dimensional anti-de Sitter space-time may be described as a hypersurface embedded in $\mathbb{R}^{2,n-1}$. Let $X^{\Lambda} = (X^0, X^{\bar{0}}, X^1, \ldots, X^{n-1})$ denote a coordinate basis for $\mathbb{R}^{2,n-1}$ with respect to which the flat metric has the form:

$$\eta_{\Lambda\Sigma} = \mathrm{diag}(+, +, -, \ldots, -),$$

AdS_n is defined as the following locus of points:

$$\eta_{\Lambda\Sigma} X^{\Lambda} X^{\Sigma} = 1 \tag{3.1}$$

(for simplicity we have set the "radius" of AdS_n equal to 1). This condition is invariant with respect to the group SO $(2, n-1)$ acting on X^Λ, which is the isometry group of AdS_n. By defining the light cone coordinates $X^\pm = X^0 \pm X^{n-1}$ and the coordinates x^μ by setting $X^\mu = X^+ x^\mu$ for $\mu = 0, 1, \ldots, n-2$, condition (3.1) may be rewritten in the following way:

$$X^+ X^- + (X^+)^2 \eta_{\mu\nu} x^\mu x^\nu = 1. \tag{3.2}$$

We restrict ourselves to the branch $X^+ \geq 0$. In the limit $X^+ \gg 1$ the above condition reduces to:

$$X^+ \eta_{\mu\nu} x^\mu x^\nu = -X^- \tag{3.3}$$

which defines an $(n-1)$-dimensional Minkowski space $M_{(n-1)}$ spanned by the coordinates x^μ, which is stable with respect to a scaling $X^\Lambda \to \lambda X^\Lambda$. Its isometry group is ISO $(1, n-2) \subset$ SO $(2, n-1)$. Moreover if we add to it the point P at $X^+ = 0$ defining $\widetilde{M}_{(n-1)} = M_{(n-1)} \cup P(X^+ = 0)$, this compact locus of points defines the boundary of AdS_n which is stable with respect to the whole group SO $(2, n-1)$. Setting $\rho = X^+$ and computing the induced metric on the hypersurface (3.2) in terms of the coordinates (ρ, x^μ), one obtains a generalization of the Bertotti–Robinson metric:

$$ds^2 = \rho^2 dx^\mu \otimes dx^\nu \eta_{\mu\nu} - \frac{d\rho^2}{\rho^2}. \tag{3.4}$$

The anti-de Sitter space may be written alternatively as a homogeneous manifold in the following way:

$$AdS_n = \frac{\mathrm{SO}\,(2, n-1)}{\mathrm{SO}\,(1, n-1)}. \tag{3.5}$$

In order to compute the non-linear action of the isometry group SO $(2, n-1)$ on the coordinates (ρ, x^μ), we shall characterize them as coordinates in the *solvable* representation of AdS_n [12]. Using the formalism of coset manifolds, one can associate with each point Q in AdS_n a *coset representative* $L(Q)$ so that the non-linear action of the isometry group SO $(2, n-1)$ on the point Q may be represented in the following way:

$$\begin{aligned} &\forall g \in \mathrm{SO}\,(2, n-1) \quad g \cdot L(Q) = L(Q'(g, Q)) \cdot h(g, Q) \\ &h(g, Q) \in \mathrm{SO}\,(1, n-1) \end{aligned} \tag{3.6}$$

In order to define a parametrization of the manifold, or equivalently $L(Q)$, one may use the following algebra decomposition:

$$\mathrm{SO}\,(2, n-1) = \mathrm{SO}\,(1, n-1) \oplus \mathrm{Solv} \tag{3.7}$$

where Solv is an n-dimensional *solvable*[1] subalgebra of SO $(2, n-1)$. As a consequence of Eq. (3.7) the anti-de Sitter space is isomorphic to a group manifold

[1] A solvable Lie algebra Solv is a Lie algebra whose k^{th} Lie derivative $\mathcal{D}^k(\mathrm{Solv})$ vanishes for some finite k. The kth Lie derivative may be defined by induction in k: $\mathcal{D}^{k+1}(\mathrm{Solv}) = \left[\mathcal{D}^k(\mathrm{Solv}), \mathcal{D}^k(\mathrm{Solv})\right]$; $\mathcal{D}^1(\mathrm{Solv}) = [\mathrm{Solv}, \mathrm{Solv}]$

generated by Solv and may be globally identified with it. In this description the local coordinates on AdS_n are therefore parameters of Solv. The Solv algebra is constructed by considering the only non-compact Cartan generator D in the coset (3.5) and diagonalizing its adjoint action on the algebra SO $(2, n-1)$. This defines a *grading* on SO $(2, n-1)$ which is decomposed as the direct sum of eigenspaces of $Ad\, j_D$, namely $g_{(0)}$, $g_{(\pm 1)}$ corresponding to the eigenvalues 0, ± 1 respectively. Solv can be constructed as $\{D\} \oplus g_{(-1)}$, where $g_{(0)} = \{D\} \oplus \mathrm{SO}(1, n-2)$ while $g_{(-1)} = \{T_\mu\}$ consists of the $n-1$ *shift operators* corresponding to all the roots having negative values on D. The subalgebra $g_{(-1)}$ is an abelian subalgebra of SO $(2, n-1)$ since there is no generator with grading -2 with respect to $Ad\ j_D$. By associating with the generator D the parameter $\log(\rho)$ and with the generators T_μ the parameters x^μ, a coset representative can be defined as an element of the solvable group Exp(Solv) as follows:

$$L(\rho, x^\mu) = \mathrm{Exp}\left(\log(\rho)D + \sum_{\mu=0}^{n-1} x^\mu T_\mu\right). \tag{3.8}$$

Constructing the Vielbeins in this parametrization and then computing the metric on the manifold, one obtains the expression in Eq. (3.4). From Eqs. (3.8) and (2.3) one may therefore deduce the action of SO $(2, n-1)$ on the coordinates (ρ, x^μ).

In this parametrization anti-de Sitter space is represented, roughly speaking, by means of Minkowski "slices" $M_{(n-1)}$ spanned by the coordinate x^μ and fibrated along the transverse coordinate $\rho \in \mathbb{R}^+$. One may define an alternative topology on ∂AdS_n, by constructing it as the limit of $M_{(n-1)}$ as $\rho \to 0$ plus a point at $\rho \to \infty$ [20]. Thinking of AdS_n as the near horizon geometry on N coinciding $(n-2)$-branes set on the $\rho = 0$ chart of the boundary, we may define the *probe* brane, introduced in the last section, as a brane whose world-volume coincides with the $M_{(n-1)}$ Minkowski "slice" at a distance ρ from the remaining N branes.

It is possible to check that the action of the group SO $(2, n-1)$ on a generic hypersurface $M_{(n-1)}$ at $\rho \neq 0$ forms a "soft" conformal algebra, that is, a conformal algebra whose structure constants depend on the coordinate ρ (the *broken conformal transformations* of [1]). In the limit $\rho \to 0$ (on the boundary) the dependence on ρ of the conformal algebra drops out and the group SO $(2, n-1)$ acts on $M_{(n-1)}$ as the usual conformal group. Now we can interpret the generators of SO $(2, n-1)$ described above, from the world-volume theory point of view, in the following way:

- D as the generator of the *dilations*;
- T_μ as the $n-1$ *translations* on $M_{(n-1)}$;
- K_μ in $g_{(+1)}$ as the $n-1$ *special conformal; transformations*;
- $\mathrm{SO}(1, n-2) = g_{(0)}/\{D\}$ as the *Lorentz transformations* on $M_{(n-1)}$.

In what follows I shall outline, without entering into the details of the calculations, the main conceptual steps in the procedure introduced in [20] for computing the supersingleton action from the quantum fluctuations of a probe p-brane around the boundary of an anti-de Sitter space. Let us consider for simplicity the special case of an $M2$ probe brane in $AdS_4 \times S^7$. As previously pointed out this space-time is a maximally supersymmetric solution of 11-dimensional supergravity. With it we may associate a superspace $AdS^{(4/8)} \otimes S^7$ spanned by 10 bosonic coordinates X^M ($M = 0, \ldots, 9$) and 32 fermionic coordinates Θ^A_α where $A = 1, \ldots, \mathcal{N} = 8$ and $\alpha = 1, \ldots, 4$ are the indices of a 4-dimensional spinor. X^M lies in the four coordinates (ρ, x^μ) of AdS_4 and the seven coordinates y^m of S^7. The superspace has the form:

$$AdS^{(4/8)} \otimes S^7 = \frac{OSp(4/8)}{\mathrm{SO}(1,3) \times \mathrm{SO}(8)} \otimes S^7. \tag{3.9}$$

The theory on the world volume $M_{(3)}$ of the $M2$ probe brane is described by a Born–Infeld action S_{BI} which, as explained in the previous section, is a σ-model defined on $M_{(3)}$ having the background superspace $AdS^{(4/8)} \otimes S^7$ as target space. The local fields in this action are therefore $X^M(\xi)$ (bosons) and $\Theta^A_\alpha(\xi)$ (fermions), ξ^a being the 3 world-volume coordinates. By using the conventions of [20] S_{BI} may be written (in suitable units) in the following form:

$$S_{BI} = 2\int_{M_{(3)}} \sqrt{-\det(h_{ab})} + (4!)^2 \int_{M_{(3)}} A^{(3)} \tag{3.10}$$

where $h_{ab}(X^M(\xi), \Theta^A_\alpha(\xi))$ is the induced metric on the world-volume and the second term is the Wess–Zumino term describing the minimal coupling of the brane with the 3-form of 11-dimensional supergravity.

We next analyze the symmetries of this action. As previously stated, its *global* symmetries are the target space superisometries which form the group $\mathcal{SG} = OSp(4/8)$. Its *local* symmetries are, on the other hand, the *diffeomorphisms* on the world-volume ($\xi \to \xi'(\xi)$) and the fermionic κ-symmetry. The former implies that the only physical degrees of freedom are the oscillations of the membrane in the directions perpendicular to its world-volume (8 bosonic d.o.f.). The latter is a symmetry of the action with respect to local transformations acting only on half of the components of the fermion fields. These components are defined by a projector of the form $\mathcal{P}^+ = (1+\Gamma)/2$, where Γ is a 4×4 matrix depending on the background fields. The invariance of S_{BI} with respect to this local transformation requires the background fields to be a solution of the 11 dimensional supergravity, which is indeed the case. The consequence of κ symmetry is that the $\Theta^+ = \mathcal{P}^+\Theta$ components of the fermionic fields can be gauged away, leaving 16 physical fermionic degrees of freedom Θ^- which match *on-shell* the 8 bosonic ones. The super singletons indeed belong to 8 supermultiplets consisting of a boson and a Majorana fermion each. These fields will be retrieved as quantum fluctuations around a suitable solution of the 8 fermionic and 8 bosonic physical degrees of freedom of the brane. The existence of κ symmetry is therefore required in order for the theory on the world-volume to be on-shell supersymmetric.

A particular solution of the theory is the static one:

$$\begin{aligned} x^\mu(\xi) &= \xi^a\ \mu,\ a = 0, 1, 2 \\ \rho(\xi) &\equiv \overline{\rho} = \text{const} \\ \partial_a y^m &= 0 \\ \Theta^A_\alpha &= 0. \end{aligned} \tag{3.11}$$

In order to retrieve the supersingleton action the strategy is to fix the local symmetries of the theory and expand the action (3.10) for small fluctuations of the physical d.o.f. around the solution (3.11) and then take the boundary limit $\overline{\rho} \to 0$ where the full superconformal symmetry is restored. The local diffeomorphisms are fixed by fixing the coordinates x^{μ} parallel to the brane to the solution (3.11). Fixing the κ symmetry is a more complicated issue. In principle one should compute the components $\Theta^{\pm}$ on the static solution, set the unphysical components Θ^{+} to zero in the action and allow the physical components to fluctuate. This is difficult to implement in the action (3.10) initially expressed in terms of the whole spinors Θ^{A}_{α} since, on the anti-de Sitter background, this dependence is rather complicated and involves higher powers of the Θ fields.

This problem has been circumvented in [20] by using the supersolvable parametrization of the superspace (3.9) which consists in redefining the target space of our σ-model to be a subspace of the whole superspace which is a supermanifold generated by a *supersolvable* algebra (SSolv) (times the sphere S^7 spanned by y^m, of course).[2] We start with a decomposition analogous to that in (3.7):

$$Osp(4/8) = [\mathrm{SO}\,(1,3) \oplus \mathrm{SO}\,(8) \oplus \mathcal{Q}] \oplus \mathrm{Ssolv}. \tag{3.12}$$

The above decomposition is obtained by performing a grading of the superalgebra with respect to $Ad\ j_D$:

$$\begin{aligned} \mathrm{OSp}(4/8) &\to g_{(-1)} \oplus sg_{(-1/2)} \oplus g_{(0)} \oplus sg_{(1/2)} \oplus g_{(1)} \\ \mathcal{Q} &= sg_{(1/2)} \\ \mathcal{S} &= sg_{(-1/2)} \\ \mathrm{SO}\,(1,3) \oplus \mathrm{SO}\,(8) \oplus \{D\} &= g_{(0)} \\ \mathrm{SSolv} &= \{D\} \oplus g_{(-1)} \oplus sg_{(-1/2)} = \mathrm{Solv} \oplus sg_{(-1/2)}, \end{aligned} \tag{3.13}$$

the spaces $g_{(\pm 1)}$ being the same as those defined previously for the anti-de Sitter space. The subspaces $sg_{(\pm 1/2)}$ are the eigenspaces of the fermionic generators corresponding to the eigenvalues $\pm 1/2$ of $Ad\ j_D$: on the world-volume theory the $\mathcal{S}$ operators (16 components) generate the *special superconformal transformations*, while the $\mathcal{Q}$ operators (16 components) generate the *supersymmetry transformations*. In contrast with solvable representation of anti-de Sitter space, the supergroup Exp(SSolv) does not coincide with the original superspace since the generators $\mathcal{Q}$ are modded out. Nevertheless it can be shown that $Ad\ j_D$ on the fermionic generators is represented by an operator $\widetilde{\Gamma}$ which coincides with the κ symmetry operator Γ on the static solution (3.11) (it is expressed as the product of the gamma matrices along the directions of the world-volume). In other words, the κ-symmetry projectors $\mathcal{P}^{\pm}$ computed on the solution (3.11) coincide with the projectors of the supersymmetry generators into the eigenspaces $sg_{(\pm 1/2)}$ respectively. Therefore the fermionic coordinates parametrizing SSolv are already the κ gauge fixed coordinates. Indeed in this parametrization the coordinates are $(\rho, x^{\mu}, y^{m}; \Theta^{-})$. The B–I action defined on this supersolvable

[2] The definition of a supersolvable Lie algebra is the same as that of a solvable Lie algebra but with the supercommutator substituted for the commutator.

target space therefore is κ gauge fixed from the very beginning and much simpler to compute.

By expanding this action around the static solution for small fluctuations of the physical fields, rescaling the latter by suitable powers of $\overline{\rho}$, taking the order α'^0 of the action and sending $\overline{\rho} \to 0$, it was possible to retrieve the supersingleton action on the boundary as a free superconformal field theory describing 8 massless bosons and 8 massless Majorana fermions.

4 Conclusions

The aim of the present paper has been, on the one hand, to give a tentatively self-contained and hopefully elementary introduction to Maldacena's conjecture and, on the other hand, to frame within a discussion of the intense research devoted to verifying this conjecture and possibly extending it, some recent results in the study of brane dynamics in certain anti-de Sitter spaces. I emphasized how some aspects of the super singleton CFT on one side of this duality could be inferred just from the dynamics of p-branes on the boundary of a $AdS_{(p+2)} \times K^{(D-p-2)}$ space-time for a general compact Einstein manifold K. Knowledge of these aspects would provide the basis for a new stringent check of the conjecture. Up to now a method for constructing the super singleton action on the world-volume of a probe brane at the boundary of $AdS_{(p+2)} \times S^{(D-p-2)}$ has been defined by means of a *normal coordinate expansion* (small quantum fluctuations) of the B–I action on the probe brane around the static configuration in which the latter is set to lie on the boundary of the anti-de Sitter space. An extension of this method to more general internal spaces K is still work in progress.

References

1. Green M., Schwarz J., Witten E. (1987): Superstring Theory. Vols. 1, 2. Cambridge University Press, Cambridge
2. Polchinski J. (1997): String Theory. Cambridge University Press, Cambridge
3. Kallosh R., Tseytlin A. (1998): Simplifying superstring action on AdS(5) $\times$ $S(5)$. JHEP **9810**, 016;
   Metsaev R., Tseytlin A. (1998): Supersymmetric D3 brane action in AdS(5) $\times$ $S^{**}(5)$. Phys. Lett. B **436**, 281–288;
   Rayaraman A., Rozali M. (1999): On the quantization of the GS string on AdS(5) $\times$ $S(5)$. Phys. Lett. B **468**, 58;
   Pesando I. (1999): On the quantization of the GS type IIB superstring action on AdS(3) $\times$ $S(3)$ with NSNS flux. hep-th/9903086
4. Maldacena J. (1998): Adv. Theor. Math. Phys. **2**, 231–252
5. Duff M.: Anti-de Sitter space, branes, singletons, superconformal field theories and all that. hep-th/980810
6. Schwarz J. (1997): Nucl. Phys. Proc. Suppl. B **55**, 1–32;
   Sen A. (1997): Nucl. Phys. Proc. Suppl. **58**, 5–19
7. Hull C.M., Townsend P.K. (1995): Nucl. Phys. B **438**, 109

8. Polchinski J. (1996): TASI lectures on D-branes, in *Fields, Strings and Duality*, ed. by C. Efthimion, B. Greene, World Scientific, Singapore, pp. 293–356
9. Lu H., Pope C.N., Sezgin E., Stelle K.S. (1995): Stainless super p-branes. Nucl. Phys. B **456**, 669;
Lu H., Pope C.N., Stelle K.S. (1996): Vertical Versus Diagonal Dimensional Reduction for p-branes. Nucl. Phys. B **481**, 313–331;
Duff M.J., Lu H., Pope C.N. (1996): The black branes of M-theory. Phys. Lett. B **382**, 73–80;
Lu H., Pope C.N., Stelle K.S. (1996): Weyl Group Invariance and p-brane Multiplets. Nucl. Phys. B **476**, 89–117
10. Fré P. (1997): Supersymmetry and first order equations for extremal states: Monopoles, hyperinstantons, black holes and p-branes. Nucl. Phys. Proc. Suppl. **57**, 52–64
11. Kallosh R., Rajaraman A. (1998): Phys. Rev. D **58**, 125003
12. Castellani L., Ceresole A., D'Auria R., Ferrara S., Fré P., Trigiante M. (1998): Nucl. Phys. B **527**, 142–170
13. Acharya B.S., Figueroa-O'Farrill J.M., Hull C.M., Spence B. (1999): Branes at conical singularities and holography. Adv. Theor. Math. Phys. **2**, 1249–1286
14. Witten E. (1998): Adv. Theor. Math. Phys. **2**, 253
15. Freund P.G.O., Rubin M.A. (1980): Phys. Lett. B **97**, 233;
Duff M.J., Nilsson B.E.W., Pope C.N. (1986): Phys. Rep. **130**, 1;
Englert F. (1982): Phys. Lett. B **119**, 339;
Castellani L., D'Auria R., Frè P. (1984): Nucl. Phys. **239**, 610;
Castellani L., Romans L. J., Warner N.P. (1984): Nucl. Phys. B **241**, 429;
de Wit B., Nicolai H. (1984): Nucl. Phys. B **231**, 506;
de Wit B., Nicolai H. (1987): Nucl. Phys. B **281**, 211
16. Dorey N., Hollowood T. J., Khoze V.V., Mattis M.P., Vandoren S. (1999): Multi-instanton calculus and the AdS/CFT correspondence in $N = 4$ superconformal field theory. Nucl. Phys. B **552**, 88–168. hep-th/9901128;
Dorey N., Hollowood T.J., Khoze V.V., Mattis M.P., Vandoren D. (1999): Multi-instantons and Maldacena's conjecture, JHER 9906:023. hep-th/9810243
17. Dirac P.A.M. (1963): J. Math. Phys. **4**, 901
18. Fronsdal C. (1982): Phys. Rev. D **26**;
Flato M., Fronsdal C. (1981): J. Math. Phys. **22**, 1100;
Angelopoulos E., Flato M., Fronsdal C., Steinheimer D. (1981): Phys. Rev. D **23**, 1278;
Bergshoeff E., Salam A., Sezgin E., Tanii Y. (1987): Phys. Lett. B **205**, 237;
Bergshoeff E., Duff M.J., Pope C.N., Sezgin E. (1987): Phys. Lett. B **199**, 69;
Bergshoeff E., Duff M.J., Pope C.N., Sezgin E. (1989): Phys. Lett. B **224**, 71
19. Ferrara S., Fronsdal C. (1998): Phys. Lett. B **433**, 19;
Ferrara S., Fronsdal C., Zaffaroni A. (1998): Nucl. Phys. B **532**, 153;
Andrianopoli L., Ferrara S. (1999): Short and long SU(2, 2/4) multiplets in the AdS/CFT correspondence. Lett. Math. Phys. **48**, 145–161
20. Dall'Agata G., Fabbri D., Fraser C., Fre P., Termonia P., Trigiante M. (1999): Nucl. Phys. B **542**, 157–194
21. Kallosh R. (1998): Superconformal actions in killing gauge. hep-th/9807206
22. Pesando I. (1998): All roads lead to Rome: Supersolvable and supercosets. Mod. Phys. Lett. A **14**, 343–348;
Pesando I. (1998): JHEP **9811**, 002

Black Hole Entropy

L. Vanzo

Abstract. According to the generalized second law of thermodynamics, the quantity that never decreases in irreversible processes involving gravity is the sum of the entropy content of ordinary matter plus the entropy provided by the black holes. As the entropy of matter has a statistical description in terms of microscopic degrees of freedom, it seems that the entropy of black holes also should have such a description. In this contribution, several proposal for address this problem are reviewed, including quantum fields theory, strings and the very recent holographic ideas, resting on the correspondence between anti-de Sitter space and conformal field theories.

1 Introduction

Several proposals for addressing the physics of black holes are reviewed. They include descriptions in quantum field theory on a curved background, sstring theory and AdS/CFT correspondence. I will not review the canonical quantum gravity approach (cf. de Pietri's contribution to this volume), since I am not an expert in this field. The content will be informative without pretence at rigour but, hopefully, it will provide a general idea of what kind of activity is involved in this fascinating field.

2 Black holes

Classical features

A black hole can be described as a trapping region of space-time, in the sense that, within a certain distance, no particle (including photons) can escape from it. The most important features of black hole solutions in gravity theories which have general relativity in the low energy limit, (relative to a scale, e.g., the Planck or string scale) are the no-hair theorems and the formation of an event horizon. There is a future directed null hypersurface, denoted H in the figure below, bounding the region of space-time that will never be seen by observers that choose to stay outside the black hole (the outside region is the one containing spatial infinity). For instance, since the horizon is future directed and null, the particle shown in the figure can only enter the black hole. The region inside the event horizon represents the black hole interior, and generically it contains a space-time singularity. The horizon can form in gravitational collapse when the escape velocity of a particle reaches that of light. That this can happen can be seen with a Newtonian argument. For a spherically symmetric star with radius R, the escape velocity is $v^2 = 2M/R$ (we are using absolute units with $G = c = \hbar = 1$). So $v = 1$ when $R = 2M$, the correct value of the Schwarzschild

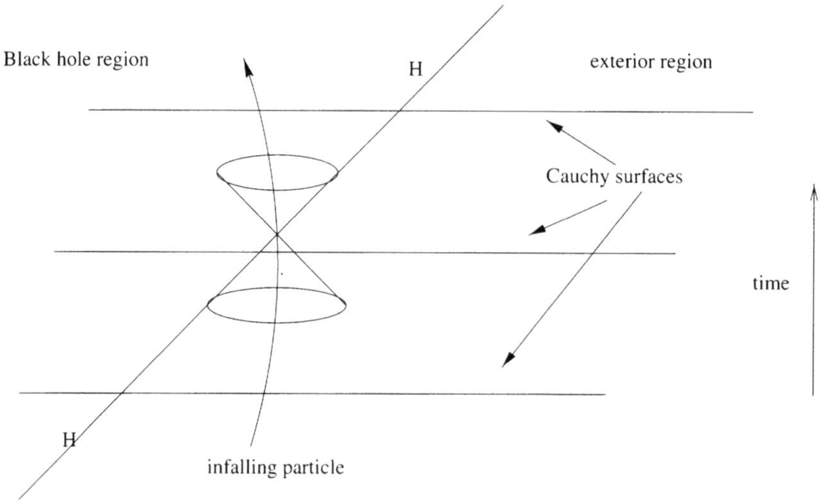

Fig. 1. The event hoizon in a spacetime containing a black hole. Note that the black hole is locally to the future of the exterior region

radius of a spherical black hole, but of course the detailed space-time picture of the black hole requires the full non-linear content of General Relativity. The event horizon must be thought of as belonging to the black hole, since no light ray within the horizon can reach infinity, or, in other words, we may say that the horizon is a property of the black hole. The horizon has at least three important attributes. First, it is believed that it has no future end point, i.e., point where the horizon stops (in quantum theory there could be an end point where the black hole disappears by Hawking evaporation). Thus a classical, isolated black hole is a perfectly stable object. The second property is that, in a world with vanishing cosmological constant, and with the dominant energy condition satisfied (this too can be violated in quantum theory), any connected component of the horizon on a spacelike surface must have the topology of a sphere. And finally, the area of any spacelike section of the horizon never decreases with time (this is Hawking's famous area theorem, which inspired the idea that horizon area is a measure of black hole entropy).

In asymptotically flat gravity ($\Lambda = 0$), the most general stationary, electrically charged black hole solution of the Einstein–Maxwell system is the Kerr–Newman solution. The horizon is located at the value of the radial coordinate $r_+ = M + \sqrt{M^2 - J^2M^{-2} - Q^2}$, where M, J and Q are the mass, angular momentum and charge of the black hole. The solution is completely specified by these three parameters, in accordance with the no-hair theorem. Thus, if a star collapses to form a black hole, all detailed information about the stellar structure is lost in the final stationary state except the mass, angular momentum and charge. In particular, a collapsed star made of anti-baryons will be indistinguishable from a collapsed baryonic star, so baryon number is really not well-defined for black holes. In terms of M, J, Q one can also express the acceleration of gravity of a stationary test particle near a black hole, in the limit where it passes onto the horizon. This is called the surface gravity κ of the black hole, and it is an important property that κ be constant on the event horizon.

Using κ, the area A of the event horizon (more exactly, of a spatial section), its angular velocity Ω and electric potential Φ, one can prove, in the classical theory of black holes, the following differential identity, connecting neighbouring black hole states

$$dM = \frac{\kappa}{8\pi} dA + \Omega\, dJ + \Phi\, dQ. \tag{2.1}$$

This identity, plus the area law and the constancy of κ, comprise the so-called three laws of black hole mechanics [1]. A fourth law has also been proved under reasonable assumptions, by Israel, and states the impossibility of attaining $\kappa = 0$ in a finite number of operations (to be performed on the black hole).

We also see that the existence of a black hole puts a bound on its parameters

$$M^2 \geq M^{-2} J^2 + Q^2 \tag{2.2}$$

which is reminiscent of various Bogomolny bounds found in soliton theory, gauge and supersymmetric field theories (indeed, the state with $J = 0$ and $M = |Q|$ is a stable BPS state in $N = 2$ supergravity). Under electromagnetic duality $F \to F^*$, a solution can also describe magnetically charged black holes and dyonic solutions are also known.

These classical black hole solutions have been generalized to de Sitter and anti-de Sitter space-time, which have positive and negative Λ respectively. In de Sitter space, there is the additional complication of the existence of a cosmological event horizon for stationary observers. Thus, in a stationary black hole solution, there is always such a cosmological horizon surrounding the event horizon, but, apart from this, the black hole's properties in de Sitter space mirror very much their asymptotically flat cousins (in particular, the topology of the horizon is a sphere and the four laws are valid, although the mathematical foundation of the fourth is less robust than the others).

In anti-de Sitter space one meets the additional surprise that even different topologies are allowed: the horizon can be a Riemann surface of any genus. Again, only a few parameters are needed to specify the solutions and the four laws are generally valid.

Black hole thermodynamics and the density of states

The four laws of black hole mechanics immediately translate into the four laws of thermodynamics, if one is willing to think of κ as proportional to temperature (in which case we can measure gravity with a thermometer!) and the area as proportional to thermodynamic entropy [2].

But the absolute temperature of a classical black hole must vanish, because the black hole is a perfect absorber from which nothing can escape. Quantum mechanically however, black holes do evaporate by a quantum emission process of pair creation by the strong gravitational field surrounding the horizon [3]. For example, a Schwarzschild black hole has a total luminosity, under emission of massless bosons, equal to

$$L = \sum_{l=0}^{\infty}(2l+1)\int_0^{\infty}|\Gamma_l(\omega)|^2 \frac{\omega}{e^{2\pi\kappa^{-1}\omega}-1}\frac{d\omega}{2\pi}. \tag{2.3}$$

The factor $|\Gamma_l(\omega)|^2$ is the transmission coefficient from the gravitational barrier around the horizon and we see that the spectrum is Planckian with a temperature, in ordinary units, equal to $\hbar\kappa/2\pi c$ (this is about 10^{-7} Kelvin degrees for a solar mass black hole).

The first law is then actually a thermodynamic law and gives the entropy of a black hole as $S = A/4$, one quarter the horizon area ($4\pi M^2$ for a Schwarzschild black hole)! This is called, nowadays, the Bekenstein–Hawking (BH) entropy of the black hole.

This is a very important result, for it tells us that the number of internal configurations of a black hole is $\sharp = \exp(A/4)$, just as if the microscopic degrees of freedom were concentrated around the horizon (in field theory one would expect this number to grow exponentially with volume).

This result can also be arrived at by assuming that black hole formation and evaporation is governed by a unitary S-matrix obeying CPT symmetry. It is then easy to see that the ratio $x = \sigma/W$ of the capture cross section of the black hole to the total emission probability of a particle with mass $m \ll M$ is

$$x = \frac{\rho(M+m)}{\rho(M)} = e^{\beta m} \tag{2.4}$$

where ρ is the density of states of the black hole and the last equality holds since the emission is thermal. From this $\rho = C\exp(4\pi M^2)$ follows, which is the entropy-area law for a Schwarzschild black hole. C is a finite constant that cannot be determined from these considerations. But if C were to be infinite, then the spectrum of the black hole's states would be strictly continuous. Then, if a black hole can form by absorption of a light particle from a heavy particle, the ratio x would be infinite so that $W = 0$. This means that there would be remnants (i.e., long living small black holes) with a tremendously high degeneracy. On the other hand, finiteness of the entropy implies a discrete spectrum for the black hole and the problem of computing this spectrum and its degeneracy, so as to recover the BH formula in the semiclassical limit, is the state counting problem which still awaits a definitive solution.

Euclidean black holes

The thermodynamic behaviour of black holes can also be inferred from the euclidean version of the various black hole solutions, without recourse to any quantum calculations [4]. By this we mean that a Wick rotation to imaginary time has been performed, after which we are left with a positive signature metric. Now it is a very significant fact that the metric so obtained is a regular solution of euclidean Einstein equations if and only if the imaginary time is a periodic variable, with period $\beta = 2\pi\kappa^{-1}$. But this is just the inverse temperature of the black hole as originally discovered by Hawking using a quantum field theory calculation! This means that a black hole will contribute to a partition function in quantum gravity as a state with that temperature. The lowest order contribution to the free energy of a Schwarzschild black hole is computed to be $F = \beta M/2$ (ignoring the instability corresponding to a positive free energy), so that the entropy will be $S = \beta(M - F) = \beta M/2 = 4\pi M^2 = A/4$.

With the euclidean procedure it is not difficult to establish the general validity of the entropy-area law for all known black holes, rotating, dyonic, magnetically charged, with or without Λ term and with different topologies. This seems, therefore, a very robust law, also valid for higher dimensional black holes (the subject of string theory).

3 The state counting problem

Recovering entropy in QFT

The identification of the microscopic states responsible for the entropy and the thermodynamic behaviour of black holes is a major problem of quantum black hole physics. Before string theory addressed this question, there were essentially four proposals for the microscopic origin of black hole entropy S:

1. $\exp S$ counts the number of configurations of the black hole in the presence of its background noise from thermal radiation [5].
2. $\exp S$ counts the number of configurations of the (massless) quantum fields (including gravitons) that can exist around the horizon and are thermalized at the Hawking temperature [6].
3. $\exp S$ counts the number of internal configurations (as opposite to horizon configurations) compatible with the given parameters of the black hole and is due to a phenomenon of quantum entanglement [7].
4. $\exp S$ counts the number of metric fluctuations of the black hole compatible with its mass, angular momentum and charge (the tones of the black hole). There is no reference here to degrees of freedom of matter [8].

Unfortunately, the first three proposals, all based on quantum field theory (QFT) in a given curved background, gave a divergent entropy that can be summarized by the formula

$$S = K\nu A_H \varepsilon^{-2} \tag{3.1}$$

where K is a numerical constant, ν the number of massless species and ε the proper distance from the horizon. So, contrary to the BH formula, in QFT S is divergent and depends on the number of massless species. To obtain the BH formula a cut-off must be invoked of order $2l_p(\nu/K)^{1/2}$, with l_p the Planck length.

It is not unlikely that quantum gravity will provide its own cut-off (this could be the scale of the horizon fluctuations) and that this will depend on the number of massless species, but no one knows how to show this since we do not have a usable quantum theory of gravity. On the other hand, the short wavelength fluctuations causing the entropy divergences are precisely the same as those which renormalize Newton's constant. But then one should start from a bare BH entropy, absorb the divergencies into a renormalized Newton constant and add the finite one-loop contribution due to matter and gravitons (which has a clearly recognizable statistical interpretation). Then we are left with a renormalized BH entropy without statistical interpretation (because the bare BH entropy has none) plus quantum corrections of statistical origin. The only sort of theory where this could work is induced gravity, for, if the Newton constant is entirely induced by a matter field (i.e., there is no such thing as a bare G), then $G^{-1} \simeq \nu$ and the entanglement entropy would be $S \simeq \nu A_H \simeq G^{-1} A_H$. This would also solve the problem of the number of species, among other things.

The fourth proposal calls into play the back reaction of the Hawking radiation on the metric, and comes very close to the BH formula. It involves the unpleasant feature that the entire history of the black hole must be taken into account, and assumes that the black hole is in a kind of thermal state, instead of deriving this. However, York's paper is a very important one, because it makes clear that one cannot ignore the back reaction. Indeed, the event horizon of a spherical black hole with back reaction included has an area which is slightly less than that of the uncorrected black hole, by an amount

$$\delta A_H \simeq 128\pi \nu \hbar \tag{3.2}$$

(ν is again the number of massless fields). This is independent of mass and so is nonzero even for very large black holes, where one expects quantum effects to be negligible. Hence there always exists a quantum ergosphere around any black hole, and York's conclusion is therefore that the fluctuations in the metric cannot be neglected.

To summarize, QFT on its own seems unable to predict the behaviour of black holes, at least in the semiclassical approximation, unless one is willing to admit a QFT of gravity itself. So now we consider such a (still incomplete) theory, that of superstrings.

Black holes in string theory

In many senses string theory is an induced theory of gravity. The low energy effective action starts with the Einstein–Hilbert term with a four-dimensional Newton constant (after dimensional reduction from $10D$) of order $G \simeq g^2\alpha'$. Here g is the (unknown) string coupling constant and $\alpha' = l_s^2$ the string slope. Hence, when considering string interactions, one cannot ignore gravity and we have to consider string propagation in a curved background. Here too, essentially four apparently distinct approaches have been considered. These were:

1. to reconsider the entangled state idea in the framework of string theory [9];
2. to study the correspondence: black holes $\leftrightarrow$ elementary string states (Susskind's heretical proposal 1993) and count BH states by counting string states;

3. to study the correspondence: black holes $\leftrightarrow$ non-perturbative string states, such as D-branes, to identify the BH constituents and count them;
4. the recently prososed AdS/CFT correspondence, which for black holes aims to explain entropy by holography with a conformal field theory (CFT) living on the boundary of anti-de Sitter space (AdS).

The first proposal was of course aimed at clarifying and eventually solving the information loss paradox and the associated entropy divergence. As was first realized by 't Hooft and Susskind, entropy divergence is related to information loss, since a system with a finite entropy has necessarily a finite storing capacity, so that, when a black hole evaporates, it has no choice than to reemit the original information.

To zeroth order in α', the Schwarzschild black hole is a solution of string theory (the heterotic string has the Kerr solution, for example), so it makes sense to consider the string partition function on such a black hole background. In flat space there is no tree-level contribution to the string free energy, because space-time is of the form $\mathcal{M} \times S^1$, where $\mathcal{M}$ is a $9D$ flat manifold and S^1 is a circle with circumference β. Now a tree-level partition function in string theory is given as a path integral over all embeddings of the sphere into space-time. Being simply connected, the sphere can only be embedded into the $\mathcal{M}$ factor, so it does not feel the temperature (it is just a vacuum contribution). As in QFT, the temperature dependence of the free energy in string theory begins to one-loop order in a perturbative expansion. The topology of an asymptotically flat black hole is instead $\mathbb{R}^2 \times S^2$, so now there is a tree-level contribution. This is seen to be just the off-shell (i.e., $\beta \neq \beta_H$, the black hole's temperature) euclidean Einstein–Hilbert action on a conical background with deficit angle $\beta - \beta_H$ (since perturbative string theory is defined only on-shell, this has to be verified by special arguments; see [9]). The Gauss–Bonnet's theorem gives

$$\log Z^{(0)} = \frac{1}{16\pi G}\int R = (\beta_H - \beta)\frac{A_H}{4G\beta_H} \tag{3.3}$$

from which $S = A_H/4$, so that the tree-level entropy is finite! Moreover, as pointed out by Susskind–Uglum, it has a statistical interpretation. Clearly, the only spheres involved in this result are those intersecting the horizon. A string world-sheet intersecting the horizon as in the figure below, is seen by an external observer in Lorentzian space-time, as an open string with the ends frozen on the horizon. The oscillations of such strings are what is counted by the black hole entropy. If we send $\alpha' \to 0$, the string world sheet reduces to a point particle, the oscillations are lost and with them the statistical interpretation.

Unfortunately, this is not the end of the story. As one goes close to the horizon, the local temperature increases without bound (the red shift). This means that, even at small string coupling, one cannot ignore the higher loop contributions. But these are infrared divergent, signaling the possibility of the Hagedorn phase transition. So a stretched horizon must be invoked (like the membrane paradigm), so as to keep the temperature below the Hagedorn threeshold. In other words string theory, like QFT, needs a horizon cut-off, or else a better understanding of the Hagedorn transition.

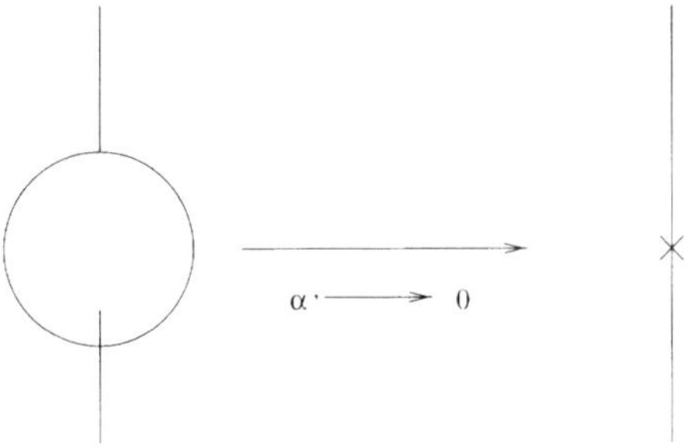

Fig. 2. A vacuum fluctuation of closed string theory, with the world sheet that interacts the horizon. To an external observer, this appears as an open string excitation with end points frozen on the event horizon

The correspondence principle

There is not much one can do with the Hagedorn transition, so string theorists have analyzed the possibility that black holes are elementary string states. At first sight this cannot work, for the following reason. The mass spectrum of elementary strings in flat space is given by $M = N(\alpha')^{-1/2}$, where N is an integer, so the radius of a black hole is of order $R = MG = g^2\sqrt{\alpha'}\,N$. For very massive black holes one needs highly excited string states, and for these the degeneracy can be computed as

$$\rho_N \simeq e^{cN} = e^{c\sqrt{\alpha'}M} \tag{3.4}$$

where c is a numerical constant depending on the string theory being considered. By contrast, the black hole's degeneracy grows exponentially with M^2. The quantity $c\sqrt{\alpha'}$ is nothing but the Hagedorn temperature, at which the string canonical ensemble is ill-defined.

This means that one cannot compare a black hole with elementary string states, without considering the effects of the gravitational field. Many arguments were put forward by Susskind and others to describe these 'dressing' gravitational effects, but they were not conclusive simply because there were no detailed calculations.

Instead, consider a black hole which is slowly evaporating. One can argue that, sooner or later, the Hawking temperature will reach the Hagedorn temperature. At this point

$$e^S = \exp(4\pi M^2) = \exp(\beta_H M/2) \simeq \exp(c\sqrt{\alpha'}M) \tag{3.5}$$

and we see that the black hole degeneracy is comparable with the string degeneracy! This led Polchinsky and Horowitz [10] to state, as a principle, that the comparison between black holes and string states should be done at a particular value of the string coupling, the value at which the size of the black hole is comparable with the string scale l_s. At this point the black hole description is no longer valid because the black hole has become an excited string state. Equating the string scale l_s with the black hole radius, one finds that the mass at the transition point is

$$M = (g^2 l_s)^{-1} = l_s^{-1}\sqrt{N}. \tag{3.6}$$

Thus, at the correspondence point, $g = N^{-1/4} \ll 1$, and so it is consistent to compare the black hole degeneracy with the string degeneracy. But we see once more the appearence of the Hagedorn transition. This argument has been extended to many other black holes in various dimensions, and always gives the correct qualitative answer. The only exceptions are the AdS black holes [11], in accordance with the fact that in AdS there seems to be no Hagedorn transition. The puzzling aspect of how the cosmological constant can change so drastically the behaviour of strings and black holes, remains a problem.

D-branes as black hole constituents

What has been said so far is very appealing from a physical point of view, but lacks almost completely the necessary mathematical details. A different approach has been offered by Polchinsky's discovery that certain extended configurations in $10D$ in supergravity theories have a weak coupling description as D-branes, hyperplanes where strings can end. To be specific, both type IIA and IIB supergravity have in the massless multiplet the metric $g_{\mu\nu}$, dilaton ϕ, antisymmetric Kalb-Ramond field $B_{\mu\nu}$ and several antisymmetric tensor fields $C^{(n)}$ coming from the Ramond–Ramond (RR) sector of both theories. As a vector couples to a particle (0-brane), an antisymmetric two-index tensor couples to a string and, more generally, a p-form field couples to a $(p-1)$-brane. These are the extended objects of superstring theory.

The branes that saturate the supersymmetric Bogomolny bound are called BPS states, and have masses proportional to their charge, generically denoted Z. Now while the fundamental string, coupling to $B_{\mu\nu}$, has $M = |Z|$, it turns out that RR-BPS-branes have masses $M \simeq |Z|/g$. The gravitational field of these objects is determined by the product $G_{10}M$, where the ten-dimensional Newton constant is of order g^2. So $G_{10}M \simeq g$ and goes to zero at weak string coupling. The novelty here is that we have a solitonic object that nevertheless has weak gravity at weak coupling (the solitonic 5-brane, for example, has $M = |Z|/g^2$, so this would not be true).

The weak coupling description of RR-BPS-branes are D-branes. There are two important results for these objects. First, the degeneracy of a given D-brane configuration is perturbatively calculable by means of supersymmetric Yang–Mills theory (this is because the open strings frozen on the D-brane world-volume have Yang–Mills theory in the massless sector). Second, this number is independent of g because it is protected by supersymmetry. Now it has long been known that supergravity admits black membrane solutions which, upon compactification, describe extreme charged black holes, provided that the Kaluza–Klein reduction is done to four and five dimensions (otherwise the horizon area vanishes). The charges carried by the black holes are the numbers of constituent RR-branes before dimensional reduction. One can then compute the BH entropy of the black holes, and use the independence of the degeneracy from g to extrapolate the result to weak coupling, where this entropy is computable from the degeneracy of the corresponding D-brane configuration. They agree perfectly, if one is careful to compute the degeneracy in a semiclassical limit (which does not always correspond to large mass black holes) of large charges (i.e., large number of constituents). To give a flavor of what is going on, consider a 5-dimensional BPS (i.e., extreme) black hole with three charges Q_1, Q_5 and Q_p [12]. This can be obtained by toroidal compactification down to $5D$ of a configuration made of Q_5 coincident $D5$-branes (this means a D-brane with 5 dimensions) with Q_1 $D1$-branes inside. The branes can be superposed because there is no net force between them, due to a cancellation of the vacuum energy between bosons and fermions as a consequence of supersymmetry. This is why the BPS condition is so crucial. Finally, the Q_p charge is carried by massless open strings with ends on the branes and has momentum $P = Q_p$. This black hole turns out to have an entropy $S = 2\pi\sqrt{Q_1 Q_5 Q_p}$.

Now the gauge theory which describes the open strings moving on the $D1$, $D5$-brane system has $n_b = 4Q_1 Q_5$ bosonic and $n_f = n_b$ fermionic degrees of freedom with energy Q_p. The partition function is a standard one

$$Z = \left[\prod_{Q_p=1}^{\infty} \frac{1+w^{Q_p}}{1-w^{Q_p}}\right]^{4Q_1 Q_5} . \tag{3.7}$$

The degeneracy of states with given Q_p is the coefficient of w^{Q_p} in the expansion of Z in powers of w. For large Q_p, this is $\exp(2\pi\sqrt{Q_1 Q_5 Q_p})$, leading to an entropy which agrees precisely with the black hole entropy.

Quite ironically, there are strong arguments within General Relativity acccording to which the entropy of extreme black holes must vanish!

To most people this D-brane stuff may seem rather artificial. After all, black holes form from the collapse of ordinary baryonic matter and it is hard to see why such exotic degrees of freedom should suddenly arise around a black hole horizon, where gravity can still be weak (e.g., in a 10^8 solar mass black hole). The calculation has nevertheless the virtue of showing that string theory has the necessary mathematical machinery to reproduce statistically the BH entropy, which must be considered a significant result.

AdS/CFT correspondence

Another piece of important recent work in string theory is the correspondence with conformal field theories in anti-de Sitter space. In asymptotically flat space, the black hole temperature increases with decreasing mass, which is precisely the reason that black holes have a finite lifetime. In AdS space, on the contrary, the temperature increases with increasing mass, making black holes virtually stable objects. In $4D$ AdS, for example, T increases as $M^{1/3}$ and S increases as T^3. This is the behaviour of massless radiation in one less dimension (the same is recovered in other dimensions, too).

These are results which are derived within semiclassical gravity in AdS, just by looking at the classical euclidean action for a black hole background. We now know that this behaviour is no coincidence. Since Maldacena's conjecture [13], it has been confirmed many times that superstring theory on $AdS_5 \times S^5$ is equivalent to $\mathcal{N} = 4$,

$SU(N)$-SUYM theory on the boundary of AdS_5, which is the $4D$ Minkowski space. The correspondence holds for AdS_d as well, possibly with different boundary CFT. Witten [14] showed how to use this correspondence explicitly, and this is formalized in the semiclassical formula (here semiclassical means taking the large N-limit of the gauge theory with fixed $g_{YM}^2 N$)

$$Z_{CFT}(S^1 \times M^{d-2}) = \sum e^{-I(X)} \tag{3.8}$$

where the sum is all over Einstein manifolds X with negative curvature, which have as boundary the manifold $S^1 \times M^{d-2}$ on which we are computing the partition function of some CFT, and $I(X)$ is the classical type IIB supergravity action for X. The formula can be generalized for higher loops, but the leading term of black hole entropy is already contained in the semiclassical result. The correspondence formula is already remarkable at this level. We see that, for an anti-de Sitter black hole,

$$-I(X) = \log Z_{CFT}(S^1 \times M^{d-2}). \tag{3.9}$$

The left-hand side is the gravitational, tree-level partition function. So the tree-level partition function for gravity, the one for which it is difficult to find a statistical interpretation, can be computed as a one-loop partition function in an ordinary, albeit intricate, superconformal field theory. Conversely, knowing $I(X)$ gives precious information about supersymmetric YM theory, that cannot be arrived at easily otherwise. So, for example, it was known after the work of Hawking–Page, that the free energy of a Schwarzschild–de Sitter black hole at sufficiently high temperature is always less than that of AdS space at the same temperature, while at low temperatures it is AdS that dominates over the black hole configuration. This means that there is a phase transition between the two spaces, the black hole and pure AdS, so there must be such a phase transition in $\mathcal{N} = 4$ SUYM theory. Indeed, the low energy supergravity action $I(X)$ is proportional to $1/g^2$ (coming from $1/G$) and the string coupling $g \simeq g_{YM}^2$. So in the large N-limit $I(X) \simeq N^2$, and the free energy also scales with N^2. As there are $N^2 - 1$ gluons in the gauge theory, this means that, when the black hole dominates, the gauge theory is in a deconfined phase. If AdS dominates, the gauge theory is close to the ground state (in a confined phase). Nevertheless small black holes can form and completely evaporate since they have higher free energy than AdS. So the gauge theory, which is unitary and has no information loss, should govern the Hawking evaporation process of small black holes in a unitary way (the 't Hooft dream!).

As for black holes, let us see how the entropy must scale with the horizon area under the rules of CFT. The temperature of a Schwarzschild black hole in AdS_d is

$$\beta = \frac{4\pi \ell^2 R}{(d-1)R^2 - (d-3)\ell^2} \simeq \frac{\ell^2}{(d-1)R}. \tag{3.10}$$

Here ℓ is the AdS radius of curvature (related to Λ), R the black hole radius and we consider the limit for large R. The CFT on the boundary lives in $(d-2)$ spatial dimensions, so its entropy scales as (for large temperatures, i.e., for large black holes as above)

$$S_{CFT} \simeq \left(\frac{\ell}{\beta}\right)^{d-2} . \tag{3.11}$$

So, using the first formula for β, we have $S_{CFT} \simeq (R/\ell)^{d-2} \simeq A_H$, and the entropy scales as the area of the event horizon! This explanation is a holographic one, as degrees of freedom living on the boundary of space-time are correctly counted by the black hole entropy. Of course one would like to locate these degrees of freedom on the horizon rather than at infinity, but in the case of AdS_3 one can do better and show that, in fact, the boundary CFT can be mapped onto the horizon of the BTZ black hole.

These results are so appealing that a negative cosmological constant, albeit an extremely tiny one, would have the approval of black holes.

4 Conclusion: where do we go from here?

Despite the large and non-trivial amount of new ideas flowing into black hole physics from sstring theory and other approaches like canonical quantum gravity, it seems that we are not yet at a conclusive stage. Each black hole seems to need a separate calculation. Extreme or near-extreme black holes can be described to a certain extent by D-brane ideas, others by the string correspondence principle, and AdS black holes by CFT. On the other hand, D-brane ideas do not work for black holes far from extremality, the correspondence principle seems invalid in AdS space and there is no CFT correspondence with asymptotically flat black holes. So the time has come to try to produce a unifying description, and one which should make contact with all other approaches, eventually disproving some of them! Otherwise we cannot say that we have a theory of black holes.

References

1. Bardeen J.M., Carter B., Hawking S.W. (1973): Commun. Math. Phys. **31**, 161
2. Bekenstein J.D. (1973): Phys. Rev. D **7**, 2333
3. Hawking S.W. (1975): Commun. Math. Phys. **43**, 199
4. Gibbons G.W., Hawking S.W. (1977): Phys. Rev. D **15**, 2572
5. Zurek W.H., Thorne K.S. (1985): Phys. Rev. Lett. **54**, 2171
6. 't Hooft G. (1985): Nucl. Phys. B **256**, 727
7. Bombelli L., Koul R.K., Lee J., Sorkin R.D. (1986): Phys. Rev. D **34**, 373
8. York J.W. Jr. (1983): Phys. Rev. D **28**, 2929
9. Susskind L., Uglum J.R. (1994): Phys. Rev. D **50**, 2700
10. Horowitz G.T., Polchinsky J. (1997): Phys. Rev. D **55**, 6189
11. Klemm D., Vanzo L. (1998): Phys. Rev. **58**, 104025
12. Strominger A., Vafa C. (1996): Phys. Lett. B **379**, 99
13. Maldacena J. (1998): Adv. Theor. Math. Phys. **2**, 231
14. Witten E. (1998): Adv. Theor. Math. Phys. **2**, 253

Gravitational Wave Astronomy from Space

A. Vecchio

Abstract. Space-borne interferometers – LISA and OMEGA – are expected to be robust gravitational wave observatories. They could easily detect radiation emitted by massive black hole binary systems at high redshift. We discuss deep surveys of massive black holes and how accurately one can expect to extract information about sources, and we compare the performances of both instruments. We show that black hole spins are likely to play a crucial role for gravitational wave astronomy from space: relativistic frame-dragging leaves peculiar "finger-prints" on the emitted radiation, which allows us to reconstruct the source parameters precisely.

1 Introduction

The emission and detection of gravitational waves (GW's) is a long-standing problem that dates back to the second decade of our century: in 1916 Einstein computed the gravitational radiation emitted by a laboratory-scale object [1] but only in the late sixties was the controversy about the "real nature" of gravitational wave solved [2–4]. The discovery in 1974 of the now-famous binary pulsar PSR1913+16 [5] – and its two decades long monitoring – and, in the early nineties, of the significantly stronger and more favorably oriented PSR1534+12 [6], has allowed us to test General Relativity to an unprecedented level of accuracy by measuring the system orbital decay due to GW emission [7]. However, as of this writing, *direct* detection of gravitational radiation has not yet been achieved. GW's are a fundamental building block of our theoretical picture of the Universe and one of the qualitative predictions of General Relativity not yet tested; however, it is likely that the most important scientific payoff from their direct observation will be the opening of a new window on the universe for astronomy.

The frequency band covered by gravitational radiation emitted by astrophysical objects extends from $\sim 10^{-18}$ Hz to $\sim 10^4$ Hz [8]; two main windows are currently explored experimentally: the *high frequency band* (10 Hz – 1kHz) and the *low frequency band* ($10^{-5} - 10^{-1}$ Hz). In the high-frequency regime a network of powerful km-size laser interferometric devices – LIGO [9], VIRGO [10], GEO [11] and TAMA [12] – is expected to come on line by 2000–2002; in the kHz range the sensitivity of narrow band acoustic devices is continuously improving [13]. Earth-based detectors can monitor objects in the mass range $\sim 1\, M_\odot - 10^3\, M_\odot$ and possibly relic radiation from the early universe (Fig. 1; cf. [8, 14] also). Close binary systems of solar-mass compact objects and massive black holes (MBH's) with mass $\sim 10^4\, M_\odot - 10^7\, M_\odot$ are fundamentally out of reach for Earth-based detectors and require observations from space.

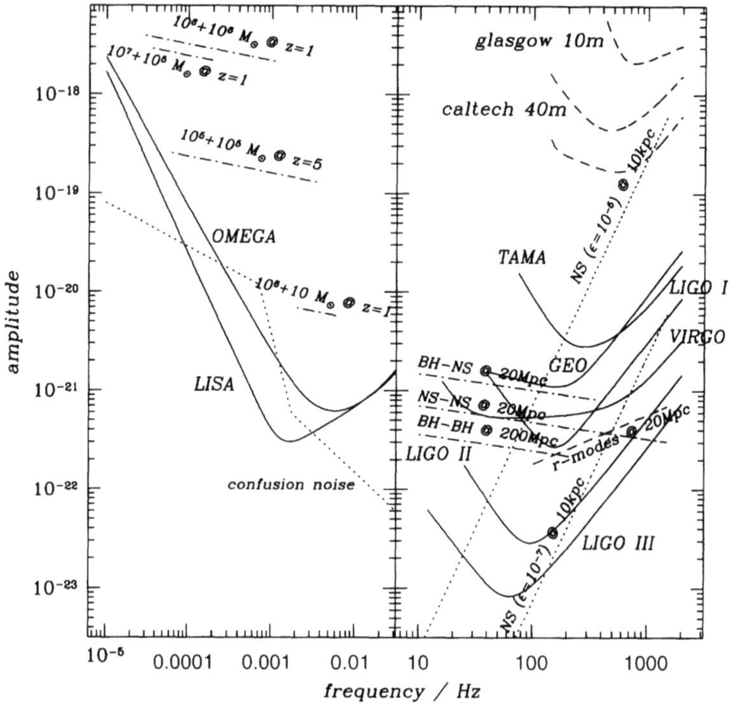

Fig. 1. The strength of gravitational radiation from some of the expected sources is compared with the sensitivity of laser interferometric detectors. The plot (*left*: low frequency band; *right*: high frequency band) shows the characteristic amplitude of the signals and the rms value of the instrument noise against frequency. The top-right *dashed lines* refer to the 10 meter prototype at Glasgow University and the 40 meter prototype at Caltech. The *solid lines* refer to the other instruments, as labelled; LIGO I, II, and III refer to the so-called "initial", "enhanced" and "advance" configurations, respectively. The dotted-dashed lines represent the in-spiral signal emitted by binary systems of neutron stars (NS) and/or black holes (BH); the masses of neutron stars and black holes are assumed to be $1.4\,M_{\odot}$ and $10\,M_{\odot}$, respectively. The dotted lines in the high frequency region are the radiation emitted by rotating neutron stars with ellipticity $\epsilon = 10^{-5}$ and 10^{-7} at 10 kpc ($\sim$ the distance from our Galactic Centre). The dashed line labelled "r-modes" refers to radiation emitted by hot-young neutron stars undergoing the so-called r-mode instability during their first year of life [15]. The dotted line labelled "confusion noise" is the background signal produced by the incoherent superposition of unresolved short-period solar-mass binaries [16]

The low-frequency band is currently being explored by low-sensitivity experiments involving Doppler tracking of interplanetary spacecraft [17, 18]; the first coincidence experiment was carried out in 1993 with the probes GALILEO, MARS-OBSERVER and ULYSSES and longer and more sensitive searches are scheduled in 2001–2004 with the space-probe CASSINI [19, 20], currently travelling toward Saturn. However radical change of performance will come with the launch of space-borne interferometers, LISA [21] or OMEGA [22]. Thanks to their high sensitivity, such instruments will allow us to undertake deep observational campaigns on a wide variety of sources, in particular, massive black holes; see Fig. 1. The latter will be detectable at a very large signal-to-noise ratio (S/N) at cosmological distances (see Fig. 2); if the event rate is reasonable [23], such observations could give precise *direct measurements* of the BH parameters and independent estimates of the fundamental cosmological parameters [21, 24] and possibly could improve our understanding of the scenario of structure formation in the universe, as the formation and evolution of MBH's might provide an independent trace of the history of their merging host galaxies.

LISA is currently identified as an ESA Cornerstone mission in the Horizon 2000-plus program; however, with NASA participation, which is under study, this instrument could fly as early as 2009. As presently conceived, LISA will consist of a constellation of three drag-free spacecraft placed at the vertices of an ideal equilateral triangle with sides of length $\simeq 5 \times 10^6$ km, forming a three-arm interferometer. The LISA orbit is designed to keep the configuration as stable as possible during the mission: the barycenter of the instrument is inserted into a heliocentric orbit, following the Earth by 20°; the detector plane is tilted by 60° with respect to the ecliptic and the instrument counter-rotates around the normal to the detector plane with the same 1-yr period.

OMEGA is a similar, although smaller-scale/lower-cost, project proposed by Ron Hellings. Six free-flying mini-probes, two at each vertex of an equilateral triangle, with sides $\simeq 10^6$ km, would be placed in a Earth orbit. The key difference between the two missions is the orbit of the detector: the OMEGA satellites lie approximately in the ecliptic plane and the configuration rotates around the line perpendicular to the detector plane once every ~ 55 days.

Here we discuss deep surveys of massive black holes by monitoring the in-spiral phase of MBH binaries; we regard LISA and OMEGA as GW observatories and carry out a comparison between the two detectors for the problem at hand. Both LISA and OMEGA would operate far above the so-called marginal detection threshold, which is likely to characterize the first generation of Earth-based interferometers. Binary systems are expected to be extremely clean and strong sources in the low frequency gravitational wave band, which will provide detailed information about the source, "orthogonal" to what one might get with telescopes. In fact, recent X-ray observations have given noteworthy examples of possible observations of general relativistic effects, such as the presence of a black hole horizon, relativistic "frame-dragging" (see [25] also) and the mapping of circular orbits around black holes (see [26], and its references, where these results are compared and contrasted with future results to be provided by LISA). However none of these results are as yet on a solid foundation, as the interpretation of X-ray data relies on the understanding and modelling of complex physical processes and because of instrument limitations.

Some of the issues discussed here have been addressed in [27–32]; we refer the reader to those papers for further details.

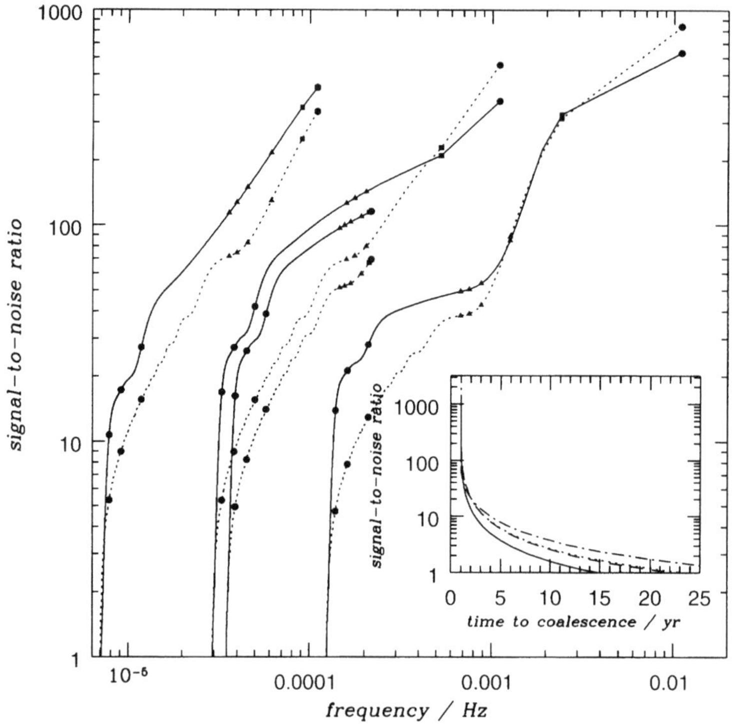

Fig. 2. The signal-to-noise ratio produced in the output of LISA and OMEGA in observation of MBH's in binary systems. The plot shows the S/N (LISA: solid lines; OMEGA: dashed lines) as a function of frequency during the final year of in-spiral for the following mass choices: $m_1 = 10^7\,M_\odot$ and $m_2 = 10^7\,M_\odot$, $m_1 = 10^6\,M_\odot$ and $m_2 = 10^6\,M_\odot$, $m_1 = 10^7\,M_\odot$ and $m_2 = 10^5\,M_\odot$ and $m_1 = 10^5\,M_\odot$ and $m_2 = 10^5\,M_\odot$, from left to right, respectively. The dots mark S/N after an integration time of 1/4 yr, 1/2 yr, 3/4 yr and 1 yr; the squares and the triangles identify S/N three hours before final coalescence and the four final days of binary's life, respectively. The fiducial source is at $z = 1$ ($H_0 = 75\,\mathrm{km\,s^{-1}\,Mpc^{-1}}$, $\Omega_0 = 0$, $\Lambda_0 = 0$, corresponding to $D_L = z/H_0 = 3.997$ Gpc) with $\cos\theta_N = 0.3$, $\phi_N = 5$, $\cos\theta_J = 0.8$ and $\phi_J = 2$ (see Sect. 3 for a definition of $\widehat{\mathbf{N}}$ and $\widehat{\mathbf{J}}$). The small box shows the angle-averaged SNR as a function of the time to coalescence for one year of integration of systems of mass $m_1 = 10^7\,M_\odot$ and $m_2 = 10^7\,M_\odot$ (solid line), $m_1 = 10^7\,M_\odot$ and $m_2 = 10^5\,M_\odot$ (dotted line), $m_1 = 10^6\,M_\odot$ and $m_2 = 10^6\,M_\odot$ (dashed line) and $m_1 = 10^5\,M_\odot$ and $m_2 = 10^5\,M_\odot$ (dotted-dashed line). In both cases spins are assumed negligible and the noise is given by the sum of the instrumental and "confusion" contributions; cf. Fig. 1

2 Massive black holes

The low frequency band is populated by a *plethora* of GW sources that could easily be detectable by space-borne interferometers (see [21, 22, 33] and their references);

they include guaranteed sources such as: known galactic short-period binary stars; solar-mass compact objects falling into a massive black hole in galactic nuclei; inspiral, merger and "ring-down" of massive black hole binary systems at cosmological distance; stochastic backgrounds of cosmological and/or generated origin.

The strongest sources, and possibly the most exciting, are likely to be binary systems of massive black holes; see Fig. 1. Compelling arguments suggest the presence of MBH's in the nuclei of most galaxies, as the near-inevitable by-product of the infall of gas in their potential well [34–36] and MBH's are invoked to explain a number of phenomena, such as the activity of quasars and active galactic nuclei [37–40]. However, the observational evidences of their existence come mainly from observations of relatively nearby galaxies, whose nuclei do not show significant activity. Central dark masses – that are believed to be "dead quasars" [41] – are mainly inferred by studying the spatial distribution and velocity of gas and/or stars in galactic cores. About twenty candidate massive black holes, with masses in the range $\sim 10^6 - 10^9\, M_\odot$, are known today [42]. The most striking evidence comes from the mapping of gas motion via the 1.35 cm water maser emission line in NGC 4258 and observations in the near infrared band of star motion in our Galactic Centre; the inferred mass of the massive dark objects is $\simeq 3.6 \times 10^6\, M_\odot$ [43] and $\simeq 2.5 \times 10^6\, M_\odot$ [44], respectively. In both cases only exotic (and highly implausible) alternatives to an MBH can still be considered [45].

Many galaxies have experienced at least one merger since the epoch $z > 2$ [46, 47]. If an MBH is present in each core of both interacting galaxies, the two massive objects would fall into the common potential well, forming a pair which loses energy and angular momentum through dynamical friction; eventually a binary forms and, driven by radiation reaction, progresses toward the final coalescence. This scenario provides, therefore, quite a natural way of producing massive binary black holes (MBHB's) [23]. The apparent precession and bending of jets in active galaxies [48] and Doppler-shifted broad emission line peaks in quasars [49] give possible indirect evidence of the existence of such systems.

MBH's in the centers of galaxies are surrounded by a dense cluster of ordinary stars, whose tidal disruption provides the gas to refuel the central object, white dwarfs, neutron stars and low mass black holes, probably in the mass range $\sim 5 - 100\, M_\odot$. In contrast to ordinary stars, compact objects are not disrupted by the hole and can occasionally be captured, presumably on highly eccentric orbits, forming binary systems with life shorter than the Hubble time. Moreover the orbits of objects with mass $\gtrsim 5M_\odot$ around the central massive body would not be perturbed by other stars in the core which would provide therefore, clean gravitational wave signals [28, 50].

The main uncertainty regarding binary systems containing MBH's concerns the event rate. According to recent estimates, several events per year emitted by high mass ratio systems would be detectable. Regarding MBHB's, the event rate is largely unknown, and viable estimates range around the value $\sim 1\ \mathrm{yr}^{-1}$ by several orders of magnitude, depending on theoretical prejudice and model assumptions (we refer the reader to [23] and its references for a thorough discussion).

3 The signal and its extraction

The gravitational radiation emitted during the in-spiral of a binary system is described by 17 independent parameters; indeed, the waveform can be very complex, in particular, for general values of orbit eccentricity and BH spins. Here we shall make the following simplifying assumptions: (a) the orbit is circular (which is probably quite realistic for binaries composed of two MBH's that have undergone a common evolution inside a galactic core) and (b) if spins are present, either the BH masses are roughly equal, or one of the BH's carries a negligible spin; in this case the system undergoes so-called *simple precession* [51].

We consider a binary source of masses m_1 and m_2 (m, μ and $\mathcal{M}$ are the total, reduced and chirp mass, respectively) and spins $\mathbf{S}_1$ and $\mathbf{S}_2$ (we define $\mathbf{S} = \mathbf{S}_1 + \mathbf{S}_2$ and $\mathbf{J} = \mathbf{L} + \mathbf{S}$) at a luminosity distance D_L; with respect to an appropriate ecliptic reference frame, the location of the source in the sky is given by $\widehat{\mathbf{N}} = (\theta_N, \phi_N)$ and the direction of the angular momentum is $\widehat{\mathbf{J}} = (\theta_J, \phi_J)$, where θ and ϕ are polar angles.

On the assumption that the noise in the arms is "totally symmetric" [29], the three read-outs can be combined in such a way as to produce two linearly-independent data streams, $h^{(I)}$ and $h^{(II)}$, over the whole sensitivity window of the instrument; they are equivalent to the outputs of two 60° interferometers in the same location and on the same plane, simply rotated by $\pi/4$ (with respect to the axis perpendicular to both arms) one with respect to the other.

In the Fourier space, the signal $\widetilde{h}^{(\iota)}(f)$ $(\iota = I\,, II)$ reads [29, 51, 53, 54]:

$$\widetilde{h}^{(\iota)}(f) \simeq \mathcal{A}\, A_p^{(\iota)}[t(f)]\, f^{-7/6}\, e^{\left\{i\left[\Psi(f)-\varphi_p^{(\iota)}[t(f)]-\varphi_D[t(f)]-\delta_p\phi[t(f)]\right]\right\}}\,, \tag{1}$$

where: $\mathcal{A}$ and $\Psi(f)$ are the gravitational wave amplitude and phase, respectively; $A_p^{(\iota)}(t)$ and $\varphi_p^{(\iota)}(t)$ are the polarization amplitude (see Fig. 3) and phase, respectively; $\varphi_D(t)$ is the Doppler phase modulation and $\delta_p\phi(t)$ is the so-called Thomas precession phase [51]. Of course, the signal must be cut off at some finite frequency, of the order $f_{\rm isco} = 1/(6^{3/2}\,\pi\, m)$. $\widetilde{h}^{(\iota)}(f)$ is phase and amplitude modulated by:

(i) the change of orientation of the detector;
(ii) the orbital motion of the instrument around the Sun;
(iii) the change of $\widehat{\mathbf{L}}$ during the in-spiral, if the observed binary undergoes precession, which causes *intrinsic* modulations on the GW signal (in this case the wave's phase carries the additional contribution $\delta_p\phi$).

The detector output is the sum of gravitational radiation $\widetilde{h}^{(\iota)}(f;\boldsymbol{\lambda})$ – where $\boldsymbol{\lambda}$ is the vector of signal parameters – and noise $\widetilde{n}(f)$, with spectral density $S_n(f)$, that we shall assume to be Gaussian and stationary. For space-borne laser interferometers, two main terms contribute to $S_n(f)$: instrumental noise and the so-called confusion noise, due to the incoherent superposition of unresolved short-period stellar-mass binaries (see Fig. 1).

The optimal signal-to-noise ratio produced by a signal h is given by $S/N = (h|h)^{1/2}$, where $(.|.)$ is the usual inner product [53]. If $\boldsymbol{\lambda}_0$ is the true value of the

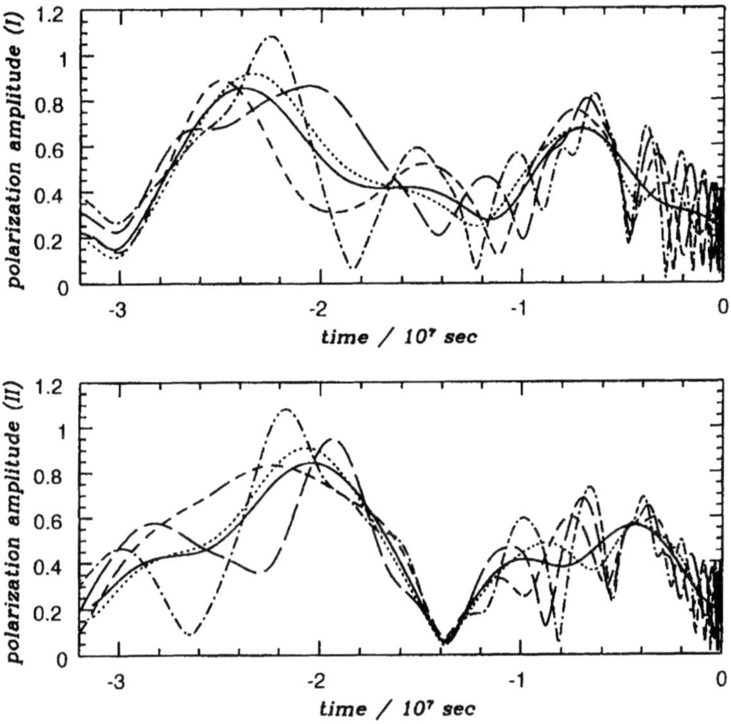

Fig. 3. Polarization amplitudes in the final year of binary in-spiral in LISA observations. The plot shows the polarization amplitude $A_p^{(\iota)}$ as a function of time in the LISA detector $\iota = I$ and $\iota = II$ for a system of mass $m_1 = 10^7\,M_\odot$ and $m_2 = 10^6\,M_\odot$ and different values of the spin-orbit parameter β (solid line: $\beta = 0$; dotted line: $\beta = 1$; short-dashed line: $\beta = 3$; long-dashed line: $\beta = 5$; dotted-dashed line: $\beta = 7$) [52]. The source location in the sky is given by $\cos\theta_N = 0.3$ and $\phi_N = 5$, the constant direction of $\widehat{\mathbf{J}}$ is $\cos\theta_J = 0.8$ and $\phi_J = 2$, and $\widehat{\mathbf{S}} \cdot \widehat{\mathbf{L}} = 0.9$. Similar or even stronger oscillations are observable in polarization phase $\varphi_p^{(\iota)}$

parameters, in general the presence of noise will induce an error $\Delta\boldsymbol{\lambda}$ in their determination; in the limit of high S/N [53, 55, 56], $\Delta\boldsymbol{\lambda}$ can be estimated easily by computing the variance-covariance matrix

$$\left\langle \Delta\lambda^j\, \Delta\lambda^k \right\rangle = \left[\left(\Gamma^{(\iota)}\right)^{-1}\right]^{jk} + \mathcal{O}(S/N)^{-1}\,, \qquad (2)$$

where

$$\Gamma^{(\iota)}_{jk} \equiv \left(\frac{\partial h^{(\iota)}}{\partial \lambda^j} \middle| \frac{\partial h^{(\iota)}}{\partial \lambda^k} \right) \qquad (3)$$

is the Fisher matrix for the ι-th instrument; the diagonal elements of Γ^{-1} are the mean square errors associated to the parameter measurements, and we define the angular resolution of the instrument as:

$$\Delta\Omega_{(\iota)}^{(N)} = 2\pi \left\{ [\Delta(\cos\theta_N)\,\Delta\phi_N]^2 - \langle\Delta(\cos\theta_N)\,\Delta\phi_N\rangle^2 \right\}^{1/2} . \quad (4)$$

For the case of the analysis with two detectors, I and II, the Fisher matrix is simply $\Gamma_{jk} = \Gamma_{jk}^{(I)} + \Gamma_{jk}^{(II)}$.

4 Deep surveys with space-borne interferometers

In this section we discuss whether space-based interferometers can indeed provide a deep census of massive black holes in the Universe by observing the in-spiral signal before the deadly merger phase[1]; in particular, we will investigate: (i) how well the physical parameters, in particular, masses, can be measured, therefore providing an accurate demography of black holes; (ii) whether a source (or its host galaxy/galaxy cluster) can be identified; this issue is of paramount importance for carrying out observations of a source and its environment in other bands.

LISA and OMEGA can study a great number and variety of systems at the same time; the key to extracting information is the different time evolutions of their signals, due both to the physics of the source and to the orbit of the detector; LISA and OMEGA have very different orbital configurations, which affect their performances as GW observatories in a critical way.

In order to establish reference figures and also to compare the two missions, we begin by assuming that we know a priori that the observed source is composed of BH's with negligible spins, or that $\widehat{\mathbf{S}}_1$, $\widehat{\mathbf{S}}_2$ and $\widehat{\mathbf{L}}$ are oriented in such a way that the direction of the angular momenta is constant (equivalently, one could assume that, for simplicity, we take into account the spin contributions only to the GW phase $\Psi(f)$); this assumption greatly simplifies the structure of the signal, and therefore the computation of the Fisher matrix, as the waveform is not intrinsically modulated. We model the radiation through the post^2-Newtonian order [52], while the amplitude is retained at the lowest Newtonian quadrupolar approximation: $\tilde{h}(f)$ depends on 11 independent parameters.

Figure 2 and Table 1 clearly show that both LISA and OMEGA can observe massive black hole binary systems up to $z \sim 5$ and sources will be visible several years before merging. A word of caution is necessary here. In our analysis, we have considered the whole final year of in-spiral; however, at some frequency (between 10^{-4} Hz and 10^{-5} Hz), instrumental noise will completely dominate the signal, killing the detection; at present there is little understanding of the real low frequency cut-off

[1] Here we are disregarding all information that one could get by also monitoring the merger and "ring-down" phase, both easily detectable [57]. At present, the merger waveform is very poorly understood and therefore impossible to use in our calculations. The ring-down signal structure is reasonably well-known and one could improve the results presented here by including it in the analysis; at the moment, for sake of simplicity, we neglect it.

Table 1. The signal-to-noise ratio (S/N) and error in the estimation of the time of coalescence Δt_c for the final year of in-spiral of MBHB's observed by LISA and OMEGA. The fiducial source is at $z = 1$ ($H_0 = 75\,\mathrm{km\,s^{-1}\,Mpc^{-1}}$, $\Omega_0 = 0$, $\Lambda_0 = 0$), corresponding to $D_L = z/H_0 = 3.997$ Gpc. For LISA we present the results corresponding to observations with only one instrument (I) and with the combination of both detectors ($I + II$); for OMEGA only the values for I are given. The waveform is modelled through the post^2-Newtonian approximation and spin effects are taken for account only for the GW phase $\Psi(f)$, with $\beta = 0$ and $\sigma = 0$ [52]. $S_n(f)$ is the sum of the instrumental and confusion noise; see Fig. 1. For each choice of m_1 and m_2, four different locations/orientations of the binary are considered; from top to bottom the values of $\widehat{\mathbf{N}}$ and $\widehat{\mathbf{L}}$ are: (1) $\cos\theta_N = 0.3$, $\phi_N = 5$, $\cos\theta_L = 0.8$, $\phi_L = 2$; (2) $\cos\theta_N = -0.6$, $\phi_N = 4$, $\cos\theta_L = 0.5$, $\phi_L = 3$; (3) $\cos\theta_N = 0.6$, $\phi_N = 1$, $\cos\theta_L = -0.2$, $\phi_L = 3$; (4) $\cos\theta_N = 0.2$, $\phi_N = 3$, $\cos\theta_L = 0.1$, $\phi_L = 1$

		LISA		OMEGA	LISA		OMEGA
m_1	m_2	$(S/N)_I$	$(S/N)_{I+II}$	$(S/N)_I$	$(\Delta t_c)_I$	$(\Delta t_c)_{I+II}$	$(\Delta t_c)_I$
$M_\odot$					(sec)	(sec)	(sec)
10^7	10^7	435	590	336	1116.676	762.866	1760.536
10^7	10^7	188	826	432	2182.218	647.838	1642.501
10^7	10^7	164	768	499	2464.082	616.135	1090.568
10^7	10^7	1638	2121	409	417.743	298.101	1659.239
10^7	10^6	273	371	229	1041.758	717.615	1531.179
10^7	10^6	134	499	297	2136.337	606.557	1238.531
10^7	10^6	132	473	357	2265.331	571.650	937.523
10^7	10^6	977	1275	286	387.675	281.225	1267.220
10^7	10^5	116	160	69	8186.841	5466.539	26738.330
10^7	10^5	81	175	85	18879.680	4944.766	24505.270
10^7	10^5	106	195	134	14253.280	4385.870	15100.990
10^7	10^5	308	420	91	3438.439	2361.636	23433.090
10^6	10^6	377	516	555	36.792	24.751	368.814
10^6	10^6	169	713	688	74.763	20.409	356.822
10^6	10^6	172	671	742	83.750	19.764	231.615
10^6	10^6	1408	1823	626	12.358	8.787	363.450
10^6	10^5	586	803	607	26.282	17.258	194.496
10^6	10^5	238	1138	749	54.371	14.289	186.892
10^6	10^5	223	1056	792	57.911	13.495	120.517
10^6	10^5	2265	2919	675	8.613	6.087	188.933
10^5	10^5	634	868	840	18.139	1.708	92.246
10^5	10^5	253	1234	1037	16.743	6.094	92.654
10^5	10^5	226	1142	1093	3.953	1.210	59.925
10^5	10^5	2460	3169	935	4.530	1.837	95.763

Table 2. Relative errors $(\Delta\mathcal{M}/\mathcal{M})$ and $(\Delta\mu/\mu)$ associated to the measurement of chirp and reduced mass by observing the final year of MBHB in-spiral with LISA and OMEGA. Signal and instrumental parameters are as in Table 1

m_1	m_2	LISA		OMEGA	LISA		OMEGA @
$M_\odot$		$(\Delta\mathcal{M}/\mathcal{M})_I$ (10^{-3})	$(\Delta\mathcal{M}/\mathcal{M})_{I+II}$ (10^{-3})	$(\Delta\mathcal{M}/\mathcal{M})_I$ (10^{-3})	$(\Delta\mu/\mu)_I$ (10^{-2})	$(\Delta\mu/\mu)_{I+II}$ (10^{-2})	$(\Delta\mu/\mu)_I$ (10^{-2})
10^7	10^7	6.546	4.393	22.240	86.351	57.999	261.467
10^7	10^7	10.862	4.413	21.055	149.938	55.088	243.650
10^7	10^7	11.520	3.795	14.206	161.297	48.801	166.936
10^7	10^7	3.301	2.259	21.472	39.824	27.509	248.861
10^7	10^6	3.979	2.685	14.781	31.598	21.358	103.783
10^7	10^6	7.029	2.586	13.915	58.786	19.728	95.770
10^7	10^6	7.014	2.229	9.481	58.732	17.445	66.239
10^7	10^6	1.883	1.338	14.361	13.942	9.943	98.898
10^7	10^5	8.998	5.998	45.331	47.424	31.541	217.436
10^7	10^5	22.363	5.740	43.345	116.602	29.805	206.507
10^7	10^5	13.117	4.761	26.452	71.654	25.119	125.904
10^7	10^5	4.234	2.949	41.785	21.799	15.052	198.438
10^6	10^6	0.571	0.377	1.777	11.365	7.608	28.788
10^6	10^6	0.954	0.379	1.581	19.921	7.115	25.130
10^6	10^6	0.999	0.331	1.135	21.754	6.439	18.479
10^6	10^6	0.282	0.190	1.626	5.025	3.454	25.834
10^6	10^5	0.361	0.235	1.079	4.074	2.703	10.148
10^6	10^5	0.591	0.220	0.969	7.332	2.391	8.966
10^6	10^5	0.619	0.194	0.686	7.732	2.186	6.504
10^6	10^5	0.163	0.113	0.988	1.672	1.187	9.155
10^5	10^5	0.112	0.064	0.225	3.159	1.866	5.126
10^5	10^5	0.178	0.061	0.207	5.120	1.616	4.586
10^5	10^5	0.168	0.054	0.150	5.305	1.514	3.443
10^5	10^5	0.052	0.031	0.219	1.267	0.793	4.848

of LISA, which turns out to be the crucial factor preventing us from working with increasingly high z: the redshifted radiation would simply fall outside the instrument band. However, binary systems in the mass range $10^4 M_\odot - 10^5 M_\odot$ could still be detectable at redshift $\simeq 10$ with $S/N \sim 10 - 100$; this might possibly provide hints about the "first generation" of MBH's.

High-redshift surveys are feasible; however, one can be concerned about the information conveyed by the data. In order to explore this issue, we have chosen fiducial sources, all at $z = 1$, with m_1 and m_2 in the range $10^5 - 10^7\, M_\odot$, and four different locations and orientations; we have then computed the errors associated to the measurements of the source parameters with LISA and OMEGA, according to the procedure presented in Sect. 3; the data stretch is assumed to correspond to the

Table 3. The angular resolution $\Delta\Omega^{(N)}$ with which a source can be located in the sky and the error in the measurement of the luminosity distance $\Delta D_L/D_L$ for observations of MBHB in-spiral with LISA and OMEGA. Signal and instrumental parameters are as in Table 1. Here $W = (\Delta D_L/D_L)$

		LISA		OMEGA	LISA		OMEGA
m_1	m_2	$\Delta\Omega_I^{(N)}$	$\Delta\Omega_{I+II}^{(N)}$	$\Delta\Omega_I^{(N)}$	W_I	W_{I+II}	W_I
$M_\odot$		$(10^{-5}$ srad)	$(10^{-5}$ srad)	$(10^{-5}$ srad)	(10^{-2})	(10^{-2})	(10^{-2})
10^7	10^7	185.460	2.255	58539250.000	3.377	1.586	457.883
10^7	10^7	135.526	11.047	29471530.000	2.937	0.463	1602.141
10^7	10^7	9.144	0.786	12109940.000	3.026	1.523	350.462
10^7	10^7	564.729	81.265	92511610.000	1.880	0.441	1733.560
10^7	10^6	157.922	2.855	15382000.000	2.786	1.363	245.398
10^7	10^6	92.486	8.891	7780560.000	2.361	0.359	833.578
10^7	10^6	7.228	0.919	3222578.000	2.517	1.310	185.279
10^7	10^6	379.685	58.202	24712640.000	1.562	0.384	839.201
10^7	10^5	251.544	15.292	10028390.000	3.012	1.619	203.600
10^7	10^5	110.835	23.197	5023249.000	3.212	0.768	525.433
10^7	10^5	17.254	5.910	1957627.000	2.816	1.558	128.962
10^7	10^5	359.974	69.164	15186930.000	1.872	0.614	424.104
10^6	10^6	111.798	1.551	1942327.000	2.329	1.179	66.112
10^6	10^6	64.347	6.663	935501.000	1.866	0.253	247.209
10^6	10^6	5.238	0.642	400485.700	1.898	1.122	54.139
10^6	10^6	261.194	41.944	2969048.000	1.319	0.312	343.647
10^6	10^5	177.278	1.893	525192.400	2.865	1.553	37.122
10^6	10^5	88.488	10.324	253030.900	2.012	0.296	134.931
10^6	10^5	9.552	0.797	107466.000	2.382	1.455	29.668
10^6	10^5	326.613	63.469	797446.500	1.585	0.354	175.836
10^5	10^5	366.500	4.739	114370.600	4.222	2.505	13.784
10^5	10^5	199.040	26.700	57298.780	3.009	0.467	54.637
10^5	10^5	16.752	1.880	24547.440	3.869	2.392	11.654
10^5	10^5	748.524	159.610	184845.200	2.551	0.548	84.561

final year of in-spiral. A short summary of the results is given in Tables 1, 2 and 3, where we display S/N, $\Delta\Omega^{(N)}$, $\Delta D_L/D_L$, Δt_c, $\Delta\mathcal{M}/\mathcal{M}$ and $\Delta\mu/\mu$ (t_c is the time at coalescence).

The first striking feature is that the accuracy of parameter extraction depends crucially (it varies by orders of magnitude) on the actual values of the source parameters: this was expected as LISA and OMEGA are all-sky monitors, whose ability in extracting information depends on the structure of the signal itself. LISA can provide a rather detailed MBH demography: $\mathcal{M}$ and μ can be measured at roughly the 0.1% and 10% level, respectively; however, we must point out that the determination of the masses improve dramatically, by a factor 10-to-100, if one includes other harmonics into the waveform [58] and not only the lowest-order quadrupole harmonic. LISA time resolution is also remarkable: the instant of black hole merger can be identified within $\sim$ 10 sec-1 hour (this depends crucially on m_1 and m_2), giving a real chance of follow-up observations with other telescopes, if one is able to pin down the source position with sufficient accuracy (this issue is discussed below). OMEGA, due to its larger instrumental noise, would provide performances worse by a factor of 2-to-10 in measuring $\mathcal{M}$, μ and t_c, depending on the mass of the binary. However it would still act as a fairly sensitive instrument for deep surveys.

The key difference between LISA and OMEGA as GW observatories lies in their angular resolutions. In order to observe, with other telescopes a region where a source is detected by using LISA/OMEGA, the "error-box" in the sky given by GW measurements should be small enough to contain, say, only a few galaxies or galaxy clusters. A crucial point in reconstructing the source position is the modulation of the signal at the detector output induced by the time-dependent antenna pattern (due to the change of orientation of the detector) and the Doppler effect produced by the motion of the instrument's centre-of-mass (this is exactly the same problem that one encounters with searches for quasi-monochromatic signals from rotating neutron stars with Earth-based detectors [59, 60]). On the Earth one can always use standard time-of-flight techniques with the LIGO/VIRGO network to pin down the source location; there is (almost) no chance to do that in space, as the typical wavelength of the radiation is $\sim$ 1 AU, which would require us to fly detectors separated by several astronomical units. The LISA orbital configuration is such that the detector sweeps the sky in a very peculiar way (see Fig. 3); OMEGA, on the contrary, has arms that are always very close to the plane of the ecliptic and all information is essentially conveyed by Doppler phase modulation[2]. Such a difference of configuration has a huge impact on the angular resolution of the instrument and the errors that affect distance measurements: for LISA, $\Delta\Omega_I^{(N)} \sim 10^{-4}$ srad and $(\Delta D_L/D_L) \sim 1\%$, though these are very crude figures as different positions and orientations of the source strongly affect both $\Delta\Omega^{(N)}$ and $(\Delta D_L/D_L)$. Note that, as was pointed out in [29], $\Delta\Omega_I^{(N)} \sim 10\,\Delta\Omega_{I+II}^{(N)}$, as the use of two detectors allows us to read out the two polarization states of the waveform independently. The situation for OMEGA is completely different, in fact, much worse. OMEGA can not locate a source with mass $\sim 10^7\,M_\odot$ at all: in fact, such a binary radiates at frequencies $\lesssim 4.4\,[10^7 M_\odot/m\,(1+z)] \times 10^{-4}$ Hz and each frequency component is shifted by $\Delta f \simeq 10^{-8}(f/10^{-4})$ Hz, which is comparable to the frequency resolution for one year of integration. The angular resolution of OMEGA increases as the mass of the binary decreases: the radiation is at higher frequencies and the Doppler effect is larger; however, $\Delta\Omega^{(N)}$ remains much worse than that for LISA. Indeed OMEGA can clearly detect a massive black hole binary, but can give very little information

[2] The OMEGA detector plane does not remain constantly in the ecliptic plane; it actually deviates from the ecliptic by several degrees, and precesses with a period of $\sim$ 7 years. But, since the precession of OMEGA is both small-amplitude and slow, as a first attempt we have assumed that the detector plane is fixed in the ecliptic plane.

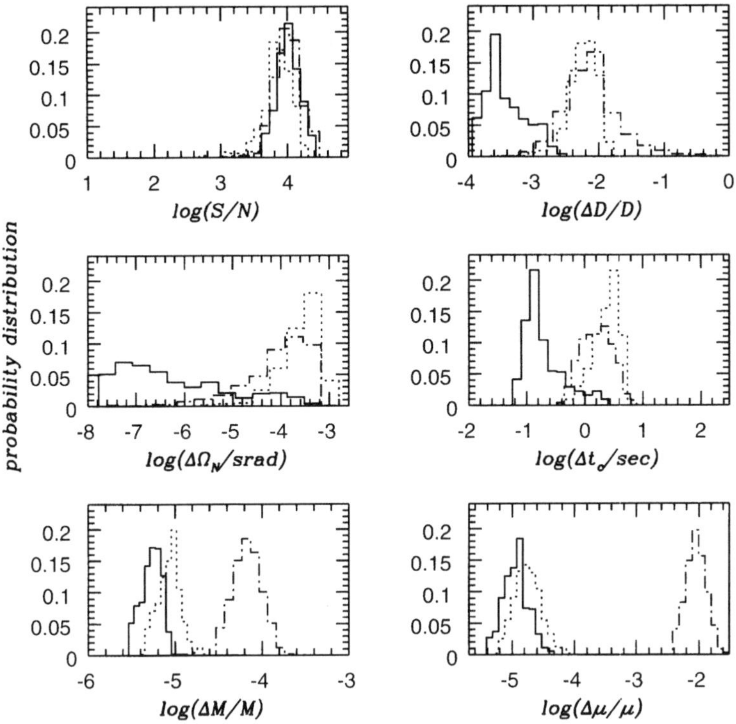

Fig. 4. Probability distribution of the signal-to-noise ratio and the errors associated to relevant source parameters for LISA observations of the final year of in-spiral of two $10^6\ M_\odot$ black holes in a binary system. The histograms show the distribution of signal-to-noise ratio (S/N), the angular resolution ($\Delta\Omega_N$) and the errors associated to the measurement of the luminosity distance ($\Delta D_L/D_L$), the time at coalescence (Δt_c), the chirp mass ($\Delta\mathcal{M}/\mathcal{M}$) and the reduced mass ($\Delta\mu/\mu$) for 1000 sources randomly distributed and oriented in the sky. All objects are at the fiducial redshift $z = 1$ ($H_0 = 75\,\mathrm{km\,s^{-1}\,Mpc^{-1}}$, $\Omega_0 = 0$, $\Lambda_0 = 0$). The plots compare the estimated errors in measurements of the parameters assuming two different physical situations: a binary undergoing simple precession, with $|\mathbf{S}|/m^2 = 0.9$ (solid line: combined observations with detector I and II; dotted line: observations with only detector I) and a binary where the BH's are not spinning, with $\beta = 0$ and $\sigma = 0$. The "raggedness" of the plots is probably due to small-number statistics

about where it is. LISA would have much better angular resolution, but still the "error box" is likely to be too large to identify the source host galaxy cluster with optical telescopes.

In the discussion up to now, we have assumed negligible BH spins for simplicity. Several observations suggest that MBH's do carry substantial intrinsic angular momentum. Indeed, a more realistic model of the GW signal should contain such information. When spins are included – in the simple precession approximation, $\widetilde{h}^{(\iota)}(f)$ depends on 12 independent parameters – the orbital plane precesses in space and the signal carries even more peculiar "finger-prints" (cf. Fig. 3), due to the stronger modulations on A_p and φ_p. In order to explore the role of black hole spins – and to consider more realistic astrophysical situations – we have performed a short numerical simulation, fixing the source distance ($z = 1$) and its mass ($m_1 = m_2 = 10^6 M_\odot$) and varying randomly the location and orientation of the binary. We have repeated this simulation by adopting two different signal models: first we consider that $|\mathbf{S}|/m^2 = 0.9$ and then that spins are negligible (which is assumed to be known a priori). In Fig. 4 we show how LISA measurements are affected. The S/N is basically the same in both cases, but the accuracy on information extraction changes drastically: if the black holes are fast spinning, all parameters are determined much more precisely than in the no-spin case. One could expect to be able to locate a source within ~ 1 arcmin^2 (the geometry of the binary system can be reconstructed through the intrinsic GW modulations caused by the precession of the source orbital plane) and to measure the distance with a $\sim 0.1\%$ error. The instant of coalescence could be determined within a second (or better); chirp and reduced mass could be estimated at the level of one part in 10^5; in fact, $\mathcal{M}$ and μ regulate not only the rate of binary in-spiral, but also the rate of precession of $\widehat{\mathbf{S}}$ and $\widehat{\mathbf{L}}$, leaving, therefore, a trace on A_p and φ_p.

Therefore, spins are likely to have a great impact on observations of massive black holes. Of course a more complete investigation is required as the parameter space is large and the results above depend strongly on $|\mathbf{S}|$ and $\widehat{\mathbf{S}} \cdot \widehat{\mathbf{L}}$.

5 Conclusions

LISA and OMEGA will be able to carry out deep surveys of massive black holes. However, due to the orbital configuration, OMEGA will be severely limited by poor angular resolution. LISA, on the contrary, will be able to measure the source parameters with high precision, expecially for black holes carrying substantial intrinsic angular momentum.

The results presented here should be regarded only as a first attempt at a much wider and more complex issue; a thorough exploration of the parameter space and more accurate GW models need to be implemented, in particular for the very interesting case – not addressed here – of high-mass ratio sources, where orbits are expected to be highly eccentric.

References

1. Einstein A. (1916): Preuss. Akad. Wiss. Berlin Sitzber. 688 (Note that the analysis was carried out by Einstein in the linearized formalism and the computed energy flux was off by a factor of two, as pointed out by Eddington in: Eddington A. S. (1924): The Mathematical Theory of Relativity. 2nd ed. Cambridge University Press, Cambridge)
2. Bondi H. (1957): Nature **179**, 1072;
Bondi H., van der Burg M.G.J., Metzner A.W.K. (1962): Proc. R. Soc. (London) A **269**, 21

3. Sachs R.K. (1962): Proc. R. Soc. (London) A **270**, 103
4. Isaacson R.A. (1968): Phys. Rev. **166**, 1263;
   Isaacson R. A. (1968): Phys. Rev. **166**, 1272
5. Hulse R.A., Taylor J.H. (1975): Astrophy. J. L **195**, 51;
   Hulse R.A. (1994): Rev. Mod. Phys. **66**, 699;
   Taylor J.H. (1994): Rev. Mod. Phys. **66**, 711
6. Wolszczan A. (1991): Nature **350**, 688
7. Will C.M. (1993): Theory and Experiment in Gravitational Physics, 2nd edition. Cambridge University Press, Cambridge;
   Will C.M. (1998): The Confrontation between General Relativity and Experiment: a 1998 update, gr-qc/9811036;
   Damour T. (1995): Gravitation and Experiments, in *Gravitation and quantizations, Les Houches, Section LVII*, ed. by B. Julisa, J. Zinn-Justin. Elsevier, Amsterdam, pp. 1–61;
   Damour T. (1997): Gravitation and Experiment, gr-qc/9711061;
   Damour T. (1999): Experimental Tests of Relativistic Gravity, gr-qc/990457
8. Thorne K.S. (1995): Gravitational Waves, in *Proc. Snowmass 95 Summer Study on Particle and Nuclear Astrophysics and Cosmology*, ed. by E.W. Kolb, R. Peccei, World Scientific, Singapore, p. 398;
   Thorne K.S. (1998): Probing Black Holes and Relativistic Stars with Gravitational waves, in *Black Holes and Relativistic Stars*, ed. by R. Wald, University of Chicago Press, Chicago, pp. 41–77
9. Abramovici A., Althouse W.E., Drever W.P. (1992): Science **256**, 325
10. Bradaschia C., Giazzotto A., Vinet J.-Y. (1990): Nucl. Instrum. Methods Phys. Res. A **289**, 518
11. Danzmann K. (1995): GEO600: A 600 meter laser interferometric gravitational wave antenna, in *Gravitational Wave Experiments*, ed. by E. Coccia, G. Pizzella, F. Ronga, World Scientific, Singapore, pp. 100–111
12. Tsubono K. (1995): 300 meter laser interferometric gravitational wave detector (TAMA 300) in Japan, in *Gravitational Wave Experiments*, ed. by E. Coccia, G. Pizzella, F. Ronga, World Scientific, Singapore, pp. 112–114
13. Cerdonio M. (1998): Cryogenic resonant detectors of gravitational waves: current operation and prospects, in *Gravitation and Relativity: At the Turn of the Millennium*, ed. by N. Dahich, J. Narlikar, IUCAA, Pune, pp. 211–230
14. Flanagan E. (1998): Sources of gravitational radiation, in *Gravitation and Relativity: At the Turn of the Millennium*, ed. by N. Dahich, J. Narlikar, IUCAA, Pune, pp. 177–198
15. Owen B.J., Lindblom L., Cutler C., Schutz B.F., Vecchio A., Andersson N. (1998): Phys. Rev. D **58**, 084020
16. Lipunov V.M., Postnov K.A. (1987): Soviet Astr. **31**, 228;
    Lipunov V.M., Postnov K.A., Prokhorov M.E. (1987): Soviet Astr. **31**, 228;
    Hils D., Bender P.L., Webbink R.F. (1990): Astrophys. J. **360**, 75;
    Bender P.L., Hils D. (1997): Class and Quantum Grav. **14**, 1439
17. Estabrook F.B., Wahlquist H.R. (1975): Gen. Rel. Grav. **6**, 439
18. Bertotti B. (1997): Gravitational wave detection from space, in *Proc. 14th Int. Conf. Gen. Rel. Grav.*, ed. by M. Francaviglia et al., World Scientific, Singapore, p. 79
19. Comoretto G., Bertotti B., Iess L., Ambrosini R. (1992): Nuovo Cim. C **15**, 1193
20. Bertotti B., Vecchio A., Iess L. (1999): Phys. Rev. D **59**, 082001
21. Bender P., Danzmann K., Schutz B.F. (1998): USA Pre-Phase A Report; Second Edition, MPQ 233
22. Hellings R.W. (1995): Ω Orbiting Medium Explorer for Gravitational Astrophysics, Mixed Proposal

23. Begelman M.C., Blandford R.D., Rees M.J. (1980): Nature **287**, 307;
Haehnelt M.G. (1994): Mont. Not. Roy. Astron. Soc. **269**, 199;
Rees M.J. (1998): Astrophysical Evidence for Black Holes, in *Black Holes and Relativistic Stars*, ed. by R. Wald, University of Chicago Press, Chicago, pp. 79–101;
Vecchio A. (1997): Class. and Quantum Grav. **14**, 1431;
Sigurdsson S., Rees M.J. (1996): Mont. Not. Royal Astron. Soc. **284**, 318;
Sigurdsson S. (1997): Class. and Quantum Grav. **14**, 1425;
Richstone D. (1998): Supermassive Black Holes Then and Now, in *Laser Interferometer Space Antenna*, ed. by W.M. Falkner, AIP Conference Proceedings **456**, pp. 41–44;
Haehnelt M.G. (1998): Supermassive Black Holes as sources for LISA, in *Laser Interferometer Space Antenna*, ed. by W.M. Falkner, AIP Conference Proceedings **456**, pp. 45-49;
Sigurdsson S. (1998): in *Laser Interferometer Space Antenna*, ed. by W.M. Falkner, AIP Conference Proceedings **456**, pp. 53–56
24. Schutz B.F. (1986): Nature **232**, 675
25. Ciufolini I. et al. (1998): Science **279**, 2100
26. Blaes O. (1998): General Relativity as seen in X-rays: what can LISA tell us that we don't already know, in *Laser Interferometer Space Antenna*, ed. W.M. Falkner, AIP Conference Proceedings **456**, pp. 57–60
27. Peterseim M., Jennrich O., Danzmann K. (1997): Class. and Quantum Grav. **13**, 279
28. Poisson E. (1996): Phys. Rev. D **54**, 5939
29. Cutler C. (1998): Phys. Rev. D **57**, 7089
30. Cutler C., Vecchio A. (1998): LISA angular resolution for monochromatic sources, in *Laser Interferometer Space Antenna*, ed. by W.M. Falkner, AIP Conference Proceedings **456**, pp. 95–100
31. Vecchio A., Cutler C. (1998): LISA: parameter estimation for massive black hole binaries, in *Laser Interferometer Space Antenna*, ed. by W.M. Falkner, AIP Conference Proceedings **456**, pp. 101–109
32. Vecchio A. (1998): Directional Information from LISA Observations, in *Third Gravitational Waves Data Analysis Workshop*, ed. by S. Finn, available at http://fiji.nirvana.phys.psu.edu/gwdaw98/Proceedings/Vecchio/index.html
33. Schutz B.F. (1997): Low Frequency Sources of Gravitational Waves, in *Proc. of the 1997 Alpbach Summer School on Fundamental Physics in Space*, ed. by A. Wilson, ESA and preprint AEI-044, pp. 229–240
34. Rees M.J. (1993): Proc. Nat. Acad Sci. **90**, 4840
35. Haehnelt M., Rees M.J. (1993): Mont. Not. Royal Astr. Soc **263**, 168
36. Loeb A., Rasio F.A. (1994): Astrophys. J. **432**, 52
37. Zel'dovich Y.B., Novikov I.D. (1964): Sov. Phys. Dokl. **158**, 811
38. Salpeter E.E. (1964): Astrophys. J. **204**, L1
39. Blandford R.D. (1990): Physical processes in Active Galactic Nuclei, in *Active Galactic Nuclei*, ed. by T. Courvoisier, M. Mayor, Springer, Berlin Heidelberg New York, pp. 161–270
40. Osterbrock D.E. (1993): Astrophys. J. **404**, 551
41. Rees M.J. (1990): Science **247**, 817
42. Richstone D. et al. (1998): Nature A **395**, 14
43. Miyoshi M., Moran J., Herrnstein J., Greenhill L., Nakai N., Diamond P., Inoue M. (1995): Nature **373**, 127
44. Eckart A., Genzel R. (1996): Nature **383**, 415
45. Maoz E. (1996): Astrophys. J. L **447**, 91;
Maoz E. (1998): Astrophys. J. L **494**, 181

46. Barnes J.E., Hernquist L. (1992): Ann. Rev. Astron. Astrophys. **30**, 705
47. Fritze-von Alvensleben U. (1996): Interacting and Merging Galaxies, in *From Stars to Galaxies*, ed. by C. Leitherer, U. Fritze-von Alvensleben, J. Huchra, ASP Conference Series **98**, p. 496
48. Roos N. (1981): Astron. Astrophys. **104**, 218
49. Gaskell C.M. (1996): Supermassive Binaries and Extragalactic Jets, in *Jets from Stars and Galaxies*, ed. by W. Kundt, Springer, Berlin, pp. 165–196
50. Ryan F.D. (1997): Phys. Rev. D **56**, 1845
51. Apostolatos T.A., Cutler C., Sussman G.S., Thorne, K.S. (1994): Phys. Rev. D **49**, 6274
52. Blanchet L., Damour T., Iyer B.R., Will C.M., Wiseman A.G. (1995): Phys. Rev. Lett. **74**, 3515
53. Cutler C., Flanagan E.E. (1994): Phys. Rev. D **49**, 2658
54. Cutler C., Vecchio A., Observing massive black hole binaries with LISA, in progress
55. Balasubramanian R., Sathyaprakash B.S., Dhurandhar S.V. (1996): Phys. Rev. D **53**, 3033
56. Nicholson D., Vecchio A. (1998): Phys. Rev. D **57**, 4588
57. Flanagan E., Hughes S. (1998): Phys. Rev. D **57**, 4535
58. Sintes A., Vecchio A.: LISA observations of massive black hole binaries using post-Newtonian waveforms, to appear
59. Schutz B.F. (1989): Data processing, analysis, and storage for interferometeric antennas, in *The Detection of Gravitational Waves*, ed. by D. Blair, Cambridge University Press, Cambridge, pp. 406–427
60. Brady P.R., Creighton T., Cutler C., Schutz B.F. (1998): Phys. Rev D **57**, 2101

Bar Mode Instability in Relativistic Rotating Stars

S. Zane

Abstract. Rapidly spinning neutron stars, with angular velocity higher than a given threshold, may become secularly unstable to an $l = m = 2$ mode and deform into a bar configuration. This instability sets in only in the presence of a suitable dissipative mechanism, such as viscosity or gravitational wave emission, and in Newtonian theory its point of onset does not depend on the nature of dissipation. The first numerical investigations, however, seem to suggest that, in general relativity, the instability is weakened when driven by viscous dissipation and strengthened when driven by gravitational radiation.
After a review of the physical problem and of previous investigations, we schematically present an analytic treatment carried out in the framework of post–Newtonian (PN) gravitation. In this analysis, we model rotating stars by homogeneous, rigidly rotating, triaxial ellipsoids employing an energy variational principle to construct relativistic equilibrium sequences and locate the point of onset of the bar mode instability along each sequence. The spacetime metric is obtained by solving Einstein's equations of general relativity in 3+1 ADM form, and we focus on the viscosity-driven instability.
We find that the value of the eccentricity, as well as related ratios like $\Omega^2/(\pi\varrho_0)$ and $T/|W|$ (= rotational kinetic energy / gravitational potential energy), all increase at the onset of instability as the star becomes more relativistic. Since higher degrees of rotation are required to trigger a viscosity-driven bar mode instability as the star becomes more compact, the effect of general relativity is to weaken the instability, even to PN order.

1 Introduction

The identification of nonaxisymmetric modes of instability in rapidly rotating equilibrium configurations is a classical problem, which has been widely studied in Newtonian theory [1] and only recently has been readdressed in the context of general relativity (GR). The motivations are manifold, from both the theoretical and the observational points of view. The issue is closely related to the problem of determining an upper limit for the spin velocity of a relativistic star, and the study of oscillations and pulsations may yield important information about both the global properties and the internal composition of the star. Moreover, rotating configurations subject to nonaxisymmetric instabilities could become important sources of emission of gravitational waves (GWs) and represent possible candidates for detection by laser interferometers like LIGO, GEO and VIRGO [2–5].

A rough estimate of the gravitational luminosity emitted by a star of mass M, mean radius R, and internal velocity $\sim V$ can be obtained from the quadrupole expression [6]

$$L \sim \frac{c^5}{G} s^2 \left(\frac{R_s}{R}\right)^2 \left(\frac{V}{c}\right)^6 \sim 3.6 \times 10^{52} s^2 \left(\frac{R_s}{R}\right)^2 \left(\frac{V}{c}\right)^6 \text{ W}, \tag{1}$$

where R_s is the Schwarzschild radius and s is an asymmetry factor which vanishes for spherical objects. Since, for typical values of R/R_s and V/c, neutron stars (NSs) turn out to be valuable candidates for such emission, the crucial point consists in quantifying the parameter s by studying deformations of the stellar shape.

As is well-known, a rigidly rotating, axisymmetric body does not emit GWs; thus a star must deviate from axisymmetry in order to have emission. In general, symmetry breaking can arise either spontaneously or due to an external process and, for most kinds of instability, spontaneous breaking sets in when the ratio $T/|W|$ between the kinetic and the gravitational potential energies exceeds a given threshold.

There are a number of plausible astrophysical situations in which this effect can set in. Numerical simulations of coalescing binary NSs show that the most probable end point of close binary evolution is merger into a single object [7–13]. The new configuration may be rapidly rotating and might be driven secularly unstable [14]. Young NSs created after a supernova explosion may accrete part of the material which is not ejected by the shock wave, increasing their kinetic energy as well. Phases characterized by substantial emission of GWs are also expected during core collapse in massive stars, accretion-induced collapse of white dwarfs [4] or coalescence of white dwarf binaries [15–17]. In all these cases, the instability arises in a young, newly born configuration. Alternatively, stationary accretion onto NSs in X-ray binary systems or in Thorne–Zytkow objects can spin up the compact star until the critical value of $T/|W|$ is reached, at which point the instability starts to power radiation. This mechanism, known as "forced gravitational emission", is particularly appealing for its capability to maintain steady, periodic emission, with the total amount of accreted angular momentum balancing that radiated away via GWs [5, 18].

In all these scenarios, independent of the evolutionary state of the NS, when the spin velocity takes or becomes larger than the critical value, the star can undergo a secular or dynamical evolution, breaking its axisymmetry. One of the most important differences is related to the temperature of the star and to the efficiency of viscous dissipation in its interior: since a newly born object is hotter than an accreting one, the mechanism responsible for the instability is different in the two cases.

From the theoretical point of view, the problem therefore consists in identifying, among all possible linear perturbations, the most relevant for the growth of the instabilities and in establishing suitable criteria which allow us to determine the point of onset of the instability.

In the rest of this paper we focus our attention on the $l = m = 2$ polar mode. In Sect. 2 we review the physical problem and previous investigations. In Sect. 3 we briefly report the main lines of an analytic formalism, recently presented by Shapiro & Zane 1998, [19], which allows us to locate the point of onset of a given instability along an equilibrium sequence. The approach consists in the generalization to post–Newtonian (PN) gravitation of an energy variational principle earlier introduced by LRS in Newtonian gravity. Results and discussion will follow.

2 Previous investigations

As is well-known, in the Newtonian framework the problem of the equilibrium shape of a rigidly rotating, incompressible body admits an exact analytical solution [1]. On these assumptions, a self-gravitating configuration of density ϱ_0 takes the shape of a Maclaurin spheroid, which is an axisymmetric ellipsoid characterized by a fixed relation between angular velocity, Ω, and eccentricity, e (Fig. 1),

$$\left(\frac{\Omega^2}{\pi \varrho_0}\right)_{\text{Mac}} = \frac{2\left(1-e^2\right)^{1/2}}{e^3}\left(3-2e^2\right)\sin^{-1} e - \frac{6}{e^2}\left(1-e^2\right). \tag{2}$$

These solutions, however, are not stable in the full parameter space. At the critical value $T/|W| = 0.27$, through the $l = 2$ mode an instability induced by hydrodynamical and gravitational effects grows in a dynamical time scale and rapidly deforms the star into a bar configuration. In addition, at lower velocities ($T/|W| = 0.1375$) the Maclaurin sequence bifurcates into two other branches of triaxial equilibria, the Jacobi and the Dedekind ellipsoids. The bifurcation point is dynamically stable, but, in the presence of a suitable dissipative mechanism such as viscosity or gravitational radiation, it becomes secularly unstable to an $l = m = 2$ bar mode. Secular instability, therefore, may play an important role in limiting the maximum rotation of NSs.

The evolution of a rapidly spinning Maclaurin spheroid towards one or the other triaxial sequence depends on the dissipative mechanism which drives the instability.

Viscosity dissipates mechanical energy E, but preserves angular momentum J. This suggests that a Maclaurin spheroid undergoing a viscosity-driven instability evolves toward a Jacobi ellipsoid, which is characterized by a lower value of E but the same value of J (and rest mass M_0). The Jacobi state is rigidly rotating, so viscous dissipation stops once it is formed. On the other hand, a rapidly spinning configuration can be also unstable to the emission of GWs [20, 21]. Since this process does not conserve J but conserves circulation, $\mathcal{C}$, under a CFS instability the growth of the bar mode leads to the deformation of a Maclaurin spheroid into a Dedekind ellipsoid, which is characterized by a lower E for a given $\mathcal{C}$. At this final state, which is stationary, the emission of GWs stops.

It is worth noticing that the relative importance of the two evolutionary scenarios, in representing possible mechanisms for the generation of GWs from rotating NSs, is not obvious. In fact, the Jacoby configuration is not stationary. Once it is created, it loses energy and angular momentum through emission of GWs until the critical point is reached and the axisymmetry is restored. As pointed out by Bonazzola and Gourgoulhon 1996 [22], the Jacobi-like instability may represent the most interesting process for the emission of gravitational radiation.

In general, the competition between the Jacobi-like and the Dedekind-like modes is governed by the ratio of the strengths of the viscous stress to the gravitational radiation reaction force, which depends crucially on the internal properties of the star. The determination of a realistic equation of state (EOS) for NSs is still an open issue, and a number of expressions for the relevant timescales associated with bulk

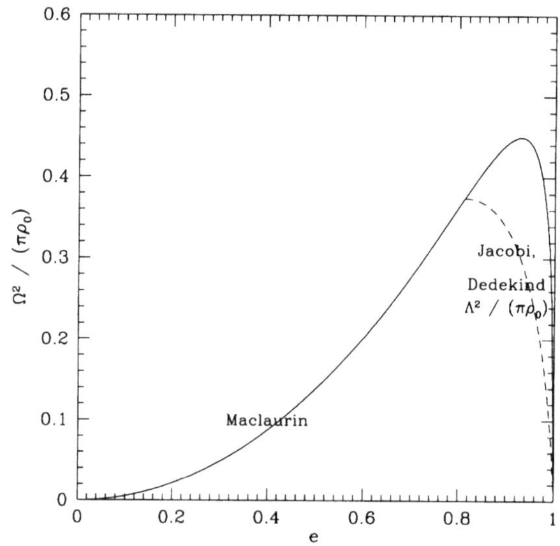

Fig. 1. Angular velocity versus eccentricity along the Maclaurin (solid line), Jacobi and Dedekind sequences (*dashed line*)

and shear viscosity have been presented in the literature (for a recent treatment see [23]). Broadly speaking, in the present state of knowledge, viscosity is thought to be more important for old NSs, and gravitational radiation in the hottest, newly born objects. The situation is more complicated when realistic models for the interior of the NS are considered [4, 24–26] or when both viscosity and gravitational radiation act together, since they tend to cancel each other [4, 27].

Much effort in Newtonian physics has been devoted to the extension of Chandrasekhar's results to more realistic, compressible fluids [1] (see, e.g., [9, 25, 28–35]. Two results are particularly important for the present discussion. First, it has been found that, for rigidly rotating polytropes, the bifurcation point can only exist when the adiabatic index exceeds a critical value, $\gamma_{crit} = 2.238$ [36–38]. This is because polytropic sequences have a natural end point at the so-called mass shedding limit, where the centrifugal force balances the gravitational force at the equator, and the EOS must be stiff enough to make the angular velocity at the bifurcation point lower than this limit. Moreover, along uniformly rotating, polytropic sequences the $m = 2$ Jacobi-like bifurcation point and the $m = 2$ Dedekind-like point coincide, as in the incompressible case [4, 9, 39].

A powerful tool for modelling rotating polytropic stars in Newtonian gravity was developed by [9], who presented equilibrium sequences for compressible analogues of most classical incompressible objects. The technique is based on an ellipsoidal energy variational principle, and is essentially a generalization to the triaxial case of an approach earlier introduced by Zel'dovich & Novikov 1971 [40], (see also [41]) in the axisymmetric case, to investigate stability against gravitational collapse. The main advantage of this approach comes from its simplicity: the method deals directly with global, conserved quantities and all results are analytic or quasi–analytic, lending themselves to straightforward physical interpretation.

Although many results are well-established for Newtonian configurations, much less is known in the context of full GR. In the axisymmetric case, the structure of a relativistic, rotating star has been investigated numerically by a number of authors [42–47], but the first fully relativistic computations of nonaxisymmetric instabilities have been presented only very recently. Three dimensional (3D) perturbation computations to identify nonaxisymmetric instabilities driven by emission of GWs have been carried out by Stergioulas & Friedman [48, 49] and Stergioulas [49], while a numerical investigation of the effects of GR on the viscosity-driven bar mode instability was presented by Bonazzola, Frieben & Gourgoulhon [25] and Bonazzola, Frieben & Gourgoulhon [50]. Stergioulas & Friedman [48] solved the coupled set of perturbed field equations, using a polytropic EOS and adopting the criterion of Friedman & Schutz [51] to determine the point of onset of the instability. According to this criterion, a nonaxisymmetric mode becomes unstable when its frequency, as measured by an observer at infinity, vanishes. They find that relativistic models are unstable for smaller degrees of rotation than for corresponding Newtonian models (for a PN analysis see also [52, 55]). The destabilizing effect of relativity is most striking in the case of the $m = 2$ mode, which can become unstable even for soft polytropes of index $\gamma \geq \gamma_{\rm crit} = 1.77$.

However, as noted by Stergioulas and Friedman [48], the results of Bonazzola, Frieben and Gourgoulhon [25] seem to suggest the opposite effect. By investigating the $m = 2$, viscosity-driven bar mode, these authors found that the critical adiabatic index becomes higher than the value of James as the relativistic character of the configuration is increased. This suggests that relativistic effects tend to stabilize the configurations. A possible explanation is that GR acts in a different way on the two instabilities, so that the viscosity-driven and the gravitational radiation-driven $m = 2$ modes may no longer coincide as they do in Newtonian theory. On the other hand, both Stergioulas' and Friedman's finding [48] and Bonazzola's, Frieben's &Gourgoulhon's [25] findings are based on a numerical solution and represent the first attempts to solve the relativistic problem in 3D. For these reasons we have recently reconsidered the problem from an analytic point of view, in PN approximation. PN effects on the equilibrium of uniformly rotating, homogeneous bodies have been studied in a series of paper by Chandrasekhar (see [19]) using the tensor virial formalism, and Bardeen 1971 [56]. However, no analytic investigations of the location of the secular stability point in GR were provided in these earlier studies or, to our knowledge, elsewhere in the literature. By generalizing the variational principle introduced by Lai, Rario and Shapiro [9], we developed a formalism which allows us to construct analytic models of incompressible, rigidly rotating stars and to evaluate their stability against nonaxisymmetric bar modes. We neglect any deviation from the ellipsoidal shape, and present the first results for the $l = m = 2$ polar mode, viscosity-driven instability.

3 A PN treatment

The energy variational principle introduced by Lai, Rasio and Shapiro [9] is based on a simple scheme. Consider a Newtonian, self-gravitating, isolated system. Each configuration, in equilibrium or not, can be specified by the total energy E and a number of global quantities, such as M_0, J, ... On the other hand, E can always be written in terms of the density and velocity fields $\varrho(\mathbf{x})$, $v(\mathbf{x})$; thus the equilibrium state can be determined by keeping the global quantities constant and extremizing E with respect to all variations in both $\varrho(\mathbf{x})$ and $v(\mathbf{x})$. Direct application of such an approach to a multidimensional system is a computationally challenging task, but a great simplification arises in the presence of symmetries when the infinite number of degrees of freedom contained in $\varrho(\mathbf{x})$ and $v(\mathbf{x})$ can be replaced by a limited number of free parameters $\alpha_1, \alpha_2, \ldots$. In this case, the equilibrium state can be determined by extremizing E according to

$$\frac{\partial}{\partial \alpha_i} E\left[\alpha_1, \alpha_2, \ldots; M_0, J, \ldots\right] = 0, \qquad i = 1, 2 \ldots \qquad \text{(equilibrium)} \quad (3)$$

under the constraint that M_0, J, ... are conserved. Along the equilibrium sequence, the point of onset of an instability can then be determined from

$$\det\left(\frac{\partial^2 E}{\partial \alpha_i \partial \alpha_j}\right)_{eq} = 0, \quad i, j = 1, 2, \ldots \qquad \text{(onset of instability).} \qquad (4)$$

Clearly, the kind of instability which is identified is that driven by a dissipative mechanism which preserves the conservation laws assumed in the construction of the equilibrium model. Accordingly, the instability triggered by viscosity can be located by assuming uniform rotation and performing all variations while holding fixed M_0 and J (but not circulation).

In order to apply the same method in PN gravity, we need to construct a variational expression for the total mass–energy M of a rotating ellipsoid. A major complication arises because, in GR, evaluation of all the integral quantities M, M_0, J requires a knowledge of the metric, which must be determined self–consistently with the matter profile by solving the Einstein field equations (EFE). Following the discussion by Wilson & Matthews [57, 58] and Wilson [59], Cool, Shapiro and Tendsky [44] provided a simplified set of EFE based on a 3+1 ADM splitting of the metric [60]. This form, which is exact to PN order, is widely used in numerical investigations and is particularly well-suited to studying quasi-equilibrium systems, i.e., systems with dynamical timescale much shorter than the timescale for emission of GWs [14, 61]. Examples include not only weak field, slow velocity systems, but also strong field, high velocity systems that are only slightly perturbed from stationary equilibrium. At present, numerical relativity in 3D is only in its infancy, and in particular lacks a large body of known solutions which it can attempt to reproduce. Our solution is derived by solving the ADM equations [60] to PN order; thus, in this respect, it should provide an important test-bed calculation for numerical studies of 3D relativistic rotating configurations.

We consider an isolated, self–gravitating, homogeneous ellipsoid with semiaxis a_1, a_2, a_3, rest mass density ϱ_0, uniformly rotating with angular velocity Ω parallel to the x_3 axis. By choosing a conformally flat decomposition of the spatial part, the most general expression for the metric in a 3+1 form is [19, 47]

$$ds^2 = -\alpha^2 dt^2 + \Psi^4 f_{ij}\left(dx^i + \beta^i dt\right)\left(dx^j + \beta^j dt\right), \tag{5}$$

where Ψ, α and β^i are respectively the conformal factor, the lapse and the shift functions, while f_{ij} is the Euclidean metric. The three functions α, β^i, Ψ can be derived as solutions of a system of non–linear, partial differential equations (PDEs) [see Eqs. (8), (14) and (18) in [47]]. However, as demonstrated in [19], to PN order these equations can be reduced to a system of second order, elliptic PDEs which can be solved by exploiting the corresponding Green functions and, under the assumptions at hand, admit an analytic solution (up to well-known elliptic integrals). The solution for the metric can then be used in the expressions for the total baryon rest mass M_0 ([44], Eq. (50)), the ADM mass M and the total angular momentum J [62]

$$M_0 = \varrho_0 \int_V \left(\alpha U^t\right) \Psi^6 d^3x, \tag{6}$$

$$M = -\frac{1}{2\pi}\oint_{S_\infty} \nabla^i \Psi d^2 S_i = -\frac{1}{2\pi}\int_V \nabla^2 \Psi d^3x. \tag{7}$$

$$J_3 = \frac{1}{16\pi}\epsilon_{3jk}\oint_{S_\infty}\left(x^j K^{km} - x^k K^{jm}\right) d^2 S_m = \int \Psi^{10}\left(x^1 S^2 - x^2 S^1\right) d^3x, \tag{8}$$

where U^ν, S^i are the fluid four-velocity and the momentum source (see [19] for the complete notation). Again, after some complicated algebra, it is possible to demonstrate that all integrals appearing in Eqs. (6), (7), (8) can be evaluated analytically (up to elliptic integrals). This yields three expressions for the functionals M, M_0, J that only depend on ϱ_0, Ω and the two axial ratios $\lambda_i = (a_3/a_i)^{2/3}$, $i = 1, 2$. The final step consists in performing the first and second variations and, due to the complexity of the calculations, was partially carried out by using an algebraic manipulator (MAPLE). Results, however, can be expressed in a simple form.

Each sequence of equilibrium can be parametrized by the value M/a_1, which is related to the compaction parameter M/R_s (R_s is the Schwarzschild radius) of the spherical, nonrotating member. These configurations yield a PN generalization of Maclaurin spheroids and were originally investigated by Chandrasekhar [63], by using the tensor virial formalism. Sequences of relativistic, numerical models at fixed $M_0\sqrt{\varrho_0}$ have been also published by Butterworth and Ipser [42]. Following the same notation as in [63], the equilibrium relation can be written as

$$\frac{\Omega^2}{\pi\varrho_0} = \left(\frac{\Omega^2}{\pi\varrho_0}\right)_{\text{Mac}} + \frac{2M_c}{a_1}E(e) \tag{9}$$

where $E(e)$ is a function of the eccentricity and the parameter $2M_c/a_1$ measures the strength of relativity. The first term on the right-hand side formally coincides with the Newtonian expression (2), although it must be noted that the eccentricity which enters in the PN formalism (and in [63]) is defined in terms of the ratio of the coordinate quantities a_3 and a_1 and is not a gauge-invariant parameter.

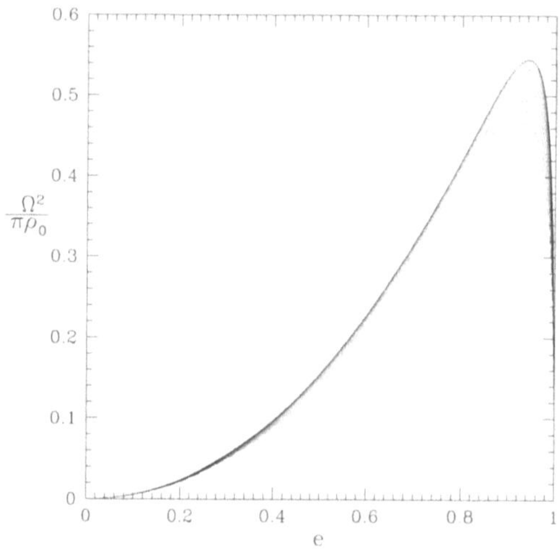

Fig. 2. The ratio $\Omega^2/(\pi\varrho_0)$ plotted again the eccentricity for PN equilibrium sequences of constant rest mass (solid lines). The different curves correspond to 11 equally spaced values of the compaction parameter M/R_s in the range [0., 0.275]. The dashed line is the Newtonian Maclaurin sequence ($M/R_s = 0$). From [19]

The function $E(e)$ was compared with the corresponding function derived by [63], and turned out to be in close agreement, with a maximum fractional error of a few percent up to $e \sim 0.8$. A comparison with numerical results by Butterworth and Ipser [42] also gives fractional differences of at most 10%. Equilibrium sequences are reported in Fig. 2 for different values of M/a_1. As can be seen, the effect of general relativity is to attribute to a star of a given eccentricity a larger value of $\Omega^2/(\pi\varrho_0)$.

The condition for stability follows from the second derivatives and gives

$$\frac{\Omega^2}{\pi\varrho_0} = 2B_{11} + \frac{2M_c}{a_1}C(e), \tag{10}$$

where, again, the leading term $2B_{11}$ formally coincides with the Newtonian expression and $C(e)$ is a function of the eccentricity.

Finally, the value of the eccentricity at the secular instability point, $e_{\rm sec}$, is calculated by equating the two expressions (9) and (10) and is reported in Table 1.

Figure 3 shows the relation between the two gauge-invariant quantities $\Omega^2/\pi\varrho_0$ and $T/|W|$, for the same equilibrium sequences as in Fig. 2. Squares mark the secular instability points, and thus separate the regions of stable and unstable configurations.

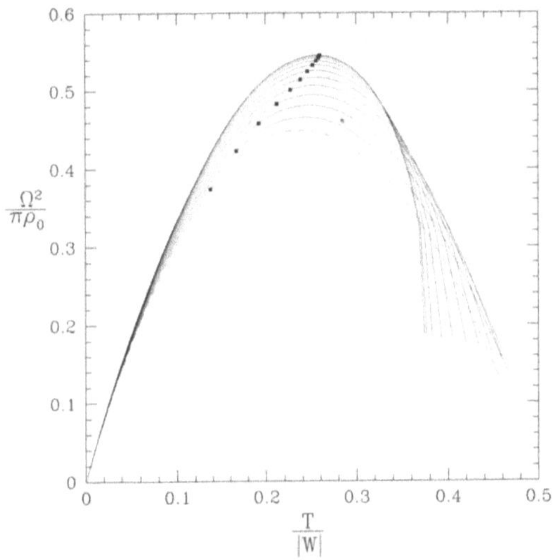

Fig. 3. The ratio $\Omega^2/(\pi\varrho_0)$ versus $T/|W|$ for the same sequences as in Figure 2 (solid lines); the dashed line is the Newtonian Maclaurin sequence. Squares mark the secular instability point. From [19]

Table 1. Location of the PN secular Instability Point

$\frac{M}{R_s}$	e_{sec}	$\left(\frac{\Omega^2}{\pi\varrho_0}\right)_{\text{sec}}$	$\left(\frac{T}{\lvert W\rvert}\right)_{\text{sec}}$	$\frac{M}{R_s}$	e_{sec}	$\left(\frac{\Omega^2}{\pi\varrho_0}\right)_{\text{sec}}$	$\left(\frac{T}{\lvert W\rvert}\right)_{\text{sec}}$
0.[a]	0.8127	0.3742	0.1375	0.150	0.9325	0.5247	0.2458
0.025	0.8561	0.4230	0.1673	0.175	0.9372	0.5327	0.2516
0.050	0.8850	0.4582	0.1924	0.200	0.9403	0.5384	0.2554
0.075	0.9042	0.4831	0.2120	0.225	0.9425	0.5424	0.2578
0.100	0.9171	0.5011	0.2268	0.250	0.9436	0.5446	0.2591
0.125	0.9261	0.5144	0.2378	0.275	0.9441	0.5457	0.2597

[a] = Newtonian limit

As we can see, the effect of general relativity is to move the instability point to an eccentricity larger than which Newtonian theory predicts. This also corresponds to larger values of the two ratios $\Omega^2/(\pi\varrho_0)$ and $T/|W|$, showing, therefore, that relativistic gravitation tends to stabilize a star against secular instability to a Jacobi-like, nonaxisymmetric bar mode.

4 Discussion

We presented here the main lines of a new PN formalism devoted to the determination of the point of onset of nonaxisymmetric instabilities for homogeneous, rigidly rotating ellipsoidal configurations. In agreement with previous findings, we found that the rotational velocity along the PN equilibrium sequence is not much different from that which is derived in the Newtonian limit, although a star of a given eccentricity has a slightly larger value of $\Omega^2/(\pi \varrho_0)$. The discrepancies between our results and those of by [63] may be ascribed mainly to our coordinate ellipsoidal approximation for the deformation and the equilibrium shape. This approximation is only exact in the Newtonian limit, altough it turns out to be quite robust to PN order. On the other hand, this is the key approximation that allows us to carry out an analytic investigation of stability properties. We would like to mention that, more recently, the validity of the ellipsoidal approximation has been tested successfully by Taniguchi, Asada & Shibata [64] and Taniguchi [65] in the context of irrotational, isolated stars and compressible, irrotational binaries respectively.

We found that GR is most crucial in influencing stability properties. According to our PN treatment, the critical value of the eccentricity for the onset of the viscosity–driven bar instability increases as the star becomes more relativistic. Both invariant ratios, $\Omega^2/(\pi \varrho_0)$ and $T/|W|$, also increase at the onset of instability above the Newtonian values. Our results show the presence of a stabilizing effect due to GR on the Jacobi-like bar mode instability, in agreement with the Bonanzzola, Frieben and Gourgoulhon [26] and Bonazzola, Frieben & Gourgoulhon [50] analyses. We also confirm a reverse behaviour between the viscosity-driven and the CFS instability investigated by Stergioulas and Friedman [49]. The Stergioulas' and Friedman's calculations are fully relativistic, so this bimodal behaviour may be explained in terms of strong field effects. However, although at present no firm conclusion can be reached, there is still the likelihood that, in GR, the viscosity-driven and the gravitational radiation-driven $m = 2$ modes may no longer coincide and that the gravitational field plays a different role in each of them.

References

1. Chandrasekhar S. (1969): Ellipsoidal Figures of Equilibrium. Yale Univ. Press, New Haven
2. Bonazzola S., Marck J.A. (1994): Ann. Rev. Nucl. Part. Sci. **45**, 655
3. Thorne K.S. (1987): Gravitational Radiation, in *300 Years of Gravitation*, ed. by S.W. Hawking, W. Israel, Cambridge University Press, Cambridge
4. Lai D., Shapiro S.L. (1995): ApJ **442**, 259
5. Schutz B.F. (1997): in *Mathematics of Gravitation. II. Gravitational Wave Detection*. ed. by A. Krolak (Banach Center, Warsaw), to appear
6. Blanchet L. (1996): in *Les Houches School "Astrophysical Sources of Gravitational Radiation" (Les Houches, France, Sept. 26–Oct. 6, (1995)*. ed. by J.-A. Marck, J.-P. Lasota
7. Rasio F.A., Shapiro S.L. (1992): ApJ **401**, 226
8. Rasio F.A., Shapiro S.L. (1994): ApJ **432**, 242
9. Lai D., Rasio F.A., Shapiro S.L. (1993a): ApJs **88**, 205
10. Lai D., Rasio F.A., Shapiro S.L. (1993b): ApJ **406**, L63
11. Lai D., Rasio F.A., Shapiro S.L. (1994a): ApJ **420**, 811
12. Lai D., Rasio F.A., Shapiro S.L. (1994b): ApJ **423**, 344
13. Lai D., Rasio F.A., Shapiro S.L. (1994c): ApJ **437**, 742

14. Baumgarte T.W., Cook G.B., Scheel M.A., Shapiro S.L., Teukolsky S.A. (1997): PRD submitted
15. Iben I., Tutukov A.V. (1984): ApJs **54**, 335
16. Yungelson L.R., Livio M., Tutukov A.V., Saffer R.A. (1994): ApJ **420**, 336
17. Chen K., Leonard P.J.T. (1993): ApJ **390**, 486
18. Wagoner R.V. (1984): AJ **278**, 345
19. Shapiro S.L., Zane S. (1998): ApJs **117**, 531
20. Chandrasekhar S. (1970): PRL **24**, 611
21. Friedman J.L., Schutz B.F. (1978): ApJ **222**, 281
22. Bonazzola S., Gourgoulhon E. (1996): Proceedings of the Les Houches School "Astrophysical Sources of Gravitational Radiation" ed. by J.-A. Marck, J.-P. Lasota, Les Houches, France, Sept. 26–Oct. 6, 1995
23. Kokkotas K., Stergioulas N. (1999): A&A **341**, 110
24. Lindblom L., Mendell G. (1994): ApJ **421**, 689
25. Bonazzola S., Frieben J., Gourgoulhon E. (1996): ApJ **460**, 379
26. Andersson N., Kokkotas K.D., Stergioulas N. (1999): ApJ **516**, 307
27. Lindblom L., Detweiler S.L. (1977): ApJ **444**, 804
28. Ostriker J.P., Tassoul J.-L. (1969): ApJ **155**, 987
29. Bodenheimer P., Ostriker J.P. (1973): ApJ **180**, 159
30. Durisen R.H. (1975): ApJ **199**, 179
31. Tassoul J.-L. (1978): Theory of Rotating Stars. Princeton University Press, Princeton
32. Ipser J.R., Managan R.A. (1981): ApJ **250**, 362
33. Hachisu I., Eriguchi Y. (1982): Prog. Theor. Phys. **68**, 206
34. Hachisu I. (1986): ApJs **61**, 479
35. Hachisu I. (1986b): ApJs **62**, 461
36. Jeans J.H. (1919): Problems of Cosmogony and Stellar Dynamics. Cambridge University Press, Cambridge
37. Jeans J.H. (1928): Astronomy and Cosmogony. Cambridge University Press, Cambridge
38. James R.A. (1964): ApJ **140**, 552
39. Ipser J.R., Managan R.A. (1985): ApJ **292**, 517
40. Zel'dovich Y.A.B., Novikov I.D. (1971): Relativistic Astrophysics, Vol. 1. Chicago University Press, Chicago
41. Shapiro S.L., Teukolsky S.A. (1983): Black Holes, White Dwarfs and Neutron Stars. Wiley, New York
42. Butterworth I.M., Ipser J.R. (1976): ApJ **204**, 200
43. Friedman J.L., Ipser J.R., Parker L. (1986): ApJ **304**, 115
44. Cook G.B., Shapiro S.L., Teukolsky S.A. (1992): ApJ **398**, 203
45. Cook G.B., Shapiro S.L., Teukolsky S.A. (1994a): ApJ **422**, 227
46. Cook G.B., Shapiro S.L., Teukolsky S.A. (1994b): ApJ **424**, 823
47. Cook G.B., Shapiro S.L., Teukolsky S.A. (1996): PRD **53** 10, 5533
48. Stergioulas N., Friedman J.L. (1998): ApJ **492**, 301
49. Stergioulas N. (1997): Structure and stability of rotating relativistic stars. Ph.D. Thesis, University of Wisconsin–Milwaukee
50. Bonazzola S., Frieben J., Gourgoulhon E. (1998): A&A **331**, 280
51. Friedman J.L., Schutz B.F. (1995): ApJ **200**, 204
52. Cutler C. (1991): ApJ **374**, 248
53. Cutler C., Lindblom L. (1992): ApJ **385**, 630
54. Lindblom L. (1995): ApJ **438**, 265
55. Yoshida S., Eriguchi Y. (1997): ApJ **490**, 779

56. Bardeen J.M. (1971): ApJ **167**, 425
57. Wilson J.R., Matthews G.J. (1989): Frontiers in Numerical Relativity. Cambridge University Press, Cambridge
58. Wilson J.R., Matthews G.J. (1995): PRL **75**, 4161
59. Wilson J.R. (1990): in *Texas Symposium on 3-Dimensional Numerical Relativity*, ed. by R.A. Matzner, University of Texas, Austin, Texas
60. Arnowitt R., Deser S., Misner C.W. (1962): The Dynamics of General Relativity, in *Gravitation: An Introduction to Current Research*. ed. by L. Witten, J. Wiley, New York
61. Baumgarte T.W, Cook G.B., Scheel M.A., Shapiro S.L., Teukolsky S.A. (1997): PRL **79**, 1182
62. Bowen J.M., York J.W.Y. (1980): PRD **21** (8), 2047
63. Chandrasekhar S. (1965): ApJ **142**, 1513
64. Taniguchi K., Asada H., Shibata M. (1998): Irrotational and Incompressible Ellipsoids in the First Post-Newtonian Approximation of General Relativity, gr-qc/9809039
65. Taniguchi K. (1999): http://xxx.lanl.gov/abs/gr-qc/9901048; Prog. Theor. Phys. in press

Recent Results

General Relativity and Theory of Gravitation

Theoretical Relativity: Classical Topics

F. Giannoni

List of the talks

[1] M. Palese, R. A. Leo, G. Soliani: The prolongation problem for the heavenly equation.
[2] S. Jhingan, G. Magli: Gravitational collapse of fluid bodies and cosmic censorship: analytic insights.
[3] P. Piccione: Time extremizing trajectories of massive and massless objects in general relativity.
[4] L. Martina: black holes and solitons.
[5] A. Spallicci: Analytic solution of the Regge–Wheeler differential equation for black hole perturbations in radial coordinate and time domains.
[6] A. M. Candela: Multiplicity of timelike geodesics in splitting Lorentzian manifolds.
[7] A. Salvatore: Null geodesics joining two submanifolds in stationary Lorentzian manifolds.
[8] V. Benci and D. Fortunato: Why is the gravitational mass equal to the inertial mass?

In [1] the so-called heavenly equation

$$u_{xx} + u_{yy} + (e^u)_{zz} = 0 \qquad (*)$$

($u = u(x, y, z)$) is considered. It occurs in the study of Einstein spaces with one rotational Killing vector (also called heavenly spaces) and in the study of extended conformal symmetries. The authors study the integrability properties of (*) using a prolongation technique and give the exact solutions (which are regular at the instant $t = 0$) of the prolongation system associated with (*) . This is done in terms of a series expansion of operators which can be interpreted as generalized Bessel coefficients (cf. Remark 3.3 and Eqs. (20) and (21) in the paper), also using a Baker–Campbell–Hausdorff expansion (cf. formula (31)) which is a very useful tool for expressing the solution of the prolongation system.

Paper [2] deals with the study of graviational collapse of an initially regular distribution of matter using Einstein's theory of gravitation. This is presently one of the most active fields of research in general relativity generated by the very famous cosmic censorship conjecture due to Roger Penrose. The study of this problem is delicate even in simple systems such as spherically symmetric ones. Paper [2] gives an overview of models satisfying the weak energy condition and the existence of a singularity-free initial data surface. Moreover the authors consider only solutions of the Einstein field equations which are physically meaningful in terms of the equation of state of the matter, focusing their attention only on analytical models describing the gravitational collapse of massive stars.

Paper [3] is a review article about recent results concerning variational problems in Lorentzian geometry under intrinsic assumptions. They can be applied to the study of trajectories of massive and massless objects in general relativity. First presented is an extension of the Fermat Principle of Classical Optics to a relativistic context where massive particles and photons are considered. It also discusses how Morse relations for light rays can be obtained as limits of Morse relations for massive particles. The paper deals also with the relativistic brachistochrone problem in a stationary spacetime, pointing out the difference between travel time brachistocrones and arrival time brachiostocrones.

Paper [4] deals with the Jackiw–Teitelboim gravity model describing a 2-dimensional reduction of the Einstein–Hilbert action or a spherical reduction of the Einstein–Hilbert–Maxwell action and its application to the description of black holes. The Euclidean version of Jackiw–Teitelboim gravity, where the de Sitter group is replaced by the orthogonal group allows one to obtain the nonlinear Schrödinger equation. Its soliton-like solutions characterize the black holes of this gravity model. Alternative descriptions of the model in terms of other completely integrable systems are also mentioned.

In [5] an analytic solution of the homogeneous Regge–Wheeler–Zerilli–Moncrief equation (with a source term) is given. The Regge–Wheeler equation represents the first order perturbations of the Schwarzschild metric. Zerilli studied emitted radiation via the radial wave equation for polar perturbations, while Moncrief showed the gauge invariant significance of the wave equations. Solutions in the time domain of the Regge–Wheeler–Zerilli–Moncrief radial wave equation give the metric tensor describing the polar perturbations. The author of the paper recently showed that radiation reaction can be identified by subtraction of the zeroth-order terms in the geodesic equations of the above metric.

In [6] timelike geodesics in splitting Lorentian manifolds (with metric coefficient explicitly depending on time) are studied. Multiplicity results for time like trajectories which are periodic with respect to the space variable and for trajectories joining two given submanifolds are obtained by variational techniques and by absolute and relative Lusternik–Schnirelmann category theory.

In [7] the number of light rays joining two given submanifolds in a splitting stationary Lorentzian manifold is studied. This is done by using the functional introduced by Fortunato, Giannoni and Masiello (cf. ref. [5] of the paper) and by applying to it Lusternik and Schnirelmann theory.

Paper [8] deals with a variational principle introduced by the authors in a paper published in 1998 in Foundations in Physics (cf. [1]). The principle states that the world-line $z = z(\sigma)$ of a massive particle can be parameterized by a parameter σ which carries physical information related to the rest mass and the charge. In particular the inertial mass is not a property of the material point, but is a constant of the motion which is determined by the initial conditions with respect to the parameter σ and the variational principle allows one to deduce the equality between the inertial mass and gravitational mass.

Why is the Gravitational Mass Equal to the Inertial Mass?

V. Benci, D. Fortunato

Abstract. In [1] a variational principle for the fundamental equations of Classical Physics was introduced. The basic point of this principle is that the world-line $z = z(\sigma)$ of a material point is parametrized by a parameter σ which carries some physical information, namely, it is related to the rest mass and to the charge. In particular the inertial mass will not be a property of a material point, but will be a constant of the motion which is determined by the initial conditions with respect the parameter σ. One of the main features of this principle is the fact that it permits us to deduce the equality between inertial and gravitational mass. Here we briefly review some of the results contained in [1].

1 The variational principle

Let us suppose that the world is described by the four-dimensional space-time variable

$$z = (z^0, z^1, z^2, z^3).$$

We may assume that $z \in \mathbf{R}^4$ or that it lives in a more general four-dimensional manifold $\mathcal{M}$.

The world-line of a material point is a function

$$\sigma \longmapsto z(\sigma);$$

the physical meaning of the parameter σ will be discussed later.

We want to state a variational principle which describes the world-lines $z = z(\sigma)$. To do this, we have to define a Lagrangian

$$\mathcal{L} = \mathcal{L}\left(\sigma, z, \frac{dz}{d\sigma}\right).$$

We assume that $\mathcal{L}$ is independent on σ. At this point, in view of the similarity with Classical Mechanics, the simplest Lagrangian which we may consider, probably is a second order polynomial in $\frac{dz}{d\sigma}$:

$$\mathcal{L} = g_{ij}(z)\frac{dz^i}{d\sigma}\frac{dz^j}{d\sigma} + a_i(z)\frac{dz^i}{d\sigma} + c(z),$$

where the $g_{ij}(z)$, $a_i(z)$ and $c(z)$ are fields which determine the dynamics of the material point. Assuming that

$$\mathcal{L}\left(z, \frac{dz}{d\sigma}\right)_{\frac{dz}{d\sigma}=0} = 0$$

we see that $c(z) = 0$.

Thus,

$$\mathcal{L} = g_{ij}(z)\frac{dz^i}{d\sigma}\frac{dz^j}{d\sigma} + a_i(z)\frac{dz^i}{d\sigma}; \tag{1}$$

we shall consider all the "physical laws" which can be deduced from it. (We adopt the Einstein summation convention if the indices go from 0 to 3; when they go from 1 to 3 we will write the sum explicitly).

The action functional takes the form

$$\mathcal{S} = \int \left(g_{ij}(z)\frac{dz^i}{d\sigma}\frac{dz^j}{d\sigma} + a_i(z)\frac{dz^i}{d\sigma} \right) d\sigma \tag{2}$$

Since $\mathcal{S}$ is an action, and z has the dimension of a length, it turns out that σ has the dimension of $\frac{\text{time}}{\text{mass}}$.

Next we write the Euler–Lagrange equations relative to the functional (2). In order to simplify the notation we define the following bilinear form

$$\langle \mathbf{v}, \mathbf{w} \rangle = g_{ij}(z)v^i w^j, \ \mathbf{v} = v^i \mathbf{e}_i, \ \mathbf{w} = w^i \mathbf{e}_i.$$

If we assume that the bilinear form $\langle \cdot, \cdot \rangle$ is not degenerate, it can be considered as a pseudoriemannian structure. When we give a physical interpretation to the structure $\langle \cdot, \cdot \rangle$, it will be the usual Lorentzian structure of General Relativity, but for the moment $\langle \cdot, \cdot \rangle$ can be considered as a generic pseudoriemannian structure which helps to simplify the computation of the variation of the functional (2).

Then (2) becomes

$$\mathcal{S} = \int < \frac{dz}{d\sigma}, \frac{dz}{d\sigma} > d\sigma + \int < \mathcal{A}(z), \frac{dz}{d\sigma} > d\sigma \tag{3}$$

where $\mathcal{A}(z)$ is the vector defined by

$$\langle \mathcal{A}(z), \mathbf{v} \rangle = a_i(z)v^i \text{ for every } \mathbf{v} = v^i \mathbf{e}_i \in T_z\mathcal{M}.$$

Taking the first variation of $\mathcal{S}$ we get the following equation for the world-lines:

$$D_\sigma \frac{dz}{d\sigma} = \frac{1}{2}\left[\left(\nabla^L \mathcal{A}\right)^T - \nabla^L \mathcal{A} \right] \cdot \frac{dz}{d\sigma}. \tag{4}$$

Here D_σ is the covariant derivative with respect the pseudoriemannian structure $\langle \cdot, \cdot \rangle$, $\nabla^L \mathcal{A}$ denotes the gradient of the vector field $\mathcal{A}$ and $\left(\nabla^L \mathcal{A}\right)^T$ is its transpose. Notice that $\nabla^L \mathcal{A}$ is formally a matrix and that it is defined by means of the formula

$$\left\langle \nabla^L \mathcal{A} \cdot \mathbf{v}, \mathbf{w} \right\rangle = \langle d\mathcal{A}[\mathbf{v}], \mathbf{w} \rangle \text{ for every } \mathbf{v}, \mathbf{w} \in T_z\mathcal{M}.$$

Equation (4) has a first integral since the Lagrangian is independent of σ. It is easy to check that it is given by

$$h = \left\langle \frac{dz}{d\sigma}, \frac{dz}{d\sigma} \right\rangle. \tag{5}$$

Observe that, if we reparametrize a critical point $z = z(\sigma)$ of (2) with an affine parameter $\sigma' = a\sigma$ (a real), we get a new curve which is no longer a stationary point of (2). Thus σ is a privileged parameter and, unlike what happens for geodesics, the constant of the motion h in (5) has a "real" physical meaning in the sense that solutions with different h cannot be obtained one from the other by affine reparametrizations. We shall see that h in (5) will be interpreted, up to a multiplicative constant, as the square of the rest mass of the particle.

2 The dynamics

Now we consider trajectories such that

$$\frac{dz^0}{d\sigma} \neq 0.$$

Then $z^0 = z^0(\sigma)$ is invertible and these trajectories can be parametrized by

$$t = \frac{z^0}{c} \tag{6}$$

with $c =$ constant.

In the following t will be interpreted as time; thus c is a constant which has the dimension of a velocity.

In this section, we will write the action (3) and the relative Euler equations using t as independent variable and we will see the mathematical consequences of this fact. In the next sections we will give a physical interpretation to them.

Using the "$\cdot$" to denote the derivative with respect to t, Eq. (5) becomes

$$h = \langle \dot{z}, \dot{z} \rangle \left(\frac{dt}{d\sigma} \right)^2. \tag{7}$$

Next we assume that there exist trajectories with $h < 0$. We shall consider only these trajectories, which seem to be the only ones which have physical meaning. We set

$$h = -m_0^2 c^2, \tag{8}$$

where c is the constant defined by (6) and m_0 (which is assumed positive) has the dimension of a mass and it depends on the initial conditions.

Thus from (7) and (8) we get

$$\frac{dt}{d\sigma} = \frac{q m_0 c}{\sqrt{-\langle \dot{z}, \dot{z} \rangle}} \tag{9}$$

where q takes the value $+1$ or -1 depending on the sign of $\frac{dt}{d\sigma}$.

Now, using (9), we can write the action (3) as follows:

$$\mathcal{S} = -m_0 c \int_{t_0}^{t_1} \sqrt{-\langle \dot{z},\ \dot{z}\rangle}\, dt + q \int_{t_0}^{t_1} \langle \mathcal{A}(z),\ \dot{z}\rangle\, dt, \tag{10}$$

where

$$t_0 = \min\left\{t(\sigma_0), t(\sigma_1)\right\} \quad \text{and } t_1 = \max\left\{t(\sigma_0), t(\sigma_1)\right\}.$$

The next step will be to separate the t variable from the other variables which will be interpreted as space variables. We set

$$t = \frac{z^0}{c};\ x = (z^1, z^2, z^3) \tag{11}$$

Using these new coordinates, we shall write the bilinear form $\langle \cdot,\ \cdot\rangle$ as follows:

$$\langle dz,\ dz\rangle = (dx \mid dx) + c\,(W(x,t) \mid dx)\, dt - c^2 g_{00}(x,t) dt^2$$

where

$$W = 2(g_{01}, g_{02}, g_{03})$$

and

$$(dx \mid dx) = \sum_{i,j=1}^{3} g_{ij}(x,t)\, dx^i dx^j.$$

Here it is not yet necessary to assume that the bilinear form $(\cdot \mid \cdot)$ and the function $g_{00}(x,t)$ are positive. However, if this is the case, $(\cdot \mid \cdot)$ is a Riemannian structure and $\langle \cdot,\ \cdot\rangle$ is a Lorentzian structure and this is the physically meaningful case. Using the above formula, we have

$$\langle \dot{z},\ \dot{z}\rangle = (\dot{x} \mid \dot{x}) + c\ (W \mid \dot{x}) - c^2 g_{00}. \tag{12}$$

Also the 4-vector $\mathcal{A} = a^i \mathbf{e}_i = g^{ij} a_j \mathbf{e}_i$ will be decomposed into its "time" and "space" components:

$$\varphi = -a_0;\ \mathbf{A} = c\sum_{i=1}^{3} a^i \mathbf{e}_i \tag{13}$$

Now substitute (13) and (12) in (10) and choose for simplicity a reference frame where $W = 0$. Then we can write the action (10) as follows:

$$\mathcal{S} = -\int_{t_0}^{t_1} m_0 c^2 \sqrt{g_{00}(x,t) - \frac{1}{c^2}(\dot{x} \mid \dot{x})}\, dt + q \int_{t_0}^{t_1} \left[\frac{1}{c}(\mathbf{A} \mid \dot{x}) - \varphi(x,t)\right] dt.$$

Taking the first variation we get the equation of motion:

$$D_t\left(\frac{m_0}{\sqrt{g_{00}-\frac{1}{c^2}(\dot{x}\mid\dot{x})}}\dot{x}\right)=-\frac{1}{2}\frac{m_0c^2}{\sqrt{g_{00}-\frac{1}{c^2}(\dot{x}\mid\dot{x})}}\nabla g_{00}$$
$$+q\left(-\nabla\varphi-\frac{1}{c}\frac{\partial\mathbf{A}}{\partial t}\right)+\frac{q}{c}\dot{x}\times(\nabla\times\mathbf{A}). \tag{14}$$

Here D_t is the covariant derivative with respect to the bilinear form $(\cdot\mid\cdot)$. If we set

$$m=\frac{m_0}{\sqrt{g_{00}-\frac{1}{c^2}(\dot{x}\mid\dot{x})}}, \tag{15}$$

$$\mathbf{E}=-\nabla\varphi-\frac{1}{c}\frac{\partial\mathbf{A}}{\partial t}, \tag{16}$$

$$\mathbf{H}=\frac{1}{c}\nabla\times\mathbf{A}, \tag{17}$$

the Euler–Lagrange equation takes the following form:

$$D_t\,(m\dot{x})=-\frac{1}{2}mc^2\nabla g_{00}+q\,\mathbf{E}+q\,\dot{x}\times\mathbf{H}. \tag{18}$$

3 Interpretation

Of course we interpret the variable t as time; hence the Eq. (18) can be interpreted as the equation of the motion of a material point parametrized by t.

Then the variables, which appear in (18), can obviously be interpreted as follows:

m = mass
q = electric charge
$\mathbf{E}$ = electric field
$\mathbf{H}$ = magnetic field.

If we consider special cases, we obtain the fundamental equations of classical physics.

First assume $g_{00}=1$ and the bilinear form $(v\mid w)$ equal to the euclidean product. Then the covariant derivative D_t reduces to the usual derivative $\frac{d}{dt}$ and (18) reduces to the usual equation of a relativistic particle in a electromagnetic field:

$$\frac{d}{dt}\left(\frac{m_0}{\sqrt{1-\frac{\dot{x}^2}{c^2}}}\dot{x}\right)=q\,\mathbf{E}+q\,\dot{x}\times\mathbf{H}.$$

Obviously, this equation allows us to interpret the quantity m_0 as rest mass.

Now assume that the electromagnetic field is null. Then in this case Eq. (18) reduces to

$$D_t\left(\frac{1}{\sqrt{g_{00}-\frac{1}{c^2}(\dot{x}\mid\dot{x})}}\dot{x}\right)=-\frac{1}{2}\frac{c^2}{\sqrt{g_{00}-\frac{1}{c^2}(\dot{x}\mid\dot{x})}}\nabla g_{00}. \tag{19}$$

Notice that we have divided both sides by m_0. So the equation becomes independent of m_0, and this fact has the physical interpretation that all heavy bodies fall in the same way in a gravitational field; only the presence of electromagnetic forces gives a physical meaning to m_0 as inertial mass (at rest).

Equation (19) is the equation of a geodesic with respect to the pseudoriemannian metric

$$\langle dz,\ dz\rangle=(dx\mid dx)-c^2 g_{00}(x,t)dt^2.$$

If we assume that the bilinear form $(dx \mid dx)$ is positive definite and that $g_{00}(x,t)>0$, then we get a Lorentzian metric and (19) is the usual equation in the theory of General Relativity of a particle moving in a gravitational field.

Then the $\{g_{ij}\}$ appearing in Eq. (1) can be interpreted as the components of the metric tensor of the General Relativity. Now, the interpretation of the vector W which appears in (12) is obvious: it describes the Coriolis forces and other forces depending on the velocity in the framework of General Relativity.

By this interpretation, we can set

$$g_{00}(x,t)=1+\frac{2}{c^2}U(x,t) \tag{20}$$

where $U(x,t)$ is the potential of the gravitational field in the gauge which makes $U(x,t)\cong 0$ at ∞ (i.e., far away from gravitational masses). In this setting, (18) becomes

$$D_t\,(m\dot{x})=-m\nabla U+q\,\mathbf{E}+q\ \dot{x}\times\mathbf{H}. \tag{21}$$

Of course, the above equation could have been obtained directly from (4) using (5) with the above definitions of $m,\ U,\ q,$ **E** and **H**.

Equation (4) determines a unique trajectory in space-time once 8 parameters are given for the initial conditions:

$$z^i(\sigma_0),\ \frac{dz^i}{d\sigma}(\sigma_0),\ i=1,2,3;\ t(\sigma_0),\ \frac{dt}{d\sigma}(\sigma_0). \tag{22}$$

Equations (9) and (15) imply that $\frac{dt}{d\sigma}(\sigma_0)=qm(\sigma_0)$, and hence $\frac{dt}{d\sigma}(\sigma_0)$ can be replaced if qm_0 is known.

Equation (21) determines a unique trajectory once the 6 parameters for the initial conditions are given:

$$x^i(t_0),\ \dot{x}^i(t_0),\ i = 1, 2, 3.$$

The extra information provided by (22) consists of two real parameters: one is the initial time $t(\sigma_0) = t_0$ which has no physical meaning since it is a convention on the setting of the clocks; the other is the rest mass m_0 and the charge q which are contained explicitly in Eq. (21).

4 Consequences

Finally we examine the main consequences of the theory, which we have presented so far.

First of all, note that, on the left-hand side of (21), m denotes the inertial mass while, on the right-hand side, we need to interpret m as the gravitational mass. Thus, in this theory, the equality

inertial mass = gravitational mass

is not assumed as an empirical postulate, but is a theorem.

The relation between the mass and the rest mass is given by (15) and it is the usual relativistic formula (it coincides with that of Special Relativity on taking $g_{00} = 1$). However, in this theory, (15) is not a consequence of the Principle of Special Relativity. It is a consequence of the form of the Lagrangian (1). Equation (1), in general, is not Lorentz invariant. Hence, we have deduced (15) without making use of the Principle of Relativity.

Taking account of (20), we get

$$m = \frac{m_0}{\sqrt{1 + \frac{2}{c^2}\left[U - \frac{1}{2}(\dot{x} \mid \dot{x})\right]}}.$$

This formula predicts that the inertial mass depends not only on the velocity, but also on the gravitational field U as in the theory of General Relativity.

Using (x, t) coordinates and the variables U, φ and $\mathbf{A}$, the Lagrangian (1) takes the following form:

$$\mathcal{L} = -m_0c^2\sqrt{1 + \frac{2}{c^2}\left[U - \frac{1}{2}(\dot{x} \mid \dot{x})\right]} + \frac{q}{c}(\mathbf{A} \mid \dot{x}) - q\varphi. \tag{23}$$

If $c \to \infty$, and $(\dot{x} \mid \dot{x})$ reduces to the usual dot product, we get

$$\mathcal{L} = \frac{1}{2}m_0|\dot{x}|^2 - m_0U - m_0c^2 + \frac{q}{c}\mathbf{A} \cdot \dot{x} - q\varphi + O\left(\frac{1}{c^2}\right)$$

which is the Lagrangian of nonrelativistic Mechanics except for the term m_0c^2 which does not affect the dynamics.

Equation (23) allows us to define the energy of our material point as follows:

$$E = \sum_{i=1}^{3} \frac{\partial \mathcal{L}}{\partial \dot{x}^i} \dot{x}^i - \mathcal{L}$$

which, by Noether's theorem, is a constant of the motion if $\mathcal{L}$ is independent of t. It is easy to see that

$$E = mc^2 + 2mU + q\varphi.$$

Using (23) we can define the momentum

$$\mathbf{P} = \frac{\partial \mathcal{L}}{\partial \dot{x}^i} = m\dot{x} + \frac{q}{c}\mathbf{A} = \mathbf{p} + \frac{q}{c}\mathbf{A}$$

where $\mathbf{p} = m\dot{x}$ is the usual momentum.

Next we discuss another important point of this theory. The rest mass m_0 and the charge q are not extra parameters imposed upon the material point, but are parameter depending by the initial conditions; by (5) and (8), m_0 can be defined by the following formula

$$m_0 = \frac{1}{c^2}\sqrt{-\langle \frac{dz}{d\sigma}, \frac{dz}{d\sigma} \rangle}.$$

Thus, it can be interpreted as "the speed with respect to σ" of a particle moving in space-time.

Thus, the time-like 4-vector $\frac{dz}{d\sigma}$ carries all information about the material point. Using (9), (12), and (15) we have

$$\frac{dt}{d\sigma} = qm \tag{24}$$

and then

$$\frac{dz}{d\sigma} = \left(c\frac{dt}{d\sigma}, \frac{dx}{d\sigma} \right) = \left(c\frac{dt}{d\sigma}, \dot{x}\frac{dt}{d\sigma} \right) = q(mc, m\dot{x}).$$

Thus, in the absence of the gravitational and the electromagnetic fields,

$$\frac{dz}{d\sigma} = q\left(\frac{E}{c}, \mathbf{p} \right)$$

is the 4-vector energy-momentum times the charge. Its relativistic length is the rest mass; if q is positive, then $\frac{dz}{d\sigma}$ is future pointing; if q is negative $\frac{dz}{d\sigma}$ is past pointing. Notice that the charge may assume only the values ± 1 unlike the rest mass; and this fact is consistent with our knowledge of elementary particles.

Next we consider the variable

$$\tau = m_0 q \sigma \tag{25}$$

since

$$< \frac{dz}{d\tau}, \frac{dz}{d\tau} >= -c^2; \tag{26}$$

it turns out that τ is proper time as defined in General Relativity: it represents the time measured by an observer moving with $z(t)$; notice that we have added the factor q so that τ is then increasing in the direction towards the future. Thus Eq. (25) allows us to give a physical interpretation to σ: it represents proper time divided by the rest mass and the charge. Notice that in this theory the charge is dimensionless since it is a pure number.

By (25) and (24), we get the following formulas which nicely connect m_0 to the proper time τ and m to the universal time t

$$m_0 = \left| \frac{d\tau}{d\sigma} \right|; \quad m = \left| \frac{dt}{d\sigma} \right|. \tag{27}$$

References

1. Benci V., Fortunato D. (1998): A new variational principle for the fundamental equations of classical physics. Found. Phys. **28**, 333–352

Multiplicity of Timelike Geodesics in Splitting Lorentzian Manifolds

A.M. Candela

1 Introduction

In the theory of General Relativity models of gravitational fields are particular examples of Lorentzian manifolds, the so–called space–times, and the trajectory of a free falling particle on which only gravity acts is just a timelike geodesic in such a manifold [1, 2].

For this reason, it is important to know everything possible information about the geometry of Lorentzian manifolds and the existence of its timelike geodesics; hence, in this short note we state some multiplicity results for timelike geodesics in an orthogonal splitting Lorentzian manifold in two cases: if they are periodic trajectories; if they join two given submanifolds.

The idea for solving both of these problems is to introduce a suitable variational approach, and, hence, to prove the existence of critical points by means of a generalization of the classical Lusternik–Schnirelman Theory which uses the relative category [3].

To this aim, we recall the main definitions.

Definition 1.1. *A Lorentzian manifold* $(\mathcal{M}, \langle\cdot,\cdot\rangle_z)$ *is said to be of the orthogonal splitting type if there exists a connected finite dimensional smooth Riemannian manifold* $(\mathcal{M}_0, \langle\cdot,\cdot\rangle)$ *such that* $\mathcal{M} = \mathcal{M}_0 \times \mathbf{R}$ *and* $\langle\cdot,\cdot\rangle_z$ *has the following form:*

$$\langle\zeta,\zeta\rangle_z = \langle\alpha(z)\xi,\xi\rangle - \beta(z)\tau^2, \tag{1.1}$$

for any $z = (x,t) \in \mathcal{M} = \mathcal{M}_0 \times \mathbf{R}$ *and* $\zeta = (\xi,\tau) \in T_z\mathcal{M} \equiv T_x\mathcal{M}_0 \times \mathbf{R}$, *where* $\alpha(z)$ *is a smooth symmetric linear strictly positive operator from* $T_x\mathcal{M}_0$ *into itself and* $\beta : \mathcal{M} \to \mathbf{R}$ *is a smooth and positive scalar field.*

From now on, let $(\mathcal{M}, \langle\cdot,\cdot\rangle_z)$ be an orthogonal splitting Lorentzian manifold.

It is well-known that the geodesics in $\mathcal{M}$ are the smooth curves $z : I \to \mathcal{M}$, I a real interval, such that

$$D_s\dot{z}(s) = 0 \quad \text{for all } s \in I;$$

moreover, since there is a constant E_z such that

$$E_z = \langle\dot{z}(s), \dot{z}(s)\rangle_z \quad \text{for all } s \in I,$$

z is called timelike (spacelike, lightlike) if E_z is negative (positive, null). This classification is the so-called causal character of geodesics [4].

Let us assume that $I = [0, 1]$.

Definition 1.2. *Let $T > 0$ be fixed. A geodesic $z : I \to \mathcal{M}$, $z(s) = (x(s), t(s))$, is a T–periodic trajectory if the following conditions hold*

$$\begin{cases} x(0) = x(1), & \dot{x}(0) = \dot{x}(1); \\ t(1) = t(0) + T, & \dot{t}(0) = \dot{t}(1). \end{cases}$$

Two T-periodic trajectories z_1 and z_2 are geometrically distinct if $z_1(I) \neq z_2(I)$.

Definition 1.3. *Let $\mathcal{N}_0$ and $\mathcal{N}_1$ be two submanifolds of $\mathcal{M}$. A normal geodesic joining $\mathcal{N}_0$ to $\mathcal{N}_1$ is any geodesic $z : I \to \mathcal{M}$ such that*

$$\begin{cases} z(0) \in \mathcal{N}_0 \\ \dot{z}(0) \in T_{z(0)}\mathcal{N}_0^{\perp} \end{cases}, \qquad \begin{cases} z(1) \in \mathcal{N}_1 \\ \dot{z}(1) \in T_{z(1)}\mathcal{N}_1^{\perp} \end{cases},$$

where $T_{z(i)}\mathcal{N}_i^{\perp} = \{v \in T_{z(i)}\mathcal{M} : \langle v, w\rangle_z = 0 \ \forall w \in T_{z(i)}\mathcal{N}_i\}$, for $i = 0, 1$.

Here, we assume that $\mathcal{N}_0$ and $\mathcal{N}_1$ are such that

$$\mathcal{N}_0 = P_0 \times \{t_0\} \quad \text{and} \quad \mathcal{N}_1 = P_1 \times \{t_1\},$$

where P_0 and P_1 are closed submanifolds of $\mathcal{M}_0$ and $t_0, t_1 \in \mathbf{R}$ are fixed.
The following results are valid.

Theorem 1.4. *Let $(\mathcal{M}, \langle\cdot,\cdot\rangle_z)$ be an orthogonal splitting manifold such that the following hypotheses hold:*

(A_1) *$\mathcal{M}_0$ is a compact Riemannian manifold such that the fundamental group $\pi_1(\mathcal{M}_0)$ is finite;*
(A_2) *there exists $T_0 > 0$ such that α and β are T_0–periodic in the t–variable, i.e.,*

$$\begin{cases} \alpha(x, t + T_0) = \alpha(x, t) \\ \beta(x, t + T_0) = \beta(x, t) \end{cases} \quad \text{for all } x \in \mathcal{M}_0,\ t \in \mathbf{R}.$$

For any $k \in \mathbf{N}$, $k \geq 1$, let $\gamma(k)$ be the number of the geometrically distinct timelike kT_0-periodic trajectories in $\mathcal{M}$. Then

$$\lim_{k \to +\infty} \gamma(k) = +\infty.$$

A complete proof of Theorem 1.4 is in [3]; other results on the existence of one timelike trajectory on a splitting Lorentzian manifold were obtained in [5] and [6].

Theorem 1.5. *Let $(\mathcal{M}, \langle\cdot,\cdot\rangle_z)$ be an orthogonal splitting Lorentzian manifold satisfying the following hypotheses:*

(H_1) *$(\mathcal{M}_0, \langle\cdot,\cdot\rangle)$ is a complete Riemannian manifold not contractible in itself;*

(H_2) *there exist constants* $\lambda,\ \lambda_1,\ \nu,\ N,\ K > 0$ *such that, for all* $z \in \mathcal{M}$, $z = (x, t)$, $\xi \in T_x\mathcal{M}_0$, *we have*

$$\lambda \langle \xi, \xi \rangle \ \le\ \langle \alpha(z)\xi, \xi \rangle, \quad \langle \alpha(x, 0)\xi, \xi \rangle \ \le\ \lambda_1 \langle \xi, \xi \rangle, \quad \nu \le \beta(z) \le N,$$

$$|\langle \alpha_t(z)\xi, \xi \rangle| \le K \langle \xi, \xi \rangle, \quad |\beta_t(z)| \le K;$$

moreover, we have

$$\limsup_{t\to+\infty} \langle \alpha_t(x, t)\xi, \xi \rangle \ \le\ 0, \quad \liminf_{t\to-\infty} \langle \alpha_t(x, t)\xi, \xi \rangle \ \ge\ 0$$

uniformly in $x \in \mathcal{M}_0$ *and* $\xi \in T_x\mathcal{M}_0$, $\langle \xi, \xi \rangle = 1$. *Let* P_0 *and* P_1 *be two subsets of* $\mathcal{M}_0$ *such that*

(H_3) P_0 *and* P_1 *are closed submanifolds of* $\mathcal{M}_0$ *such that one of them is compact;*
(H_4) P_0 *and* P_1 *are both contractible in* $\mathcal{M}_0$.

Then, if $N(P_0, P_1, t_1 - t_0)$ *is the number of timelike normal geodesics joining* $\mathcal{N}_0$ *to* $\mathcal{N}_1$, *it follows that*

$$\lim_{|t_1 - t_0|\to+\infty} N(P_0, P_1, t_1 - t_0) = +\infty.$$

A complete proof of Theorem 1.5 and a result on the existence of at least one timelike normal geodesic joining $\mathcal{N}_0$ to $\mathcal{N}_1$ are in [7].

Remark 1.6. In general, the causal character of normal geodesics joining $\mathcal{N}_0$ to $\mathcal{N}_1$ is not prescribed. However, if $P_1 = \mathcal{M}_0$ and $t_1 \neq t_0$, then all normal geodesics from $\mathcal{N}_0$ to $\mathcal{M}_0 \times \{t_1\}$ are timelike.

Remark 1.7. Let P_0 and P_1 both be reduced to a single point. Then Theorem 1.5 gives a multiplicity result of timelike geodesics joining two fixed points. Other results on geodesic connectedness are in [8, 9].

Remark 1.8. In general, if the Lorentzian metric does not satisfy assumptions (H_1) and (H_2), Theorem 1.5 may not hold. Counterexamples are provided by the de Sitter and anti–de Sitter space–times [10].

2 Variational setting

In order to introduce a suitable variational approach for both the given problems, we have to define some functional manifolds. To this aim, for any $n \in \mathbf{N}$, $n \ge 1$, we consider $H^1(I, \mathbf{R}^n)$, the Sobolev space of absolutely continuous curves whose derivative is square summable, endowed with the norm

$$\|x\|_1^2 = \int_0^1 |\dot{x}|^2 ds + \int_0^1 |x|^2 ds.$$

By the Nash Embedding Theorem $\mathcal{M}_0$ is a submanifold of a certain Euclidean space $\mathbf{R}^N$; hence, the Hilbert manifold $H^1(I, \mathcal{M}_0)$ is such that

$$H^1(I, \mathcal{M}_0) \equiv \{x \in H^1(I, \mathbf{R}^N) : x(I) \subset \mathcal{M}_0\}.$$

It can be proved that

$$\Lambda^1 = \{x \in H^1(I, \mathcal{M}_0) : x(0) = x(1)\}$$

and

$$\Omega = \{z \in H^1(I, \mathcal{M}_0) : x(0) \in P_0,\ x(1) \in P_1\}$$

are both smooth submanifolds of $H^1(I, \mathcal{M}_0)$ [11, 12]).
Moreover, for any $\bar{t}_0, \bar{t}_1 \in \mathbf{R}$, the following closed affine submanifold can be defined

$$W(\bar{t}_0, \bar{t}_1) = \{t \in H^1(I, \mathbf{R}) : t(0) = \bar{t}_0,\ t(1) = \bar{t}_1\}. \tag{2.1}$$

If

$$F(z) = \int_0^1 \langle \dot{z}, \dot{z} \rangle_z \, ds = \int_0^1 (\langle \alpha(z)\dot{x}, \dot{x} \rangle - \beta(z)\dot{t}^2) \, ds$$

is the action integral in $H^1(I, \mathcal{M}_0) \times H^1(I, \mathbf{R})$, the following propositions hold.

Proposition 2.1. *Assume $Z_k = \Lambda^1 \times W(0, kT_0)$ and $f_k = F|_{Z_k}$ for any $k \in \mathbf{N}$, $k \geq 1$. If $z \in Z_k$ is such that $f_k'(z) = 0$ and $f_k(z) < 0$, then z is a timelike kT_0–periodic trajectory in $\mathcal{M}$.*

(For the proof, cf. [3, Lemma 3.3]).

Proposition 2.2. *Assume $Z_* = \Omega \times W(t_0, t_1)$ and $f_* = F|_{Z_*}$. If $z \in Z_*$ is such that $f_*'(z) = 0$ and $f_*(z) < 0$, then z is a timelike normal geodesic joining $\mathcal{N}_0$ to $\mathcal{N}_1$.*

(For the proof it is enough to reason as in the Riemannian case; cf. [12]).

Note that both f_k and f_* are indefinite functionals; hence, we have to introduce the notion of relative category in order to obtain a multiplicity theorem for unbounded functionals (cf. [13–15]).

Definition 2.3. *Let Y and A be closed subsets of a topological space Z. The category of A in Z relative to Y, $\mathrm{cat}_{Z,Y}(A)$, is the least integer n such that there exist $n+1$ closed subsets of Z, $A_0, A_1, \ldots, A_n$, $A = A_0 \cup A_1 \cup \cdots \cup A_n$, and $n+1$ functions, $h_j \in C([0,1] \times A_j, Z)$, $j = 0, 1, \ldots, n$, such that*

(a) $h_j(0, z) = z$ for $z \in A_j$, $0 \leq j \leq n$;
(b) $h_0(1, z) \in Y$ for $z \in A_0$, and $h_0(\sigma, y) \in Y$ for all $y \in A_0 Y$, $\sigma \in [0, 1]$;
(c) $h_j(1, z) = z_j$ for $z \in A_j$ and some $z_j \in Z$, $1 \leq j \leq n$.

If a finite number of such sets does not exist, we set $\mathrm{cat}_{Z,Y}(A) = +\infty$.

It is the case that $\text{cat}_Z(A) = \text{cat}_{Z,\emptyset}(A)$ is the classical Lusternik–Schnirelman category of A in Z (for more details about this theory, [16]).

Aside from technical differences, both the given problems can be solved by applying the following multiplicity theorem for strongly indefinite functionals (for the proof, see [3, Theorem 1.4]).

Theorem 2.4. *Let Z be a C^2 complete Riemannian manifold modelled on a Hilbert space and let $f : Z \to \mathbf{R}$ be a C^1 functional. Let f satisfy the Palais–Smale condition, i.e., every sequence $(z_n)_{n\in\mathbf{N}} \subset Z$ such that*

$$\sup_{n\in\mathbf{N}} |f(z_n)| < +\infty \quad \textit{and} \quad \lim_{n\to+\infty} \|f'(z_n)\| = 0$$

converges up to subsequences.
Let us assume that there exist two subsets V and C of Z such that C is a closed strong deformation retract of $Z\backslash V$, i.e., there exists a continuous map $\mathcal{R} : I \times (Z\backslash V) \to Z$ such that

$$\begin{cases} \mathcal{R}(0, z) = z \;\; \textit{for all } z \in Z\backslash V, \\ \mathcal{R}(1, z) \in C \;\; \textit{for all } z \in Z\backslash V, \\ \mathcal{R}(\sigma, z) = z \;\; \textit{for all } z \in C,\ \sigma \in I. \end{cases}$$

If

$$\textit{inf}_{z\in V} f(z) > \sup_{z\in C} f(z) \quad \textit{and} \quad \text{cat}_{Z,C}(Z) > 0,$$

then f has at least $\text{cat}_{Z,C}(Z)$ critical points in Z whose critical levels are greater than or equal to $\textit{inf} f(V)$.

3 Sketch of the proofs

Unfortunately, we cannot apply Theorem 2.4 directly to both f_k and f_* since the Palais–Smale compactness condition does not hold and the relative category of the spaces Z_k and Z_* may be trivial. Thus, two approximation schemes are needed: a penalization of the action functional overcomes the lack of compactness and a Galerkin approximation allows us to consider a space of curves whose relative category is infinite. Moreover, a priori estimates of the critical levels of such approximating functionals permit us to obtain critical points for the action integral.

In order to penalize the action functional, for any $\varepsilon > 0$ let $\psi_\varepsilon \in C^2(\mathbf{R}_+, \mathbf{R}_+)$ be the "cut–function"

$$\psi_\varepsilon(s) = \begin{cases} 0 & \text{if } 0 \le s \le \dfrac{1}{\varepsilon} \\ \displaystyle\sum_{n=3}^{+\infty} \frac{\left(s - \frac{1}{\varepsilon}\right)^n}{n!} & \text{if } s > \dfrac{1}{\varepsilon}. \end{cases}$$

Then the two families of penalized functionals are

$$f_{k,\varepsilon}(z) = f_k(z) - \psi_\varepsilon(\|\dot{t}\|^2), \; z \in Z_k,$$
$$f_{*,\varepsilon}(z) = f_*(z) - \psi_\varepsilon(\|\dot{t}\|^2), \; z \in Z_*,$$

where $\|\dot{t}\|^2 = \int_0^1 |\dot{t}|^2 ds$.

It can be proved that

(i) for all $\varepsilon > 0$ both $f_{k,\varepsilon}$ and $f_{*,\varepsilon}$ satisfy the Palais–Smale condition in Z_k and Z_* respectively;
(ii) $K > 0$ and $\varepsilon_0 > 0$ exist such that, for all $\varepsilon \in]0, \varepsilon_0]$, we have

$$z \in Z_k, \; f'_{k,\varepsilon}(z) = 0, \; f_{k,\varepsilon}(z) < 0 \implies f'_k(z) = 0, \; f_k(z) < 0;$$
$$z \in Z_*, \; f'_{*,\varepsilon}(z) = 0, \; f_{*,\varepsilon}(z) < 0 \implies f'_*(z) = 0, \; f_*(z) < 0$$

(for the proofs, cf. [3, Sect. 4] and [7, Sects. 4 and 5]).

Now, we have to introduce a Galerkin approximation in order to have non–trivial relative categories. To this aim, note that, for any pair of real numbers $(\bar{t}_0, \bar{t}_1)$, the corresponding affine submanifold $W(\bar{t}_0, \bar{t}_1)$ in (2.1) is such that

$$W(\bar{t}_0, \bar{t}_1) = H_0^1 + \mathbf{R}\, j_{\bar{t}_0,\bar{t}_1},$$

where

$$H_0^1 = \{t \in H^1(I, \mathbf{R}) : t(0) = t(1) = 0\},$$
$$j_{\bar{t}_0,\bar{t}_1} : s \in I \longmapsto s\bar{t}_1 + (1-s)\bar{t}_0 \in \mathbf{R}.$$

We can consider the orthonormal basis $\{\sin(i\pi s)\}_{i\in\mathbf{N}}$ of H_0^1 and the finite dimensional subspaces

$$H_m = span\,\{\sin(i\pi s) : \; i = 1, 2, \dots, m\}$$

for all $m \geq 1$. Thus, we can assume that

$$W_m(\bar{t}_0, \bar{t}_1) = H_m + \mathbf{R}\, j_{\bar{t}_0,\bar{t}_1}.$$

The following submanifolds

$$Z_{k,m} = \Lambda^1 \times W_m(0, kT_0), \quad Z_{*,m} = \Omega \times W_m(t_0, t_1),$$

are well-defined and we can introduce

$$f^m_{k,\varepsilon} = f_{k,\varepsilon}|_{Z_{k,m}}, \qquad f^m_{*,\varepsilon} = f_{*,\varepsilon}|_{Z_{*,m}}.$$

To conclude the Galerkin approximation, we need the following lemma (the main tools of its proof are in [8]).

Lemma 3.1. *Let* $\varepsilon \in]0,1]$. *Suppose that, for all* $m \in \mathbf{N}$, $m \geq 1$, *there exists* $z_m \in Z_{k,m}$ *a critical point, of* $f^m_{k,\varepsilon}$. *If two constants,* c_1 *and* c_2, *exist independent of* m *such that*

$$c_1 \leq f^m_{k,\varepsilon}(z_m) \leq c_2 \quad \textit{for all } m \geq 1,$$

then, up to subsequences, $(z_m)_{m\geq 1}$ *converges to an element* $z \in Z_k$ *such that*

$$f'_{k,\varepsilon}(z) = 0, \qquad \lim_{m\to+\infty} f^m_{k,\varepsilon}(z_m) = f_{k,\varepsilon}(z), \quad \textit{and} \quad c_1 \leq f_{k,\varepsilon}(z) \leq c_2.$$

A similar lemma can be stated if we put $f^m_{*,\varepsilon}$, $f_{*,\varepsilon}$, $Z_{*,m}$, and Z_* in place of $f^m_{k,\varepsilon}$, $f_{k,\varepsilon}$, $Z_{k,m}$, and Z_k respectively.

Since more information about the topology of the spaces of curves $Z_{k,m}$ and $Z_{*,m}$ can be obtained by some results due to Fadell and Husseini [17, 18], it can be proved that Theorem 2.4 applies to $f^m_{k,\varepsilon}$, or $f^m_{*,\varepsilon}$, and suitable subsets of $Z_{k,m}$, respectively $Z_{*,m}$.

Hence, a priori estimates of the critical levels found imply the existence of timelike geodesics and complete the proofs of Theorems 1.4 and 1.5.

Acknowledgement. The author was supported by M.U.R.S.T. (research funds ex 40% and 60%).

References

1. Beem J.K., Ehrlich P.E., Easley K.L. (1996): Global Lorentzian Geometry. Monographs Textbooks Pure Appl. Math. **202**, Dekker, New York
2. Hawking S.W., Ellis G.F.R. (1973): The Large Scale Structure of Space–Time. Cambridge University Press, Cambridge
3. Candela A.M., Giannoni F., Masiello A.: Multiple critical points for indefinite functionals and applications. J. Differential Equations, to appear
4. O'Neill B. (1983): Semi-Riemannian Geometry with Applications to Relativity. Academic Press Inc., New York
5. Greco C. (1990): Periodic trajectories for a class of Lorentz metrics of a time–dependent gravitational field. Math. Ann. **287**, 515–521
6. Masiello A. (1995): On the existence of a timelike trajectory for a Lorentzian metric. Proc. Roy. Soc. Edinburgh A **125**, 807–815
7. Candela A.M., Masiello A., Salvatore A.: Existence and multiplicity of normal geodesics in Lorentzian manifolds. J. Geom. Anal. to appear
8. Benci V., Fortunato D., Masiello A. (1994): On the geodesic connectedness of Lorentzian manifolds. Math. Z. **217**, 73–93
9. Giannoni F., Masiello A. (1995): Geodesics on product Lorentzian manifolds. Ann. Inst. H. Poincaré Anal. Non Linéaire **12**, 27–60
10. Masiello A. (1994): Variational Methods in Lorentzian Geometry. Pitman Res. Notes Math. Ser. **309**, Longman Sci. Tech., Harlow
11. Klingenberg W. (1978): Lectures on Closed Geodesics. Springer, Berlin
12. Klingenberg W. (1982) Riemannian Geometry. de Gruyter, Berlin

13. Fadell E. (1985): Lectures in cohomological index theories of G–spaces with applications to critical point theory. Raccolta di seminari, Università della Calabria
14. Fournier G., Willem M. (1990): Relative category and the calculus of variations, in *Variational problems*, ed. by H. Beresticky, J.M. Coron, I. Ekeland, Birkhäuser, Basel, pp. 95–104
15. Szulkin A. (1990): A relative category and applications to critical point theory for strongly indefinite functionals, Nonlinear Anal. TMA **15**, 725–739
16. Palais R.S. (1966): Lusternik–Schnirelman theory on Banach manifolds. Topology **5**, 115–132
17. Fadell E., Husseini S. (1991): Category of loop spaces of open subsets in Euclidean space. Nonlinear Anal. TMA **17**, 1153–1161
18. Fadell E., Husseini S. (1994): Relative category, products and coproducts. Rend. Sem. Mat. Fis. Univ. Milano **LXIV**, 99–117

Gravitational Collapse of Fluid Bodies and Cosmic Censorship: Analytic Insights

S. Jhingan, G. Magli

Abstract. The present analytical understanding of the nature of the singularities which form at the end state of gravitational collapse of massive fluid bodies ("stars") is reviewed. Special emphasis is given to the issue of the physical reasonableness of the models.

1 Introduction

The investigation of the "final fate" of gravitational collapse of initially regular distributions of matter is one of the most active fields of research in contemporary general relativity. It is, indeed, known that, under fairly general hypotheses, solutions of the Einstein equations with "physically reasonable" matter can develop into singularities [1]. The key problem that still remains unsolved is the *nature* of such singularities. The main open issue is whether the singularities, which arise as the end point of collapse, can actually be observed.

Roger Penrose [2] was the first to propose the idea, known as the cosmic censorship conjecture: *does there exist a cosmic censor who forbids the occurrence of naked singularities, clothing each one in an absolute event horizon?* This conjecture can be formulated in a "strong" sense (in a "reasonable" spacetime we cannot have a naked singularity) or in a weak sense (even if such singularities occur they are safely hidden behind an event horizon, and therefore cannot communicate with far-away observers). Since Penrose's proposal there have been various attempts to prove the conjecture (see [3] and references therein). Unfortunately, no such attempts have been successful so far. As a consequence, research in this field turned to more tractable objectives. In particular, one would like to understand what happens in simple systems, like spherically symmetric ones (interestingly enough, even this apparently innocuous problem is far from being solved completely, although, as we shall see, a general pattern does seem to arise).

Our aim here is to survey only those models which have a clear physical interpretation. Therefore, we shall require satisfaction of the weak energy condition as well as existence of a singularity-free initial data surface. Moreover, we shall take into consideration solutions of the Einstein field equations which are physically meaningful in terms of a (phenomenological) equation of state of matter. In this respect it is worth mentioning that material schemes having a well-defined *microscopical* interpretation, like the Vlasov–Einstein system, would be closer to such a requirement [4]. However, very little is known about the dynamics of such models (a numerical investigation has been recently carried out [5]).

Given excellent general reviews on gravitational collapse and cosmic censorship [6–11], we have focused our attention on a specific issue, namely, the investigation of analytical models describing the gravitational collapse of massive stars. Therefore we are not going to address here other related and important topics. These include Vaidya spacetimes [3], radiation shells (see [9] and references therein), gravitational collapse of scalar fields [12, 13], critical behaviour in numerical relativity [14], stability of Cauchy horizon in Reissner–Nordström spacetimes [15], the Hoop conjecture [16, 17], among others [6, 10, 18, 19].

2 Einstein equations for spherical collapse

What is known *analytically* in gravitational collapse is essentially restricted to spherical symmetry (one exception is given by the Szekeres "quasi-spherical" spacetimes [20], but the results [21, 22] are very similar to those holding in Tolman–Bondi models; see Sect. 4.2.4). Therefore we discuss the mathematical structure of the Einstein field equations describing collapse of a deformable body only in the spherically symmetric case. For perfect fluids this structure is well-known [23]; we present here a more general case which takes into account anisotropic materials as well [24]. Also, we consider only non-dissipative processes since very little is known in the dissipative case [25].

The general, spherically symmetric, non-static line element in comoving coordinates t, r, θ, φ can be written in terms of three functions ν, η, Y of r and t:

$$ds^2 = -e^{2\nu}dt^2 + \eta^{-1}dr^2 + Y^2(d\theta^2 + \sin^2\theta d\varphi^2). \tag{1}$$

Throughout this paper, we shall assume that the collapsing body is "materially spherically symmetric" in the sense that all the physical "observables" do not depend on angles. The matter density of the material (baryon number density) is then given by

$$\varrho = \varrho_0(r)Y^{-2}\sqrt{\eta}, \tag{2}$$

where $\varrho_0(r)$ is an arbitrary (positive) function. As in any theory of continuous media, to choose a specific material one has to specify the *internal energy* ϵ. This function depends on the parameters characterizing the state of strain of the body (for a recent review of relativistic elasticity theory, see [26]). It can be shown that such parameters can be identified with Y and η in the comoving frame (in other words, any deformation is described "gravitationally"). Therefore, we introduce as *equation of state* of the material a positive function

$$\epsilon = \psi(r, Y, \eta).$$

Here the explicit dependence on r takes into account possible inhomogeneities.

The energy-momentum tensor can be readily calculated and the result is a diagonal tensor of the form $T_\nu^\mu = \mathrm{diag}(-\epsilon, \Sigma, \Pi, \Pi)$. Here the stress-strain relations (i.e., the relations giving the radial stress Σ and the tangential stress Π in terms of the constitutive function) are given by

$$\Sigma = 2\eta \frac{\partial \psi}{\partial \eta} - \psi, \tag{3}$$

$$\Pi = -\frac{1}{2} Y \frac{\partial \psi}{\partial Y} - \psi. \tag{4}$$

We shall always use the word stress rather than pressure since both Σ and Π can be, in principle, negative (tensions) without violating the energy conditions.

Different materials correspond to different choices of the function ψ. It is, however, worth mentioning that the material scheme most widely used in astrophysical applications is the perfect fluid, which can be characterized as a material whose function of state depends on the number density only ($\epsilon = \widetilde{\psi}(\varrho)$). In this case both stresses coincide:

$$\Pi = \Sigma = \varrho \frac{d\widetilde{\psi}}{d\varrho} - \widetilde{\psi} := p,$$

where p is the isotropic pressure. Two particular cases are worth mentioning in this scheme. One is that of the linear pressure-density relation ($p = k\epsilon$); the corresponding equation of state is $\widetilde{\psi}(\varrho) = \varrho(1 + A\varrho^k)$ where A is a constant. The other is the *dust model* for which $p = 0$. In this case energy is distributed proportionally to the mass ($\widetilde{\psi}(\varrho) = \varrho$).

As soon as one allows anisotropy to occur, other interesting models arise (see [27] for a review of the role of anisotropy in relativistic astrophysics). Recently, a particular anisotropic model has been singled out [24, 28] (for previous investigations on this kind of models see the references in [28]). In this model, one assumes that the radial stress identically vanishes. The key role is played by the Eq. (3) which shows that the dependence of ψ on η must be a multiplicative dependence from $\sqrt{\eta}$ only. Therefore, materials with vanishing radial stresses can be characterized, using Eq. (2), *via* equations of state of the form

$$\psi(r, \eta, Y) = \varrho h(r, Y), \tag{5}$$

where h is a positive, but otherwise arbitrary, function.

Once an equation of state has been chosen, the Einstein field equations become a closed system; in spherical symmetry there are three independent equations for the three variables ν, η and Y. It has proven, however, to be very useful to write the field equations as a system of *four* differential equations. This is done by introducing the *mass function* [23, 29] defined as

$$m(r, t) = \frac{Y}{2}\left(1 - Y'^2\eta + \dot{Y}^2 e^{-2\nu}\right), \tag{6}$$

where the dash and the dot denote derivatives with respect to r and t respectively. The mass function is arbitrary (positive) and allows us to write the field equations in the following form (four compatible equations in three variables)

$$m' = 4\pi\epsilon Y^2 Y', \tag{7}$$

$$\dot{m} = -4\pi \Sigma Y^2 \dot{Y}, \tag{8}$$

$$Y' \dot{\eta} = -2\eta(\dot{Y}' - \dot{Y}\nu'), \tag{9}$$

$$\Sigma' = -(\epsilon + \Sigma)\nu' - 2(\Sigma - \Pi)(Y'/Y). \tag{10}$$

3 Physical reasonableness and initial data

It is easy to produce new "solutions" of the Einstein field equations in "matter". Indeed, just pick a metric at will and claim that this is a "solution" referring to the *calculated* energy momentum tensor. Of course, what one has to do to remove the quotation marks from the above statements is to check the physical reasonableness of the results.

First of all, the weak energy condition must be imposed on the energy momentum tensor. This condition requires $T_{\mu\nu}u^\mu u^\nu \geq 0$ for any non-spacelike u^μ and implies, besides positivity of ϵ, non-negativeness of $\epsilon + \Sigma$ and $\epsilon + \Pi$. Due to Eqs. (3) and (4), such conditions are equivalent to differential inequalities on the function ψ, namely, $\frac{\partial \psi}{\partial \eta} \geq 0$ and $\frac{\partial \psi}{\partial Y} \leq 0$.

Imposing the weak energy condition *per se* does not assure physical reasonableness, since there is no guarantee that the energy momentum tensor can be deduced from a field theoretic description of matter. What is needed is the satisfaction of a suitable equation of state. We shall require, in addition, the equation of state to be locally stable [24] (this last requirement could be relaxed in the presence of rotationally-induced stress). The local stability condition requires the (local) equilibrium state of the material to be unstrained. In the "comoving picture" this amounts to saying that the energy must have an absolute minimum at the flat-space values of the metric.

We collect the functions describing admissible equations of state in spherical symmetry as set

$$\Psi = \left\{ \psi \in C^2(\mathrm{R}^3_+, \mathrm{R}_+) : \psi(r, r, 1) = \min \psi(r, Y, \eta),\ \frac{\partial \psi}{\partial \eta} \geq 0,\ \frac{\partial \psi}{\partial Y} \leq 0 \right\}.$$

We shall always assume that the value of ψ at the minimum (which in general can be a function of r) has been rescaled to unity (this can be done without loss of generality).

A solution of the Einstein field equations describes the collapse of an initially regular distribution of matter only if the spacetime admits a spacelike hypersurface ($t = 0$, say) which carries *regular* initial data. This means that the metric, its inverse, and the second fundamental form all have to be continuous at $t = 0$.

On the initial hypersurface we use the scaling freedom of the r coordinate to set $Y(r, 0) = r$. We call a set of initial data *complete* if it is minimal, in the sense that no part of its content can be gauged away by a coordinate transformation. We will now prove that a complete set of initial data for Eqs. (7)–(10), at fixed equation of state, is composed of a pair of functions. Physically, such functions describe the initial distribution of energy density $\epsilon_0 = \epsilon(r, 0)$ and of velocity $V_0 = e^{-\nu(r,0)}\dot{Y}(r, 0)$. It is sometimes convenient to parameterize these two distributions in terms of two other functions $\{F(r), f(r)\}$ where $F(r) = m(r, 0)$ is the initial distribution of mass and $f(r)$ is called the "energy function".

The relationship between the two sets $\{F, f\}$ and $\{\epsilon_0, V_0\}$ is given by the following formulae

$$F' = 4\pi r^2 \epsilon_0, \quad f = V_0^2 - 2F/r. \tag{11}$$

The function F has to be non-negative with $\epsilon_0(0) = \lim_{r\to 0} F(r)/r^3$ finite and non-vanishing, while f has to be greater than -1 to preserve the signature of the metric (see Eq. (13) below) with $\lim_{r\to 0} f(r) = 0$.

The proof that only two arbitrary functions of r give a complete set of initial data at fixed equation of state is implicitly contained in many papers on spherical collapse (see, e.g., [23, 30]). It seems, however, that a complete proof has never been published in detail, so we take this occasion to give it. We denote by a subscript, the initial value of each quantity appearing in the Einstein field equations. We know $F = m_0$, $V_0 = e^{-\nu_0}\dot{Y}_0$. Since $Y_0 = r$ and ψ is a known function of r, Y and η, its initial value ψ_0 is also known – as well as the initial values of the stresses due to formulae (3) and (4) – as a function of r and η_0 ($\psi_0 = \psi(r, r, \eta_0)$). Evaluation of Eq. (7) at $t = 0$ now gives $\psi_0 = F'/(4\pi r^2)$, from which the value of η_0 can be extracted algebraically. At this point, the remaining three field equations can be used to evaluate the remaining data, i.e., $\dot{m}_0$, ν_0 and $\dot{\eta}_0$:

$$\dot{m}_0 = -4\pi \Sigma_0 r^2 V_0, \quad \nu'_0 = -\frac{\Sigma'_0}{\psi_0 + \Sigma_0} - 2\frac{\Sigma_0 - \Pi_0}{r(\psi_0 + \Sigma_0)}, \quad \dot{\eta}_0 = -2\eta_0 e^{\nu_0} V'_0. \tag{12}$$

This completes the proof.

In what follows, we consider only solutions which can be interpreted as models of collapsing stars, i.e., isolated objects rather than "universes". This is possible only if the metric matches smoothly with the Schwarzschild vacuum solution (the matching between two metrics is smooth if both the first and the second fundamental forms are continuous on the matching surface).

4 Classification and nature of singularities

To understand the collapsing scenarios as well as the nature of the singularities, one would like to analyse exact solutions of the Einstein field equations. However, it goes without saying that the non-linearity of such equations makes them essentially untractable, even in spherical symmetry, without further simplifying hypotheses (such as, e.g., vanishing shear or acceleration; see [31, 32]). In view of such difficulties, one would like to extract information from the Einstein equations without solving them completely.

The mathematical structure of spherical collapse discussed in the previous section shows that there is a one-to-one correspondence between solutions and choices of triplets (s, say) of functions $s = \{F(r), f(r), \psi(r, \eta, Y)\}$ (we are, of course, identifying solutions *modulo* gauge transformations). We denote the set of solutions parameterized in this way by $\mathcal{S}$. Since in this parameterization we have already taken into account regularity as well as physical admissibility, the whole physical content of the cosmic censorship problem can be translated in the mathematical terms of predicting the end state of any choice of $s \in \mathcal{S}$.

Non-singular solutions

The existence of non-singular, non-static solutions of the Einstein field equations is not forbidden by the singularity theorems (we refer the reader to [33] for a recent review on singularity theorems and further references) and a gravitational collapse can lead to a bouncing back, at a finite area radius, without singularity formation. This phenomenon can be recursive, producing an eternally oscillating, globally regular solution [34–36]. Speaking very roughly, one can prepare regular initial data in such a way that the region of possible trapped surface formation is disconnected from the data. In this way the remaining hypotheses of the singularity theorems can be satisfied without singularity formation (see also [24]).

Much more exotic than oscillating solutions, are globally regular solutions that describe non-singular blackholes [33]. These are matter objects "sitting" inside their Schwarzschild radii. Their finite extension replaces the central singularity with a matter-filled, non-singular region. In this case trapped surfaces are obviously present and therefore the strong energy condition must be violated (a simple argument shows that the dominant energy condition also is necessarily violated). In any case, such strange objects are not ruled out by general relativity in as much as only the weak energy condition and an equation of state are required [37–39]. However, as far as we are aware, it is not known if a fully dynamic solution exists which could eventually lead to this exotic end state.

Singular solutions

We now move to the case in which a singularity is formed in the future of a regular initial data set. The singularities of spherically symmetric matter-filled spacetimes can be recognized from the divergence of the energy density and curvature scalars, such as e.g., the Kretschmann scalar $R^{\mu\nu\varrho\sigma}R_{\mu\nu\varrho\sigma}$. Essentially, these singularities can be of two kinds. We shall call *shell crossing* singularities those at which Y' vanishes ($Y \neq 0$), and *shell focusing* singularities those at which Y vanishes. To the two kinds of singularities correspond two curves $t_{sc}(r)$ and $t_f(r)$ in the r-t plane, defined by $Y'(r, t_{sc}(r)) = 0$ and by $Y(r, t_f(r)) = 0$ respectively. Physically, the shell crossing curve gives the time at which two neighbouring shells of matter intersect each other, while the shell focusing curve identifies at which time the shell labeled r "crushes to zero size". Of these two kinds of singularities, the one of physical interest is obviously that occurring first in the sense that, at fixed r, one has $t_f > t_{sc}$ or vice versa. It would be very interesting, therefore, to carry out a study of the field equations in order to obtain conditions for shell crossing avoidance in spherical spacetimes. This study has been carried out up to now only for dust spacetimes [40–42].

From the point of view of censorship, the nature of a singularity in an asymptotically flat, initially regular spacetime can be one of the following. First of all, a singularity can be *spacelike*, like, e.g., the Schwarzschild singularity or the singularity occurring at the end state of the collapse of a Oppenheimer–Snyder dust cloud (see Sect. 4.2.2 below). These singularities lie in the future of all possible observers and therefore are strongly censored (i.e., allowed by strong cosmic censorship). If a singularity is not strongly censored, then it is naked, i.e., visible to some observer. However, two different cases can occur, namely, the singularity can be *locally* or *globally* naked. A singularity is locally naked if light signals can emerge from it but fall back without reaching any asymptotic observer. A singularity of this kind will be visible only to observers who have crossed the horizon; therefore, weak cosmic censorship holds for such end states. An example of a spacetime containing a locally naked singularity is provided by the Kerr spacetime with mass greater than the angular momentum per unit mass (or by the Reissner–Nordström spacetime with mass greater than charge). Finally, a singularity is globally naked if light-rays emerging from it can reach an asymptotic observer.

4.2.1 Shell crossing singularities

The first explicit example showing formation of a naked singularity was found as a shell crossing singularity in a spherical dust cloud [43]. It can be shown that these singularities are timelike and are always locally naked.

Some definitions have been proposed to put singularities "in the order of increasing seriousness" [44]. Essentially, what is done is to check the behaviour of the invariants of the Riemann tensor in the approach to the singularity. According to such criteria, the shell crossing singularities turn out to be "weak" at least as compared with the shell focusing singularities. This "weakness" is considered by some authors as a hint of a possible extension of the spacetime [45]. However, in spite of their "weakness", there is at the moment no general proof available of extendibility of spacetimes through a shell cross (although some encouraging results exist; see [46]). The unique exception is a paper by Papapetrou and Hamoui [47]. In this paper the authors claim to have explicitly found the extension in the case of "degenerate" shell crossing singularities, i.e., when the curve $t_{sc}(r)$ degenerates to an "instant of time" $t_{sc} = T = \text{const.}$ In this case, it is easy to check that the crossing occurs at a "point" $r = \text{const.}$ rather than at a "3-space" and this is the key to their treatment. However, some results of this paper are unclear from the physical point of view.

What is actually available is only that a *continuous* extension of shell crossing singularities exists in the dust case [41]. We shall show this in a slightly more general case. Integrating equation (9) formally with respect to time we can write

$$\eta^{-1} = \frac{Y'^2}{1+f}\Omega^2, \tag{13}$$

where f is the energy function and $\Omega(r,t) = \exp\left(-\int_0^t \frac{\nu'\dot{Y}}{Y'}d\tilde{t}\right)$. Changing variable from r to $Y = Y(r,t)$ one gets the metric

$$ds^2 = \frac{\Omega^2}{1+f}\left[\left(\dot{Y}^2 - (1+f)\Omega^{-2}e^{2\nu}\right)dt^2 - 2\dot{Y}dYdt + dY^2\right] + Y^2(d\theta^2 + \sin^2\theta d\varphi^2).$$

If Ω is finite and non-vanishing at $t = t_{sc}(r)$, the above metric is continuous with continuous inverse at such a surface. This happens if ν' vanishes (dust case) or, more generally, if ν' goes to zero at least as Y' does at the shell cross (this happens, for instance, in the case of vanishing radial stresses).

In a recent paper [48], Szekeres and Lun have shown that there exists a system of coordinates in which the metric is of class C^1 as $t \to t_{sc}^-$. However, again, this result *per se* does not show extendibility of the spacetime (see also [49]).

4.2.2 Non-central shell focusing singularities

It is important to distinguish the *central* shell focusing singularity, i.e., that occurring at $r = 0$, from the other focusing singularities, since in many cases it is easy to prove that non-central singularities are censored.

A necessary condition for the visibility of a "point" r is that the condition $1 - \frac{2m(r,t)}{Y(r,t)} > 0$, implying absence of trapped surfaces (see next subsection), is satisfied. Since the apparent horizon is the boundary of the region containing trapped surfaces, the above condition implies that the time of formation of the apparent horizon $t_{ah}(r)$, defined as $2m(r, t_{ah}(r))Y(r, t_{ah}(r))$, must not be *before* $t_f(r)$, i.e., $t_{ah}(r) \geq t_f(r)$. Now suppose $m(r, t_f(r))$ to be different from zero (as we have seen, this is not the case at the central singularity where m has to vanish as r^3). Then $1 - 2m(r,t)/Y(r,t)$ goes to minus infinity as t tends to $t_f(r)$ so that the singularity is covered. This shows that all naked shell focusing singularities are necessarily massless, in the sense that m has to vanish there [50]. It follows that any non-central singularity will certainly be censored if the mass is an increasing function with respect to t. Now Eq. (8) gives $\dot{m} = -4\pi\Sigma Y^2\dot{Y}$ and, since $\dot{Y} < 0$ during collapse, we conclude that non-central singularities are always covered if the radial stress Σ is non-negative [51] (in particular, all non-central singularities occurring in dust as well as in models with only tangential stress are covered, since mass does not depend on time in this case). In the presence of radial tensions, the question is still open. It is known that a perfect fluid with $p = k\epsilon$ exhibits naked non-central singularities if $k < -1/3$ [51].

4.2.3 Central singularities: the root equation

The first explicit examples of the formation of a naked shell focusing singularity were provided by Eardley [52], Eardley and Smarr [53] and by Christodoulou [54]. Since then the techniques to study the nature of the singularities in spherically symmetric spacetimes have been developed by many authors (see the references in [42]) and finally settled by Dwivedi and Joshi [55].

The key idea is the following: if the singularity is visible, at least locally, there must exist light signals coming out from it. Therefore, by investigating the behaviour of radial null geodesics near the singularity, one can try to find out if outgoing null curves meet the singularity in the past. On such radial null geodesics, the derivative of $Y(r, t)$ reads

$$\frac{dY}{dr} = Y' + \eta^{-1/2}\dot{Y}e^{-\nu}.$$

Using (6) and (4) in the above equation we obtain

$$\frac{dY}{dr} = Y'\left[1 - \sqrt{1 + \frac{\Omega^2}{1+f}\left(\frac{2m}{Y} - 1\right)}\right], \tag{14}$$

where Y' has to be understood as a known function of Y and r. If the singularity is naked, Eq. (14) must have at least one solution with definite, outgoing tangent at $r = 0$, i.e., a solution of the kind $Y = X_0 r^\alpha$ where $\alpha > 1$ and X_0 is a positive constant. Clearly, this behaviour is possible only if the necessary condition $1 - 2m/Y > 0$ is satisfied. Indeed Y' is equal to one, and therefore positive, on the initial data surface. If no shell crossing occurs, it remains positive, so that the right-hand side of Eq. (14) cannot remain positive if $2m/Y - 1$ changes sign.

Once the necessary condition is satisfied, one has to check if both X_0 and α exist such that the solution of Eq. (14) is of the specified form near the singularity. On using l'Hôpital's rule we have

$$X_0 = \lim_{r\to 0}\left(\frac{1}{\alpha r^{\alpha-1}}\frac{dY}{dr}\right)_{Y=X_0 r^\alpha}. \tag{15}$$

Using again (14), this equation becomes an *algebraic* relation for X_0 at fixed α. If a positive definite X_0 exists the singularity is naked.

4.2.4 The dust case

The general exact solution of the Einstein field equations is known in the most simple case of vanishing pressure (dust) [29, 56, 57]. In this case, from Eq. (10), one gets $\nu = 0$ (more precisely, e^ν is an arbitrary function of t only which can be rescaled to unity without loss of generality). Then $\Omega = 0$ and it follows from Eq. (4) that $\eta^{-1} = Y'^2/(1+f)$. The mass is constant in time ($m = F(r)$) due to Eq. (8) with $\Sigma = 0$. Therefore (6) can be written as a Kepler-like equation ($\dot{Y}^2 = f + 2F/Y$), which is integrable in parametric form for $f \neq 0$ and in closed form for $f = 0$. Finally, the density can be read off from (7) as $\epsilon = F'/(4\pi Y^2 Y')$.

Great effort has been spent in trying to understand the nature of the central singularity in this solution [41, 42, 53–55, 58–62] and we now know the complete spectrum of possible end states of the dust evolution in dependence on the initial data. We recall here what happens in the case of marginally bound solutions ($f = 0$) [62] since it

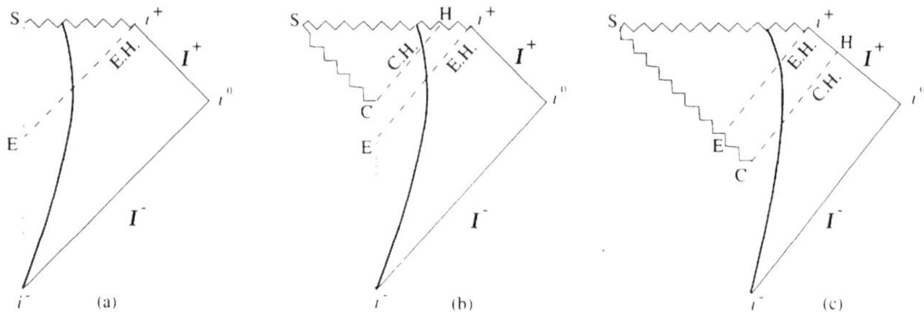

Fig. 1. Penrose diagrams for collapsing dust clouds. The dotted line represents the center of the star, and the bold jagged line represents the singularity. The continuous line connecting i^- to the singularity represents the boundary of the collapsing cloud. Three different situations can occur, depending on the choice of initial data: (a) "generalized" Oppenheimer–Snyder collapse: the singularity is covered; (b) locally naked singularity; (c) globally naked singularity

is sufficiently general to illustrate a tendency and simple enough to be recalled in a few lines.

For marginally bound dust, the solutions $s = \{F, 1, 1\}$ can be uniquely characterized by the expansion of the function $F(r)$ at $r = 0$ or, what is the same, by the expansion of $\epsilon_0 = F'/4\pi r^2$. Using this expansion in the root equation, it is not difficult to check the following results.

- If the first non-vanishing term corresponds to $n = 1$ or $n = 2$ Eq. (15) always has a real positive root: the singularity is naked.
- If the first non-vanishing term is $n = 3$, the root equation reads

$$2x^4 + x^3 + \xi x - \xi = 0, \tag{16}$$

  where $x^2 = X$ and $\xi = F_3/F_0{}^{5/2}$. From the theory of quartics one can show that this equation admits a real positive root if $\xi < \xi_c = -(26 + 15\sqrt{3})/2$. Therefore, ξ_c is a critical parameter: at $\xi = \xi_c$ a "phase transition" occurs and the end state of collapse turns from a naked singularity to a black hole.
- If $n > 3$, the limit in Eq. (14) diverges: the singularity is covered. In particular, this case contains the solution first discovered by Oppenheimer and Snyder [58] describing a homogeneous dust cloud.

The naked singularities mentioned above are *locally* naked. It can, however, be shown that if locally naked singularities occur in dust spacetimes, then spacetimes containing globally naked singularities can be build up from these by matching procedures. The Penrose diagrams corresponding to the three different cases are shown in Fig. 1.

The above results can be extended to the general case of collapsing dust clouds, so that the final fate of the dust solutions $s = \{F, f, 1\}$ is completely known. The final fate depends on a parameter which is a combination of coefficients of the expansions of F and f near $r = 0$, and a structure similar to that of marginally bound collapse arises (see [42] for details).

4.2.5 Vanishing radial stresses

Recently, the general solution for spherically symmetric dust has been extended to the case in which only the radial stress vanishes [24, 28]. The solution can be reduced to quadratures using a system of coordinates first introduced by Ori [63] for charged dust. One of the new coordinates is the mass m which is constant in time (due to (8) with $\Sigma = 0$), the other coordinate is the "area radius" Y. In such coordinates the metric reads

$$ds^2 = -\Gamma^2\left(1 - \frac{2m}{Y}\right)dm^2 + 2\sqrt{1+f}\frac{\Gamma}{hu}dY\,dm - \frac{1}{u^2}dY^2 \\ + Y^2(d\theta^2 + \sin^2\theta\, d\varphi^2), \tag{17}$$

where

$$u = -\sqrt{-1 + \frac{2m}{Y} + \frac{1+f}{h^2}}, \quad \Gamma = g(m) + \int \frac{h}{u^2\sqrt{1+f}}\frac{\partial u}{\partial m}dY, \tag{18}$$

and the function $g(m)$ is arbitrary.

The problem of understanding the nature of the singularities for such solutions is essentially still open. It is, indeed, possible to write the root equation in explicit form, but this equation contains a sort of "non-locality" due to the integral entering in the definition of Γ. Cosequently only a few special cases have been analysed so far [28, 64, 65].

Among the solutions with tangential stresses a particularly interesting one is the *Einstein cluster* [66–68]. This is a spherically symmetric cluster of rotating particles. The motion of the particles is sustained by the angular momentum L whose average effect is to introduce a non-vanishing tangential stress in the energy-momentum tensor. The corresponding equation of state has the form $h(m, Y) = \sqrt{1 + L^2(m)/Y^2}$. Therefore, a solution is uniquely identified by the choice of three arbitrary functions of m alone, namely, F, f and L^2 (for $L = 0$ one recovers dust). It turns out that the final state "at fixed dust background" (i.e., for fixed F, f) depends on the expansion of L^2 near $m = 0$ ($L \approx \beta m^y$, say) [69, 70]. Considering, for simplicity, the marginally bound case, one finds that for $4/3 < y < 2$ the singularity does not form (the system bounces back), while for $4/3 < y < 7/3$ a naked singularity is always formed. At $y = 7/3$ a sort of transition takes place and the evolution of the model is such that only the critical branch is changed, un-covering a part of the blackhole region in the corresponding dust spacetime; the non-critical branch is the same as in dust spacetimes. Finally for $y > 7/3$ the evolution always leads to the same end state of the corresponding dust solution.

4.2.6 Self-similar collapse

A spherically symmetric spacetime is self-similar if it admits an homothetic vector ξ, i.e., a vector satisfying $\mathcal{L}_\xi g_{\mu\nu} = 2g_{\mu\nu}$. In the comoving frame the dimensionless variables ν, λ and Y/r depend only on the "similarity variable" $z := r/t$, and the Einstein field equations become *ordinary* differential equations (we refer the reader to [71] for a complete treatment of self-similar solutions). Being governed by ordinary differential equations, self-similar spherical collapse can be analyzed with the powerful techniques of dynamical systems theory [72]. The analysis of the singularities forming in self-similar spacetimes has been carried out by many authors for different equations of state, such as dust [52, 53, 74–77], barotropic perfect fluids [75, 78, 79], radiation (Vaidya) shells [80–82] and in general cases [83, 84].

The picture arising resembles the dust case in the sense that both naked singularities and black holes can form depending on the values of the parameters characterizing the solution. A thorough review of these and other features of self-similar solutions can be found in [73]; here we limit ourselves to stressing the fact that naked singularities exist in self-similar solutions with pressure, thereby showing that the phenomenon of naked singularity formation cannot be considered as an artifact of dust solutions (i.e., vanishing stresses).

4.2.7 General stresses

The problem of predicting the final fate of an initially regular distribution of matter supported by an arbitrary distribution of stress-energy (including, e.g., the case of isotropic perfect fluids but also anisotropic crystalline structures which are thought to form at extremely high densities), is still open even in the spherically symmetric case. First of all, one has to take into account the fact that there is a high degree of uncertainty in the properties of the equation of state at very high densities. Recently, Christodoulou [85] initiated the analysis of a simple model composed of a dust ("soft") phase for energy density below a certain value $\overline{\epsilon}$ and a stiff ("hard") phase for $\epsilon > \overline{\epsilon}$ (in the hard phase the pressure is given by $p = \epsilon - \overline{\epsilon}$).

Although the details of the collapse in the presence of general matter fields are still largely unknown, it is very unlikely that the "embarrassing" examples of naked singularity formation like those occurring in dust can eventually be eliminated with the "addition" of stresses. Indeed, it is reasonable to think that a sector of naked singularities exists in the choice of initial data for any fixed equation of state [50]. The main issues that have, therefore, to be addressed are the *genericity* and the *stability* of naked singularity formation.

Both the above italicized terms have a somewhat intuitive meaning that is, however, difficult to express in mathematical terms. Regarding genericity, one can mean that the set of initial data leading to naked singularities is not of measure zero. For instance, it has been shown [42] that among the dust solutions $s = \{F, f, 1\}$ naked singularities are generic in the sense that, at a fixed density profile F, one can always choose energy functions f leading to black holes or naked singularities. A generalization of such a result would be that the naked singularities are generic – in this specific sense – at a fixed, but arbitrary, equation of state ψ (there is some convincing evidence for this; see [30]).

The issue of stability is even more delicate than that of genericity. Indeed, any exact solution of a physical theory must survive to small but arbitrary perturbations in order to serve as a candidate for describing nature. In the case of exact solutions of the Einstein field equations, the notion of stability is very delicate due to to the gauge invariance of the theory with respect to spacetime diffeomorphisms. Recently, some evidence of stability of dust naked singularities against perturbations has been obtained [86].

5 A picnic on the side of the road

It is well-known that, if the mass of a collapsing object does not fall below the neutron star limit ($\sim 3M_{\odot}$), no physical process is able to produce enough pressure to balance the gravitational pull so that continued gravitational collapse must occur. It is widely believed that the final state of this process is a black hole. However, what general relativity actually predicts, in the cases which have been analyzed so far, is that either a black hole or a naked singularity is formed, depending on the initial distribution of density and velocity and on the constitutive nature of the collapsing matter. One may raise the objection that most of the known analytical results could be an artifact of spherical symmetry. However, from the numerical point of view, some evidence that this is not the case is showing up. Therefore, as singularity theorems showed that the singularities occurring in collapse are generic and not any artifact of symmetry, a similar situation may hold for the nature of singularities as well.

One may ask at this point if and when, a cosmic censorship theorem holds in nature. An answer to such a question could be in the negative. However, it remains to understand the physics underlying the end state of gravitational collapse with respect to the choice of initial data in a fixed matter model. In a famous novel [87] (that inspired the film *Stalker* by A. Tarkovsky) a short visit by extraterrestrial beings to the earth occurs. The gap between the two civilizations is so high that human beings are, with respect to the "garbage" left by the visitors, like ants exploring what remains on the side of the road after a human picnic. Something they find is useful, something useless, something even dangerous, but anyway everything is obscure and difficult to understand, looking like the weak shadow of a wonderful abyss of knowledge. Our present understanding of cosmic censorship resembles this situation. Indeed, we are getting a variety of mathematical hints with somewhat obscure physical meanings. For instance: the condition on ξ recalled in Sect. 4.2.4, the constraints arising in the gravitational collapse of Einstein clusters, the dimensionless numbers arising in Choptuik's numerical results [14], the condition on the radiation flux which arises in Vaidya collapse [3]. Such "numbers" should presumably be the remnant weak shadow of a general theorem when its hypotheses are enormously restricted by the choice of the equation of state and of the symmetries adopted.

To get rid of this puzzle appears to be one of the most exciting objectives of future research in classical relativity theory.

Acknowledgements

Many discussions with Elisa Brinis and Pankaj Joshi are gratefully acknowledged. S.J. thanks the ICSC World Laboratory (Lausanne, Switzerland) for the Chandrasekhar Memorial Fellowship (1998–99).

References

1. Hawking S.W., Ellis G.F.R. (1973): The Large Scale Structure of Space-Time. Cambridge University Press, Cambridge
2. Penrose R. (1969): Nuovo Cimento **1**, 252
3. Joshi P.S. (1993): Global Aspects in Gravitation and Cosmology. Clarendon press, Oxford
4. Rendall A.D. (1996): An introduction to the Einstein–Vlasov system (Lectures given at the Banach centre, Warsaw). gr-qc/9604001
5. Rein G., Rendall A.D., Schaeffer, J. (1998): Phys. Rev. D **58**, 044007
6. Israel W. (1984): Found. Phys. **14**, 1049
7. Eardley D.M. (1987): Cosmic Strings and Origin of Structure in the Universe, in *Gravitation and Astrophysics, Cargese Lectures*, ed. by B. Carter, J. Hartle, Plenum, New York–London, pp. 293–305
8. Clarke C.J.S. (1993): Class. Quantum Grav. **10**, 1375
9. Wald R.M. (1998): Gravitational Collapse and Cosmic Censorship, in *Black Holes, Gravitational Radiation and the Universe, Esssays in the honour of C. V. Vishveswara*, ed. by R. Bala, B. Iyer and Biplab Bhawal, pp. 69–85
10. Joshi P.S. (1997): Gravitational Collapse, in *Singularities, Black Holes and Cosmic Censorship*, ed. by P.S. Joshi, IUCAA publication, Pune, India, pp. 33–54
11. Singh T.P. (1996): Gravitational Collapse and Cosmic Censorship, in *Proc. of the XVIIIth IAGRG meeting*, ed. by G. Date, B. Iyer, Inst. of Math. Sci. Report **117**, 57–76
12. Christodoulou D. (1994): Ann. Math. **140**, 607
13. Christodoulou D. (1999): Ann. Math. **149**, 183
14. Gundlach C. (1998): Adv. Theor. Math. Phys. **2**, 1
15. Burko L., Ori A. eds. (1998): The Internal Structure of Black Holes and Spacetime Singularities. Ann. Isr. Phys. Soc. **13** I.O.P., Bristol
16. Thorne K.S. (1972): Nonspherical Gravitational Collapse: A Short Review, in *Magic without Magic*, ed. by J. R. Klauder, Freeman, San Francisco, pp. 231–258
17. Lemos J.P.S. (1999): Phys. Rev. D **59**, 044020
18. Israel W. (1986): Canad. J. Phys. **64**, 120
19. Israel W. (1986): Phys. Rev. Lett. **56**, 789
20. Szekeres P. (1975): Phys. Rev. D **12**, 2941
21. Joshi P.S., Królak A. (1996): Class. Quantum Grav. **13**, 3069
22. Deshingkar S.S., Jhingan S., Joshi P.S. (1997): Gen. Rel. Grav. **27**, 1071
23. Misner C.W., Sharp D.H. (1964): Phys. Rev. B **136**, 571
24. Magli G. (1997): Class. Quantum Grav. **14**, 1937
25. Herrera L., Martinez J. (1998): Gen. Rel. Grav. **30**, 445
26. Kijowski J., Magli G. (1998): Class. Quantum Grav. **15**, 3891
27. Herrera L., Santos N. (1997): Phys. Rep. **286**, 2
28. Magli G. (1998): Class. Quantum Grav. **15**, 3215
29. Lemaitre G. (1933): Ann. Soc. Sci. Bruxelles A **53**, 51; (1997): Gen. Rel. Grav. **29**, 641
30. Joshi P.S., Dwivedi I.H. (1999): Class. Quantum Grav. **16**, 41

31. Kramer D., Stephani H., Herlt E., MacCallum M. (1980): Exact Solutions of Einstein's Field Equations. Cambridge University Press, Cambridge
32. Krasiński A. (1997): Inhomogeneous Cosmological Models, Cambridge University Press, Cambridge
33. Senovilla, J.M.M. (1998): Gen. Relativity Gravitation **30**, 701
34. Bonnor W.B., Faulkes M.C. (1967): Mon. Not. Roy. Astr. Soc. **137**, 239
35. Thompson I.H., Whitrow G.J. (1968): Mon. Not. Roy. Astr. Soc. **139**, 499
36. Bondi H. (1969): Mon. Not. Roy. Astr. Soc. **142**, 333
37. Mars M., Martín-Prats M., Senovilla J. (1996): Class. Quantum Grav. L **13**, 51
38. Ayón-Beato E., Garcia A. (1998): Phys. Rev. Lett. **80**, 5056
39. Magli G. (1999): Rep. Math. Phys. **44**, 407–412
40. Hellaby C., Lake K. (1985): Ap. J. **290**, 381
41. Newman R.P.A.C. (1986): Class. Quantum Grav. **3**, 527
42. Jhingan S., Joshi P.S. (1997): The Structure of Singularity in Spherical Inhomogeneous Dust Collapse, in *The internal structure of black holes and spacetime singularities*, ed. by L. Burko, A. Ori, Ann. Isr. Phys. Soc. **13**, pp. 357–389
43. Yodzis P., Seifert H., Müller zum Hagen H. (1973): Comm. Math. Phys. **34**, 135
44. Clarke C.J.S., Królak A. (1986): J. Geom. and Phys. **2**, 127
45. Clarke C.J.S. (1993): The Analysis of Space-Time Singularities, Cambridge University Press, Cambridge
46. Clarke C.J.S., O'Donnell N. (1993): Ren. Sem. Mat. Un. Pol. Torino **50**, 39
47. Papapetrou A., Hamoui A. (1967): Ann. Inst. Henri Poincaré **6**, 343
48. Szekeres P., Lun A. (1997): What is a shell crossing singularity? Pre-print
49. Szekeres P., Iyer V. (1993): Phys. Rev. D **47**, 4362
50. Dwivedi I.H., Joshi P.S. (1994): Comm. Math. Phys. **166**, 117
51. Cooperstock F.I., Jhingan S., Joshi P.S., Singh, T.P. (1997): Class. Quantum Grav. **14**, 2195
52. Eardley D.M. (1974): Phys. Rev. Lett. **33**, 442
53. Eardley D.M., Smarr L. (1979): Phys. Rev. D **19**, 2239
54. Christodoulou D. (1984): Comm. Math. Phys. **93**, 171
55. Joshi P.S., Dwivedi I.H. (1993): Phys. Rev. D **47**, 5357
56. Tolman R.C. (1934): Proc. Nat. Acad. Sci. **20**, 169
57. Bondi H. (1947): Mon. Not. Roy. Astr. Soc. **107**, 410
58. Oppenheimer J., Snyder H. (1939): Phys. Rev. **56**, 455
59. Gorini V., Grillo G., Pelizza M. (1989): Phys. Lett. A **135**, 154
60. Grillo G. (1991): Class. Quantum Grav. **8**, 739
61. Lake K. (1992): Phys. Rev. Lett. **68**, 3129
62. Singh T.P., Joshi P.S. (1996): Class. Quantum Grav. **13**, 559
63. Ori A. (1990): Class. Quantum Grav. **7**, 985
64. Singh T.P., Witten L. (1997): Class. Quantum Grav. **14**, 3489
65. Barve S., Singh T.P., Witten L.: Spherical gravitational collapse: tangential pressure and related equations of state. gr-qc/9901080
66. Einstein A. (1939): Ann. Math. **40** (4), 922
67. Datta B. K. (1970): Gen. Rel. Grav. **1**, 19
68. Bondi H. (1971): Gen. Rel. Grav. **2**, 321
69. Harada T., Iguchi H., Nakao K. (1998): Phys. Rev. D **58**, R041502
70. Jhingan S., Magli G.: Black holes vs. naked singularities formation in collapsing Einstein's clusters, Phys. Rev. D, in press
71. Carr B., Coley A.A.: A complete classification of spherically symmetric perfect fluid similarity solutions, Phys. Rev. D, in press

72. Bogoyavlenski O.I. (1985): Methods in the Qualitative Theory of Dynamical Systems in Astrophysics and Gas Dynamics. Springer, Berlin–Heidelberg–New York
73. Carr B.J. (2000): Spherically symmetric similarity solutions, in *Proc. 7th Canadian Conference on General Relativity and Relativistic Astrophysics*, ed. by D. Hobill, in press
74. Waugh B., Lake K. (1988): Phys. Rev. D **38**, 1315
75. Ori A., Piran T. (1990): Phys. Rev. D **42**, 1068
76. Lemos J.P.S. (1991): Phys. Lett. A **158**, 279
77. Lemos J.P.S. (1992): Phys. Rev. Lett. **68**, 1447
78. Ori A., Piran T. (1987): Phys. Rev. Lett. **59**, 2137
79. Joshi P.S., Dwivedi I.H. (1992): Comm. Math. Phys. **146**, 333
80. Kuroda Y. (1984): Prog. Theor. Phys. **72**, 63
81. Papapetrou A. (1985): Formation of Singularity and Causality, in *A Random Walk in Relativity and Cosmology*, ed. by N. Dadhich, J. K. Rao, J. V. Narlikar, C. V. Vishveshwara, Wiley Eastern, New Delhi, pp. 184–191
82. Lake K. (1986): Phys. Lett. A **116**, 17
83. Waugh B., Lake K. (1989): Phys. Rev. D **40**, 2137
84. Lake K., Zannias T. (1990): Phys. Rev. D. **41**, 3866
85. Christodoulou D. (1993): Relativistic fluids and gravitational collapse, in *Global Structure and Evolution in General Relativity*, Lecture Notes in Phys. **460**, Springer, Berlin Heidelberg New York, pp. 33–54
86. Iguchi H., Nakao K., Harada T. (1998): Phys. Rev. D **57**, 7262
87. Strugackij A., Strugackij B. (1980): Picnic na občine, Molodaja gvardija, Moscow

Black Holes and Solitons

L. Martina

Abstract. The Two-dimensional Jackiw–Teitelboim gravity is represented as a completely integrable nonlinear reaction – diffusion system, whose Euclidean version leads to the non-linear Schrödinger equation. The soliton-like solutions, to such systems called dissipatons, characterize completely the black holes of the gravity model under consideration (the black hole horizon, the Hawking temperature and the causal structure). Collision of black holes is described in terms of elastic scattering of dissipatons, which shows a novel transmissionless character, creating a metastable state with a specific lifetime. Finally, alternative descriptions of the model in terms of other completely integrable systems are discussed.

The wonderful properties of black holes have long made it attractive to have available a lower-dimensional analog, which would exhibit the key features without the unnecessary complications. This is true in the Jackiw–Teitelboim (JT) model [1, 2], obtained as a 2-dimensional reduction of the (2+1) Einstein–Hilbert action, or also as a spherical reduction of a 4D dilaton-gravity Einstein–Hilbert–Maxwell action [3]. Recently, the study of this low-dimensional gravity model has received a great deal of attention because of the discovery of black hole (BH) solutions [4, 5]. Moreover, the existence of a black hole implies a non-trivial causal structure, which is related to the Hawking radiation phenomenon [6] and to interesting thermodynamical properties [3, 7–9]. However, the non-trivial causal structure associated with a BH is hardly described in the conformal gauge. Hence, one is encouraged to look for alternative formulations. For instance, the connection between JT gravity and the sine-Gordon equation is studied in [10] . More directly, and in the context of the gauge formulation of the JT model, we have investigated [11, 12] a noncovariant gauge choices, related to the SO (2, 1) Heisenberg model, which leads to a multicomponent completely integrable system for the Zweibein fields. This system is the real analog of the nonlinear Schrödinger (NLS) equation and the 1-soliton solutions obtained, called "dissipatons" because of the exponential dependence of their shape on time, are related to the BH of the JT theory. In this sense we can describe the horizon position and the causal structure of the solution in terms of solitonic parameters. Moreover, since the dissipaton amplitude scales as the inverse of the cosmological constant, strong nonlinear effects are no longer negligible at large scales. Furthermore, only dissipatons moving at velocities less than a critial value lead to BH solutions. Thus, the relativistic bound on the velocities allowed is recovered in the present non-relativistic picture. Furthermore, we deal with the Euclidean version of the previous treatment, which enables us to compute the Hawking temperature in terms of the NLS soliton parameters. Finally, we analyze a type of collision of two black holes in terms of elastic scattering of dissipatons. This scattering has a transmissionless character (in contrast with the reflectionless solitons) and creates a metastable state of BH's with a specific lifetime.

We first review the gauge formulation of the JT model. The main idea of a gauge field theory of gravity is that the general coordinate transformations are implemented by gauge transformations. These do not act on the original metric tensor, but on the Vielbein and spin connections. Thus, the local gauge group induces the metric tensor on a base space [13]. This approach has been used for the JT model [14–17], given by

$$S = \int_{\mathcal{M}} \sqrt{-g}\; V_0 (R - \Lambda) dx^1\, dx^2, \tag{1}$$

where R is the scalar curvature, Λ the cosmological constant and V_0 is a world scalar Lagrange multiplier, or dilaton field. Then one introduces the "rotated" Zweibein fields $q_a^{\pm}$ $(a = 1, 2)$, defined by

$$g_{ab} = -\frac{4}{\Lambda}(q_a^+ q_b^- + q_b^+ q_a^-) \tag{2}$$

and a spin connection V_a, taken as independent variables and combined into the connection 1-form

$$J = J_a dx^a, \quad J_a = \frac{i}{2}\left(q_a^+ + q_a^-\right)\tau_1 - \frac{i}{2}\left(q_a^+ - q_a^-\right)\tau_2 + \frac{i}{4} V_a \tau_0, \tag{3}$$

where τ_i $(i = 0, 1, 2)$ are the basis elements of the SL$(2, \mathbf{R})$ algebra, satisfying the set of relations

$$\tau_i \tau_j = h_{ij} + i\, c_{ijk} \tau_k, \quad \left(h_{ij}\right) = \mathrm{diag}(-1, -1, 1)\,, \quad c_{ijk} = -\epsilon_{ijl} h_{lk}. \tag{4}$$

This parametrization realizes a $\mathbf{Z}_2$-grading of the connection algebra with isotropy group O$(1, 1)$. Therefore, τ_a $(a = 1, 2)$ and τ_0 are the generators of translations and of the Lorentz transformations, respectively, generating the de Sitter group SO$(1, 2)$.

Now, from the vanishing of the curvature 2-form $F = dJ + J \wedge J \equiv 0$, one gets certain equations, involving the covariant derivative $D_a^{\pm} \equiv \partial_a \pm (1/2) V_a$, which provide the curvature of the connection (3) and the torsionless condition. Such a system can be derived also from the action functional

$$S = \int_{\mathcal{M}} \epsilon_{ab} \left[q_0^+\, D_a^+\, q_b^- + q_0^-\, D_a^-\, q_b^+ + \frac{1}{8} V_0 (\partial_a V_b - 4\, q_a^+\, q_b^-) \right] dx^1\, dx^2,$$

where $q_0^{\pm}$ are new Lagrange multipliers and J_0 is defined by (3). The BF theory [18] action (5) is invariant with respect to the infinitesimal SL$(2, \mathbf{R})$ gauge transformations. These are equivalent to general coordinate transformations if the equation of motion are used [19, 20]. Moreover, the equation for the multiplier J_0 is given by

$\mathcal{D}_a J_0 = 0$, where $\mathcal{D}_a = \partial_a - [J_a, \ \ ]$. The total set of field equations arising from the variation of the action (5) can be written in the form

$$D_\mu^\mp q_\nu^\pm = D_\nu^\mp q_\mu^\pm, \qquad \partial_\mu V_\nu - \partial_\nu V_\mu = 4(q_\mu^+ q_\nu^- - q_\nu^+ q_\mu^-), \tag{5}$$

with $\mu, \nu = 0, 1, 2$, and under the assumption that all the derivatives of $q_a^\pm$ and V_a $(a = 1, 2)$ with respect to an auxiliary variable x^0 vanish. Then, Eq. (5) is again the zero - curvature condition in 3D for the connection J. Actually, Eq. (5) represents the Euler–Lagrange equation for the Chern–Simons (CS) action on the SL(2, **R** group [19–21], which can be seen as a subgroup of the corresponding Poincaré group. In 2D the local symmetry group identifies with the de Sitter group. Thus, the structure obtained is a relict of the 3D CS theory under the effects of the dimensional reduction. Of course, the space of all solutions of the classical field equations, modulo SL(2, **R**) gauge transformations, is finite dimensional. Since the general solution of the system (5) can be given in the form of the right-invariant chiral current $J_\mu = G^{-1}\partial_\mu G$, where G is a differentiable mapping on SL (2, **R**) .

Now, we introduce the so-called moving trihedrals frame $\{\mathbf{n}_i\}$ [22] by the local adjoint representation of the algebra

$$G\tau_i G^{-1} = n_i^k \tau_k, \qquad \mathbf{n}_i = \left(h_{jj}\, n_j^i\right). \tag{6}$$

It satisfies the orthonormal conditions

$$\left(\mathbf{n}_i \mathbf{n}_j\right) = h_{ij}, \ \ \mathbf{n}_i \wedge \mathbf{n}_j = c_{ijk}\mathbf{n}_k \tag{7}$$

and the moving frame changes as

$$\partial_\mu \mathbf{n}_i = (J_\mu)_{ik}^{(ad)} \mathbf{n}_k, \tag{8}$$

where the $J_\mu^{(ad)}$ are the chiral currents J_μ in the adjoint representation. Its integrability is assured by the zero curvature condition, namely, by Eq. (5). Assigning to $\mathbf{n}_0 = \mathbf{s}$ the special role of (pseudo)-spin variable, we see thet it has as phase space the 1-sheeted hyperboloid $(\mathbf{s}, \mathbf{s}) = -1$, and its tangent plane is defined by the vectors $\mathbf{n}_1$, $\mathbf{n}_2$. In the new basis $\mathbf{n}_\pm = \mathbf{n}_1 \pm \mathbf{n}_2$, the linear system (8) takes the form

$$D_\mu^\mp \mathbf{n}_\pm = \mp 2 q_\mu^\pm \mathbf{s}, \qquad \partial_\mu \mathbf{s} = q_\mu^+ \mathbf{n}_- - q_\mu^- \mathbf{n}_+, \tag{9}$$

where one can perform the local gauge transformation

$$\mathbf{s} \to \mathbf{s}, \ \mathbf{n}_+ \to e^{+\alpha}\, \mathbf{n}_+, \mathbf{n}_- \to e^{-\alpha}\, \mathbf{n}_- \tag{10}$$

and, correspondingly, for the potentials we have

$$V_\mu' = V_\mu + 2\partial_\mu \alpha, \ q_\mu^{+\,\prime} = e^\alpha q_\mu^+, \ q_\mu^{-\,\prime} = e^{-\alpha} q_\mu^-. \tag{11}$$

Finally, in establishing the relation between the local trihedrals and the JT metric, we easily obtain the expression

$$g_{ab} = -\frac{2}{\Lambda}\left(\partial_a \mathbf{s}, \partial_b \mathbf{s}\right). \tag{12}$$

This formula enables us to give a gravitational interpretation of the σ- model we are going to discuss.

The idea we follow is to add to (9) a differential constraint in the (x^1, x^2) space for $\mathbf{s}$, such that a completely integrable dynamics is introduced in order to partially fix the gauge freedom in a controlled fashion and allowing a residual local Lorentz covariance. Moreover, for the moment we forget all that concerns the variable x^0 and the current J_0. Precisely, we consider as a constraint the classical continuous Heisenberg model realized on the SL $(2, \mathbf{R})$ /SO $(1, 1)$ coset space

$$\partial_2 \mathbf{s} = C\mathbf{s} \wedge \partial_1^2 \mathbf{s}, \tag{13}$$

where C is an arbitrary dimensional constant, which is set equal to 1. Substitution of (13) into (9) yields

$$q_2^+ = D_1^- q_1^+, \qquad q_2^- = -D_1^+ q_1^-. \tag{14}$$

Taking account of these relations, we can wirte the field equations (5) as

$$D_2^{\mp} q_1^{\pm} \mp (D_1^{\mp})^2 q_1^{\pm} = 0, \qquad \partial_2 V_1 - \partial_1 V_2 = 4\partial_1 (q_1^+ q_1^-).$$

Defining the flat connection

$$A_2 = V_2 + 4(q_1^+ q_1^- - a), \qquad A_1 = V_1, \tag{15}$$

where a is an arbitrary real constant, and gauging it out by a local SO $(1, 1)$ transformation $A_j = 2\partial_j \lambda$, $q_1^{\pm} = q^{\pm} e^{\pm\lambda}$ for a regular real function λ, we get the nonlinear reaction-diffusion (RD) system

$$\partial_2 q^{\pm} \mp \partial_1^2 q^{\pm} \pm 2(q^+ q^- - a) q^{\pm} = 0. \tag{16}$$

Here only the global SO $(1, 1)$ invariance $q^{\pm} \to e^{\pm\alpha} q^{\pm}$ survives. Equation (16) looks like a 2-component reactive - diffusive system, playing an important role in synergetics [23]. However, the unusual negative value for the second diffusion coefficient is crucial for the existence of Hamiltonian structure and the integrability of the model. The main issues concerning the integrability structures associated with system (16) [24] can be obtained by a proper treatment of those for the NLS equation. However, although most of the algebraic forms are slightly generalized with respect the NLS case, the analytical aspects are less trivially extended, because of the reality of the fields $q^{\pm}$ and of their boundary conditions. The Lax pair for (16) is of Zakharov–Shabat type [25], where a rotation of $\pi/2$ in the complex plane of the spectral parameter is required. However, we observe that, in general, the Galilei transformations allowed by (16) $x^1 \to x^1 + 2Vx^2$, $x^2 \to x^2$, $q^{\pm} \to e^{\mp\alpha(x^{1'}, x^{2'})} q^{\pm}(x^{1'}, x^{2'})$, where $\alpha(x^1, x^2) = V^2 x^2 + Vx^1$, do not preserve the boundary conditions. Thus, solutions with nice asymptotics $q^{\pm} \to 0$ for $x^1 \to \pm\infty$ may become unbounded in reference frames moving at sufficiently high velocity.

In the gauge fixed by (13), the metric tensor (2) takes the form

$$g_{00} = \frac{8}{\Lambda} \partial_1 q^+ \partial_1 q^-, \quad g_{11} = -\frac{8}{\Lambda} q^+ q^-, \tag{17}$$

$$g_{01} = g_{10} = -\frac{4}{\Lambda}(q^-\partial_1 q^+ - q^+\partial_1 q^-).$$

We observe that the components g_{11} and g_{01} are densities of the simplest conserved quantities for (16), i.e., the "mass" and the "momentum", respectively.

In the chosen gauge, the equations for the dilaton field are

$$\partial_1 V_0 = 4\left(q_1^+ q_0^- - q_0^+ q_1^-\right), \qquad \partial_2 V_0 = 4\left(\partial_1 q_1^+ q_0^- + q_0^+ \partial_1 q_1^-\right),$$
$$\partial_1 q_0^\pm = \frac{1}{2} V_0 q_1^\pm, \qquad \left(\partial_2 \pm 2 q_1^+ q_1^-\right) q_0^\pm = -\frac{1}{2} V_0 \partial_1 q_1^\pm. \tag{18}$$

The integrability condition of this subsystem does not provide any further constraint and it has to be solved for known $q_1^\pm$.

By using the Bäcklund transformations, one can find several types of solutions [11]. Here we will consider only a few of them. In particular, we will consider the analog of the bright soliton solution for the NLS equation ($a = 0$). In a moving frame coordinate $\xi = x^1 - vx^2 + \xi_0$, such a solution is given by

$$q^\pm(x^1, x^2) = \pm k e^{\pm\left[\left(k^2 - \frac{1}{4}v^2\right)x^2 - \frac{1}{2}v(\xi - \xi_0)\right]} \operatorname{sech} k\xi, \tag{19}$$

depending on the two real parameters k and v. But, in contrast with the bright soliton of the NLS equation, the above solution does not preserve the amplitude for the components $q^\pm$ independently. During the time evolution one of the fields is exponentially growing, while the other one is decaying. At the same time, the product q^+q^- has the usual solitonic shape. Since this is similar to the pattern formation in the context of dissipative structures, for the solution (19) the name of "dissipaton" is suggested. For the N-dissipaton solution see [11]. Note that the parameter v corresponds to the real part of the spectrum of the NLS solitons. Furthermore, in the space of the parameters (v, k), there exists the critical value $v_{\text{crit}} = 2k$. For the solution (19) obtained with $v < v_{\text{crit}}$, at infinity one has $q^\pm \to 0$. At the critical value the solution is a steady state in the moving frame $q^\pm = \pm k e^{\pm k\xi_0}(1 \mp \tanh k\xi)$, with constant asymptotics $q^\pm \to \pm 2ke^{\pm k\xi_0}$ for $x^1 \to \mp\infty$ and $q^\pm \to \pm 0$ for $x^1 \to \pm\infty$. In the super-critical case $v > v_{\text{crit}}$, we are led to $q^\pm \to \pm\infty$ for $x^1 \to \mp\infty$ and $q^\pm \to \pm 0$ for $x^1 \to \pm\infty$.

We first analyze the dissipaton (19) at rest, which describes the stationary metric of JT gravity (we will use the simple notation $x^1 \to x$ and $x^2 \to t$)

$$ds^2 = -\frac{8k^2}{\Lambda}\cosh^{-2} kx[k^2 \tanh^2 kx(dt)^2 - (dx)^2]. \tag{20}$$

This metric is regular everywhere, except at $x = 0$, and is vanishing for $x \to \infty$. However, one can promptly check that it satisfies everywhere the equation provided by the action (1). Hence, the dissipaton maximum position $x = 0$ has to be interpreted as the event horizon.

Introducing a "tortoise"-like coordinate defined by

$$x' = \frac{1}{k}x + \frac{1}{k^2}\ln|1 - e^{-2kx}|, \tag{21}$$

the metric (20) takes the form

$$ds^2 = -\frac{8k^4}{\Lambda} \cosh^{-2} kx \tanh^2 kx \, [(dt)^2 - (dx')^2]. \tag{22}$$

Now, introducing the system of Kruskal–Szekeres (KS) coordinates

$$2v = e^{k^2(x'+t)}\ , \ 2u = -e^{k^2(x'-t)}, \tag{23}$$

we get the conformal metric

$$ds^2 = -\frac{8}{\Lambda}\frac{du dv}{(1-uv)^2}, \tag{24}$$

which can be analytically extended. In the coordinate system

$$v = U + V\ , \ u = U - V, \tag{25}$$

we have the diagram reported in Fig. 1. The original constant curved (x, t) space-time corresponds to the region I. The soliton maximum position $x = 0$ corresponds to the diagonal coordinates lines $u = 0$ $(U = V)$ and $v = 0$ $(U = -V)$. It is clear that the maximally extended KS coordinates admit a physical singularity at $uv = 1$, represented by the past and future hyperbola $U^2 - V^2 = 1$. In the original variables this singularity corresponds to the analytical extension to the complex x plane. Since $uv = -\sinh^2 kx$, one easily finds the singularity for pure imaginary $x = \pm i\frac{\pi}{2k}$. This suggests that singular solutions of the NLS type equation have a physical interpretation as singularities in gravitational theories. Furthermore, we can introduce the "Schwarzschild" coordinates type, defining the space-like variable

$$z = \frac{4k^2}{\Lambda^2}\left(2\,\mathrm{sech}^2 kx - 1\right), \tag{26}$$

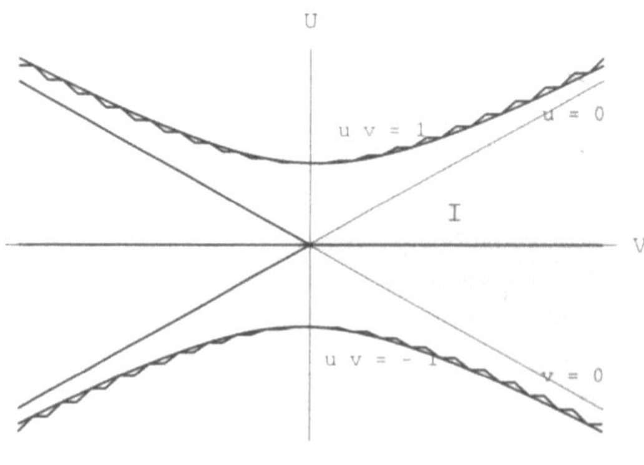

Fig. 1. Standard KS diagram for a static BH. Physical region is I

which takes values on the finite interval $-\frac{4k^2}{\Lambda^2} \leq z \leq \frac{4k^2}{\Lambda^2}$. The metric takes the standard form

$$ds^2 = -\left(\frac{|\Lambda|}{2}z^2 - M\right)(dt)^2 + \frac{1}{\left(\frac{|\Lambda|}{2}z^2 - M\right)}(dz)^2, \tag{27}$$

where the mass M of the black hole is

$$M = \frac{2\,k^4}{|\Lambda|}. \tag{28}$$

By using known formulas [4] for the entropy and temperature of the BH, we get

$$\mathcal{S} = \frac{8\pi k^2}{|\Lambda|}, \qquad T = \frac{k^2}{2\,\pi}. \tag{29}$$

We notice here that the solution has to be completed with the corresponding dilaton field, provided by (18). In the case under consideration one finds

$$V_0 = v_0\left(1 - 2\,\mathrm{sech}^2 kx\right), \qquad q_0^{\pm} = \frac{v_0}{2}\mathrm{sech}kx\ \tanh kx e^{\pm\left[\left(k^2 - \frac{1}{4}v^2\right)x^2\right]}, \tag{30}$$

where v_0 is an arbitrary real constant. Recalling that the dilaton field V_0 is the $\theta\theta$ component of the 3-dimensional metric [4], which is singular to the zeroes of that function. From the previous expression one sees that there are two (symmetric) such points. However, when we use the "Schwarzschild" coordinates (27), the dilaton field takes the familiar linear expression

$$V_0 = -v_0 \frac{\Lambda^2}{4k^2}\, z. \tag{31}$$

When we choose $v \neq 0$ in the solution (19), the line element becomes

$$ds^2 = -\frac{8}{\Lambda}k^2 \cosh^{-2}\varrho \times \left[(k^2 \tanh^2\varrho - \frac{1}{4}v^2)(dt)^2 - \frac{1}{k^2}(d\varrho)^2 - \frac{v}{k}d\varrho dt\right], \tag{32}$$

where we have introduced the moving frame coordinate of the soliton $\varrho = k(x - vt)$. In this system of coordinates the time t cannot be defined globally because of the cross term $d\varrho dt$. Nevertheless, we can find the "synchronized" system of coordinates (ϱ, T) defined by

$$dT = dt - \frac{v}{2k(k^2\tanh^2\varrho - v^2/4)}d\varrho, \tag{33}$$

which leads to

$$\begin{aligned} ds^2 = &-\frac{8}{\Lambda}k^2\cosh^{-2}\varrho \\ &\times\left[(k^2\tanh^2\varrho - \frac{1}{4}v^2)(dT)^2 - \frac{\tanh^2\varrho}{(k^2\tanh^2\varrho - v^2/4)}(d\varrho)^2\right]. \end{aligned} \tag{34}$$

This metric shows a horizon singularity at

$$\tanh \varrho = \pm \frac{v}{2k} \quad \text{only if} \quad |v| \leq 2|k| \equiv |v_{\max}| \tag{35}$$

for the dissipaton velocity. Consequently, a BH cannot move faster than the maximal value of the velocity $|v_{\max}| = 2|k|$. If we increase the dissipaton velocity to a value $v > v_{\max} \equiv v_{crit}$, the synchronization of clocks is still possible, but the metric (34) becomes regular everywhere except at $|\varrho| \to \infty$.

On the other hand, for $v < v_{\max} \equiv v_{\text{crit}}$ (or $k^2 > \frac{v^2}{4}$), we can introduce the analog of the transformation (27)

$$z = \frac{|2r^2 - r_H^2|}{r_H^4}, \qquad r = |k| \cosh^{-1} \varrho, \qquad r_H = \left(k^2 - v^2/4\right)^{\frac{1}{2}}.$$

This new variable runs over the interval

$$0 \leq z \leq z_{\max} = \frac{1}{r_H^2} \frac{4k^2 + v^2}{4k^2 - v^2}, \tag{36}$$

and the singular point r_H corresponds to $z_H = 1/r_H^2 \leq z_{\max}$. The metric of Schwarzschild type reads

$$ds^2 = -\frac{2r_H^4}{\Lambda} \left[-\left(r_H^4 z^2 - 1\right) (dT)^2 + \frac{(dz)^2}{\left(r_H^4 z^2 - 1\right)} \right], \tag{37}$$

which is singular at z_H. We could interpret this result as saying that, for $z \in [0, z_H)$, the metric (37) describes the region inside the event horizon, while, for $z \in (z_H, z_{\max}]$, it describes the outside region. In order to make our argument more clear, we refer to Fig. 2. From this diagram one can see that the central part of the dissipaton given by $|\varrho| < \varrho_H$ corresponds to the outside of the BH. Vice versa, the outside of the dissipaton, i.e., $|\varrho| > \varrho_H$, gives the inside of the BH.

Now, the metric of KS type (23) is obtained by the change of variables

$$v = e^{r_H^2 (R+T)}, \quad u = \text{sign}\,(r - r_H)\, e^{r_H^2 (R-T)} \quad R = \frac{1}{2r_H^2} \ln |1 - \frac{r_H^2}{r^2}|. \tag{38}$$

The KS diagram (see Fig. 3) is obtained by using again the variables (25). However, the diagonal lines $u = 0$ and $v = 0$ correspond now to the BH horizon $r = r_H < |k|$. Region I corresponds to the inner part of the BH, i.e., for $0 < r < r_H$ or equivalently $\varrho > \varrho_H$. Region II below the hyperbola $U^2 - V^2 = v^2/4k^2$ describes the outside of region the BH, with $r > r_H$ or $\varrho < \varrho_H$. It approaches the singularity of the KS metric given by the hyperbola $uv = U^2 - V^2 = 1$, when the dissipaton velocity reaches the critical value $|v|_{\max}$.

Analytic continuation of the previous black hole space times is important to understand the quantum and thermodynamical aspects of the proposed 2D gravity theory.

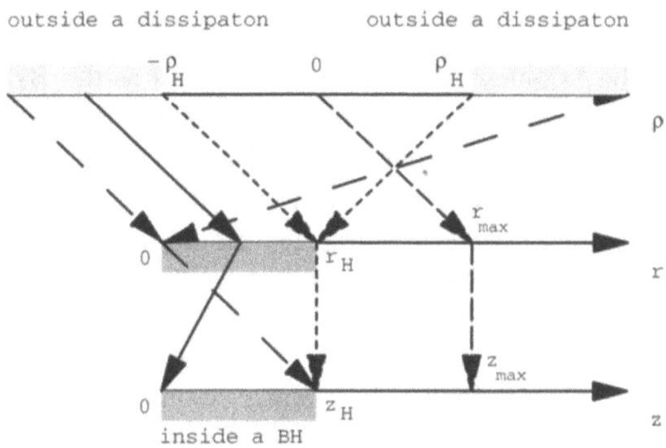

Fig. 2. Mapping a moving dissipaton into a Black Hole

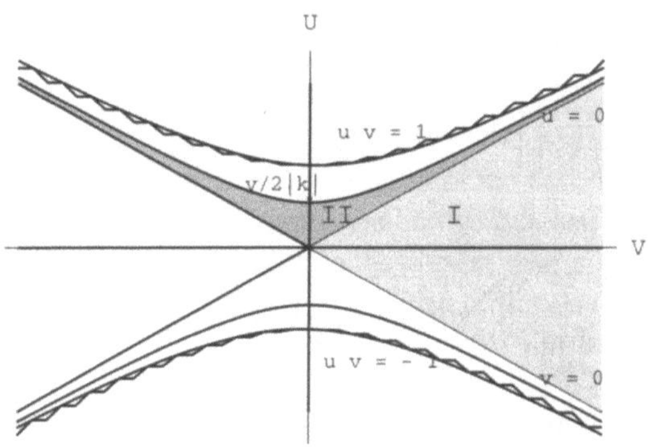

Fig. 3. Standard KS diagram for a moving BH. Physical regions are I and II

This is now more relevant because of the strict relation with some well-known completely integrable systems. Indeed, the Euclidean version of JT gravity, where the de Sitter group SL(2, **R**) is replaced by the orthogonal group SO(3), can be obtained by following the above construction with small changes. In brief, we obtain the NLS equation

$$i\partial_\tau \psi + \partial_x^2 \psi + 2|\psi|^2 = 0, \tag{39}$$

where we have set $x_1 \to x$ and $x_2 \to \tau$, which is the "analytic" continuation of system (16) through

$$t = i\tau, \ \psi = iq^+, \ \overline{\psi} = iq^-. \tag{40}$$

As is well-known [25], this equation admits the "bright" soliton solution

$$\psi = ik\, e^{-i(V^2/4-k^2)\tau + iVx/2} \operatorname{sech} k(x - V\tau), \tag{41}$$

where k and V are real quantities, such that $\lambda = -\frac{V}{2} + ik$ plays the role of the spectral parameter in the Inverse Spectral Transform Method. Now, we saw above that the dissipaton admits an interpretation in terms of a gravitational BH in the pseudo-Euclidean metric. Then the bright soliton can be considered as the BH in the corresponding Euclidean gravity, usually interpreted as the gravitational instanton. This is a complete nonsingular positive definite metric solution of the vacuum Einstein equation. Moreover, those metrics which are asymptotically flat in spatial directions and periodic in the imaginary time direction, contribute to the thermal canonical ensemble and are relevant for the thermodynamical properties of the BH [6]. First let us consider the static bright soliton ($V = 0$ in (41)). One easily sees that it is a periodic function in time, with period $\frac{2\pi}{k^2}$. Moreover, the corresponding metric

$$ds_E^2 = \xi_E^2 d\tau^2 + \frac{dx^2}{P} = -\frac{8k^2}{\Lambda}\cosh^{-2} kx[k^2 \tanh^2 kx(d\tau)^2 + (dx)^2] \tag{42}$$

defines the Euclidean BH spacetime with the horizon fixed at $x = 0$. As in [26, 27] we introduce the proper distance from the Euclidean horizon

$$l = 4\sqrt{\frac{2}{|\Lambda|}}\left[\operatorname{arctg} e^{kx} - \frac{\pi}{4}\right],$$

which is defined by the equation $dl = dx/P^{1/2}$. Then, near the horizon, the metric is given approximately by

$$ds_E^2 \approx \kappa_{gr}^2 l^2 d\tau^2 + dl^2, \; \kappa_{gr} = \lim_{l\to 0} P^{1/2}\frac{d\xi}{dx} = k^2, \tag{43}$$

where κ_{gr} is the so-called surface gravity. Furthermore, the Euclidean time is ranging over $\tau \in (0, 2\pi/\kappa)$ and the Hawking temperature of the black hole is

$$T_H = \frac{k^2}{2\pi}, \tag{44}$$

in agreement with the previous result (29). Repeating the analogous calculations for the moving dissipaton, one obtains the Hawking temperature

$$T_H = \frac{r_H^2}{2\pi} = \frac{k^2 - V^2/4}{2\pi}. \tag{45}$$

The integrability of Eq. (16) implies the existence of N-dissipaton solutions, the superposition formula of which is given in [11]. It is well-known [25] that the N-soliton is asymptotically decomposed into individual solitons when $t \to \pm\infty$. Furthermore, the collision of two individual solitons is elastic and, asymptotically, the only effect of the interaction is a shift of phase and position. We could expect a similar behaviour for the dissipatons also. But this suggests that the horizon of the individual black hole, which is related to the dissipaton, will shift as well. Actually, the main difficulty is due to the nonstationary character of the metric corresponding to the colliding dissipatons. This fact makes the synchronization problem highly non-trivial. However,we can describe here the particular 2-dissipaton solution of (16)

$$q^{\pm}(x,t) = \pm\frac{2}{\Delta}\left(\frac{k_1 + k_2}{k_1 - k_2}\right) \times \left(k_1 \cosh\theta_2 e^{\pm k_1^2 t} + k_2 \cosh\theta_1 e^{\pm k_2^2 t}\right), \tag{46}$$

where

$$\Delta = \cosh(\theta_1 + \theta_2) + \left(\frac{k_1 + k_2}{k_1 - k_2}\right)^2 \cosh(\theta_1 - \theta_2) + \frac{4k_1k_2}{(k_1 - k_2)^2} \cosh\left(k_1^2 - k_2^2\right) t, \tag{47}$$

with $\theta_i = k_i (x - x_{0i})$ $i = 1, 2$. At any fixed time t this solution is exponentially decaying at space infinity.

Next we put the dissipaton labeled by 1 at the origin of the coordinates (i.e., $x_{01} = 0$ in (47)) and the second displaced by a certain amount, say, $x_{02} = d$ with $d > 0$. Then, we look at the limit $d \to \infty$ in a fixed bounded domain of (x, t). We find that, while the dissipaton labeled by 2 goes far apart on the right, the first suffers a negative shift of position. But this type of solution corresponds to a static BH with horizon at the dissipaton location. Since the second dissipaton is far apart, it is located inside the event horizon. However, in this case it cannot escape outside, that is, overcome the horizon position. Consequently, the interaction is repulsive and it induces a shift of the horizon of the first. Because of the left-right symmetry, a horizon shift of the same amount will occur if the second dissipaton is sent to $x \to -\infty$, but to the right. In the present case, similar to the bound states of two off-phase solitons with equal amplitudes [28], two dissipatons start to move from $\pm\infty$ with opposite velocities $v_1 = -v_2 = k_1 - k_2$ and, after a repulsive interaction, go back to $\pm\infty$. In Fig. 4 we consider a dissipaton - dissipaton collision for a particular choice of the amplitude parameters $k_1 = 2$ and $k_2 = 1$. In this case we always have $v = k_- < 2k = k_+$. In particular, we can compute the position of the horizons x_H , at least in the asymptotics $t \to \pm\infty$. Our parameters are chosen in an interval in which $x_H \ll \frac{1}{k}$, where the latter quantity estimates the coherence length of the dissipaton. This means that the horizon is located very close to the dissipaton central position $x_{\text{diss}} = x_{0+} \pm (\overline{x}_0 + k_- t)$. Moreover, since we defined the BH interior by $x > |x_H - x_{\text{diss}}|$, the horizon surface is now given by two disconnected parts, which never can overcome each other. This is a sort of hard core interaction, which drastically changes the dissipatons interaction.

Finally, after reaching a minimal distance $\sim d$, the dissipatons form a metastable bound state, with the lifetime

$$\Delta T \approx \frac{2}{\left(k_1^2 - k_2^2\right)} \left(k_2 d + \ln\left[\frac{k_1\left(k_1^2 - k_2^2\right)}{2k_2}\right]\right). \tag{48}$$

This formula is in good agreement with the computer calculations represented in Fig. 4.

We have investigated the JT model of gravity in the context of a gauge field formulation. At first sight the conditions of gauge fixing (14) which we chose, may appear quite unusual; however, they definitely lead to wide classes (hierarchies of countable many equations) of completely integrable systems. In particular, we have studied a sort of NLS equation whose soliton-like solutions, dissipaton, can be interpreted as

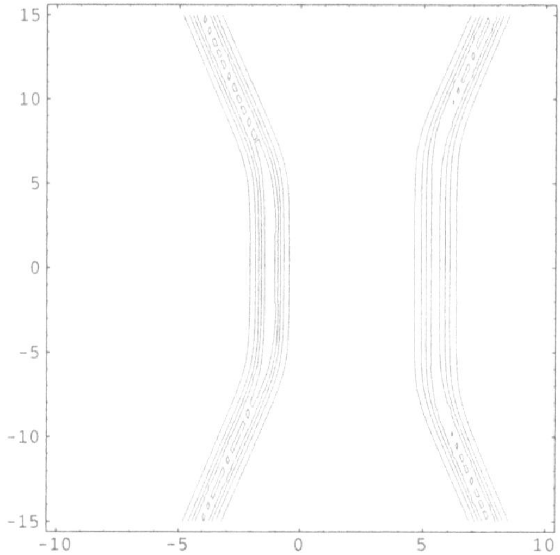

Fig. 4. Two-dissipaton collision for $d = 4$

BH's. Among several cases, the most interesting is given by the soliton moving with a limited velocity. In such a case the metric associated with the dissipaton describes the inner and the outer part of a BH in a compact space. The position of the event horizon is determined in terms of the two dissipaton parameters. Finally, surface gravity and Hawking temperature within the Euclidean version of the theory have been related to the soliton parameters. Thus, the novelty of our reformulation lies in the possibility not only of describing analytically the BH solution, which was known before, but also of deriving explicitly the N-BH solutions and, in particular, the interaction of two dissipatons. This has been reinterpreted in terms of the scattering of the event horizons of the two BH's.

Acknowledgements

This work was supported in part by MURST of Italy and by INFN - Sezione di Lecce.

References

1. Jackiw R. (1984): Liouville Field Theory: A Two-Dimensional Model for Gravity, in *Quantum Theory of Gravity*, ed. by S. Christensen, Adam Hilger, Bristol;
   Jackiw R. (1985): Lower dimensional Gravity. Nucl. Phys. B **252**, 343
2. Teitelboim C. (1983): Gravitation and Hamiltonian Structure in Two Space-time Dimensions. Phys. Lett. B **126**, 41; The Hamiltonian Structure of Two-Dimensional Space Time and its Relation with the Conformal Anomaly, in *Quantum Theory of Gravity*, ed by S. Christensen, Adam Hilger, Bristol, pp. 327–344
3. Cadoni M., Mignemi S. (1995): Phys. Rev. D **51**, 4319
4. Bañados M., Teitelboim C., Zanelli J. (1992): Phys. Rev. Lett. **69**, 1849

5. Bañados M., Teitelboim C., Zanelli J. (1993): Phys. D **48**,1506
6. Hawking S.W. (1975): Comm. Math. Phys. **43**, 199
7. Kumar A., Ray K. (1995): Phys. Lett. B **351**, 431
8. Elizalde E., Fosalba-Vela P., Natulin S., Odintsov S.D. (1995): Phys. Lett. B **352**, 235
9. Lemos J.P.S. (1996): Phys. Rev. D **54**, 6206
10. Gegenberg J., Kunstatter G. (1997): Solitons and Black Holes. hep-th/9707181
11. Martina L., Pashaev O.K., Soliani G. (1997): Class. Quant. Grav. **14**, 3179
12. Martina L., Pashaev O.K., Soliani G. (1998): Phys. Rev. D **58**, 84025
13. Eguchi T., Gilkey P.B., J. Hanson A. (1980): Phys. Rev. **66**, 213–393
14. Fukuyama T., Kamimura K. (1985): Phys. Lett. B **160**, 259
15. Isler K., Trugenberger C.A. (1989): Phys. Rev. Lett. **63**, 834
16. Chamseddine A.H., Wyler D. (1990): Nucl. Phys. B **340**, 595
17. Montano D., Sonnenschein J. (1989): Nucl. Phys. B **324**, 348
18. Birmingham D., Blau M., Rakowski M., Thompson G. (1991): Phys. Rep. **209**, 129
19. Achucarro A., Townsend P. (1986): Phys. Lett. B **180**, 89
20. Witten E. (1988/89): Nucl. Phys. B **311**, 46
21. Fernando S., Mansouri F.: Internal structure of Black Holes and Chern–Simons Theory in $(2+1)$-Dimensions. hep-th 9804147
22. Cartan E. (1937): Théorie des Groupes Finis et Continus et la Géométrie Différentielle traitées par la Méthode du Repér Mobile. Gauthier-Villars, Paris
23. Prigogine I. (1980): From Being to Becoming. Time and Complexity in the Physical Sciences. W. H. Freeman and C., San Francisco
24. Faddeev L., Takhtajan L. (1987): Hamiltonian Methods in the Theory of Solitons. Springer, Berlin Heidelberg New York
25. Zakharov V.E., Shabat A.B. (1971): Zh. Eksp. Teor. Fiz. **61**, 118
26. Frolov V.P. (1992): Phys. Rev. D **46**, 5383
27. Lemos J.P.S. (1996): Phys. Rev. D **54**, 6206
28. Hasegawa A., Kodama Y. (1995): Solitons in Optical Communications. Clarendon Press, Oxford

The Prolongation Problem for the Heavenly Equation

M. Palese, R.A. Leo, G. Soliani

Abstract. We provide an exact regular solution of an operator system arising as the prolongation structure associated with the heavenly equation. This solution is expressed in terms of operator Bessel coefficients.

1 Introduction

It is well-known that the so-called heavenly equation

$$u_{xx} + u_{yy} + (e^u)_{zz} = 0, \tag{1}$$

where $u = u(x, y, z)$ and subscripts mean partial derivatives, occurs in the study of heavenly spaces (Einstein spaces with one rotational Killing vector) [1–5] and in the context of extended conformal symmetries [6]. Reduced versions of Eq. (1) have been found in [7] via the symmetry approach, providing instanton and meron–like configurations. Furthermore, an algebra of the Virasoro type without central charge was associated with the heavenly equation by using to its invariance under conformal transformations [7].

To investigate the algebraic aspects of the integrability properties of Eq. (1), we shall apply the prolongation technique, which can lead to the relative linear spectral problem [8–11]. In more than two independent variables, the extension of the prolongation procedure is generally nontrivial and some aspects remain to be explored (see, for example, [9, 10]). However, the main result of this note is represented by the exact solution of the prolongation system associated with Eq. (1). This system consists of three operator equations in the form of commutator relations which can be written as second order (operator) differential equations, formally resembling equations of Bessel type. Our solution is achieved through a series expansion defining operator Bessel coefficients.

2 The prolongation structure

We introduce on a manifold with local coordinates (x, y, z, u, p, q, r) the closed differential ideal defined by the set of 3-forms:

$$\theta_1 = du \wedge dx \wedge dy - r dx \wedge dy \wedge dz, \tag{2}$$

$$\theta_2 = du \wedge dy \wedge dz - p dx \wedge dy \wedge dz, \tag{3}$$

$$\theta_3 = du \wedge dx \wedge dz + q dx \wedge dy \wedge dz, \tag{4}$$

$$\theta_4 = dp \wedge dy \wedge dz - dq \wedge dx \wedge dz + e^u dr \wedge dx \wedge dy + e^u r^2 dx \wedge dy \wedge dz, \quad (5)$$

where $\wedge$ stands for the exterior product.

It is easy to verify the following

Proposition 2.1. *On every integral submanifold defined by* $u = u(x, y, z)$, $p = u_x$, $q = u_y$, $r = u_z$, *with* $dx \wedge dy \wedge dz \neq 0$, *the ideal* (2)–(5) *is equivalent to Eq.* (1).

Next we consider the 2-forms:

$$\Omega^k = H^k(u, u_x, u_y, u_z; \xi^m)dx \wedge dy + F^k(u, u_x, u_y, u_z; \xi^m)dx \wedge dz + G^k(u, u_x, u_y, u_z; \xi^m)dy \wedge dz + A^k_m d\xi^m \wedge dx + B^k_m d\xi^m \wedge dz + d\xi^k \wedge dy, \quad (6)$$

where $\xi = \{\xi^m\}$, $k, m = 1, 2, \ldots, N$ (N arbitrary), and H^k, F^k and G^k are, respectively, the pseudopotential and functions to be determined. Furthermore, the quantities A^k_m and B^k_m denote the elements of two $N \times N$ constant regular matrices, and the summation convention over repeated indices is understood.

Definition 2.1. The forms Ω^k are called the *prolongation forms* associated with Eq. (1). Let $\mathcal{I}$ be the *prolonged* ideal generated by the forms θ_j and Ω^k. We say that $\mathcal{I}$ is closed if $d\Omega^k \in \mathcal{I}\,(\theta_j, \Omega^k)$.

Lemma 2.1. *The closure condition for* $\mathcal{I}$ *yields*

$$H^k = e^u u_z L^k(\xi^m) + P^k(u, \xi^m), \quad (7)$$

$$F^k = -u_y L^k(\xi^m) + N^k(\xi^m), \quad (8)$$

$$G^k = u_x L^k(\xi^m) + M^k(u, \xi^m), \quad (9)$$

where L^k, P^k, N^k, M^k *are functions of integration.*

Proof. The closure condition is equivalent to the following constraints:

$$H^k_{u_z} - e^u G^k_{u_x} = 0, \qquad F^k_{u_x} + G^k_{u_y} = 0, \quad (10)$$

$$H^k_{u_x} = H^k_{u_y} = F^k_{u_x} = F^k_{u_y} = G^k_{u_y} = G^k_{u_z} = 0, \quad (11)$$

$$u_z H^k_u - u_y F^k_u + u_x G^k_u - e^u u_z^2 G^k_{u_x} + [G, H]^k = 0, \quad (12)$$

$$F^k_{\xi^l} - G^k_{\xi^m} A^m_l - H^k_{\xi^m} B^m_l = 0, \quad (13)$$

$$F^k_{\xi^m} (B^{-1})^m_n G^n - G^k_{\xi^m} (B^{-1})^m_n F^n = 0, \quad (14)$$

$$[A, B] = 0, \quad (15)$$

where $[G, H]^k = G^j H^k_{\xi^j} - H^j G^k_{\xi^j}$ (Lie bracket), $H^k_{\xi^l} = \frac{\partial H^k}{\partial \xi^l}$, $H^k_u = \frac{\partial H^k}{\partial u}$, and so on. Eqs. (10)–(12) provide the result. □

In the following we shall omit the indices k, m for simplicity.

Proposition 2.2. *The following prolongation equations hold.*

$$P_u = e^u[L, M], \tag{16}$$

$$M_u = -[L, P], \tag{17}$$

$$[M, P] = 0. \tag{18}$$

Proof. This is a straightforward consequence of Lemma 2.1 (Eq. (12)). □

Hereafter, L, P, M will be regarded as (regular) operators, in the sense that

$$L \to L^j \frac{\partial}{\partial \xi^j}, \quad M \to M^j \frac{\partial}{\partial \xi^j}, \quad P \to P^j \frac{\partial}{\partial \xi^j},$$

while the Lie brackets become commutators.

3 Exact solution of the prolongation equations

Next we seek an exact solution to Eqs. (16)–(18).

Definition 3.1. For any $A = A^j \frac{\partial}{\partial \xi^j}$, we define $\mathcal{L}[\mathcal{A}]$ by

$$\mathcal{L}[A] = [L, A]. \tag{19}$$

Remark 3.1. The prolongation Eqs. (16) and (17) can be written as

$$P_{tt} - \frac{1}{t} P_t + \mathcal{L}^2[P] = 0, \tag{20}$$

$$M_{tt} + \frac{1}{t} M_t + \frac{1}{t} \mathcal{L}^2[M] = 0, \tag{21}$$

where $t = 2e^{\frac{u}{2}}$.

Remark 3.2. Notice that the above equations resemble formally conventional Bessel equations of the type

$$\kappa_{tt} - \frac{1}{t}\kappa_t + \omega^2 \kappa = 0, \qquad \chi_{tt} + \frac{1}{t}\chi_t + \frac{1}{t}\omega^2 \chi = 0,$$

with ω a constant, whose regular solutions at $t = 0$ are given by

$$\kappa(t) = p_0 \frac{t}{2} J_1(t\omega), \qquad \chi(t) = m_0 J_0(t\omega),$$

with p_0, m_0 constants of integration.

We are interested in solutions of Eqs. (20) and (21) which are regular at $t = 0$. We proceed in a heuristic way, generalizing the scheme used in the solution of the operator equations

$$\widetilde{P}_{tt} + \mathcal{L}^2[\widetilde{P}] = 0, \quad \widetilde{M}_{tt} + \mathcal{L}^2[\widetilde{M}] = 0, \tag{22}$$

whose formal solution is

$$\widetilde{P}(t, \xi) = e^{it\mathcal{L}}[A_0(\xi)] + e^{-it\mathcal{L}}[B_0(\xi)], \tag{23}$$

$$\widetilde{M}(t, \xi) = e^{it\mathcal{L}}[C_0(\xi)] + e^{-it\mathcal{L}}[D_0(\xi)], \tag{24}$$

where A_0, B_0, C_0, D_0 can be determined from the initial conditions.

Definition 3.2. We define the n-th power of the operator $\mathcal{L}$ by setting

$$\mathcal{L}^n[A] = [L, [L, \ldots, [L, A] \ldots], \tag{25}$$

where L appears $n$times, and $\mathcal{L}^0[A] = A$.

Definition 3.3. Let $\mathbf{J_0}(\cdot)$ and $\mathbf{J_1}(\cdot)$ be defined formally by

$$\mathbf{J_0}(t\mathcal{L}[M_0]) = \sum_{m=0}^{\infty} \frac{(-1)^m}{(m!)^2} \left(\frac{t}{2}\right)^{2m} \mathcal{L}^{2m}[M_0], \tag{26}$$

$$\mathbf{J_1}(t\mathcal{L}[P_0]) = \sum_{m=0}^{\infty} \frac{(-1)^m}{m!(m+1)!} \left(\frac{t}{2}\right)^{1+2m} \mathcal{L}^{1+2m}[P_0]. \tag{27}$$

The following result holds; its proof is straightforward.

Proposition 3.1. *A solution of Eqs.* (20) *and* (21) *regular at* $t = 0$ *is given by*

$$P = \frac{t}{2}\mathbf{J_1}(t\mathcal{L}[P_0]), \qquad M = \mathbf{J_0}(t\mathcal{L}[M_0]), \tag{28}$$

where $M_0 \equiv M_0(\xi) = M(t;\xi)\,|_{t=0}$ *and* $P_0 \equiv P_0(\xi)$ *is such that* $[L, P_0] = [L, M_0]$.

Remark 3.3. Here $\mathbf{J_0}$ and $\mathbf{J_1}$ are borrowed from the series expansion of the Bessel functions J_ν, with ν an integer:

$$J_\nu(\tau) = \sum_{m=0}^{\infty} \frac{(-1)^m}{m!(m+\nu)!} \left(\frac{\tau}{2}\right)^{\nu+2m} \mathcal{L}^{\nu+2m}[P_0], \tag{29}$$

for $\nu = 0$ and $\nu = 1$, respectively (see [12]).

Remark 3.4. Inserting the operators (28) into Eqs. (20) and (21) yields the condition $[\mathcal{L}[M_0], M_0] = 0$.

Remark 3.5. We note that the search of solutions to Eqs. (20) and (21), regular at $t = 0$, is equivalent to solving the corresponding Cauchy problem with initial conditions $P(0;\xi) = 0$, $P_t(t,\xi)\,|_{t=0} = 0$, $M(0;\xi) = M_0$, $M_t(t,\xi)\,|_{t=0} = 0$.

For practical purposes, e.g., to determine the spectral problem and Bäcklund transformations associated with Eq. (1), we shall rewrite the solutions (28) in a form which contains the operator L and not the operator $\mathcal{L}$. We shall give a regular solution of Eqs. (20) and (21) following a scheme similar to that used for (22).

By induction it is easy to prove the following lemma.

Lemma 3.1. *For* $n > 0$, (25) *takes the form*

$$\mathcal{L}^n[A] = \sum_{k=0}^{n} (-1)^k \binom{n}{k} L^{n-k} A L^k. \tag{30}$$

As a straightforward consequence we have

$$e^{it\mathcal{L}}[A_0] = \sum_{n=0}^{\infty} \frac{(it)^n}{n!} \sum_{k=0}^{n} (-1)^k \binom{n}{k} L^{n-k}[A_0]L^k$$
$$= \sum_{j=0}^{\infty} \frac{(it)^j}{j!} L^j [A_0] \left(\sum_{K=0}^{\infty} \frac{(it)^k}{k!} L^k \right) \equiv e^{itL}[A_0]e^{-itL}, \tag{31}$$

which is just the *Baker–Campbell–Hausdorff expansion* (see, e.g., [13]).

Thus, we can rewrite the solutions of the operator equations (22) in a more suitable form.

Operator Bessel coefficients

In order to express the solutions (28) of Eqs. (20) and (21) in terms of the operator L, we introduce operator Bessel coefficients by means of a formal expansion analogous to that used in the case of Bessel functions.

Definition 3.4. Let X be a regular operator. We define the *operator Bessel coefficients* $\mathbf{J_m}(tX)$ as the coefficients of the formal expansion:

$$e^{\frac{t}{2}X(z-1/z)} = \sum_{m=-\infty}^{\infty} z^m \mathbf{J_m}(tX). \tag{32}$$

Remark 3.6. We stress that the Laurent series on the right is uniformly convergent.

First we prove a technical lemma.

Lemma 3.2. *Operator Bessel coefficients satisfy the following recurrence and derivation formulae.*

$$\mathbf{J_{-k}}(tL) = (-1)^k \mathbf{J_k}(tL), \tag{33}$$
$$2k\mathbf{J_k}(tL) = tL[\mathbf{J_{k-1}}(tL) + \mathbf{J_{k+1}}(tL)], \tag{34}$$
$$2\frac{d}{dt}[\mathbf{J_k}(tL)] = L[\mathbf{J_{k-1}}(tL) - \mathbf{J_{k+1}}(tL)], \tag{35}$$
$$\frac{d}{dt}[t^k \mathbf{J_k}(tL)] = Lt^k \mathbf{J_{k-1}}(tL), \tag{36}$$
$$\frac{d}{dt}[t^{-k} \mathbf{J_k}(tL)] = -Lt^{-k} \mathbf{J_{k+1}}(tL). \tag{37}$$

Proof. By differentiating the formal expansion (32) with respect to z, with $X = L$, we obtain

$$\frac{1}{2}tL(1+\frac{1}{z^2}) \sum_{k=-\infty}^{\infty} z^k \mathbf{J_k}(tL) = \sum_{k=-\infty}^{\infty} kz^{k-1} \mathbf{J_k}(tL).$$

Then, by equating coefficients of z^{k-1} in the above identity, we obtain formula (34). Furthermore, if we differentiate the formal expansion with respect to t, we have

$$\frac{1}{2}L(z-1/z)\sum_{k=-\infty}^{\infty} z^k \mathbf{J_k}(tL) = \sum_{k=-\infty}^{\infty} z^k \frac{d}{dt}\mathbf{J_k}(tL).$$

By equating coefficients of z^k on either side of this identity, we obtain formula (35). Formulae (36) and (37) can be obtained by adding and subtracting (34) and (35), while Eq. (33) follows directly from (32) and from the hypothesis that L is a regular operator. □

A form of the solution of the prolongation equations in terms of L

In the following we shall provide an equivalent solution to Eqs. (20) and (21) in terms of L which is in some sense an analogue of formula (31).

Proposition 3.2. *We can rewrite the solution* (28) *in terms of L as follows:*

$$P = \frac{t}{2}\sum_{k=-\infty}^{\infty} \mathbf{J_{k+1}}(tL)P_0\mathbf{J_k}(tL), \qquad M = \sum_{k=-\infty}^{\infty} \mathbf{J_k}(tL)M_0\mathbf{J_k}(tL). \tag{38}$$

Proof. To verify that the operators (38) obey Eqs. (20) and (21), we refer to Lemma 3.2. In fact, by using (36), (37) and (33), we have

$$P_{tt} = \frac{1}{2}\sum_{k=-\infty}^{\infty} [\mathbf{J_k}(tL)\mathcal{L}[P_0]\mathbf{J_k}(tL)] + \frac{t}{2}\sum_{k=-\infty}^{\infty} [\mathbf{J_k}(tL)\mathcal{L}^2[P_0]\mathbf{J_{k+1}}(tL)].$$

On the other hand, since

$$\mathcal{L}^2[P] = -\frac{t}{2}\sum_{k=-\infty}^{\infty} (-1)^k[\mathbf{J_k}(tL)L^2P_0\mathbf{J_{1-k}}(tL) - 2\mathbf{J_k}(tL)LP_0L\mathbf{J_{1-k}}(tL)$$
$$+\mathbf{J_k}(tL)P_0L^2\mathbf{J_{1-k}}(tL)] = -\frac{t}{2}\sum_{k=-\infty}^{\infty} [\mathbf{J_k}(tL)\mathcal{L}^2[P_0]\mathbf{J_{k+1}}(tL)],$$

we obtain the result.

In a similar way, by virtue of (34), we can write

$$L\mathbf{J_{k+2}}(tL) = \frac{2(k+1)}{t}\mathbf{J_{k+1}}(tL) - L\mathbf{J_k}(tL). \tag{39}$$

This, with the help of Lemma 3.2, gives the following expression for M_{tt}:

$$-\sum_{k=-\infty}^{\infty} [\mathbf{J_k}(tL)\mathcal{L}^2[M_0]\mathbf{J_k}(tL)]$$
$$+\sum_{k=-\infty}^{\infty} (-1)^k 2(k-1)[\mathbf{J_{k-1}}(tL)LM_0\mathbf{J_{-k}}(tL)$$
$$-2\mathbf{J_{k-1}}(tL)M_0L\mathbf{J_{-k}}(tL) + \mathbf{J_{-k}}(tL)M_0L\mathbf{J_{k-1}}(tL)].$$

Then, using the expression

$$\mathcal{L}^2[M] = \sum_{k=-\infty}^{\infty} [\mathbf{J_k}(tL)\mathcal{L}^2[M_0]\mathbf{J_{-k}}(tL)],$$

and taking into account (33) and (34), we have

$$tM_{tt} + M_t + \mathcal{L}^2[M] = \sum_{k=-\infty}^{\infty} [2k\mathbf{J_{k-1}}(tL)LM_0\mathbf{J_k}(tL) +$$
$$-2(k-1)\mathbf{J_{k-1}}(tL)M_0L\mathbf{J_k}(tL)] = 0.$$

Hence Eq. (21) is satisfied. This completes the proof. □

4 Conclusions

We have solved the prolongation problem for Eq. (1) in terms of a series expansion of operators which can be interpreted as generalized Bessel coefficients. The operators (38) have been derived under the hypothesis that P and M are regular at $t = 0$, i.e., at $u \to -\infty$. The "key" for our result is based on Eqs. (20) and (21), which constitute an extended form (of operator Bessel type) of the operator equations (22). We note that the introduction of the operator (30) has strongly facilitated our task. In this regard, a useful step is represented by Eq. (31), in which a correspondence is established between the operators $\mathcal{L}$ and L via the Baker–Campbell–Hausdorff formula.

In theory, knowledge of H, F and G (see (7)–(9)) may be exploited to find the spectral problem related to Eq. (1). However, in contrast with what happens in other cases, the determination of the spectral problem of Eq. (1) within the prolongation scheme offers notable difficulties, mainly owing to the fact that we have only been able to obtain a solution of the prolongation equations (16)–(18) which is regular at $t = 0$. At present the construction of a prolongation algebra whose elements depend uniquely on the pseudopotential variables remains an open problem.

Acknowledgments

One of us (M. P.) would like to thank M. Ferraris and M. Francaviglia for useful discussions.

References

1. Boyer C., Finley J.D. (1982): Killing vectors in self-dual, Euclidean Einstein spaces. J. Math. Phys. **23**, 1126–1128
2. Gegenberg J.D., Das A. (1984): Stationary Riemannian space-times with self-dual curvature. Gen. Relativity and Gravitation **16**, 817–829
3. Gibbons G.W., Hawking S.W. (1978): Gravitational Multi–Instantons. Phys. Lett. B **78**, 430–432

4. Lebrun C. (1991): Explicit self-dual metrics on $CP_2\#...\#CP_2$. J. Diff. Geom. **34**, 223
5. Plebanski J.F. (1975): Some solutions of complex Einstein equations. J. Math. Phys. **16**, 2395–2402
6. Park Q-Han (1990): Extended conformal symmetries in real heavens. Phys. Lett. B **236**, 429–432
7. Alfinito E., Soliani G., Solombrino L. (1997): The symmetry structure of the heavenly equation. Lett. Math. Phys. **41**, 379–389
8. Estabrook F.B., Wahlquist H.D. (1976): Prolongation structures of nonlinear evolution equations. II, J. Math. Phys. **17**, 1293–1297
9. Morris H.C. (1976): Prolongation structures and nonlinear evolution equations in two spatial dimensions. J. Math. Phys. **17**, 1870–1872
10. Tondo G.S. (1985): The eigenvalue problem for the Three-Wave Resonant Interaction in (2+1) dimensions via the prolongation structure. Lett. Nuovo Cimento **44**, 297–302
11. Whalquist H.D., Estabrook F.B. (1975): Prolongation structures of nonlinear evolution equations. J. Math. Phys. **16**, 1–7
12. Watson G.N. (1952): A Treatise on the Theory of Bessel Functions. Cambridge University Press, Cambridge
13. Baker H.F. (1904): Alternant and continuous group. Proc. London Math. Soc. (2) **3**, 24–47;
    Campbell J.E. (1898): On a law of combination of operators. Proc. London Math. Soc. **29**, 14–32;
    Hausdorff F. (1906): The symbolic exponential formula in group theory. Ber. Verh. Sächs. Gess. Wiss. Leipzig. Math. Phys. Kl. **58**, 19–48

Time Extremizing Trajectories of Massive and Massless Objects in General Relativity

P. Piccione

Abstract. This is a review article about recent results concerning one-dimensional variational problems in Lorentzian geometry see [1–12]. We will discuss from a mathematical point of view a general-relativistic version of Fermat's principle, that characterizes the trajectories of massive and massless (photons) objects freely falling under the action of the gravitational field. We obtain two variational problems whose solutions are (future or past pointing) causal geodesics joining a spacelike submanifold P and a timelike submanifold Γ of $\mathcal{M}$. Moreover, we will present two general-relativistic versions of the classical brachistochrone problem. The solutions of the brachistochrone variational problem represent trajectories of massive objects subject to the gravitational field and also to some *constraint* forces, and so they are *not* geodesics in the spacetime metric. We will distinguish between the *travel time* and the *arrival time* brachistochrones, which are curves extremizing the time measured respectively by a watch which is traveling together with the massive object and by a watch fixed at the arrival point in space.

1 Introduction

We consider an arbitrary Lorentzian manifold $(\mathcal{M}, g)$. More precisely, we assume that $\mathcal{M}$ is a finite-dimensional real C^∞ manifold of dimension greater than two whose topology satisfies the second countability axiom and the Hausdorff separation axiom, and we assume that g is a semi-Riemannian C^∞ metric of signature $(+, ..., +, -)$ on $\mathcal{M}$. The physically interesting case is, of course, $\dim(\mathcal{M}) = 4$, in which case $\mathcal{M}$ represents an arbitrary spacetime in the sense of general relativity.

The symbol $\langle \cdot, \cdot \rangle$ will denote the bilinear form induced by g on the tangent spaces of $\mathcal{M}$; the usual nabla symbol ∇ will denote the covariant derivative relative to the Levi–Civita connection of g. Given a smooth function ϕ on $\mathcal{M}$, for $q \in \mathcal{M}$ we denote by $\nabla\phi(q)$ the gradient of ϕ at q with respect to g, which is the vector in $T_q\mathcal{M}$ defined by $\langle \nabla\phi(q), \cdot \rangle = \mathrm{d}\phi(q)[\,\cdot\,]$; the Hessian $H^\phi(q)$ of ϕ at q is the symmetric bilinear form on $T_q\mathcal{M}$ given by $H^\phi(q)[v_1, v_2] = \langle \nabla_{v_1} \nabla\phi, v_2 \rangle$, for $v_1, v_2 \in T_q\mathcal{M}$.

At each point $p \in \mathcal{M}$, we denote the tangent space to $\mathcal{M}$ by $T_p\mathcal{M}$. We call a linear subspace $\mathcal{W}_p$ of $T_p\mathcal{M}$ *spacelike* if g is positive definite on $\mathcal{W}_p$, *lightlike* if g is positive semidefinite but not positive definite on $\mathcal{W}_p$, and *timelike* otherwise. This implies that $\mathcal{W}_p$ is spacelike if and only if the orthocomplement of $\mathcal{W}_p$ is timelike, and vice versa (cf., e.g., [13, p. 20]). A vector $v \in T_p\mathcal{M}$ is called *spacelike*, *lightlike* or *timelike* if the linear subspace $\{sv | s \in \mathbb{R}\}$ has the respective property. Equivalently, v is *spacelike* if $g(v, v) > 0$ or $v = 0$, *lightlike* if $g(v, v) = 0$ but $v \neq 0$, and *timelike* if $g(v, v) < 0$. Finally, we call a submanifold Σ of $\mathcal{M}$ *spacelike*, *lightlike* or *timelike* if at all points $p \in \Sigma$ the tangent space $T_p\Sigma$ has the respective property. Note that, according to this definition, a single point is a (zero-dimensional) spacelike submanifold.

We are interested in the spacetime trajectory of massive or massless objects from a given place in space at a given time, typically represented by a spacelike submanifold P of $\mathcal{M}$, to another place in space at *some* time in the future, represented by a timelike submanifold Γ of $\mathcal{M}$. In an absolutely similar fashion, one could consider instead trajectories starting at *some* time at a given place and ending at a given place at a given instant; this corresponds to reversing the orientation of the curve parameterizations. From a mathematical point of view, the difference between the two problems is not significant, and we will restrict our attention to the first case.

In the mathematical setup, some interesting technical differences arise between the timelike and the lightlike cases. The theory of timelike geodesics (and of the brachistochrones) in a Lorentzian manifold is very similar to the corresponding theory for Riemannian geodesics. They are obtained as stationary points for suitable functionals: the set of trial curves for the variational problems, satisfying a suitable Sobolev regularity, has the structure of a smooth infinite dimensional manifold and the functionals are smooth in this structure. Hence, the variational problems can be studied with powerful techniques of Critical Point theory and Global Analysis on manifolds, like Lusternik–Schnirelman theory and Morse theory on Hilbert manifold. On the other hand, the theory of lightlike geodesics is more complicated, due to the fact that, for the Fermat principle, the set of trial paths is *not* a regular manifolds. In applications, to overcome this problem a possible approach is to build an approximation scheme, as in [2,4], to obtain existence and multiplicity results by taking limits in the timelike case. One of the issues discussed in this article is the approximation problem in Fermat's principle (see Sect. 4); among the main results of the section we discuss the *stability* of the geometric index of lightlike geodesics by perturbations with causal geodesics (Theorem 4.5).

For our variational setup, we introduce the following spaces. If $I \subseteq \mathbb{R}$ is any interval, we will denote by $H^2(I, \mathbb{R}^n)$ the Sobolev space (see [14]) of C^1 curves $z : I \longmapsto \mathbb{R}^n$ having absolutely continuous derivative and such that the integral $\int_I |\ddot{z}|^2 \, dt$ is finite, where the dots on the symbol of the curve z mean derivatives and $|\cdot|$ denotes the Euclidean norm in $\mathbb{R}^n$. We also denote by $a \cdot b$ the Euclidean inner product between vectors in $\mathbb{R}^n$.

As to the topology of H^2, observe that a sequence z_n is H^2-convergent if and only if it is uniformly convergent up to the first derivative, and the sequence of the second derivatives $\ddot{z}_n$ is L^2-convergent.

The space $H^2(I, \mathbb{R}^n)$ is an infinite dimensional separable Hilbert space, with inner product given by

$$\langle z_1, z_2 \rangle_2 = \int_0^1 (z_1 \cdot z_2 + \dot{z}_1 \cdot \dot{z}_2 + \ddot{z}_1 \cdot \ddot{z}_2) \, dt.$$

Given *any* differentiable manifold N, the set $H^2([0,1], N)$ is defined as the set of all C^1-curves $z : [0,1] \longmapsto N$ such that, for every local chart (V, φ) on N, with

$\varphi : U \longmapsto \mathbb{R}^n$ a diffeomorphism, and for every closed subinterval $I \subseteq [0,1]$ such that $z(I) \subset V$, we have $\varphi \circ z \in H^2(I, \mathbb{R}^n)$.

It is not difficult to see that this definition of $H^2([0,1], N)$ may be given equivalently in the following two ways:

- a curve $z : [0,1] \longmapsto N$ belongs to $H^2([0,1], N)$ if and only if there exists a finite sequence $I_1, \dots, I_k$ of closed subintervals of $[0,1]$ and a finite number of charts $\varphi_i : U_i \longmapsto \mathbb{R}^n$ on N, $i = 1, \dots, k$, such that $\bigcup_{i=1}^k I_k = [0,1]$, $z(I_i) \subset U_i$, and $\varphi_i \circ z \in H^2(I_i, \mathbb{R}^n)$ for all $i = 1, \dots, k$;
- a C^1-curve $z : [0,1] \longmapsto N$ is in $H^2([0,1], N)$ if and only if $\dot z$ is absolutely continuous and for one (hence, for every) Riemannian metric g_R on N, the integral $\int_0^1 g_R(\nabla^{(R)}_{\dot z} \dot z, \nabla^{(R)}_{\dot z} \dot z)\,dt$ is finite, where $\nabla^{(R)}$ is the Levi–Civita connection of g_R.

A classical result of Global Analysis (see [15]) states that, for every differentiable manifold N, the Sobolev space $H^2([0,1], N)$ has the structure of an infinite dimensional manifold, modeled on the Hilbert space $H^2([0,1], \mathbb{R}^n)$.

Let $T\mathcal{M}$ denote the tangent bundle of $\mathcal{M}$, with $\pi : T\mathcal{M} \longmapsto \mathcal{M}$ the canonical projection. For a curve $z : I \subseteq \mathbb{R} \longmapsto \mathcal{M}$, a *vector field along* z is a map $\zeta : I \longmapsto T\mathcal{M}$ such that $\pi \circ \zeta = z$.

For the background material assumed in this paper, we recommend the following references: [16, 17] for the basics of Lorentzian geometry, [8, 13] for the main facts of General Relativity, [18] for infinite dimensional differential geometry and [19] for the main tools of the Calculus of Variations in Lorentzian manifolds.

2 Fermat's principle for freely falling massive objects

In this section we present a variational principle that characterizes the timelike geodesics between P and Γ.

We will denote by $\Omega^{(2)}_{P,\Gamma}$ the set of curves in $H^2([0,1], \mathcal{M})$ that join P and Γ:

$$\Omega^{(2)}_{P,\Gamma} = \left\{ z \in H^2([0,1], \mathcal{M}) \;\middle|\; z(0) \in P,\; z(1) \in \Gamma \right\}. \tag{2.1}$$

It is well-known [15] that $\Omega^{(2)}_{P,\Gamma}$ is a submanifold of $H^2([0,1], \mathcal{M})$, with the same regularity as the manifolds P and Γ. For all $z \in \Omega^{(2)}_{P,\Gamma}$, the tangent space $T_z\Omega^{(2)}_{P,\Gamma}$ is given by:

$$T_z\Omega^{(2)}_{P,\Gamma} = \left\{ \zeta \in H^2([0,1], T\mathcal{M}) \;\middle|\; \zeta \text{ a vector field along } z \right\}. \tag{2.2}$$

In order to construct an *arrival time* functional for curves joining P and Γ, we need a function that *measures the time of events* of Γ. To this aim, we assume now that $T : \Gamma \longmapsto \mathbb{R}$ is a smooth *time function* on Γ, i.e., the gradient ∇T of T is a timelike vector field on Γ. The existence of such a function T is the motivation for requiring that Γ be timelike, as mentioned above.

If Γ is one-dimensional, i.e., if Γ is a timelike curve embedded in $\mathcal{M}$, then such a function T is just a parameterization of Γ.

If $\dim(\Gamma) > 1$, then a time function $T : \Gamma \longmapsto \mathbb{R}$ exists if and only if Γ, viewed as a Lorentzian manifold in its own right, is *stably causal* (see [20, p. 198]).

Let T be a fixed time on Γ. Since ∇T is timelike, then, for all $c \in \mathbb{R}$, the inverse image $T^{-1}(c)$ is a possibly empty spacelike submanifold of Γ, which will be denoted by Γ_c.

We consider the smooth timelike vector field Y on Γ, given by a suitable normalization of ∇T:

$$Y = \frac{\nabla T}{\langle \nabla T, \nabla T \rangle}. \tag{2.3}$$

The following clearly holds:

$$\mathrm{d}T[Y] = \langle \nabla T, Y \rangle \equiv 1, \quad \text{and} \quad Y \perp \Gamma_c. \tag{2.4}$$

The existence of a time function on Γ allows us to define an *arrival time functional* τ on the space of all curves with final endpoint in Γ: for all $z \in \Omega^{(2)}_{P,\Gamma}$, let $\tau(z)$ be defined by

$$\tau(z) = T(z(1)). \tag{2.5}$$

Proposition 2.1. *The functional τ is smooth on $\Omega^{(2)}_{P,\Gamma}$. For all $z \in \Omega^{(2)}_{P,\Gamma}$ and all $\zeta \in T_z\Omega^{(2)}_{P,\Gamma}$, the Gateaux derivative $\mathrm{d}\tau(z)[\zeta]$ is given by*

$$\mathrm{d}\tau(z)[\zeta] = \langle \nabla T(z(1)), \zeta(1) \rangle. \tag{2.6}$$

Proof. See [21, Proposition 2.1]. □

Given any C^1-curve $z : [0, 1] \longmapsto \mathcal{M}$ having endpoint $z(1)$ in Γ, we denote by U_z the C^1-vector field on z obtained by parallel transport of $Y(z(1))$ along z. This means that U_z is the unique vector field along z satisfying

$$\nabla_{\dot z} U_z = 0, \quad U_z(1) = Y(z(1)).$$

Observe that, since parallel transport is an isometry, the quantity

$$\langle U_z, U_z \rangle = \langle Y(z(1)), Y(z(1)) \rangle$$

is constant (negative) along z.

If z is a causal curve with final endpoint in Γ, we say that z is *future pointing* with respect to the time orientation of Γ induced by Y if, for all t, the tangent vector $\dot z(t)$ and the vector $U_z(t)$ belong to the same light half-cone, i.e., if $\langle \dot z(t), U_z(t) \rangle < 0$ everywhere. A similar definition is given for *past pointing* curves.

For our variational principle, we want to consider the set of timelike curves in $\Omega^{(2)}_{P,\Gamma}$ which are parameterized by a constant multiple of the proper time and which are future pointing with respect to the time orientation of Γ induced by the timelike vector field Y. To this aim, for all $\varepsilon > 0$ we introduce the set

$$\mathcal{R}^{+}_{P,\Gamma,\varepsilon} = \left\{ z \in \Omega^{(2)}_{P,\Gamma} : \langle \dot{z}, \dot{z} \rangle \equiv -\varepsilon^2 \text{ and } \langle \dot{z}, U_z \rangle < 0 \text{ on } [0,1] \right\}. \tag{2.7}$$

Proposition 2.2. *For all $\varepsilon > 0$, the set $\mathcal{R}^{+}_{P,\Gamma,\varepsilon}$ is a smooth submanifold of $\Omega^{(2)}_{P,\Gamma}$, and, in particular, the restriction of the functional τ to $\mathcal{R}^{+}_{P,\Gamma,\varepsilon}$ is smooth. For all $z \in \mathcal{R}^{+}_{P,\Gamma,\varepsilon}$, the tangent space $T_z\mathcal{R}^{+}_{P,\Gamma,\varepsilon}$ can be identified with the Hilbert subspace of $T_z\Omega^{(2)}_{P,\Gamma}$ given by:*

$$T_z\mathcal{R}^{+}_{P,\Gamma,\varepsilon} = \left\{ \zeta \in T_z\Omega^{(2)}_{P,\Gamma} \;\middle|\; \langle \nabla_{\dot{z}}\zeta, \dot{z} \rangle = 0 \textit{ on } [0,1] \right\}. \tag{2.8}$$

Proof. See [21, Proposition 2.2]. □

We are ready to state our variational principle for timelike geodesics between P and Γ.

Theorem 2.3 (Fermat Principle for Timelike Geodesics). *Let $\varepsilon > 0$ be fixed and z a curve in $\mathcal{R}^{+}_{P,\Gamma,\varepsilon}$. Then, z is a stationary point for τ in $\mathcal{R}^{+}_{P,\Gamma,\varepsilon}$ if and only if z is a geodesic such that $\dot{z}(0)$ is perpendicular to P and $\dot{z}(1)$ is perpendicular to the time slice $\Gamma_{\tau(z)}$ of Γ.*

Proof. See [21, Theorem 3.2]. □

Remark 2.4. Observe that a vector orthogonal to a causal vector, i.e., lightlike or timelike, is always spacelike. For this reason, since solutions to our variational problem are timelike geodesics orthogonal to P, we must assume that P be spacelike.

A complete variational theory for timelike geodesics can be found in [1, 3, 5, 21].

3 Fermat's principle for light rays

In this section we discuss a general-relativistic Fermat principle for lightlike geodesics. From a mathematical point of view, the problem is more delicate, due to the fact that the set of lightlike geodesics joining P and Γ is not a C^1-submanifold of $H^2([0,1], \mathcal{M})$. For this reason, we do not introduce an infinite dimensional differentiable structure on our set of "trail paths", and we define our variational setup as follows.

As the trial paths for our variational principle we want to consider all piecewise smooth lightlike curves from P to Γ, i.e., all possible ways to go from P to Γ at the (vacuum) speed of light. More precisely, we introduce the space of trial paths in the following way.

Definition 3.1. *Let $\mathcal{L}_{P,\Gamma}$ denote the set of all piecewise C^∞ maps $\beta : [0,1] \to \mathcal{M}$ with:*

(a) $\langle \beta', \beta' \rangle = 0$;
(b) $\beta(0) \in P$ *and* $\beta(1) \in \Gamma$;
(c) $\langle \beta', U \rangle < 0$.

Here and in the following, $\beta' : [0,1] \to T\mathcal{M}$ denotes the tangent field of β. In (c), we have introduced the vector field $U : [0,1] \longrightarrow T\mathcal{M}$ along β which is defined by parallel-transporting the vector $Y_{\beta(1)}$. (Note that the vector field Y was introduced by (2.3)).

Allowing for piecewise smooth curves, rather than just for smooth curves, is of advantage in view of the Morse index theorem for null geodesics (see [12]. Condition (c) of Definition 3.1 guarantees that $\beta'(1)$ is future pointing with respect to the time function $T : \Gamma \to \mathbb{R}$ and that at the (possible) break points of β the tangent vector field β' does not jump from one half of the light cone to the other.

In the curve space $\mathcal{L}_{P,\Gamma}$ we consider the following kind of variations.

Definition 3.2. *For $\beta \in \mathcal{L}_{P,\Gamma}$, an* allowed variation *of β is a map*

$$\eta :]-\varepsilon_0, \varepsilon_0[\times [0,1] \longrightarrow \mathcal{M},$$

for some $\varepsilon_0 > 0$, that satisfies the following properties:

(a) *there is a subdivision $0 = s_0 < s_1 < \dots < s_{N-1} < s_N = 1$ such that η is a C^∞-map on each $]-\varepsilon_0, \varepsilon_0[\times [s_i, s_{i+1}]$;*
(b) *for all $\varepsilon \in]-\varepsilon_0, \varepsilon_0[$ the curve $s \longmapsto \eta(\varepsilon, s)$ is in $\mathcal{L}_{P,\Gamma}$;*
(c) *$\eta(0, s) = \beta(s)$ for all $s \in [0,1]$.*

For $\beta \in \mathcal{L}_{P,\Gamma}$, we define the space $T_\beta \mathcal{L}_{P,\Gamma}$ of *variational vector fields* along β as

$$T_\beta \mathcal{L}_{P,\Gamma} = \Big\{ X{:}[0,1] \longrightarrow T\mathcal{M} \;\Big|\; X \text{ a piecewise smooth vector field along } \beta, \\ \langle \nabla_{\beta'} X, \beta' \rangle = 0,\ X(0) \in T_{\beta(0)} P,\ X(1) \in T_{\beta(1)} \Gamma \Big\}. \tag{3.1}$$

Then the following proposition holds.

Proposition 3.3. *Let β be any curve in $\mathcal{L}_{P,\Gamma}$. Then every allowed variation η of β defines a variational vector field $X \in T_\beta \mathcal{L}_{P,\Gamma}$ by*

$$X(s) = \eta(\cdot, s)'(0). \tag{3.2}$$

Conversely, to every $X \in T_\beta \mathcal{L}_{P,\Gamma}$, there is an allowed variation η of β such that (3.2) holds.

Proof. See [12, Proposition 3.3]. □

Clearly, the set $T_\beta \mathcal{L}_{P,\Gamma}$ has the structure of an infinite dimensional real vector space. As suggested by our notation, the reader may view $T_\beta \mathcal{L}_{P,\Gamma}$ as the tangent space of $\mathcal{L}_{P,\Gamma}$ at the point β.

The functional we want to extremize is the *arrival time* $\tau : \mathcal{L}_{P,\Gamma} \longrightarrow \mathbb{R}$, defined by (2.5).

Without a differentiable structure on $\mathcal{L}_{P,\Gamma}$ we cannot speak of the derivative of τ. However, for any $\beta \in \mathcal{L}_{P,\Gamma}$ and any allowed variation η of β, the map $]-\varepsilon, \varepsilon[\longrightarrow \mathbb{R}$, $\varepsilon \longmapsto \tau(\eta(\varepsilon, \cdot))T(\eta(\varepsilon, 1))$ is the composition of two C^∞-maps. By the chain rule, its derivative at $\varepsilon = 0$ is given by

$$\frac{d}{d\varepsilon}\tau(\eta(\varepsilon, \cdot))\Big|_{\varepsilon=0} = \mathrm{d}T(\beta(1))[X(1)] \tag{3.3}$$

where X is defined by (3.2). This suggests that we view the linear map $\mathrm{d}\tau(\beta) : T_\beta \mathcal{L}_{P,\Gamma} \longrightarrow \mathbb{R}$, which is defined by

$$\mathrm{d}\tau(\beta)[X] = \mathrm{d}T(\beta(1))[X(1)], \tag{3.4}$$

as the differential of τ at the point β. Although we have no differentiable structure on $\mathcal{L}_{P,\Gamma}$, we can now define what we mean by a stationary point of τ.

Definition 3.4. *A curve $\beta \in \mathcal{L}_{P,\Gamma}$ is a* stationary point *of τ if*

$$\frac{\mathrm{d}}{\mathrm{d}\varepsilon}\,\tau(\eta(\varepsilon, \cdot))\Big|_{\varepsilon=0} = 0$$

for all allowed variations η of β. Equivalently, β is a stationary point of τ if $(d\tau)_\beta(X) = 0$ for all $X \in T_\beta \mathcal{L}_{P,\Gamma}$.

Here we make use of the above-mentioned fact that every $X \in T_\beta \mathcal{L}_{P,\Gamma}$ can be written in the form of Eq. (3.2) with an allowed variation η of β. The variational problem we want to solve is to determine the stationary points of τ. We have the following variational principle.

Theorem 3.5 (Fermat Principle for Lightlike Geodesics). *Let β be a curve in the space $\mathcal{L}_{P,\Gamma}$. Then β is a stationary point of the arrival time functional τ if and only if $\beta \circ \phi^{-1}$ is an affinely parameterized geodesic, $\beta'(0)$ is orthogonal to $T_{\beta(0)}P$ and $\beta'(1)$ is orthogonal to $T_{\beta(1)}\Gamma_{\tau(\beta)}$. Here $\phi : [0, 1] \longrightarrow [0, 1]$ denotes the piecewise C^∞ diffeomorphism defined by*

$$\phi(s) = \frac{\int_0^s \langle \beta', U\rangle\, \mathrm{d}\tilde{s}}{\int_0^1 \langle \beta', U\rangle\, \mathrm{d}\tilde{s}}. \tag{3.5}$$

Proof. See [12, Theorem 3.6]. □

The original formulations of Fermat's principle in General Relativity, as presented in this paper, are to be found in [9, 22]; its formulation to the case of varying endpoints is in [12]. A Lusternik–Schnirelman theory and a Morse theory for light rays based on Fermat's principle are presented in [2, 4, 5, 19].

4 Lightlike geodesics are limits of timelike geodesics

In this section we will discuss briefly a method of approximating the lightlike geodesics between P and Γ with timelike geodesics, for which the results of the previous sections apply. The idea of such approximation is not new in General Relativity (see T. Levi–Civita [23]).

We recall that, for any *non-degenerate* submanifold Σ of our Lorentzian manifold $\mathcal{M}$, i.e., such that the restriction of the Lorentzian metric g to the tangent bundle $T\Sigma$ is non-degenerate, the *second fundamental form* S^Σ (also known as the *shape tensor* of Σ) can be defined in analogy to the Riemannian case. Namely, for each $p \in \Sigma$ and each vector $n \in T_p\mathcal{M}$ which is perpendicular to $T_p\Sigma$, the second fundamental form of Σ in the direction of n is the bilinear form $S_n^\Sigma : T_p\Sigma \times T_p\Sigma \longrightarrow \mathbb{R}$ defined by

$$S_n^\Sigma(v_1, v_2) = \langle n, \nabla_{v_1} V_2 \rangle \tag{4.1}$$

where V_2 is any C^∞-vector field on Σ which takes the value v_2 at p. As in Riemannian geometry, one can show that S_n^Σ is, indeed, well-defined (i.e., independent of the extension V_2 of v_2) and symmetric. For details we refer the reader to [16] and [17]. Observe that every spacelike submanifold of $\mathcal{M}$ is non-degenerate.

We give the following definition of P-focal point along a lightlike geodesic.

Definition 4.1. *Let $z : [0, 1] \longmapsto \mathcal{M}$ be a lightlike geodesic with $z(0) \in P$ and $\dot z(0) \perp T_{z(0)}P$. A P-*Jacobi field *along z is a smooth vector field J along z satisfying:*

1. $\nabla_{\dot z}^2 J - R(\dot z, J)\,\dot z = 0$;
2. $J(0) \in T_{z(0)}P$;
3. $\langle \nabla_{\dot z(0)} J, v \rangle + S^P_{\dot z(0)}(J(0), v) = 0$ *for all* $v \in T_{z(0)}P$.

*A point $z(t_0)$ of z is said to be a P-*focal point *if there exists a non-zero P-Jacobi field along z such that $J(t_0) = 0$; the* multiplicity *of the P-focal point $z(t_0)$ is the dimension of the vector space of all P-Jacobi fields vanishing at t_0. We denote by $\mathcal{J}_z^P$ the vector space of all P-Jacobi fields along z. It is known that the set of P-focal points along a causal geodesic is discrete if P is a non-degenerate submanifold of $\mathcal{M}$; the* geometric index *of z is the sum of the dimensions of all the P-focal points along z.*

Equation 1 is the Jacobi equation, whose geometric meaning is that J is the variational field corresponding to a (smooth) variation z_s of z consisting of geodesics; condition 2 means that, in first approximation, each z_s starts at some point of P; condition 3 means that $\dot z_s(0)$ is orthogonal to P for all s.

A vector field that satisfies the differential equation 1 is said to be a *Jacobi field* along z; a Jacobi field J is uniquely determined by the initial conditions $J(0)$ and $\nabla_{\dot z(0)} J$. There is a linear isomorphism between the vector space $\mathcal{J}_z$ of Jacobi fields along z and the tangent space $T_{\dot z(0)}(T\mathcal{M})$ to the tangent bundle $T\mathcal{M}$ at the point $\dot z(0)$. To describe such an isomorphism, we make the following remarks.

Given any point $p \in \mathcal{M}$ and any vector $v \in T_p\mathcal{M}$, we can identify the tangent space $T_v(T\mathcal{M})$ at v of the tangent bundle $T\mathcal{M}$ with the direct sum $T_p\mathcal{M} \oplus T_p\mathcal{M}$. The first summand $T_p\mathcal{M}$ is the *horizontal* space of the Levi–Civita connection of g, and the second summand $T_p\mathcal{M}$ is the *vertical* subspace, which is tangent to the fiber $T_p\mathcal{M}$. For $\lambda \in T_v(T\mathcal{M})$, we will write $\lambda = (\lambda_h, \lambda_v)$, with $\lambda_v, \lambda_h \in T_p\mathcal{M}$ being the components of λ with respect to this decomposition; this identification will be made throughout the section.

The isomorphism between $T_{\dot{z}(0)}(T\mathcal{M}) \simeq T_{z(0)}\mathcal{M} \oplus T_{z(0)}\mathcal{M}$ and $\mathcal{J}_z$ is obtained by assigning to each $\lambda = (\lambda_h, \lambda_v) \in T_{\dot{z}(0)}T\mathcal{M}$ the unique Jacobi field $J \in \mathcal{J}_z$ satisfying $J(0) = \lambda_h$ and $\nabla_{\dot{z}(0)} J = \lambda_v$. An inverse for this map is obtained as follows. Given $J \in \mathcal{J}_z$, let z_s be a smooth variation of z consisting of geodesics and with variational vector field J. Then, the element $\lambda \in T_{\dot{z}(0)}T\mathcal{M}$ is given by $\lambda = \frac{\mathrm{D}}{\mathrm{d}s}\frac{\mathrm{d}}{\mathrm{d}t} z_s(t)|_{t=s=0}$. Observing that $z_s(1) = \exp(\frac{\mathrm{d}}{\mathrm{d}t} z_s(0)$, we have

$$J(1) = \mathrm{d}\exp_{z(0)}(\dot{z}(0))[\lambda]. \tag{4.2}$$

We denote by $TP^{\perp}$ the normal bundle of P in $\mathcal{M}$, which is a submanifold of $T\mathcal{M}$; the P-Jacobi fields along z correspond to vectors $\lambda \in T_{\dot{z}(0)}(TP^{\perp})$ tangent to the normal bundle of P.

It is now easy to prove the following

Lemma 4.2. *Let $z : [0, 1] \longmapsto \mathcal{M}$ be a (lightlike) geodesic with $z(0) \in P$ and $\dot{z}(0) \in TP^{\perp}$. If $z(1)$ is not a P-focal point along z, then the exponential map* exp *of $\mathcal{M}$ gives a local diffeomorphism between a neighborhood of $\dot{z}(0)$ in $TP^{\perp}$ and a neighborhood of $z(1)$ in $\mathcal{M}$.* □

Let A denote the open subset of $T\mathcal{M}$ on which the exponential map is defined. For $p \in \mathcal{M}$, $v \in AT_p\mathcal{M}$ and $w \in T_p\mathcal{M}$, we have

$$\mathrm{d}\exp_p(v)[w] = \mathrm{d}\exp(v)[(0, w)].$$

Moreover, if $z(t) = \exp(tv)$ is the geodesic such that $z(0) = p$ and $\dot{z}(0) = v$, then $\mathrm{d}\exp_p(v)[v] = \frac{\mathrm{d}}{\mathrm{d}t}\exp(tv)|_{t=1} = \dot{z}(1)$.

We present the following generalization of the classical *Gauss Lemma* in semi-Riemannian geometry (see [16]). We recall that the Gauss Lemma states that the differential $\mathrm{d}\exp_p$ of the exponential map at a *fixed* point p is a *radial isometry*. In our case, we need to prove a similar property for the exponential map when the basepoint is not necessarily fixed.

Proposition 4.3. *Let $p \in \mathcal{M}$, $v \in AT_p\mathcal{M}$ and $\lambda = (\lambda_h, \lambda_v) \in T_v(T\mathcal{M})$ be given. Then we have*

$$\big\langle \mathrm{d}\exp(v)[(0, v)], \mathrm{d}\exp(v)[\lambda]\big\rangle = \big\langle v, \lambda_h + \lambda_v\big\rangle. \tag{4.3}$$

Proof. See [21, Proposition 4.3]. □

Using Proposition 4.3, it is not hard to prove that the lightlike geodesics between P and Γ can be obtained as limits of timelike geodesics between P and Γ. We will state the result only for the case when Γ is a unidimensional submanifold of $\mathcal{M}$, i.e., when Γ is a timelike curve.

Proposition 4.4. *Let $\Gamma : \mathbb{R} \longmapsto \mathcal{M}$ be a (smooth) timelike curve of $\mathcal{M}$, P any spacelike submanifold of $\mathcal{M}$ and $z : [0,1] \longmapsto \mathcal{M}$ a lightlike geodesic with $z(0) \in P$, $\dot z(0) \in TP^{\perp}$ and $z(1) = \Gamma(s_0)$ for some $s_0 \in \mathbb{R}$. Assume that $z(1)$ is not a P-focal point along z. Then there exists a neighborhood V of $z([0,1])$ in $\mathcal{M}$ and a positive number ε_0 such that, for all $\varepsilon \in]-\varepsilon_0, \varepsilon_0[$, there exists a unique timelike geodesic $z_\varepsilon : [0,1] \longmapsto \mathcal{M}$ joining P and Γ, with $\dot z_\varepsilon(0) \in TP^{\perp}$. Moreover, z_ε tends to z in $C^2([0,1], \mathcal{M})$ as ε tends to 0. If z is future pointing with respect to Γ, i.e., if $\langle \dot z(1), \dot\Gamma(s_0) \rangle < 0$, then also z_ε is future pointing with respect to Γ for all ε.*

Proof. See [21, Proposition 4.4]. □

The conclusion that we draw from Proposition 4.4 is that the lightlike geodesics between P and Γ that start orthogonally at P can be obtained as limits of critical points for the arrival time functional in $\mathcal{R}^+_{P,\Gamma,\varepsilon}$ for $\varepsilon \to 0$.

In order to develop a Morse theory for lightlike geodesics as a *limit* of the Morse theory for timelike geodesics (see [5]), one needs to prove the *stability* of the geometric index for causal Lorentzian geodesics. This is established in the next

Theorem 4.5. *Suppose that $\gamma : [0,1] \mapsto \mathcal{M}$ is a lightlike geodesic joining P and Γ, with $\dot\gamma(0) \in T_{\gamma(0)}P^{\perp}$, such that $\gamma(1)$ is* not *a P-focal point along γ. Let $\gamma_n : [0,1] \mapsto \mathcal{M}$ be a sequence of causal geodesics joining P and Γ, with $\dot\gamma_n(0) \in T_{\gamma_n(0)}P^{\perp}$ and converging to γ as $n \to \infty$. Then, for n sufficiently large, the geometric index of γ_n equals the geometric index of γ.*

Proof (Sketch). The proof of Theorem 4.5 involves the notion of *Maslov* index for a semi-Riemannian geodesic (see [24]). For causal Lorentzian geodesics, the Maslov index coincides with the geometric index of the geodesic, while in the general case it is given by a sort of *algebraic* count of the multiplicities of the conjugate points along the geodesic. The Maslov index can be characterized as the intersection number between a curve and a codimension one subvariety of the *Lagrangian Grassmannian* of a symplectic space, and thus it is stable under homotopies. The stability of the geometric index can be proven in more general contexts; details of the proof may be found in [24, Theorem 5.2.1, Corollary 5.2.2]. □

This approach to the study of light rays between extended light sources and receivers is the subject of a subsequent paper ([5]).

5 The relativistic brachistochrone problems

The brachistochrone problem can be formulated on Lorentzian manifolds in the following way.

We fix a timelike smooth vector field Y on $\mathcal{M}$. For simplicity, we assume that Y is complete, i.e., its integral lines are defined over the entire real line. The integral curves of Y can be interpreted as the worldlines of *observers*. Please note that we do not require Y to be normalized, i.e., in general the worldlines of our observers are not parameterized by proper time. The reason is that, in the stationary case, i.e., if

$(\mathcal{M}, g)$ admits a timelike Killing vector field, it is convenient to choose this Killing vector field for Y and not for a renormalized version of it.

To formulate the brachistochrone problem with respect to our arbitrarily chosen observer field Y, we fix a point p in $\mathcal{M}$, a (maximal) integral curve $\gamma : \mathbb{R} \longmapsto \mathcal{M}$ of Y and a real number $k > 0$. The *trial paths* for our variational problem are all timelike smooth curves $\sigma : [0, 1] \longmapsto \mathcal{M}$ which are nowhere tangent to Y and satisfy the following conditions.

$$\sigma(0) = p; \tag{5.1}$$

$$\sigma(1) \in \gamma(\mathbb{R}); \tag{5.2}$$

$$\langle \dot\sigma(0), Y(\sigma(0)) \rangle = -k \left(-\langle \dot\sigma(0), \dot\sigma(0) \rangle \right)^{1/2}; \tag{5.3}$$

$$\langle \nabla_{\dot\sigma}\dot\sigma, \dot\sigma \rangle = 0; \tag{5.4}$$

$$\langle \nabla_{\dot\sigma}\dot\sigma, Y \rangle = 0. \tag{5.5}$$

If we interpret each integral curve of Y as a "point in space", (5.1) and (5.2) mean that all trial paths connect the same two points in space, where the starting time is fixed whereas the arrival time is not. Condition (5.3) says that all trial paths start with the same speed with respect to the observer field Y. By condition (5.4), the quantity T_σ defined by $-T_\sigma^2 = \langle \dot\sigma, \dot\sigma \rangle$ is a constant for each trial path σ (but takes different values for different trial paths). This implies that the curve parameter s along σ is related to proper time τ by an affine transformation, $\tau = T_\sigma s + \text{const}$. As a consequence, the 4-velocity along each trial path is given by $T_\sigma^{-1}\dot\sigma$, whereas the 4-acceleration is given by $T_\sigma^{-2}\nabla_{\dot\sigma}\dot\sigma$. Hence, conditions (5.4) and (5.5) require the 4-acceleration to be perpendicular to the plane spanned by $\dot\sigma$ and Y. In other words, with respect to the observer field Y there are only forces perpendicular to the direction of motion. Such forces can be interpreted as constraint forces supplied by a frictionless slide which is at rest with respect to the observer field Y.

The brachistochrone problem can now be formulated in the following way.

Among all trial paths that satisfy the above conditions, we want to find those curves for which the travel time is minimal or, more generally, stationary.

A different general-relativistic brachistochrone problem can be formulated by requiring that the solutions be stationary points for the *arrival time* functional, given by $\tau(\sigma) = \gamma^{-1}(\sigma(1))$. In other words, $\tau(\sigma)$ is the value of the time of the receiver at the arrival event; this is proper time if and only if Y is normalized along γ. In physical terms, the two brachistochrone problems differ in their methods of measuring time: in the first case, time is measured by a watch traveling along the trajectory of the mass; in the second, time is measured by the observer who receives the mass at the end of its trajectory. For a physical interpretation of our brachistochrone problem, the timelike vector field Y should be related to some observable quantities, i.e., Y should be co-moving with some bodies. For instance, if we are in the solar system and Y is comoving with the planets, the solutions to our brachistochrone problem will give worldlines of particles that minimize the travel time among all curves that have a fixed specific energy in the rest system of the planets. If Y is at rest with respect to the sun and to the distant stars, then the brachistochrones will be the worldlines of massive objects that minimize the travel time among all curves that have fixed energy in a reference system oriented at distant stars.

It is also possible to return to the original interpretation of of the brachistochrone problem and think of the body guided by a frictionless slide, in which case Y is determined by being the rest system of the slide.

We emphasize that, as in the Newtonian case, the solutions of the brachistochrone variational problem represent trajectories of massive objects subject to the gravitational field and also to some *constraint* forces. Hence, they are *not* in general geodesics in the spacetime metric.

If $(\mathcal{M}, g)$ is a stationary spacetime and Y is a *Killing* vector field, i.e., the flow of Y preserves metric g, then condition (5.5) means that the product $\langle\dot\sigma, Y\rangle$ is constant along σ. The value of this constant can be computed easily using condition (5.3), which gives $\langle\dot\sigma, Y\rangle \equiv -kT_\sigma$. Hence, in the stationary case, conditions (5.3) and (5.5) can be subsumed in the condition:

$$\langle\dot\sigma, Y\rangle = -kT_\sigma. \tag{5.6}$$

Condition (5.6) is the relativistic counterpart of the energy conservation law in the Newtonian case. Although physically meaningful, the mathematical approach to the general-relativistic brachistochrone problem in the non-stationary case presents difficulties of higher order than in the stationary case. For instance, it is not even clear whether the non-stationary brachistochrones are solutions to a second-order differential equation; in [11], the authors used a Lagrange multiplier technique to derive a system of differential equations for the travel time brachistochrones and for the Lagrange multipliers. Unfortunately, it does not seem to be possible to eliminate the Lagrange multipliers from the system without introducing integrals, unless in the stationary case. Thus, it looks as if the brachistochrones in the non-stationary case are not determined by a second-order differential equation, but rather by an integro-differential equation.

For these technical reasons, we will stick to the case of a manifold $\mathcal{M}$ with metric g which is stationary with respect to the observer field Y.

We introduce for convenience the auxiliary Riemannian metric g_{R} on $\mathcal{M}$, given by

$$g_{\mathrm{R}}(q)(v_1, v_2) = \langle v_1, v_2\rangle_{(\mathrm{R})} = \langle v_1, v_2\rangle - \frac{\langle v_1, Y(q)\rangle \cdot \langle v_2, Y(q)\rangle}{\langle Y(q), Y(q)\rangle}, \tag{5.7}$$

for $q \in \mathcal{M}$ and $v_1, v_2 \in T_q\mathcal{M}$. It is easy to see that Y is Killing also for the metric g_{R}; moreover, the restrictions of g and g_{R} to the orthocomplement of Y coincide.

Let k be a fixed positive constant, with $-k^2 < \sup_{\mathcal{M}}\langle Y(q), Y(q)\rangle$, and let U_k be the open set

$$U_k = \Big\{q \in \mathcal{M} : \langle Y(q), Y(q)\rangle + k^2 > 0\Big\}. \tag{5.8}$$

Since Y is Killing, the quantity $\langle Y, Y\rangle$ is constant along the integral lines of Y, and so U_k is invariant with respect to the flow of Y.

We will denote by p a fixed event of U_k and by $\gamma : \mathbb{R} \longmapsto U_k$ a given integral line of Y which does not pass through p. We will work in the space $\Omega^{(2)}_{p,\gamma} = \Omega^{(2)}_{p,\gamma}(U_k)$ of curves in U_k joining p and γ, defined as in (2.1). As we have observed, $\Omega^{(2)}_{p,\gamma}$ is a smooth submanifold of $H^2([0,1], U_k)$; for $\sigma \in \Omega^{(2)}_{p,\gamma}$, the tangent space $T_\sigma \Omega^{(2)}_{p,\gamma}$, defined as in (2.2), is a Hilbert space with respect to the inner product

$$\langle \zeta, \zeta \rangle_2 = \int_0^1 \left(\langle \nabla_{\dot\sigma} \zeta, \nabla_{\dot\sigma} \zeta \rangle_{(R)} + \langle \nabla^2_{\dot\sigma} \zeta, \nabla^2_{\dot\sigma} \zeta \rangle_{(R)} \right) \mathrm{d}t. \tag{5.9}$$

Finally, for all positive constant $k \in \mathbb{R}^+$, we introduce the space $\mathcal{B}^{(2)}_{p,\gamma}(k)$ by

$$\mathcal{B}^{(2)}_{p,\gamma}(k) = \left\{ \sigma \in \Omega^{(2)}_{p,\gamma} : \exists\, T_\sigma \in \mathbb{R}^+ \text{ such that } \langle \dot\sigma, Y \rangle \equiv -k\, T_\sigma \text{ and } \langle \dot\sigma, \dot\sigma \rangle \equiv -T_\sigma^2 \right\}. \tag{5.10}$$

The space $\mathcal{B}^{(2)}_{p,\gamma}(k)$ has the structure of an infinite dimensional smooth Hilbert manifold.

Proposition 5.1. *$\mathcal{B}^{(2)}_{p,\gamma}(k)$ is a smooth submanifold of $\Omega^{(2)}_{p,\gamma}$. For $\sigma \in \mathcal{B}^{(2)}_{p,\gamma}(k)$, the tangent space $T_\sigma \mathcal{B}^{(2)}_{p,\gamma}(k)$ can be identified with the Hilbert space*

$$T_\sigma \mathcal{B}^{(2)}_{p,\gamma}(k) = \Big\{ \zeta \in T_\sigma \Omega^{(2)}_{p,\gamma} : \exists\, C_\zeta \in \mathbb{R} \text{ such that } \langle \nabla_{\dot\sigma} \zeta, Y \rangle - \langle \zeta, \nabla_{\dot\sigma} Y \rangle \equiv C_\zeta \text{ and } \langle \nabla_{\dot\sigma} \zeta, \dot\sigma \rangle \equiv \frac{T_\sigma C_\zeta}{k} \Big\}, \tag{5.11}$$

endowed with the inner product $\langle \cdot, \cdot \rangle_2$ of formula (5.9).

Proof. See [25, Proposition 2.1]. □

We consider the *action* functional F on $\Omega^{(2)}_{p,\gamma}$, given by

$$F(\sigma) = \frac{1}{2} \int_0^1 \langle \dot\sigma, \dot\sigma \rangle \, \mathrm{d}t. \tag{5.12}$$

It is well-known that F is smooth; for $\sigma \in \Omega^{(2)}_{p,\gamma}$ and $V \in T_\sigma \Omega^{(2)}_{p,\gamma}$, the Gateaux derivative $\mathrm{d}F(\sigma)[V]$ is given by

$$\mathrm{d}F(\sigma)[V] = \int_0^1 \langle \nabla_{\dot\sigma} V, \dot\sigma \rangle \, \mathrm{d}t. \tag{5.13}$$

Travel time brachistochrones

If $\sigma \in \mathcal{B}^{(2)}_{p,\gamma}(k)$, then, from (5.12), we obtain immediately that

$$F(\sigma) = -\frac{1}{2} T_\sigma^2; \tag{5.14}$$

hence, in $\mathcal{B}^{(2)}_{p,\gamma}(k)$ the *travel time* functional T, given by $T(\sigma) = T_\sigma$, has the following form:

$$T(\sigma) = \sqrt{-2\,F(\sigma)}. \tag{5.15}$$

In particular, from (5.14) and (5.15), one can write the travel time in an integral form, so it is possible to study its critical points using the Euler–Lagrange formalism (see [25]).

We have the following

Proposition 5.2. *The travel time functional T on $\mathcal{B}^{(2)}_{p,\gamma}(k)$ is smooth on $\mathcal{B}^{(2)}_{p,\gamma}(k)$. For $\zeta \in T_\sigma \mathcal{B}^{(2)}_{p,\gamma}(k)$, the Gateaux derivative $\mathrm{d}T(\sigma)[\zeta]$ is given by*

$$\mathrm{d}T(\sigma)[\zeta] = -\frac{C_\zeta}{k}. \tag{5.16}$$

Proof. See [25, Corollary 2.2]. □

After setting up our variational framework, we are ready to give the following definition:

Definition 5.3. *A* travel time brachistochrone *of energy k between p and γ is a stationary point for the travel time functional T on $\mathcal{B}^{(2)}_{p,\gamma}(k)$. A travel time brachistochrone curve σ is said to be* minimal *if σ is a minimum point for T on $\mathcal{B}^{(2)}_{p,\gamma}(k)$. From Proposition 5.2 it follows easily that a curve $\sigma \in \mathcal{B}^{(2)}_{p,\gamma}(k)$ is a travel time brachistochrone if and only if, for every $\zeta \in T_\sigma \mathcal{B}^{(2)}_{p,\gamma}(k)$, we have $C_\zeta = 0$.*

Since T is strictly positive on $\mathcal{B}^{(2)}_{p,\gamma}(k)$, its critical points coincide with the critical points in $\mathcal{B}^{(2)}_{p,\gamma}(k)$ of the restriction of the action functional $F = -\frac{1}{2}T^2$. The minimal brachistochrones of energy k are *maximum* points of F on $\mathcal{B}^{(2)}_{p,\gamma}(k)$.

We now give a different description of the travel time brachistochrone curves, as curves that minimize *locally* their travel time.

If q is any point in U_k, we denote by γ_q the maximal integral line of Y through q. Moreover, if $I = [a, b] \subseteq [0, 1]$ is any interval, and if q_1, q_2 are any two points in U_k, we define $\mathcal{B}^{(2)}_{q_1,\gamma_{q_2}}(k, I)$ as the space of curves $\tau \in H^2(I, U_k)$ such that $\tau(a) = q_1$, $\tau(b) \in \gamma_{q_2}(\mathbb{R})$, and satisfying $\langle \dot{\tau}, Y \rangle \equiv -k\,T_\tau$, $\langle \dot{\tau}, \dot{\tau} \rangle \equiv -T_\tau^2$ for some $T_\tau \in \mathbb{R}^+$.

Observe that, if $\sigma \in \mathcal{B}^{(2)}_{p,\gamma}(k)$, then, for every $I = [a, b] \subseteq [0, 1]$, the restriction of σ to I is a curve in $\mathcal{B}^{(2)}_{\sigma(a),\gamma_{\sigma(b)}}(k, I)$.

Definition 5.4. *A curve $\sigma \in \mathcal{B}^{(2)}_{p,\gamma}(k)$ is said to be a* local minimizer *for the travel time if, for all $0 \le a < b \le 1$ such that $b - a$ is sufficiently small, the restriction of σ to the interval $I = [a, b]$ is a minimum point for the travel time functional in the space $\mathcal{B}^{(2)}_{\sigma(a),\gamma_{\sigma(b)}}(k, I)$.*

We will see in Proposition 5.7 that a curve $\sigma \in \mathcal{B}^{(2)}_{p,\gamma}(k)$ is a travel time brachistochrone in the sense of Definition 5.3 if and only if it is a local minimizer for the travel time.

As in the classical case, the travel time brachistochrone problem can be reduced to a geodesic problem with respect to a suitable Riemannian structure.

We denote by Δ the smooth distribution on $\mathcal{M}$ given by the orthocomplement of the vector field Y. Observe that, since Y is timelike, Δ is *spacelike*, i.e., the restriction of the Lorentzian metric g on Δ is positive definite.

Let $\psi : \mathcal{M} \times \mathbb{R} \longmapsto \mathcal{M}$ be the flow of Y, i.e., for $q \in \mathcal{M}$ and $t \in \mathbb{R}$, $\psi(q, t)$ is the value $\gamma_q(t)$, where γ_q is the maximal integral line of Y satisfying $\gamma_q(0) = q$. Since Y is Killing, $\psi(\cdot, t)$ is a local isometry for all $t \in \mathbb{R}$; moreover, it is easy to see that the distribution Δ is ψ-invariant, which means that $\psi_x(q, t_0)(\Delta_q) = \Delta_{\psi(q,t_0)}$, where $\psi_x(q, t_0)$ denotes the differential of the map $\psi(\cdot, t_0)$ at the point q. A function $\phi : \mathcal{M} \longmapsto \mathbb{R}$ is said to be Y-invariant if it is constant along each flow line of Y; if ψ is C^1, this amounts to saying that $\langle Y, \nabla\phi\rangle \equiv 0$.

We define $\Omega^{(2)}_{p,\gamma}(\Delta)$ to be the subset of $\Omega^{(2)}_{p,\gamma}$ consisting of curves whose tangent vector at each point lies in Δ:

$$\Omega^{(2)}_{p,\gamma}(\Delta) = \Big\{w \in \Omega^{(2)}_{p,\gamma} : \dot w(t) \in \Delta_{w(t)},\ \forall t \in [0, 1]\Big\}. \tag{5.17}$$

Using the language of sub-Riemannian geometry, we will call the curves in $\Omega^{(2)}_{p,\gamma}$ *horizontal*. By the same arguments as in Proposition 5.1, one checks immediately that $\Omega^{(2)}_{p,\gamma}(\Delta)$ is a smooth submanifold of $\Omega^{(2)}_{p,\gamma}$, and that, for $w \in \Omega^{(2)}_{p,\gamma}(\Delta)$, the tangent space $T_w\Omega^{(2)}_{p,\gamma}(\Delta)$ is given by

$$T_w\Omega^{(2)}_{p,\gamma}(\Delta) = \Big\{V \in T_w\Omega^{(2)}_{p,\gamma} : \langle\nabla_{\dot w} V, Y\rangle - \langle V, \nabla_{\dot w} Y\rangle = 0\Big\}. \tag{5.18}$$

Lemma 5.5. *Let ϕ be a smooth Y-invariant positive function. Then the functional*

$$E_\phi(w)\frac{1}{2}\int_0^1 \phi(w)\,\langle \dot w, \dot w\rangle_{(R)}\,\mathrm{d}t \tag{5.19}$$

on $\Omega^{(2)}_{p,\gamma}$ and its restriction to $\Omega^{(2)}_{p,\gamma}(\Delta)$ have the same critical points. These critical points are geodesics in $\mathcal{M}$ with respect to the Riemannian metric $\phi \cdot g_{\mathrm{R}}$ that join p and γ and that are orthogonal to γ.

Proof. See [25, Lemma 4.3]. □

The functional E_ϕ of (5.19) is called the *energy* functional relative to the metric $\phi \cdot g_{\mathrm{R}}$. The critical points of E_ϕ in $\Omega^{(2)}_{p,\gamma}$ (or, equivalently, in $\Omega^{(2)}_{p,\gamma}(\Delta)$) will be called *horizontal geodesics* between p and γ with respect to the Riemannian metric $\phi \cdot g_{\mathrm{R}}$.

In order to state our variational principle properly, we introduce an operator $\mathcal{D}$ that *deforms* curves in $\Omega^{(2)}_{p,\gamma}$ into horizontal curves using the flow of Y.

Let $\mathcal{D}$ be the map

$$\mathcal{D} : \Omega^{(2)}_{p,\gamma} \longmapsto \Omega^{(2)}_{p,\gamma}(\Delta)$$

defined by $\mathcal{D}(\sigma) = w$, where

$$w(t) = \psi(\sigma(t), \tau_\sigma(t)), \tag{5.20}$$

and τ_σ is the unique solution on $[0, 1]$ of the Cauchy problem

$$\tau_\sigma' = -\frac{\langle \dot\sigma, Y\rangle}{\langle Y, Y\rangle}, \qquad \tau_\sigma(0) = 0. \tag{5.21}$$

Using the Killing property of Y it is easily checked that $\mathcal{D}$ is well-defined, i.e., that the maximal solution of (5.21) is defined on the entire interval $[0, 1]$ and the corresponding curve w given by (5.20) is horizontal. Moreover, the smooth dependence on σ of the solution τ_σ of (5.21) proves that $\mathcal{D}$ is a smooth map.

We have the following

Proposition 5.6. *The map $\mathcal{D}$ is smooth. For $\sigma \in \mathcal{B}^{(2)}_{p,\gamma}(k)$ and $\zeta \in T_\sigma\mathcal{B}^{(2)}_{p,\gamma}(k)$, the Gateaux derivative $\mathrm{d}\mathcal{D}(\sigma)[\zeta]$ is given by*

$$\mathrm{d}\mathcal{D}(\sigma)[\zeta] = \mathrm{d}_x\psi(\sigma, t_\sigma)\left[\zeta + \tau_\zeta \cdot Y(\sigma)\right], \tag{5.22}$$

where $\mathrm{d}_x\psi$ denotes the partial derivative of ψ with respect to the first variable and $\tau_\zeta : [0, 1] \longmapsto \mathbb{R}$ is the function

$$\tau_\zeta(t) = -\int_0^t \frac{C_\zeta\langle Y, Y\rangle + 2k\,T_\sigma\langle\nabla_\zeta Y, Y\rangle}{\langle Y, Y\rangle^2}\,\mathrm{d}r. \tag{5.23}$$

In particular, if σ is a brachistochrone, then τ_ζ takes the following form

$$\tau_\zeta(t) = -2k\,T_\sigma\int_0^t \frac{\langle\nabla_\zeta Y, Y\rangle}{\langle Y, Y\rangle^2}\,\mathrm{d}r. \tag{5.24}$$

Moreover, for all $\sigma \in \mathcal{B}^{(2)}_{p,\gamma}(k)$, the differential $\mathrm{d}\mathcal{D}(\sigma) : T_\sigma\mathcal{B}^{(2)}_{p,\gamma}(k) \longmapsto T_{\mathcal{D}(\sigma)}\Omega^{(2)}_{p,\gamma}$ is injective.

Proof. See [25, Proposition 4.4]. □

Now everything is ready to state the following

Proposition 5.7 (Variational Principle for Travel Time Brachistochrones). *Let σ be a curve in $\mathcal{B}^{(2)}_{p,\gamma}(k)$. The following are equivalent.*

1. *σ is a brachistochrone of energy k between p and γ.*
2. *σ is a local minimizer for the travel time.*
3. *$w = \mathcal{D}(\sigma)$ is a horizontal geodesic between p and γ with respect to the Riemannian metric $\phi_k \cdot g_R$, where g_R is the Riemannian metric defined in (5.7), and ϕ_k is given by*

$$\phi_k = -\frac{\langle Y, Y\rangle}{k^2 + \langle Y, Y\rangle}. \tag{5.25}$$

Moreover, if one of the conditions above is satisfied, then $E_{\phi_k}(w) = \frac{1}{2}T_\sigma^2$, *where* E_{ϕ_k} *is the energy functional relative to the metric* $\phi_k \cdot g_R$. □

Proof. See [26, Lemma 2.6] and [25, Proposition 5.7]. □

Observe that the result of Lemma 5.5 applies to the function ϕ_k; an essential point in the proof of Proposition 5.7 is the following equality:

$$F = -E_{\phi_k} \circ \mathcal{D} \quad \text{in } \mathcal{B}^{(2)}_{p,\gamma}(k). \tag{5.26}$$

As a corollary of the variational principle, we obtain the following characterization of the travel time brachistochrones:

Proposition 5.8. *A curve* $\sigma \in \Omega^{(2)}_{p,\gamma}$ *is a travel time brachistochrone of energy* k *between* p *and* γ *if and only if* σ *is smooth and there exists a* $T_\sigma > 0$ *such that* σ *satisfies the following second-order differential equation:*

$$\begin{aligned}\nabla_{\dot\sigma}\dot\sigma \,+\, 2k^2 \frac{\langle \nabla_{\dot\sigma} Y, Y\rangle}{\langle Y, Y\rangle (k^2 + \langle Y, Y\rangle)}\,\dot\sigma + \frac{2k\,T_\sigma}{\langle Y, Y\rangle}\,\nabla_{\dot\sigma} Y \\ -2k\,T_\sigma \frac{\langle \nabla_{\dot\sigma} Y, Y\rangle}{\langle Y, Y\rangle (k^2 + \langle Y, Y\rangle)}\,Y = 0.\end{aligned} \tag{5.27}$$

with initial velocity $\dot\sigma(0)$ *satisfying*

$$\langle \dot\sigma(0), \dot\sigma(0)\rangle = -T_\sigma^2, \quad \text{and} \quad \langle \dot\sigma(0), Y(p)\rangle = -k\,T_\sigma. \tag{5.28}$$

Proof. See [26, Corollary 2.7] and [25, Proposition 3.6]. □

The travel time brachistochrones are studied in details in references [7, 8, 25, 26]. In particular, a Lusternik–Schnirelman theory for travel time brachistochrones is developed in [7], and a Morse theory is discussed in [25].

Arrival time brachistochrones

Definition 5.9. *A curve* $\sigma \in \mathcal{B}^{(2)}_{p,\gamma}(k)$ *is said to be a* localized minimizer for the arrival time *if, for all* $0 \le a < b \le 1$ *such that* $b - a$ *is sufficiently small, the restriction of* σ *to the interval* $I = [a, b]$ *is a minimum point for the arrival time functional* τ *in the space* $\mathcal{B}^{(2)}_{\sigma(a),\gamma_{\sigma(b)}}(k, I)$.

As in the case of the travel time brachistochrones, a localized minimizer for the arrival time can be described as a critical point for the arrival time. A curve σ is said to be an *arrival time brachistochrone of energy* k *between* p *and* γ if it is a critical point of class C^2 for τ in $\mathcal{B}^{(2)}_{p,\gamma}(k)$.

Lemma 5.10. *A curve* $\sigma \in \mathcal{A}^{(1)}_{p,\gamma}(k)$ *of class* C^2 *is an arrival time brachistochrone of energy* k *between* p *and* γ *if and only if, for all* $\zeta \in T_\sigma\mathcal{A}^{(1)}_{p,\gamma}(k)$, *we have* $\zeta(1) = 0$.

Proof. See [8, Lemma 3.1]. □

On the basis of this fact, the following theorem is easily proven.

Theorem 5.11. *Let $k > 0$ be fixed. A curve $\sigma : [0,1] \longmapsto U_k$ of class C^2 joining p and γ is an arrival time brachistochrone of energy k between p and γ if and only if σ is a curve that satisfies the second-order differential equation*

$$\nabla_{\dot\sigma}\dot\sigma - \frac{2\,\mathcal{T}_\sigma}{k}\nabla_{\dot\sigma}Y - \frac{2}{k}\frac{\langle\nabla_{\dot\sigma}Y, Y\rangle}{k^2 + \langle Y, Y\rangle}(k\dot\sigma - \mathcal{T}_\sigma Y) = 0, \tag{5.29}$$

and whose initial tangent vector $\dot\sigma(0)$ is timelike, future pointing and satisfies the condition

$$\langle\dot\sigma(0), Y(\sigma(0))\rangle^2 = -k^2\langle\dot\sigma(0), \dot\sigma(0)\rangle. \tag{5.30}$$

Proof. See [8, Proposition 3.2]. □

A complete variational theory for arrival time brachistochrones, including existence and multiplicity results, can be found in [8].

Acknowledgements

Partially sponsored by CNPq, Brazil, Processo n. 301410/95-0 (RN).

References

1. Giannoni F. (1996): A timelike extension of Fermat's principle in general relativity. Comparisons with the lightlike case. Proceedings of the WCNA96, Athens
2. Giannoni F., Masiello A., Piccione P. (1997): A Variational Theory for Light Rays in Stably Causal Lorentzian Manifolds. Existence and Regularity Results. Commun. Math. Phys. **187**, 375
3. Giannoni F., Masiello A., Piccione P. (1998): A timelike extension of Fermat's principle in general relativity and applications. Calculus of Variations and PDE **6** (3), 263–283
4. Giannoni F., Masiello A., Piccione P. (1998): A Morse theory for light rays in stably causal Lorentzian manifolds. Ann. Inst. H. Poincaré, Physique Théorique **69**, 359–412
5. Giannoni F., Masiello A., Piccione P. (1999): A Morse theory for massive particles and photons in general relativity, in press
6. Giannoni F., Perlick V., Piccione P., Verderesi J.A. (2000): Time minimizing curves in Lorentzian geometry. The general relativistic brachistochrone problem. Matematica Contemporanea, in stampa, Proc. 10th School of Differential Geometry, Belo Horizonte, July 1998
7. Giannoni F., Piccione P. (1998): An existence theory for relativistic brachistochrones in stationary Spacetimes, J. Math. Phys. **39** (11), 6137–6152
8. Giannoni F., Piccione P. (1999): The arrival time brachistochrones in a general relativistic spacetime, in press
9. Perlick V. (1990): On Fermat's principle in general relativity: I. The general case. Class. Quantum Grav. **7**, 1319–1331

10. Perlick V. (1991): The brachistochrone problem in a stationary space-time. J. Math. Phys. **32** (11), 3148–3157
11. Perlick V., Piccione P. (1997): The brachistochrone problem in arbitrary spacetimes. RT–MAT 97-16, IME, Universidade de São Paulo. Preprint LANL math. DG/9905096
12. Perlick V., Piccione P. (1999): A general-relativistic Fermat principle for extended light sources and extended receivers. General Relativity and Gravitation **30** (10), 1461–1476
13. Sachs R., Wu H. (1977): General Relativity for Mathematicians. Springer, New York
14. Brezis H. (1983): Analyse Fonctionelle. Masson, Paris
15. Palais R. (1968): Foundations of Global Nonlinear Analysis. Benjamin, New York
16. Beem J.K., Ehrlich P.E., Easley K.L. (1996): Global Lorentzian Geometry, Marcel Dekker, New York and Basel
17. O'Neill B. (1983): Semi-Riemannian Geometry with Applications to Relativity. Academic Press, New York
18. Lang S. (1985): Differential Manifolds. Springer, Berlin–Heidelber–New York
19. Masiello A. (1994): Variational Methods in Lorentzian Geometry. Pitman Research Notes in Mathematics **309**, Longman, London
20. Hawking S.W., Ellis G.F. (1973): The large scale structure of space-time. Cambridge University Press, Cambridge
21. Piccione P. (1998): Causal Trajectories between Submanifolds in Lorentzian Geometry. 7th International Conference of Differential Geometry and Applications, Masaryk University in Brno (Czech Republic), pp. 631–644
22. Kovner I. (1990): Fermat principle in arbitrary gravitational fields. Astrophysical J. **351**, 114–120
23. Levi Civita T. (1928, 1986): Fondamenti di Meccanica Relativistica. Zanichelli, Bologna
24. Mercuri F., Piccione P., Tausk D.V. (1999): Stability of the focal and geometric index in semi-Riemannian geometry via the Maslov index. RT-MAT 99-08. Universidade de São Paulo, Brazil, Preprint
25. Giannoni F., Piccione P., Tausk D. (1998): Morse theory for the travel time brachistochrones in stationary spacetimes. Preprint
26. Giannoni F., Piccione P., Verderesi J.A. (1997): An approach to the relativistic brachistochrone problem by sub-Riemannian geometry, J. Math. Phys. **38** (12), 6367–6381

Null Geodesics Joining Two Submanifolds in Stationary Lorentzian Manifolds

A. Salvatore

Abstract. We present a variational principle, similar to the Fermat principle in Optic, concerning the existence of lightlike geodesics joining two submanifolds in stationary Lorentzian manifolds.

1 Introduction and statement of results

Let $\mathcal{M}$ a smooth finite dimensional manifold and $\langle \cdot, \cdot \rangle_z$ a Lorentz metric on it, that is, a smooth symmetric (0,2) tensor field on $\mathcal{M}$ which defines a non-degenerate bilinear form of index 1 on each tangent space $T_z\mathcal{M}$, $z \in \mathcal{M}$. A smooth curve $\gamma : [0, 1] \to \mathcal{M}$ is a geodesic in $\mathcal{M}$ if

$$D_s \dot{\gamma} = 0,$$

where D_s denotes the covariant derivative along γ induced by the Levi-Civita connection of the metric. It is well-known that, for each geodesic $z = z(s)$, the energy $E(z) = \langle \dot{z}(s), \dot{z}(s) \rangle_z$ is constant in $[0, 1]$, so a geodesic $z = z(s)$ is said to be timelike, lightlike or spacelike if $E(z)$ is negative, null or positive, respectively. In a space-time (i.e., a four-dimensional Lorentz manifold) the timelike geodesics represent the world-lines of a free falling particle, that is, a particle for which only gravitational effects are present. Lightlike geodesics represent the trajectories of light rays, while spacelike geodesics are less meaningful from a physical point of view. Lightlike geodesics are also called null geodesics or light rays.

It is well-known that the Fermat principle states that the time elapsed in the passage of light between two fixed points is an extremum with respect to all possible paths connecting the points. A similar variational principle describes the light rays in General Relativity. We present here some recent results and generalizations in this direction, concerning a stationary Lorentzian manifold (for the study of light rays in more general Lorentzian manifolds, see, e.g., [1–3]).

A Lorentz manifold $(\mathcal{M}, \langle \cdot, \cdot \rangle_z)$ is called stationary if there exists a finite dimensional Riemannian manifold $(\mathcal{M}_0, \langle \cdot, \cdot \rangle_x)$ such that $\mathcal{M} = \mathcal{M}_0 \times \mathbf{R}$ and $\langle \cdot, \cdot \rangle_z$ is given by

$$\langle \zeta, \zeta \rangle_z = \langle \xi, \xi \rangle_x + 2\langle \delta(x), \xi \rangle_x \tau - \beta(x)\tau^2 \tag{1.1}$$

for any $z = (x, t) \in \mathcal{M}_0 \times \mathbf{R}$ and $\zeta = (\xi, \tau) \in T_z\mathcal{M} = T_x\mathcal{M}_0 \times \mathbf{R}$, with $\beta : \mathcal{M}_0 \to \mathbf{R}$ a smooth and positive scalar field and $\delta : \mathcal{M}_0 \to T\mathcal{M}_0$ a smooth vector field. If $\delta(x) = 0$, the metric (1.1) is static and $(\mathcal{M}, \langle\cdot,\cdot\rangle_z)$ is called a static Lorentzian manifold.

From now on, let $\mathcal{M} = \mathcal{M}_0 \times \mathbf{R}$ be a stationary Lorentzian manifold equipped with the metric (1.1). Let P_0 and P_1 be two given submanifolds of $\mathcal{M}_0$ and let $t_0 \in \mathbf{R}$ be fixed. We are interested in the study of the existence of lightlike geodesics $z : [0,1] \to \mathcal{M}$, $z = (x,t)$, joining $\widetilde{P}_0 = P_0 \times \{t_0\}$ and $\widetilde{P}_1 = P_1 \times \mathbf{R}$, that is, such that $z(0) \in \widetilde{P}_0$ and $z(1) \in \widetilde{P}_1$, and whose space component x satisfies the orthogonal conditions

$$\begin{cases} \langle \dot{x}(0), \xi\rangle_x = 0 & \text{for all } \xi \in T_{x(0)} P_0, \\ \langle \dot{x}(1), \xi\rangle_x = 0 & \text{for all } \xi \in T_{x(1)} P_1. \end{cases} \tag{1.2}$$

More precisely, we look for smooth functions $z : [0,1] \to \mathcal{M}$, $z(s) = (x(s), t(s))$, which are solutions of the following system:

$$\begin{cases} D_s \dot{z}(s) = 0 & \text{for all } s \in [0,1], \\ E(z) = \langle \dot{z}(s), \dot{z}(s)\rangle_z = 0 & \text{for all } s \in [0,1], \\ x(0) \in P_0, \quad t(0) = t_0, \quad x(1) \in P_1, & \\ \langle \dot{x}(0), \xi\rangle_x = 0 & \text{for all } \xi \in T_{x(0)} P_0, \\ \langle \dot{x}(1), \xi\rangle_x = 0 & \text{for all } \xi \in T_{x(1)} P_1. \end{cases} \tag{1.3}$$

From a physical point of view, lightlike geodesics verifying (1.3) are the trajectories of light rays joining two celestial bodies of which one is a light source. Moreover, an example of a stationary Lorentz manifold is Kerr space-time which describes the space-time outside an axially symmetric body rotating around its axis while an example of a static Lorentz manifold is Schwarzschild space-time which represents the manifold outside a static spherically symmetric massive body (see [4, 5]).

Lightlike geodesics joining an event $\widetilde{P}_0 = \{(x_0, t_0)\}$ to a line $\widetilde{P}_1 = \{x_1\} \times \mathbf{R}$ have been studied in [6] while the existence of geodesics, not necessarily lightlike, from a point to a subspace $\widetilde{P}_1 = P_1 \times \{t_1\}$ of a static Lorentzian manifold has been established in [7]; closed light rays have been studied in [8]. The following results state that the number of lightlike geodesics joining $\widetilde{P}_0 = P_0 \times \{t_0\}$ and $\widetilde{P}_1 = P_1 \times \mathbf{R}$ depends on the topological properties of $\mathcal{M}$, P_0 and P_1 (for proof, see [9]). Analogous results on Riemannian manifolds are contained in [10] and [11].

Theorem 1.1. *Let $\mathcal{M} = \mathcal{M}_0 \times \mathbf{R}$ be a manifold equipped with a stationary Lorentz metric (1.1) such that*

(M_0) $(\mathcal{M}_0, \langle\cdot,\cdot\rangle_x)$ is a connected, complete, C^3 n-dimensional Riemannian manifold;

(M_1) there exist constants ν, N, $D > 0$ such that

$$\nu \le \beta(x) \le N \quad \textit{and} \quad \langle \delta(x), \delta(x)\rangle_x \le D \ \textit{ for all } x \in \mathcal{M}_0. \tag{1.4}$$

Let P_0 and P_1 be two disjoint closed submanifolds of $\mathcal{M}_0$ such that

(C) *P_0 or P_1 is compact;*

(O_0) $\langle\delta(x), \xi\rangle_x = 0$ *for any* $x \in P_0$, $\xi \in T_x P_0$;

(O_1) $\langle\delta(x), \xi\rangle_x = 0$ *for any* $x \in P_1$, $\xi \in T_x P_1$.

Then there exists at least one solution of (1.3). If, moreover, P_0 and P_1 are both contractible in $\mathcal{M}_0$, then problem (1.3) has at least $cat(P_0 \times P_1)$ solutions.

The following multiplicity theorem holds even if P_0 and P_1 are not disjoint:

Theorem 1.2. *Let $\mathcal{M} = \mathcal{M}_0 \times \mathbf{R}$ be a manifold equipped with the stationary Lorentz metric (1.1) such that (M_0) and (M_1) are satisfied. Let P_0 and P_1 be two closed submanifolds of $\mathcal{M}_0$ such that (C), (O_0) and (O_1) hold. If $\mathcal{M}_0$ is not contractible in itself while P_0 and P_1 are both contractible in $\mathcal{M}_0$, then problem (1.3) has infinitely many non-constant solutions $z_n(s) = (x_n(s), t_n(s))$ whose "arrival times" $(t_n(1))_{n\in\mathbf{N}}$ form a divergent increasing sequence.*

2 Variational approach

Let $(\mathcal{M}, \langle\cdot,\cdot\rangle_z)$ be a manifold equipped with a stationary Lorentz metric (1.1) such that (M_0) and (M_1) hold. As lightlike geodesics are independent, up to reparametrization, on a conformal change of the metric, according to (M_1) it is not restrictive to take $\beta(x) = 1$. Moreover, by the Nash Embedding Theorem, $\mathcal{M}_0$ is a submanifold of a Euclidean space $\mathbf{R}^N$ and its metric $\langle\cdot,\cdot\rangle_x$ is the Euclidean metric on $\mathbf{R}^N$ which will be denoted by $\langle\cdot,\cdot\rangle$, so that the Lorentz metric on $\mathcal{M}$ becomes

$$\langle\zeta,\zeta\rangle_z = \langle\xi,\xi\rangle + 2\langle\delta(x),\xi\rangle\tau - \beta(x)\tau^2 \tag{2.1}$$

for any $z = (x,t) \in \mathcal{M}_0 \times \mathbf{R}$ and $\zeta = (\xi,\tau) \in T_z\mathcal{M} = T_x\mathcal{M}_0 \times \mathbf{R}$. Let $I = [0,1]$ and $H^1(I, \mathbf{R}^N)$ be the Sobolev space of absolutely continuous curves whose derivative is square summable, endowed with the usual norm $\|\cdot\|$. It is known (see, e.g., [12]) that, if $\mathcal{M}_0$ is complete and P_0, P_1 are closed in $\mathcal{M}_0$, then the subset

$$\Gamma(P_0, P_1) = \left\{x \in H^1(I, \mathbf{R}^N) : x(I) \subset \mathcal{M}_0;\ x(0) \in P_0,\ x(1) \in P_1\right\}$$

is a complete Riemannian manifold whose tangent space in $x \in \Gamma(P_0, P_1)$ is

$$T_x\Gamma(P_0, P_1) = \{\xi \in H^1(I, T\mathcal{M}_0) : \xi(s) \in T_{x(s)}\mathcal{M}_0 \text{ for all } s \in I;$$
$$\xi(0) \in T_{x(0)}P_0,\ \xi(1) \in T_{x(1)}P_1\}.$$

If (O_0) and (O_1) hold, it is possible to prove that the solutions of (1.3) are the critical points at level zero of the functional

$$f(z) = \frac{1}{2}\int_0^1 \langle\dot z(s), \dot z(s)\rangle_z ds$$

in $\Gamma(P_0, P_1) \times H^1(I, \mathbf{R})$. As f is not bounded from above and from below, we use a variational principle which reduces the search for the lightlike geodesics joining $\widetilde{P}_0$ and $\widetilde{P}_1$ in a stationary Lorentz manifold, to the study of the critical points of a smooth functional bounded from below on a Riemannian manifold. This variational approach was introduced by Fortunato, Giannoni and Masiello in [6] (see [13] also) in the study of lightlike geodesics joining an event to a vertical line and it was extended in [9] to the case of lightlike geodesics joining two submanifolds.

Fixing $\lambda \in \mathbf{R}$, we set

$$W_\lambda = \left\{ t \in H^1(I, \mathbf{R}) : t(0) = 0, \mathrm{t}(1)=\lambda \right\},$$

a closed affine submanifold of $H^1(I, \mathbf{R})$ whose tangent space at each point is given by

$$H_0^1 = \left\{ \tau \in H^1(I, \mathbf{R}) : \tau(0) = \tau(1) = 0 \right\}$$

and we define $Z_\lambda = \Gamma(P_0, P_1) \times W_\lambda$, a Hilbert manifold such that

$$T_z Z_\lambda = T_x \Gamma(P_0, P_1) \times H_0^1$$

for each $z = (x, t) \in Z_\lambda$.

It is easy to prove that $f_\lambda = f \;|_{Z_\lambda}$ is a C^1 functional on Z_λ. Moreover, if $z = (x, t) \in Z_\lambda$ and $\zeta = (\xi, \tau) \in T_z Z_\lambda$, we have

$$\begin{aligned} \frac{\partial f_\lambda}{\partial x}(z)[\xi] &= f'_\lambda(z)[(\xi, 0)] = \\ &= \int_0^1 \langle -D_s \dot{x} + \dot{t}\delta'(x)^*[\dot{x}] - \frac{d}{ds}(\dot{t}\delta(x)), \xi \rangle ds \\ &\quad + [\langle \dot{x}, \xi \rangle]_0^1 + [\dot{t}\langle \delta(x), \xi \rangle]_0^1 \end{aligned} \tag{2.2}$$

where $\delta'(x(s))^*$ is the adjoint of $\delta'(x(s))$ for any $s \in I$, while, for all $\tau \in H_0^1$, we have

$$\frac{\partial f_\lambda}{\partial t}(z)[\tau] = f'_\lambda(z)[(0, \tau)] = \int_0^1 (\ddot{t} - \frac{d}{ds}(\langle \delta(x), \dot{x} \rangle) \tau ds. \tag{2.3}$$

Now, it can be shown that, if P_0 and P_1 satisfy the orthogonality conditions (O_0) and (O_1), then the following propositions are equivalent:

(a) z is a solution of (1.3) with "arrival time" $t(1) = \lambda$;

(b) z is a critical point of f_λ on Z_λ such that $f_\lambda(z) = 0$.

Moreover, denoting by N_λ the kernel of the map $\dfrac{\partial f_\lambda}{\partial t}$, we see from (2.3) that N_λ is the graph of the smooth function $\Phi_\lambda : x \in \Gamma(P_0, P_1) \to \Phi_\lambda(x) \in W_\lambda$ defined as

$$\Phi_\lambda(x)(s) = \int_0^s \langle \delta(x(r)), \dot{x}(r) \rangle dr + s(\lambda - \int_0^1 \langle \delta(x), \dot{x} \rangle dr) \qquad \text{for all } s \in I.$$

Now, we consider the restriction of f_λ on N_λ, that is, the functional

$$J_\lambda : x \in \Gamma(P_0, P_1) \to J_\lambda(x) = f_\lambda(x, \Phi_\lambda(x)) \in \mathbf{R}.$$

Then, for any $x \in \Gamma(P_0, P_1)$, it follows that

$$J_\lambda(x) = \frac{1}{2}\int_0^1 (\langle \dot{x}, \dot{x}\rangle + \langle \delta(x), \dot{x}\rangle^2) ds - \left(\lambda - \int_0^1 \langle \delta(x), \dot{x}\rangle ds\right)^2.$$

Arguing as in [13], we can prove the following variational principle.

Proposition 2.1. *For $z = (x, t) \in Z_\lambda$, the following propositions are equivalent:*

(a) *z is a critical point of f_λ;*
(b) *x is a critical point of J_λ and $t = \Phi_\lambda(x)$.*

Moreover, if (a) or (b) holds, then $f_\lambda(x, t) = J_\lambda(x)$.

If $\lambda \in \mathbf{R}$ is fixed, the previous proposition permits us to obtain the geodesics on Lorentz manifolds as the critical points of a "Riemannian" functional. However, in our problem, λ is strictly positive but unknown as it gives the arrival time of the lightlike geodesic on $\widetilde{P}_1$; therefore we consider λ as a new variable and introduce the functional $H : (\lambda, x) \in \mathbf{R}_+ \times \Gamma(P_0, P_1) \to 2J_\lambda(x) \in \mathbf{R}$. Then, the search for solutions of (1.3) is equivalent to finding $(\lambda, x) \in \mathbf{R}_+ \times \Gamma(P_0, P_1)$ which is a solution of the problem

$$\frac{\partial H}{\partial \lambda}(\lambda, x) = 0, \qquad H(\lambda, x) = 0, \qquad \lambda > 0. \tag{2.4}$$

By solving the first equation with respect to λ, we obtain the functional

$$F(x) = \int_0^1 \langle \delta(x), \dot{x}\rangle ds + \sqrt{\int_0^1 (\langle \dot{x}, \dot{x}\rangle + \langle \delta(x), \dot{x}\rangle^2) ds} \qquad x \in \Gamma(P_0, P_1). \tag{2.5}$$

Extending a similar result contained in [6] (see [8] also), we proved in [9] the following theorem:

Theorem 2.2. *Let P_0 and P_1 be two given submanifolds of $\mathcal{M}_0$ which satisfy the orthogonality conditions (O_0) and (O_1). If $\overline{x} \in \Gamma(P_0, P_1)$ is such that $F'(\overline{x}) = 0$ and $F(\overline{x}) > 0$, then, if $\overline{\lambda} = F(\overline{x})$, $\overline{z} = (\overline{x}, \Phi_{\overline{\lambda}}(\overline{x}))$ is a solution of problem (1.3).*

From the previous theorem it follows that solving the problem (1.3) is equivalent to finding positive critical levels of the functional F introduced in (2.5). As F is bounded from below, we will use the properties of the Lusternik–Schnirelman category (see, e.g., [14]). We recall that the Lusternik–Schnirelman category of a subset A of a topological space X, $\mathrm{cat}_X(A)$, is the least number of closed and contractible subsets of X covering A; if it is not possible to cover A with a finite number of such sets, $\mathrm{cat}_X(A) = +\infty$. In [15] Fadell and Husseini proved that, if $\mathcal{M}_0$ is not contractible in itself while P_0 and P_1 are contractible in $\mathcal{M}_0$, then $\Gamma(P_0, P_1)$ has infinite category. Moreover, if we remove the assumption that $\mathcal{M}_0$ not be contractible, we have seen in [9] that $\mathrm{cat}(\Gamma(P_0, P_1)) \geq \mathrm{cat}(P_0 \times P_1)$.

Finally, we give an idea of the proofs of Theorems 1.1 and 1.2 (for the details, see [9]).

Proof of Theorem 1.1. First of all, note that, if $P_0 P_1 = \emptyset$, then F is a C^1 strictly positive functional in $\Gamma(P_0, P_1)$ and its critical points give the solutions of (1.3). As it is possible to prove that F satisfies the classical Palais–Smale compactness condition at any strictly positive level a, by well-known results of the Lusternik–Schnirelman theory, we conclude that F has at least $cat(\Gamma(P_0, P_1)) \geq cat(P_0 \times P_1)$ critical points. □

Proof of Theorem 1.2. Arguing as above, we see that F has infinitely many strictly positive critical levels (as $cat(\Gamma(P_0, P_1)) = +\infty$). However, more careful estimates on the critical levels prove that, even if $P_0 P_1 \neq \emptyset$, there exists a sequence $\{x_n\}$ of critical points of F such that $\{F(x_n)\}$ positively diverges. □

Remark 3. Clearly, if $\delta(x) = 0$, Theorems 1.1 and 1.2 still hold, but in this case the proof can be given by a different and easier variational approach. Indeed, the Fermat principle for static space-times states that a lightlike curve $z(s) = (x(s), t(s))$ is a light ray joining $\widetilde{P}_0$ and $\widetilde{P}_1$ if and only if x is a critical point of the arrival time functional $T(z) = t(1) = \int_0^1 \langle \dot{x}(s), \dot{x}(s) \rangle ds$.

Acknowledgements

This work was supported by M.U.R.S.T. (Research funds ex 40% and 60%).

References

1. Giannoni F., Masiello A., Piccione P. (1997): A variational theory for light rays in stably causal Lorentzian manifolds. Existence and regularity results. Comm. Math. Phys. **187**, 375–415
2. Giannoni F., Masiello A., Piccione P. (1998): A Morse theory for light rays on stably causal Lorentzian manifolds. Ann. Inst. H. Poincaré, Physique thèorique **69**, 359–412
3. Masiello A., Piccione P. (1998): Shortening null geodesics in Lorentzian manifolds. Applications to closed light rays. Differential Geom. Appl. **8**, 47–70
4. Fortunato D., Masiello A. (1995): Fermat Principles in General Relativity and the Existence of Light Rays on Lorentzian Manifolds, in *Proc. Variational and Local Methods in the Study of Hamiltonian Systems ICTP*, ed. by A. Ambrosetti, G.F. Dell'Antonio, World Scientific, Singapore, pp. 34–64

5. Hawking S.W., Ellis G.F.R. (1973): The Large Scale Structure of Space-Time. Cambridge University Press, Cambridge
6. Fortunato D., Giannoni F., Masiello A. (1995): A Fermat principle for stationary space times and applications to light rays. J. Geom. Phys. **15**, 159–188
7. Molina J. (1996): Existence and multiplicity of normal geodesics on static space-time manifolds. Bull. Un. Mat. Ital. A **10**, 305–318
8. Candela A.M. (1996): Lightlike periodic trajectories in space-times. Ann. Mat. Pura Appl. **CLXXI**, 131–158
9. Candela A.M., Salvatore A. (1997): Light rays joining two submanifolds in space-times. J. Geom. Phys. **22**, 281–297
10. Grove K. (1973): Condition (C) for the energy integral on certain path spaces and applications to the theory of geodesics. J. Differential Geom. **8**, 207–223
11. Salvatore A. (1996): Trajectories of dynamical systems joining two given submanifolds. Differential Integral Equations **9**, 779–790
12. Klingenberg W. (1982): Riemannian Geometry. de Gruyter, Berlin
13. Giannoni F., Masiello A. (1991): On the existence of geodesics on stationary Lorentz manifolds with convex boundary. J. Funct. Anal. **101**, 340–369
14. Schwartz J.T. (1969): Nonlinear Functional Analysis. Gordon and Breach, New York
15. Fadell E., Husseini S. (1991): Category of loop spaces of open subsets in Euclidean space. Nonlinear Anal. TMA **17**, 1153–1161

Analytic Solution of Regge–Wheeler Differential Equation for Black Hole Perturbations in Radial Coordinate and Time Domains

A.D.A.M. Spallicci

Abstract. An analytic solution of the Regge–Wheeler (RW) equation has been found via the Frobenius method at the regular singularity of the horizon $2M$, in the form of a time and radial coordinate dependent series. The RW partial differential equation, derived from the Einstein field equations, represents the first order perturbations of the Schwarzschild metric. The known solutions are numerical in the time domain, and approximate and asymptotic for low or high frequencies in the Fourier domain. The former is of little relevance for comprehension of the geodesic equations for a body in the black hole field, while the latter is mainly useful for the description of emitted gravitational radiation. Instead a time domain solution is essential for the determination of the radiation reaction of a particle falling into a black hole, i.e., the influence of emitted radiation on the motion of the perturbing mass in the black hole field. To this end, a semi-analytic solution of the inhomogeneous RW equation with the source term (Regge–Wheeler–Zerilli equation) shall be attempted where a numerical contribution shall be present. Discussion on the features of the series shall be shown in further work.

1 Introduction

Regge and Wheeler [1] proved the vacuum stability of the Schwarzschild black hole, while Zerilli [2, 3] studied emitted radiation via the radial wave equation for polar perturbations, whose source is a test mass m freely falling towards the black hole of large mass M. Moncrief [4] showed the gauge invariant significance of the wave equations.

Past work was concerned with the Fourier analysis of emitted radiation (for a review see Ruffini [5]) and not with the motion of the particle. Along the lines of radiation emission analysis, more recent work has been developed mainly in Japan (Osaka and Tokyo), United States (Salt Lake City and Caltech), while symbolic computing on perturbations has been pursued in Canada (Kingston) and Italy (Rome, La Sapienza, ICRA).

We first examine the known approaches to solutions of the RW equations: a Fourier anti-transform from an approximate frequency solution (where low or high frequencies are neglected, and thus also the corresponding anti-transformed and yet unspecified time-parts of the trajectory of the incoming test mass) presents controversial issues of interpretation; a numerical time solution is not adequate for comprehension of the different contributions to the geodesic equations.

Therefore, an analytic solution, in the time and, obviously, radial coordinate domains, remains the only way to study the problem of motion.

Past work was confined to a particle falling in an *unperturbed* Schwarzschild metric, but radiation reaction requires a geodesic path through the emission of radiation, in a *perturbed* Schwarzschild metric.

This novel approach to the calculation of the motion and radiation reaction for the two-body problem (body plus particle, the small parameter m/M being the ratio of the masses) is based on the following concept: in the background curvature given by the Schwarzschild geometry rippled by gravitational waves, the geodesic equations also insure the presence of radiation reaction for high velocities and strong field (the expansion parameter being m/M and not v/c or ϕ/c). This concept is applicable to any orbit, but radial fall is of interest due to the non-adiabatic regime (equality of radiation reaction and fall time scales), in which *the particle has to react locally and immediately to the emitted radiation*. The energy balance hypothesis (emitted radiation is equal to the variation in kinetic energy) is only used for the determination of the 4-velocity via the Lagrangian and for the normalization of divergencies; the solution in the time domain of the Regge–Wheeler–Zerilli–Moncrief radial wave equation determines the metric tensor expressing the polar perturbations in terms of which the geodesic equations are written; finally, radiation reaction shall be identified by subtraction of the zeroth-order terms in the geodesic equations [6, 7].

The long-term aim of this work is the identification of radiation reaction in the trajectory up to the horizon, without the assumption of adiabaticity and with minimal use of the energy balance postulate.

Future developments may include: i) a general method for the determination of the motion of small objects in any orbit; ii) a contribution to the solution of inhomogeneous second-order equations, whose energy-momentum tensor is based on geodesic equations *with perturbation effects* and which is yet to be found; iii) a post-Schwarzschildian formalism.

In this paper we are concerned with the partial fulfillment of the first step, namely, the solution of the associated homogeneous equation.

Radiation reaction is a fundamental concept in the motion theory of bodies, but also has relevant implications for detector templates since the capture of stars by black holes is a source of gravitational waves.

For a brief critical review of the status of current undestanding of radiation reaction and for references, refer to [6, 7].

2 The equation

The RWZM equation for polar perturbations is

$$\frac{d^2\Psi_l(r,t)}{dr^{*2}} - \frac{d^2\Psi_l(r,t)}{dt^2} - V_l(r)\Psi_l(r,t) = S_l(r,t) \tag{1}$$

where

$$r^* = r + 2M \ln\left(\frac{r}{2M} - 1\right) \tag{2}$$

is the tortoise coordinate and the potential $V_l(r)$ is

$$V_l(r) = \left(1 - \frac{2M}{r}\right) \frac{2\lambda^2(\lambda+1)r^3 + 6\lambda^2 M r^2 + 18\lambda M^2 r + 18M^3}{r^3(\lambda r + 3M)^2}. \tag{3}$$

Further $\lambda = \frac{1}{2}(l-1)(l+2)$ and $S_l(r,t)$ is the 2^l-pole source component and, for a radially falling particle, is expressed by:

$$\begin{aligned} S_l = & \frac{\left(1-\frac{2M}{r}\right)c}{(\lambda+1)(\lambda r+3M)} r \left(1-\frac{2M}{r}\right)^2 \delta'[r - r_0(t)] \\ & - \frac{\left(1-\frac{2M}{r}\right)c}{(\lambda+1)(\lambda r+3M)} \left[(\lambda+1) - \frac{M}{r} - \frac{6Mr}{\lambda r + 3M}\right] \delta[r - r_0(t)] \end{aligned} \tag{4}$$

where $c = 4m\sqrt{(2l+1)\pi}$ and $r_0(t)$ (known geodesic equation of motion in *unperturbed Schwarzschild metric*) is the inverse of:

$$t = -4M\left(\frac{r}{2M}\right)^{1/2} - \frac{4M}{3}\left(\frac{r}{2M}\right)^{3/2} - 2M \ln\left[\frac{\sqrt{\frac{r}{2M}} - 1}{\sqrt{\frac{r}{2M}} + 1}\right] \tag{5}$$

Equation 5 shows that the time domain spans $-\infty$ to $+\infty$. The total energy radiated after a given t_0 per l mode is

$$E = \frac{1}{64\pi}\frac{(l+2)!}{(l-2)!}\int_{t_0}^{\infty} (\dot{\Psi})^2 dt. \tag{6}$$

3 An analytic solution

Equation (1) can be rewritten in terms of t and r^* (tortoise coordinate) only, resulting in a partial differential equation with constant coefficients but solely via an approximate inverse function $r(r^*)$ of Eq. (2) and thus resulting in an approximate p.d.e. Further, the solution is most interesting at $r^* = -\infty$ where a detailed analytic study is not possible. In the (t, r) domain, on the other hand Eq. (1) becomes

$$\frac{1}{A^2(r)}\frac{d^2\Psi_l}{dr^2} - \frac{1-A(r)}{rA^2(r)}\frac{d\Psi_l}{dr} - \frac{d^2\Psi_l}{dt^2} - V_l(r)\Psi_l = S_l(r,t) \tag{7}$$

where $A(r) = dr^*/dr = r/(r-2M)$. The initial value problem is well defined (Ψ and $\dot{\Psi}$ are zero, i.e., a particle at rest at infinity).The homogeneous equation associated to Eq. (7) can be reduced to an ordinary differential equation with $\Psi(r,t) = R(r)T(t)$. Dividing Eq. (7) by RT we obtain (dropping the l notation):

$$\frac{1}{RA^2}\frac{d^2R}{dr^2} - \frac{1-A}{rRA^2}\frac{dR}{dr} - \frac{1}{T}\frac{d^2T}{dt^2} - V = 0. \tag{8}$$

The first, second and fourth terms are not dependent upon t and form time constants. Thus the third term must also be a constant in time and Eq. (8) is autonomous [8]:

$$\frac{1}{T}\frac{d^2T}{dt^2} = K^2. \tag{9}$$

We set $dT/dt = u$ and hence $d^2T/dt^2 = uu_T$ where the lower subscript indicates a derivative in T. Equation (9) becomes

$$\frac{1}{T}uu_T = K^2 \tag{10}$$

whose solution is $T = e^{\pm K}$. Equation (8) becomes

$$\frac{1}{RA^2}\frac{d^2R}{dr^2} - \frac{1-A}{rRA^2}\frac{dR}{dr} - K^2 - V = 0 \tag{11}$$

or, showing the dependence of $A(r)$ on r,

$$\frac{d^2R}{dr^2} + \frac{2M}{r(r-2M)}\frac{dR}{dr} - \frac{r^2R}{(r-2M)^2}(K^2+V) = 0 \tag{12}$$

For finding a solution, it is better to recast Eq. (12) as

$$\frac{d^2R}{dr^2} + \frac{2M}{r(r-2M)}\frac{dR}{dr} - f(r)R = 0 \tag{13}$$

where $f(r)$ is given by

$$\begin{aligned} &\frac{\lambda^2K^2r^6 + 6\lambda K^2Mr^5 + (9K^2M^2 + 2\lambda^3 + 2\lambda^2)r^4}{r^2(r-2M)^2(\lambda r+3M)^2} \\ &+\frac{2\lambda^2M(1-2\lambda)r^3 + 6\lambda M^2(3-2\lambda)r^2 + 18M^3(1-2\lambda)r - 36M^4}{r^2(r-2M)^2(\lambda r+3M)^2} \end{aligned} \tag{14}$$

With $\varrho = r - 2M$, Eq. (13) is finally recast as

$$\begin{aligned} &(\varrho+2m)^2[\lambda(\varrho+2M)+3M]^2\frac{d^2R(\varrho)}{d\varrho^2} \\ &+\frac{2M(\varrho+2m)[\lambda(\varrho+2M)+3M]^2}{\varrho}\frac{dR(\varrho)}{d\varrho} + \frac{\gamma(\varrho)}{\varrho^2}R(\varrho) = 0 \end{aligned} \tag{15}$$

where

$$\begin{aligned} \gamma(\varrho) = {} & -\lambda^2K^2(\varrho+2M)^6 - 6\lambda K^2M(\varrho+2M)^5 \\ & -(9K^2M^2 + 2\lambda^3 + 2\lambda^2)(\varrho+2M)^4 \\ & -2\lambda^2M(1-2\lambda)(\varrho+2M)^3 - 6\lambda M^2(3-2\lambda)(\varrho+2M)^2 \\ & -18M^3(1-2\lambda)(\varrho+2M) + 36M^4, \end{aligned} \tag{16}$$

Equation (15) may be solved by the Frobenius method [9]. We rewrite Eq. (15) as

$$\alpha(\varrho)\frac{d^2R(\varrho)}{d\varrho^2}+\frac{\beta(\varrho)}{\varrho}\frac{dR(\varrho)}{d\varrho}+\frac{\gamma(\varrho)}{\varrho^2}R=0 \tag{17}$$

where the expressions for $\alpha(\varrho)$ and $\beta(\varrho)$ are derived straightforwardly from (15). The functions α, β and γ are regular about the expansion point $\varrho=0$ and have the form

$$\alpha(\varrho)=\sum_{n=0}^{4}\alpha_n\varrho^n \quad , \ \beta(\varrho)=\sum_{n=0}^{3}\beta_n\varrho^n \quad , \ \gamma(\varrho)=\sum_{n=0}^{6}\gamma_n\varrho^n .$$

The indicial equation ($\alpha_0 s^2+(\beta_0-\alpha_0)s+\gamma_0=0$) is

$$\begin{aligned}&4M^4(2\lambda+3)^2s^2+[4M^4(2\lambda+3)^2-4M^4(2\lambda+3)^2)s\\&-16K^2M^6(2\lambda+3)^2=0\end{aligned} \tag{18}$$

and shows that the roots are $s_{1,2}=\pm 2KM$ and differ by an integer ($4KM$). One solution, $R_a(\varrho)$, is a series of the form

$$R_a(\varrho)=\varrho^s\sum_{n=0}^{\infty}R_n\varrho^n \tag{19}$$

while the other is of the form

$$\begin{aligned}R_b(\varrho)&=\ln\varrho\left[\sum_{n=0}^{\infty}(s+2KM)R_n(s)\varrho^{n+s}\right]_{s=-2KM}\\&\qquad+\varrho^{-2KM}\sum_{n=0}^{\infty}\frac{d}{ds}[(s+2KM)R_n(s)]_{s=-2KM}\,\varrho^n,\end{aligned} \tag{20}$$

R_0 has an arbitrary value and all other terms depend upon it. The first three terms (for $L=2$, $M=1$) of the series are given with $R_0=1$, as

$$R_1=\frac{-27+14K-56K^2}{2(-7+28K)} \tag{21}$$

$$R_2=\frac{-420+210K-1428K^2-(168+1036K+784K^2)R_1}{4(-196+392K)} \tag{22}$$

$$\begin{aligned}R_3=&\frac{-168+72K-1016K^2-(348+498K+1428K^2)R_1}{4(-441+588K)}\\&+\frac{(658-2268K-784K^2)R_2}{4(-441+588K)}.\end{aligned} \tag{23}$$

Acknowledgements

Thanks are due to NIKHEF (Nationaal Instituut voor Kernfysica en Hoge Energie Fysika) and FOM (Stichting voor Fundamenteel Onderzoek der Materie) for their Research Grant in Mathematical Physics at the Theory Section of NIKHEF-Amsterdam

(Dr. J. van Holten); to Prof. I.W. Roxburgh (Astronomy Unit, School of Mathematical Sciences, Queen Mary & Westfield College, London); to Prof. M. Ferraris (Ist. Fisica Matematica J.-L. Lagrange, Univ. Torino) for assistance with symbolic computations, finally, for the access to the computing facilities of the Mathematics Dept. "G. Castelnuovo" of Roma I Università La Sapienza, is acknowledged.

References

1. Regge T., Wheeler J.A. (1957): Stability of a Schwarzschild singularity. Phys. Rev. **108**, 1063–1069
2. Zerilli F.J. (1970): Gravitational field of a particle falling in a Schwarzschild geometry analyzed in tensor harmonics. Phys. Rev. D **2**, 2141–2160
3. Zerilli F.J. (1975): in *Black Holes, Gravitational Waves and Cosmology: an Introduction to Current Research*. A-7, Gordon & Breach, New York, (errata Fr [2])
4. Moncrief V. (1974): Gravitational perturbations of spherically symmetric systems I. The exterior problem. Ann. Phys. N.Y. **88**, 323–342
5. Ruffini R. (1973): Energetics of Black Holes: Part III Gravitational Radiation in Schwarzschild Geometries, in *Black Holes*, August 1972, Les Houches, ed. by. C. DeWitt, B. DeWitt, Gordon & Breach, New York, R 528–137
6. Spallicci A.D.A.M. (1999): On the perturbed Schwarzschild geometry for determination of particle motion, in *Proc. Gravitational Waves, 2nd Eduardo Amaldi Conference*, 1–4 July 1997, CERN, Geneve, ed. by E. Coccia, G. Veneziano, G. Pizzella, World Scientific, Singapore, pp. 303–308
7. Spallicci A.D.A.M. (1999): Radiation reaction in free fall from perturbative geodesic equations in spherical coordinates, in *Proc. 8th Marcel Grossmann Meeting, 22–28 June 1997, Jerusalem*, ed. by T. Piran, R. Ruffini, World Scientific, Sinagpore, pp. 1107–1110
8. Zwillinger D. (1998): Handbook of Differential Equations, Academic Press, Boston
9. Vvedenski D. (1993): Partial Differential Equations with Mathematica. Addison-Wesley, Redwood City

Quantum Gravity

Problems in Quantum General Relativity and Gravitation

E. Sorace

From the early sixties a long-term scientific wave was initiated in the communities of the physical and mathematical disciplines. It was described very well by the title of the survey book (1993) prepared for the 65th birthday of D. Sciama: *"The Renaissance of General Relativity and Cosmology"* [1]. Six years later one can see that in this area experimental and theoretical issues are still growing in quantity and quality and cross-fertilizing in a measure conceivable only in a very fortunate period of the evolution of a science.

Even at this conference we have seen the enormous advances that have been made in experimental techniques which allow us to test GR with errors much smaller than any physicist could have imagined a few years ago, and the analysis of classical GR has progressed equally. In this explosive development, however, the propulsive role of quantum problems is the least far from as can be shown even by a terse and very short review of current research on the quantum theory of general relativity and of gravity (see, however, the comprehensive critical analysis in the contribution of C. Rovelli to GR_{15} [2]).

It is perhaps worth recalling that Einstein, at the end of his first paper on gravitational waves in 1916 [3], advocated a modification using quantum theory of *"not only the electrodynamics of Maxwell, but also the new theory of gravitation"*, again for the sake of the stability of atoms *in nature*. Recall also that the Bekenstein–Hawking entropy [4] of black holes, their evaporation and all their thermodynamical properties are interpreted and calculated in terms of quantum fluctuations of the vacuum around the horizon: in 1975 Hawking was able to recover from a field model interacting with the geometry the fact that the theory of quantum fields in a curved spacetime implies that a black hole generated by collapse will emit thermal radiation [5]; in 1976 Hartle and Hawking assimilated the evaporation to a tunneling [6].

One should, however, distinguish between the search for a theoretical solution to the general problem of constructing a quantum consistent field theory whose classical counterpart is correlated to classical GR, from that for an effective theory, based on classical GR, in terms of which one can perform semiclassical one loop calculations.

These last attempts are mainly based on the gauge invariant QFT as developed by means of the path integral framework in the successful approach to the standard model. As is well-known the standard model is renormalisable on the contrary of QG. The fields are taken to be the components of the metric tensor and perturbation theory is effected by its expansion around a classical solution. The introduction of a gauge fixing and of the BRST symmetry gives an effective action that can be regularized, by using, e.g., the Riemann ζ-function. Nonetheless, it is the case that when matter fields are present renormalizability of such effective quantum gravity encounters unsormontable difficulties (see e.g., [7]). This approach has been, and is still, very fruitful in quantum cosmology and any progress in that direction can be an important step in the comprehension of QG and in phenomenological calculations. An extension of the range of the usual ζ-function regularization by means of a generalization based on the extended use of the *local* ζ-function has been done by Moretti, whose results have been presented in this workshop. A critical analysis of the compatibility of the local invariance of the boundary conditions with the global properties of the ghost boundary operator for the fluctuations of the metric in the one loop semiclassical path integral calculations in Euclidean QG has been presented by Esposito et al.

On the other hand hypotheses, phenomenologically very relevant and anthropomorphically developed in the presence of a "*'cosmic censor' who forbids the appearance of naked singularities*" arising in gravitational collapses "*by clothing each one in an absolute event horizon*" [8], have been formulated in the framework of the analysis of black hole type solutions of classical GR. But recently it has been proposed that quantum fluctuations can avoid the creation of naked singularities. In his communication Caldarelli has presented a class of models of classical collapsing black holes in which classical singularities are hidden by geometrical effects of the quantum fluctuations of a coupled matter field.

Of course one cannot forget that we have yet to see an indusputable signature of a gravitational wave in a terrestrial laboratory (Virgo and Ligo and the like are still pending as has been well explained at this conference) and that the proportionality of entropy to the area of a black hole, which is nearly a paradigm of the contemporary research in the field – as it is recovered in all existing relativistic theories of gravity (see, e.g., [2]), is up to now a very speculative consequence of hypotheses not subjected to experiment. Even the quantum consistency of cosmic censorship cannot obviously be considered *per se* as an experimental test of the validity of QG or of a particular way of performing the quantization.

It seems, then, very reasonable to analyse, at the scale of the universe, the effects of the presence of quantum gravity. It is therefore worth noting the proposal [9] to analyse the electromagnetic signals of γ-ray bursts, whose sources are now confirmed to be at cosmological distance (actually on the order of billions of light years), with the object of testing nonconventional dispersion laws. Deformations of the dispersion law connected to a fundamental length, mainly the Planck scale, appear in many theories of QG, or in theories which modify fundamental physics, so that one thinks that the QG problem can be confronted there in a consistent way. Into this category would fall the so-called k-Poincaré deformation, once a physical theory consistent with this *quantum* group is constructed. This is not an easy task as we have to deal with more than the small breakdown of the Lie commutator structure, from which the deformation of the dispersion law derives and which could be eliminated by nonlinear change of the basis. Indeed the structure is changed fundamentally because the law of combining representations is irreversibly modified in a nonsymmetric way and, in consequence, the parameters of the group transformations no longer commute among themselves. Thus one has generated a peculiar kind of noncommutative geometry of spacetime, on which a complete and physically consistent field theory is yet to be constructed [10].

Therefore the possibility of exploiting the cosmological distance of γ-ray bursts to test a priori discriminant properties of space time is very attractive and has been analysed even in the loop representation of canonical gravity [11], where it has been shown that there is even a possible helicity effect. These results should be interpreted as a constraint on the semiclassical states representing the universe. This inversion of role arises because this last theory is an example of a non-perturbative approach so that, at the beginning, there is no background spacetime but only an arbitrary (4-dimensional Riemannian) manifold. The loop QG created by the work of Ashtekar and his collaborators, starting from a formulation of GR in terms of peculiar connections and continuing by using generalized Wilson loops of those canonical variables as states, has been very fruitful and has solved many of the mathematical problems which obstructed the original program [12]. Many bridges have thus been established with many different disciplinary areas such as, e.g., knot theory [12]: spin networks have been shown to be the correct objects for constructing an orthonormal basis of the Hilbert space states on the space. These discrete combinatorial structures labelled by sets of representations of SU(2), introduced by R. Penrose in 1971 [13], can be connected by a duality map to the triangulations of a manifold. To pass to spacetime one can use *spin foam*, the analog, with one further dimension, of spin networks (see [14] and its extensive bibliography). Classical geometrical quantities such as length, area and volume are represented by operators whose eigenvalues are discrete and the geometry is then quantized [16]. This permitted the counting of microstates and provided a statistical sum to calculate the entropy of a black hole, in agreement with the results of [4] and [5]. At this workshop Marzuoli and her collaborators have shown in their communication the calculation of invariants of spin networks with boundary as a sum over states, which, in the absence of boundary, in 3-dimensions reproduces the partition function of Ponzano–Regge: the method is extended to the deformed Hopf algebra $U_q(sl(2, \mathbf{C}))$ at a root q of unity and is in correspondence with the Turaev–Viro invariant.

Thus loop QG appears in some way to be analogous, as are quantum groups as well, with the noncommutative geometry of Connes [17]. It must be stressed, however, that the 7 axioms imposed by Connes on a spectral triple, to generate a mathematically *bona fide* noncommutative geometry, are intended to characterize noncommutative analogs of any classical geometry but that the physical goal was in principle the construction of a geometry in which the standard model should be encoded. Even the more recent, very important achievements of the spectral action principle [18] in which, from a formula for an action containing a single Dirac operator, determined by the requirement that the spectrum of the standard model be reproduced, one gets the standard model coupled to the gravitational field of an Einstein plus Weyl type, seem to be a classical theory on a noncommutative geometry which has to be further developed in order to make a new step in quantum field theory.

In this workshop string theory has not been the explicit object of any communication. But it is, on the contrary, the most popular framework in which high energy theorists – who comprise a large proportion – study QG. Some phenomenological consequences of string theory in cosmology (see e.g., [19]) have been discussed in the general sessions. Here it can only be added that string theory shares some common features with loop gravity, spectral Connes theory and deformed symmetries as regards the structure of space time at the Planck scale. On the other hand it is analogous to semiclassical quantum gravity as its working formulation is (up to now) perturbative and background dependent. The string however is actually proteiform and it is predicted that, when the horizon of a black hole has shrunk to string scale, then it becomes, in fact, an excited string state. This relation with black hole solutions of GR, which are one of the physically distinguished background for perturbative calculations, has stimulated the search for new classical solutions to nonlinear Einstein equations of black hole type. Theorems which impose a spherical symmetry to the event horizons of such solutions have been bypassed by the minimal modification of adding a negative cosmological constant. By this change a path has been opened to the construction of solutions having nontrivial topologies, with genus ≥ 1, which can rotate and even be charged. One can then test quantization effects on these new objects, called topological black holes, and look at their thermodynamical properties and even at their relation to string states, which seems however to be not as easy to understand as the usual ones. It is also possible to generalize such solutions by increasing the dimension of spacetime by even or odd coordinates and entering into the framework of p-branes or M-theory (see, e.g., [20]). Klemm has presented in this workshop an analysis of topological black holes for $N = 2$ gauged supergravity.

In a different context another way to look at such nonlinear problems, in which the topology plays an essential role in determining background configurations, is to search for soliton-like solutions of the associated static variational problem which are classified by the conserved topological number. As is necessary in all these models, to persist up to Derrick's theorem, higher powers of the derivatives of the fields must be present in the Lagrangian. A typical situation, well-known to physicists, is given by the Skyrme model, where the topological number B is the homotopy class of the map between the space domain compactified by the zero value of the field at infinity and the field of fixed modulus. This model has been recovered by Witten as the low energy limit of QCD for large colour number and, as we have heard here in the lectures of Fabbrocini, represents the main theoretical model of nuclear physics. This fundamental model has been shown to have solutions (of given continuous symmetry) only for $B = 1$, the spherical configuration of the Skyrme hedgehog, and for $B = 2$, the toroidal configurations. For higher numbers, up until a few years ago there were only expressions proposed with axial symmetry or dubious configurations recovered by numerical brute force [21]. Rather recently some configurations with finite symmetries have been definitely accepted as solutions for higher values of B [22]. Here Pisani has presented the mathematical analysis of some models of this kind, corresponding to relativistic charged particles, and he showed the existence of solutions by using present achievements of developing the calculus of variations.

As a final point one can mention that there are many unorthodox approaches. It seems natural to judge as the most radical those which seek modifications even to the usual quantum mechanics. Among them the de Broglie–Bohm theory still appears rather relevant [23]. Its main point aims to eliminate the indeterminism from quantum physics by asking for the physical reality of the wave functions, requiring them to become the so-called "pilot waves" accompanying massive particles subject to a modified Newton's second law. In fact this theory has, for a very long time, survived the hard selection of the basic theories, as impelled by experimental and theoretical confutations. Nonetheless, up to now, quantum mechanics has been completely confirmed in all the crucial experiments designed its falsification. Feoli has presented here a communication in which it is shown that gravitational waves, i.e., solutions of the linearized Einstein equations, can take the peculiar analytical form demanded by that theory for the wave functions of a free relativistic particle.

We can now conclude with the remark that the quantization of gravity has progressed vigorously and that there are now well-established and different theories in which QG is on the horizon as a clear problem. It is then possible that there are only very difficult and purely technical obstacles to overcome in order to reach the goal. Nevertheless still apposite is the following reconsideration by David Hilbert of the construction of the final general relativistic theory: *"Every boy in the streets of Göttingen understands more about four-dimensional geometry than Einstein. Yet, in spite of that, Einstein did the work and not the mathematicians."* ([24], p. 124).

References

1. Ellis G., Lanza A., Miller J., eds. (1993): Renaissance of General Relativity and Cosmology. Cambridge University Press, Cambridge
2. Rovelli C. (1997): Strings, loops and others: a critical survey of the present approaches to quantum gravity. Plenary lecture on quantum gravity at the GR_{15} conference, Poona, India. gr-qc/9803024
3. Einstein A. (1916): Näherungsweise Integration der Feldgleichungen der Gravitation. König. Preuss. Akad. der Wiss. Berlin. Sitzungberichte, 688-696. Quoted by Cattani C., De Maria M. (1993): Conservation Laws and Gravitational Waves, in *The Attraction of Gravitation*, ed. by J. Earman, M. Janssen, J.D. Norton, Birkhäuser, Boston
4. Bekenstein J.D. (1974): Generalized second law of thermodynamics in black hole physics. Phys. Rev. D **9**, 3292;
   Hawking S.W. (1974): Do black holes explode? Nature **248**, 30
5. Hawking S.W. (1975): Particle creation by black holes. Commun. Math. Phys. **43**, 199–220
6. Hartle J.B., Hawking S.W. (1976): Path-integral derivation of black-hole radiance. Phys. Rev. D **13**, 2188–2203
7. Moss I.G. (1996): Quantum Theory, Black Holes and Inflation. John Wiley, Chichester
8. Penrose R. (1969): Gravitational collapse: the role of general relativity. Riv. del Nuovo Cimento **1**, 252–276
9. Amelino-Camelia G., Ellis J., Mavromatos N.E., Nanopoulos D.V., Sarkar S. (1998): Tests of quantum gravity from observation of γ-ray bursts. Nature **393**, 763–765
10. Kosiński P., Lukierski J., Maślanka P. (1999): Local D = 4 field theory on k-deformed Minkowski space. hep-th/9902037

11. Gambini R., Pullin J. (1999): Nonstandard optics from quantum space-time. Phys. Rev. D **59**, 124021-1-4
12. Ashtekar A. (1988): New Perspectives in Canonical Gravity. Bibliopolis, Naples
13. Asthekar A. (1997): Polymer Geometry at Planck Scale and Quantum Einstein Equations, in *Proc. of the 14th International Conference on General Relativity and Gravitation*, ed. by M. Francaviglia, G. Longhi, L. Lusanna, E. Sorace, World Scientific, Singapore, pp. 8–29;
Gambini R., Pullin J. (1996): Loops, Knots, Gauge Theories and Quantum Gravity. Cambridge University Press, Cambridge
14. Penrose R. (1971): Angular Momentum: an Approach to Combinatorial Space-Time, in *Quantum theory and beyond*, ed. by T. Bastin, University Press, Cambridge, pp. 151–180
15. Baez J.C. (1999): An introduction to spin foam models of *BF* theory and quantum gravity. gr-qc/9905087
16. Asthekar A. (1999): Quantum mechanics of geometry. gr-qc/9901023
17. Connes A. (1994): Noncommutative Geometry. Academic Press, San Diego
18. Chamseddine A.H., Connes A. (1997): The spectral action principle. Commun. Math. Phys. **186**, 731–750
19. Durrer R., Gasperini M., Sakellaradiou M., Veneziano G. (1999): Seeds of large scale anisotropy in string cosmology. Phys. Rev. **59**, 043511
20. Aharony O., Gerbser S.S., Maldacen J., Ooguri H., Oz Y. (1999): Large *N* field theory, string theory and gravity. hep-th/9905111
21. Makhankov V.D., Rybakov Y.P., Sanyuk V.I. (1993): The Skyrme Model: Fundamentals, Methods, Applications. Springer, Berlin Heidelberg New York
22. Houghton C.J., Manton N.S., Sutcliffe P.M. (1998): Rational maps, monopoles and skyrmions. Nucl. Phys. B **510**, 507–537
23. Holland R.P. (1993): The de Broglie-Bohm theory of motion and quantum field theory. Phys. Rep. **224**, 95–150
24. Reid C. (1986): Hilbert-Courant. Springer, New York, quoted by Kichessamy S. (1993): Variational derivations of Einstein's equations. op. cit. [3], p. 197

A Quantum Implementation of the Cosmic Censorship Conjecture

M.M. Caldarelli

Abstract. We review some aspects of the quantum field theory of a conformally coupled scalar field on the singular background obtained in the massless limit of a class of toroidal black holes. The stress-tensor and its back reaction on the metric are computed, using point-splitting regularization, in the cases of transparent, Neumann and Dirichlet boundary conditions. We find that the quantum fluctuations generate an event horizon which hides the singularity. The resulting object can be interpreted as a long-lived remnant. We discuss the relevance of this result in the context of the cosmic censorship conjecture, and in connection to the end point of the quantum evaporation process.

The cosmic censorship conjecture states that, during gravitational collapse, naked singularities never occur, at least for asymptotically flat spacetimes [1]. It is interesting to investigate how quantum mechanics modifies the issue of the collapse; we ask whether quantum fluctuations, which cannot be neglected in the final stages of the collapse, can prevent the formation the classical naked singularities, hiding them behind an event horizon. If this turns out to be the case, semi-classical back reaction of quantum fields on the geometry would strengthen the cosmic censorship conjecture. An instance of this is the case of extreme black holes, where quantum emission drives the hole away from extremality [2]. The issue of these questions is also fundamental to the problem of the end point of evaporation. Black holes are known to evaporate by the emission of thermal radiation [3]. The end point of this evaporation process is still an unsolved issue, related to unitarity of the quantum theory of gravitation.

These questions have been analysed by various authors [4–8] in the case of the non-rotating Bañados-Teitelboim-Zanelli (BTZ) black hole [9, 10]. Taking into account the back reaction on the metric due to the propagation of quantum fields on the background of this three-dimensional black hole, it was shown that in the $M \to 0$ limit, where space-time acquires a naked singularity, quantum fluctuations produce a horizon at the semi-classical level and dress the singularity.

These results were obtained in a lower dimensional model of gravity; here we shall review their extension to a physically more relevant four-dimensional black hole [11]. These computations have been made possible by the recent discovery of topological black holes [12–16], namely, solutions of the vacuum Einstein equations with a negative cosmological constant, which are asymptotically anti–de Sitter black holes and have a compact event horizon with a nontrivial topology. These black holes can result from gravitational collapse [17, 18] and from quantum tunnelling processes [19]. They can be charged [20, 16], exist in dilaton gravity [20, 21] and also admit a rotating generalization [22, 23]. For toroidal topology, the hoop and cosmic censorship conjectures have been studied in [24]. The reason why these black holes violate the classical theorems of general relativity forbidding nonspherical topologies is the presence of the cosmological constant. A simple framework in which to study quantum effects in the presence of naked singularities is the class of toroidal black holes described by the metric

$$\begin{aligned} ds^2 = &-\left(-\frac{2\eta}{r}+\frac{r^2}{\ell^2}\right) dt^2 \\ &+\left(-\frac{2\eta}{r}+\frac{r^2}{\ell^2}\right)^{-1} dr^2 + r^2\left(|\tau|^2\, dx^2 + dy^2 + 2\mathrm{Re}\tau\, dx dy\right) \end{aligned} \tag{1}$$

where $x, y \in [0, 1[$ and τ is the Teichmüller complex parameter that determines the conformal class of the torus. If $\eta > 0$, solution (1) is a black hole of mass $M = \eta|\mathrm{Im}\,\tau|/4\pi$ with an event horizon localized in $r_+ = (2\eta\ell^2)^{1/3}$. As the mass decreases to zero the event horizon shrinks and disappears for $\eta = 0$, leaving a naked singularity. We shall review the propagation of a conformally coupled scalar field on this singular background, and show from a semi-classical analysis of its back reaction on the metric, that the quantum fluctuations of the field induce the formation of an event horizon which hides the singularity [11].

The background $\mathcal{M}$ of toroidal black holes (1) is described by the singular metric (from now on we restrict attention to the $\tau = i$ case)

$$ds^2 = -\frac{r^2}{\ell^2}\, dt^2 + \frac{\ell^2}{r^2}\, dr^2 + r^2\left(dX^2 + dY^2\right) \tag{2}$$

where t ranges over $\mathbb{R}$, $r > 0$ and $X, Y \in [0, 1[$ with the points 0 and 1 identified. A first important remark about this metric is that it is a locally anti–de Sitter (AdS) metric and can be obtained by quotienting AdS space with respect to a discrete subgroup of its isometry group. The universal covering of AdS space, $\overline{\mathrm{AdS}}$, is a homogeneous and symmetric manifold, with isometry group SO(3, 2). It solves Einstein equations with negative cosmological constant $\Lambda = -3/\ell^2$ and has a constant negative curvature $R = -12/\ell^2$. It is possible to construct AdS space as an hypersurface in a flat five-dimensional pseudo-euclidean spacetime $\mathcal{V}$ with signature $(-, +, +, +, -)$; the corresponding scalar product will be denoted $(\cdot, \cdot)_{\mathcal{V}}$. Its metric assumes the form (2) in an appropriate coordinate system; in fact it is just the universal covering of $\mathcal{M}$. To obtain the background $\mathcal{M}$ from $\overline{\mathrm{AdS}}$ we need to make the identifications $\Lambda^m : X \mapsto X + m$ and $\Gamma^n : Y \mapsto Y + n$, with $m, n \in \mathbb{Z}$. These identifications form an abelian subgroup $H \cong \mathbb{Z} \times \mathbb{Z} \subset \mathrm{SO}(3, 2)$ of the homogeneity group. Thus, $\mathcal{M}$ is regular everywhere except for $r = 0$. This is not a curvature singularity, but there are inextendible geodesics ending there as the elements of the identification group leave this point fixed. The manifold $\mathcal{M}$ being the quotient $\overline{\mathrm{AdS}}/H$, the analogy with the BTZ black hole – which is itself the quotient of three-dimentional AdS space with respect to a discrete group of isometries – is also formal. This fact simplifies the computation of the quantum stress-tensor of a conformally coupled scalar field, and leads us to interpret it as a consequence of the Casimir effect.

Next we consider the propagation of a conformally coupled scalar field ϕ on the background $\mathcal{M}$ and calculate its Green function. The total action of the theory is

$$S[g_{\mu\nu},\phi] = \frac{1}{2\kappa}\int \sqrt{-g}\,(R-2\Lambda) - \frac{1}{2}\int \sqrt{-g}\left[g^{\mu\nu}\nabla_\mu\phi\nabla_\nu\phi + \xi R\phi^2\right], \tag{3}$$

where the first term is the usual Hilbert-Einstein action and the second term describes the matter field. Here $\kappa = 8\pi G$, and $\xi = 1/6$. The equations of motion are, as usual, $G_{\mu\nu} + \Lambda g_{\mu\nu} = \kappa T_{\mu\nu}$ and $(\Box - \xi R)\,\phi = 0$.

As $\mathcal{M}$ is a quotient of $\overline{\text{AdS}}$, its Green function can be obtained from that of the covering space $\overline{\text{AdS}}$ by the method of images. But $\overline{\text{AdS}}$ is not a globally hyperbolic spacetime and therefore we have to impose boundary conditions at infinity. To this end we conformally map $\overline{\text{AdS}}$ to half of the Einstein static universe (ESU). The boundary conditions at infinity become conditions on the fields on the equator of ESU. As shown in [25], three choices of boundary conditions are possible; the transparent boundary conditions are obtained by quantizing the field using smooth modes over the whole ESU, while the Neumann and Dirichlet boundary conditions can be imposed on the equator of the ESU. The Green functions are then

$$\mathcal{G}^{\epsilon}_{AdS}(\mathbf{x},\mathbf{x}') = \frac{1}{4\pi^2}\frac{1}{(\mathbf{x}-\mathbf{x}',\mathbf{x}-\mathbf{x}')_{\mathcal{V}}} + \frac{\epsilon}{4\pi^2}\frac{1}{(\mathbf{x}+\mathbf{x}',\mathbf{x}+\mathbf{x}')_{\mathcal{V}}} \tag{4}$$

with $\epsilon = 0, +1, -1$ for transparent, Neumann and Dirichlet boundary conditions respectively. The functional form of the Green functions is the same as in Minkowski space because the background is conformally flat and the field conformally coupled. The second term in Eq. (4) corresponds to an antipodal image in $\overline{\text{AdS}}$ needed to force the boundary conditions. The same results can be obtained by a direct mode sum [26, 27].

Next we obtain the Green function with the appropriate boundary conditions simply by summing over images, obtaining

$$\mathcal{G}^{\epsilon}_{\mathcal{M}}(\mathbf{x},\mathbf{x}') = \frac{1}{4\pi^2} \cdot \sum_{m,n\in\mathbb{Z}}\left[\frac{1}{(\mathbf{x}-\Lambda^m\Gamma^n\mathbf{x}',\mathbf{x}-\Lambda^m\Gamma^n\mathbf{x}')_{\mathcal{V}}} + \frac{\epsilon}{(\mathbf{x}+\Lambda^m\Gamma^n\mathbf{x}',\mathbf{x}+\Lambda^m\Gamma^n\mathbf{x}')_{\mathcal{V}}}\right]. \tag{5}$$

The series converges in the Dirichlet case, but diverges as $\sum[1/(m^2+n^2)]$ for transparent and the Neumann boundary conditions [5]. This is an infrared divergence, due to the existence of zero modes in the Neumann and transparent cases, that render the Klein-Gordon operator non-invertible. Hence, we have to regularize the Green functions eliminating these zero modes. This can be done by defining the regularized

Green functions by

$$
\mathcal{G}^{\epsilon}_{\mathcal{M}}(\mathbf{x},\mathbf{x}'|s) = \frac{1}{4\pi^2} \cdot \sum_{m,n\in\mathbb{Z}} \left[\frac{1}{(\mathbf{x}-\Lambda^m\Gamma^n\mathbf{x}',\mathbf{x}-\Lambda^m\Gamma^n\mathbf{x}')^s_{\mathcal{V}}} + \frac{\epsilon}{(\mathbf{x}+\Lambda^m\Gamma^n\mathbf{x}',\mathbf{x}+\Lambda^m\Gamma^n\mathbf{x}')^s_{\mathcal{V}}} \right] \tag{6}
$$

which are convergent for $s > 1$ and diverge in the $s \to 1$ limit. Note that Eq. (6) is a Green function in $(2s+2)$-dimensional anti–de Sitter space: we have used dimensional regularization. In the calculation of the stress tensor we shall take derivatives of these functions with respect to the spacetime variables, that will render finite the series in $s = 1$. Here the stress tensor will be the analytic continuation of the dimensionally continued stress-tensor.

Having the Green functions at hand, we turn to the stress tensor for the field ϕ; classically it is given by

$$
\begin{aligned}
T_{\mu\nu} = {} & \frac{2}{3}\,\nabla_\mu\phi\nabla_\nu\phi - \frac{1}{3}\,\phi\nabla_\mu\nabla_\nu\phi - \frac{1}{6}\,g_{\mu\nu}g^{\alpha\beta}\nabla_\alpha\phi\nabla_\beta\phi \\
& + \frac{1}{3}g_{\mu\nu}\phi\Box\phi + \frac{1}{2\ell^2}\,g_{\mu\nu}\phi^2 .
\end{aligned} \tag{7}
$$

We observe first that the tensor $T_{\mu\nu}$ is traceless, as expected in a classical conformally invariant theory. However, as we are in an even dimension, a conformal anomaly appears and breaks this invariance at the quantum level.

Next we seek to calculate the quantum stress tensor of the theory. We cannot take the expectation value of formula (7) directly: after the quantization procedure the field $\phi(\mathbf{x})$ becomes a distribution and expectation values like $\langle 0|\,\phi(\mathbf{x})\phi(\mathbf{x})\,|0\rangle$ are formally divergent quantities; we need to define a regularization procedure. We shall use the point-splitting technique [28]; in this framework, we define the regularized stress tensor to be

$$
\begin{aligned}
T_{\mu\nu}(\mathbf{x}) = \lim_{\mathbf{x}'\to\mathbf{x}} \frac{1}{3} & \Bigg[\nabla_\mu\nabla'_\nu + \nabla'_\mu\nabla_\nu - \frac{1}{2}\,\nabla_\mu\nabla_\nu - \frac{1}{2}\,\nabla'_\mu\nabla'_\nu \\
& -\frac{1}{2}\,g_{\mu\nu}\left(g^{\alpha\beta}\nabla_\alpha\nabla'_\beta + \frac{1}{\ell^2}\right)\Bigg]\left[\mathcal{G}(\mathbf{x},\mathbf{x}') - \mathcal{G}_{\text{sing}}(\mathbf{x},\mathbf{x}')\right],
\end{aligned} \tag{8}
$$

where $\mathcal{G}_{\text{sing}}(\mathbf{x},\mathbf{x}')$ is the singular part of the Green function and we have used the equations of motion.

We have to distinguish two contributions to the Green function: the former is the direct contribution, that is, the Green function of $\overline{\text{AdS}}$ space; the latter is the contribution coming from the sum over the images. The $\overline{\text{AdS}}$ contribution is divergent in the limit $\mathbf{x}' \to \mathbf{x}$ and we have to substract its singular part to renormalize it. In fact, the stress tensor of the scalar field in $\overline{\text{AdS}}$ has to be proportional to the only tensor available, i.e., the metric tensor $g_{\mu\nu}$. The proportionality factor is one quarter of the conformal anomaly in our space [28], from which we deduce that the stress tensor contribution of $\overline{\text{AdS}}$ amounts to

$$\langle T_{\mu\nu}^{AdS} \rangle = -\frac{\hbar}{960\pi^2\ell^4}\, g_{\mu\nu}. \tag{9}$$

In contrast, the contribution of the images remains finite after the coincidence limit of the Green function. The calculation of the image contribution $\langle T_{\mu\nu}^{\pm}\rangle$ can be simplified by performing it in the flat five-dimensional embedding space of AdS; the quantum stress tensor can successively be pulled back to $\mathcal{M}$ [4, 11]. As a result, the quantum stress-tensor reads

$$\langle T_{\mu\nu}^{\epsilon}\rangle = \frac{\hbar}{2\pi^2\ell^2}\operatorname{diag}\left(-\frac{\alpha}{r^2}+\frac{\epsilon}{3\ell^2}\left(\frac{r}{2\ell}\right)^4 E\left(\frac{r}{2\ell}\right);\quad \frac{\alpha\ell^4}{r^6};\right.$$
$$\left.-\frac{\alpha}{r^2}\ell^2+\frac{\epsilon}{6}\left(\frac{r}{2\ell}\right)^4 E\left(\frac{r}{2\ell}\right); -\frac{\alpha}{r^2}\ell^2+\frac{\epsilon}{6}\left(\frac{r}{2\ell}\right)^4 E\left(\frac{r}{2\ell}\right)\right)+\langle T_{\mu\nu}^{AdS}\rangle \tag{10}$$

where we have defined the auxiliary function

$$E(\zeta) \equiv \sum_{m,n\in\mathbb{Z}} \frac{m^2+n^2}{\left(1+(m^2+n^2)\zeta^2\right)^3}, \tag{11}$$

and $\alpha = 6.0268\ldots$ is a constant. The stress tensor obtained is regular and finite, except for $r = 0$ where it diverges. Furthermore, as can be easily verified, it is conserved. Finally, we note that the stress tensor found involves a negative energy density which may be interpreted as a Casimir energy in $\mathcal{M}$ due to the identifications under the action of the group H.

Next we compute the semi-classical corrections to the geometry of $\mathcal{M}$ at order $\hbar$ due to the presence of the quantum field ϕ. These corrections arise from solving the semi-classical equations $G_{\mu\nu} + \Lambda g_{\mu\nu} = \kappa\langle T_{\mu\nu}\rangle$.

We consider first the $\overline{\text{AdS}}$ term of the stress tensor. As $\langle T_{\mu\nu}^{AdS}\rangle$ is proportional to the metric, it only contributes by changing the cosmological constant by a finite renormalization,

$$\Lambda_{\text{ren}} = \Lambda + \frac{\hbar\kappa}{960\pi^2\ell^4} = -\frac{3}{\ell^2} + \frac{\hbar\kappa}{960\pi^2\ell^4}. \tag{12}$$

Having performed this renormalization we can forget the $\overline{\text{AdS}}$ contribution in the stress tensor and consider only the contribution of the images to the corrections to the metric. In the following we take ℓ to be the renormalized constant.

We now look for a static solution, with toroidal symmetry, of the semi-classical equations. The general metric possessing these properties has the form

$$\mathrm{d}s^2 = -f^2(r)\,\mathrm{d}t^2 + f^{-2}(r)\,\mathrm{d}r^2 + e^{2A(r)}\left(\mathrm{d}X^2 + \mathrm{d}Y^2\right). \tag{13}$$

Solving the semi-classical Einstein equations for this expression and keeping only the corrections of order $\hbar$ to the metric of $\mathcal{M}$, we find that

$$f^2(r) = \frac{r^2}{\ell^2} - \hbar\left[\frac{\alpha\kappa}{2\pi^2 r^2} + \frac{\epsilon\kappa}{3\pi^2\ell^2}\left(\frac{2\ell}{r}\right) E_3\left(\frac{r}{2\ell}\right)\right] + \mathcal{O}(\hbar^2), \tag{14}$$

where

$$E_3(\zeta) = \sum_{m,n\in\mathbb{Z}}{}' \Bigg[\left(\frac{\zeta^2}{16\left(m^2+n^2\right)^{1/2}} - \frac{3}{16\left(m^2+n^2\right)^{3/2}} \right) \cdot \left(\arctan\left(\sqrt{m^2+n^2}\zeta\right) - \frac{\pi}{2} \right) + \frac{\xi}{16(m^2+n^2)} + \frac{\zeta}{8(m^2+n^2)\left(1+(m^2+n^2)\zeta^2\right)} \Bigg]. \tag{15}$$

In the last function, an arbitrary additive constant has been set equal to zero to match asymptotically the perturbed metric with its classical background. We do not give the exact form of $A(r)$, but it can be proven that $\exp(2A)$ behaves asymptotically like r^2 and is regular everywhere. From the asymptotic behavior of $E_3(\zeta)$, we deduce the asymptotic form of the lapse function:

$$f^2(r) = \frac{r^2}{\ell^2} - C_\epsilon \frac{\hbar\kappa\alpha}{\pi^2 r^2} + \mathcal{O}\left(\frac{\hbar}{r^4}\right) \tag{16}$$

with $C_\epsilon = 0, 1/18$ and $17/18$ for transparent, Neumann and Dirichlet boundary conditions respectively.

Thus the semi-classical metric acquires a curvature singularity at the origin and the perturbation due to the propagation of the quantum field ϕ develops a horizon that hides the singularity. This is easily seen in the case of transparent boundary conditions ($\epsilon = 0$), where a horizon forms in $r_+ = \left(\kappa\alpha\hbar\ell^2\right)^{1/4}$. In terms of the Planck length ℓ_p the black hole radius behaves like $\sqrt{\ell\ell_p}$ and formally has a Hawking temperature [3]

$$T = \frac{\hbar r_+}{\pi\ell^2}, \tag{17}$$

which is immensely smaller than the temperature of the initially macroscopic toroidal black hole. If the dressed singularity continues to evaporate, then it is a kind of long-lived remnant.

When Neumann or Dirichlet boundary conditions are imposed, the expression of the lapse function is more complicated, but it is possible to prove that the effect of quantum fluctuations is analogous to the transparent boundary case; only the location of the horizon changes slightly. For a Neumann field we find the horizon nearer to the singularity, while Dirichlet boundary conditions push the horizon outwards, and transparent boundary conditions give rise to a horizon located between the other two.

Lastly we observe that the validity of the semi-classical approximation could break down near the horizon, where the fluctuations of $\langle T_{\mu\nu}\rangle$ become of the same order as $\langle T_{\mu\nu}\rangle$. To cure this flaw, a possible trick consists in considering N independent scalar fields instead of one [29, 30, 5, 8]. This reduces the ratio of the fluctuations to $\langle T_{\mu\nu}\rangle$, and the semi-classical approximation may become almost exact, because higher loop terms are of the order $\mathcal{O}(1/N)$.

The quantum dressing of naked singularities could be a general feature of propagation of quantum fields on singular backgrounds generated by collapsing matter, and would imply that quantum effects would work in favor of the cosmic censorship conjecture. Furthermore, as $\mathcal{M}$ appears to be the natural end point of the evaporation process, the semi-classical insight seems to discredit the possibility of naked singularities as end points of evaporation. Yet, as the horizon enjoys thermal properties, the best we can conclude is that the dressed singularity behaves as a kind of long-lived remnant, or perhaps that the evaporation does not stop at $\eta = 0$, but would continue through a series of dressed conical spacetimes, as has been suggested in three dimensions [5]. In the final stages of evaporation, quantum effects are expected to dominate, thus no final conclusion can be drawn without a full theory of quantum gravity.

Acknowledgements

I would like to thank G. Magli, V. Moretti, L. Vanzo and S. Zerbini for many valuable discussions.

References

1. Penrose R. (1969): Riv. Nuovo Cim. **1**, 252
2. Vanzo L. (1997): Phys. Rev. D **55**, 2192
3. Hawking S.W. (1975): Commun. Math. Phys. **43**, 199
4. Steif A.R. (1994): Phys. Rev. D **49**, 585
5. Lifschytz G., Ortiz M. (1994): Phys. Rev. D **49**: 1929
6. Shiraishi K., Maki T. (1994): Phys. Rev. D **49**, 5286
7. Martinez C., Zanelli J. (1997): Phys. Rev. D **55**, 3642
8. Binosi D., Moretti V., Vanzo L., Zerbini S. (1999): Phys. Rev. D **59**, 104017
9. Banados M., Teitelboim C., Zanelli J. (1992): Phys. Rev. Lett. **69**, 1849
10. Banados M., Henneaux M., Teitelboim C., Zanelli J. (1993): Phys. Rev. D **48**, 1506
11. Caldarelli M.M. (1998): Phys. Rev. D **58**, 064008
12. Huang C.-G., Liang C.-B. (1995): Phys. Lett. A **201**, 27
13. Aminneborg S., Bengtsson I., Holst S., Peldan P. (1996): Class. Quant. Grav. **13**, 2707
14. Mann R. (1997): Class. Quant. Grav. **14**, 2927
15. Vanzo L. (1997): Phys. Rev. D **56**, 6475
16. Brill D.R., Louko J., Peldan P. (1997): Phys. Rev. D **56**, 3600
17. Smith W.L., Mann R.B. (1997): Phys. Rev. D **56**, 4942
18. Lemos J.P.S. (1998): Phys. Rev. D **57**, 4600
19. Mann R.B. (1997): Class. Quant. Grav. L **14**, 109
20. Cai R.-G., Zhang Y.-Z. (1996): Phys. Rev. D **54**, 4891
21. Cai R.-G., Ji J.-Y., Soh K.-S. (1998): Phys. Rev. D **57**, 6547
22. Klemm D., Moretti V.,Vanzo L. (1998): Phys. Rev. D **57**, 6127
23. Lemos J.P.S., Zanchin V.T. (1996): Phys. Rev. D **54**, 3840
24. Lemos J.P.S. (1999): Phys. Rev. D **59**, 044020
25. Avis S.J., Isham C.J., Storey D. (1978): Phys. Rev. D **18**, 3565
26. Burgess C.P., Lutken C.A. (1985): Phys. Lett. B **153**, 137
27. Inami T., Ooguri H. (1985): Prog. Theor. Phys. **73**, 1051

28. Birrell N.D., Davies P.C.W. (1982): Quantum Fields in Curved Space. Cambridge University Press, Cambridge
29. Toumboulis E. (1977): Phys. Lett. B **70**, 361
30. Curtis J., Callan G., Giddings S.B., Harvey J.A., Strominger A. (1992): Phys. Rev. D **45**, 1005

Foundational Problems in Quantum Gravity

I.G. Avramidi, G. Esposito

Abstract. Boundary conditions play a crucial role in the path-integral approach to quantum gravity and quantum cosmology, as well as in current attempts to understand the one-loop semiclassical properties of quantum field theories. Within this framework, one is led to consider boundary conditions completely invariant under infinitesimal diffeomorphisms on metric perturbations. These are part of a general scheme, which can be developed for Maxwell theory, Yang–Mills theory, Rarita–Schwinger fields and any other gauge theory. A general condition for strong ellipticity of the resulting field theory on manifolds with boundary is proved here, following recent work by the authors. Its relevance to Euclidean quantum gravity is also discussed.

In several branches of classical and quantum field theory, as well as in the current attempts to develop a quantum theory of the universe and of gravitational interactions, it remains very useful to describe physical phenomena in terms of differential equations for the variables of the theory, supplemented by boundary conditions for the solutions of such equations. For example, the problems of electrostatics, the analysis of waveguides, the theory of vibrating membranes, the Casimir effect, van der Waals forces, and the problem of how the universe could evolve from an initial state, all need a careful assignment of boundary conditions. In the last case, if one follows a path-integral approach, one faces two formidable tasks:

(i) the classification of the geometries occurring in the "sum over histories" and matching the assigned boundary data;

(ii) the choice of boundary conditions on metric perturbations which may lead to the evaluation of the one-loop semi-classical approximation. Indeed, while the full path integral for quantum gravity is a fascinating idea but remains a formal tool, the one-loop calculation may be put on solid ground, and appears particularly interesting because it yields the first quantum corrections to the underlying classical theory (despite the well-known lack of perturbative renormalizability of quantum gravity based on Einstein's theory). Within this framework, it is of crucial importance to understand whether the property of strong ellipticity of the boundary-value problem (see Appendix) is compatible with the imposition of local and gauge-invariant boundary conditions for a self-adjoint operator on perturbations.

For this purpose, we are now going to study gauge-invariant boundary conditions in a general gauge theory, following [1]. Given a Riemannian manifold M, a gauge theory is defined by two vector bundles, V and G, such that $\dim V > \dim G$. V is the bundle of gauge fields $\varphi \in C^{\infty}(V, M)$, and G is the bundle of parameters of gauge transformations $\epsilon \in C^{\infty}(G, M)$. Both bundles V and G are equipped with Hermitian positive-definite metrics E, $E^\dagger = E$, and γ, $\gamma^\dagger = \gamma$, and with the corresponding natural L^2 scalar products $(,)_V$ and $(,)_G$.

The infinitesimal gauge transformations

$$\delta\varphi = R\epsilon \tag{1}$$

are generated by a first-order differential operator R,

$$R: C^\infty(G, M) \to C^\infty(V, M). \tag{2}$$

Furthermore, two auxiliary operators are introduced,

$$X: C^\infty(V, M) \to C^\infty(G, M) \tag{3}$$

and

$$Y: C^\infty(G, M) \to C^\infty(G, M), \tag{4}$$

which make it possible to define the differential operators

$$L \equiv XR: C^\infty(G, M) \to C^\infty(G, M) \tag{5}$$

and

$$H \equiv \overline{X}YX: C^\infty(V, M) \to C^\infty(V, M), \tag{6}$$

where $\overline{X} \equiv E^{-1}X^\dagger\gamma$. The operators X and Y should satisfy the following conditions (but are otherwise arbitrary):

(1) The differential operators L and H have the same order.
(2) The operators L and H are formally self-adjoint (or anti-self-adjoint).
(3) The operators L and Y are elliptic.

From these conditions, two essentially different cases arise:

Case I. X is of first order and Y is of zeroth order, i.e.,

$$X = \overline{R}, \quad Y = I_G, \tag{7}$$

where $\overline{R} \equiv \gamma^{-1}R^\dagger E$. Then, of course, L and H are both *second-order* differential operators,

$$L = \overline{R}R, \quad H = R\overline{R}. \tag{8}$$

Case II. X is of zeroth order and Y is of first order. Let $\mathcal{R}$ be the bundle of maps of G into V, and let $\beta \in \mathcal{R}$ be a zeroth-order differential operator. Then

$$X = \overline{\beta}, \quad Y = \overline{\beta}R, \tag{9}$$

where $\overline{\beta} \equiv \gamma^{-1}\beta^\dagger E$, and the operators L and H are of *first* order,

$$L = \overline{\beta}R, \quad H = \beta\overline{\beta}R\overline{\beta} = \beta L\overline{\beta}. \tag{10}$$

We assume that, by suitable choice of the parameters, the second-order operator $\overline{R}R$ can be made of Laplace type and the first-order operator $\overline{\beta}R$ can be made of Dirac type, and, therefore, have non-degenerate leading symbols (here denoted by σ_L)

$$\det{}_G \sigma_L(\overline{R}R) \neq 0, \tag{11}$$

$$\det{}_G \sigma_L(\overline{\beta}R) \neq 0. \tag{12}$$

The dynamics of gauge fields $\varphi \in C^\infty(V, M)$ at the linearized (one-loop) level is described by a formally self-adjoint (or anti-self-adjoint) differential operator,

$$\Delta : C^\infty(V, M) \to C^\infty(V, M). \tag{13}$$

This operator is of second order for bosonic fields and of first order for fermionic fields. In both cases its leading symbol satisfies the identities

$$\sigma_L(\Delta R) = 0, \qquad \sigma_L(\overline{R}\Delta) = 0, \tag{14}$$

and, therefore, is degenerate.

We consider only the case in which the gauge generators are *linearly independent*. This means that the equation

$$\sigma_L(R)\epsilon = 0, \tag{15}$$

has *the unique* solution $\epsilon = 0$. In other words,

$$\text{Ker } \sigma_L(R) = \emptyset, \tag{16}$$

i.e., the rank of the leading symbol of the operator R equals the dimension of the bundle G,

$$\text{rank } \sigma_L(R) = \dim G. \tag{17}$$

We also assume that the leading symbols of the generators R are *complete* in that they generate *all* zero-modes of the leading symbol of the operator Δ, i.e., *all* solutions of the equation

$$\sigma_L(\Delta)\varphi = 0, \tag{18}$$

have the form

$$\varphi = \sigma_L(R)\epsilon, \tag{19}$$

for some ϵ. In other words,

$$\text{Ker } \sigma_L(\Delta) = \{\sigma_L(R)\epsilon \mid \epsilon \in G\}, \tag{20}$$

and hence

$$\text{rank}\ \sigma_L(\Delta) = \dim V - \dim G. \tag{21}$$

Furthermore, we take the operator H to be of the *same* order as the operator Δ and construct a formally (anti-)self-adjoint operator,

$$F \equiv \Delta + H, \tag{22}$$

so that

$$\sigma_L(F) = \sigma_L(\Delta) + \sigma_L(H). \tag{23}$$

It is easy to derive the following result:

The leading symbol of the operator F is non-degenerate, i.e.,

$$\det_V \sigma_L(F) \neq 0. \tag{24}$$

Indeed, suppose that there exists a zero-mode φ_0 of the leading symbol of the operator F, i.e.,

$$\sigma_L(F)\varphi_0 = \overline{\varphi}_0\sigma_L(F) = 0, \tag{25}$$

where $\overline{\varphi} \equiv \varphi^\dagger E$. Then we have

$$\overline{\varphi}_0\sigma_L(F)\sigma_L(R) = \overline{\varphi}_0\sigma_L(\overline{X}Y)\sigma_L(L) = 0, \tag{26}$$

and, since $\sigma_L(L)$ is non-degenerate,

$$\overline{\varphi}_0\sigma_L(\overline{X}Y) = \sigma_L(YX)\varphi_0 = 0. \tag{27}$$

But this implies that

$$\sigma_L(H)\varphi_0 = 0, \tag{28}$$

and hence

$$\sigma_L(F)\varphi_0 = \sigma_L(\Delta)\varphi_0 = 0. \tag{29}$$

Thus, φ_0 is a zero-mode of the leading symbol of the operator Δ and according to the completeness of the generators R must have the form $\varphi_0 = \sigma_L(R)\epsilon$ for some ϵ. Substituting this form into Eq. (27) we obtain

$$\sigma_L(YX)\sigma_L(R)\epsilon = \sigma_L(YL)\epsilon = 0. \tag{30}$$

Therefore, by taking into account the non-singularity of $\sigma_L(YL)$, it follows that $\epsilon = \varphi_0 = 0$, and hence the leading symbol of the operator F has no zero-modes, i.e., it is non-degenerate.

Thus, the operators L and F both have non-degenerate leading symbols. In quantum field theory the operator X is called the gauge-fixing operator, F the gauge-field operator, the operator L the (Faddeev–Popov) ghost operator and the operator Y in Case II the third (or Nielsen–Kallosh) ghost operator. The most convenient and the most important case is that in which, by suitable choice of the parameters, it turns out to be possible to make both operators F and L either of Laplace type or of Dirac type. The one-loop effective action for gauge fields is given by the functional superdeterminants of the gauge-field operator F and the ghost operators L and Y:

$$\Gamma = \frac{1}{2}\log(\mathrm{Sdet}\, F) - \log(\mathrm{Sdet}\, L) - \frac{1}{2}\log(\mathrm{Sdet}\, Y). \tag{31}$$

We now focus on bosonic fields, when Δ is a second-order formally self-adjoint operator. The gauge invariance identity (14) means, in particular, that

$$\sigma_L(\Delta)\sigma_L(R) = 0. \tag{32}$$

Now we assume that both the operators $L = \overline{R}R$ and $F = \Delta + R\overline{R}$ are of Laplace type, i.e.,

$$\sigma_L(\overline{R}R) = |\xi|^2 I_G, \tag{33}$$

$$\sigma_L(F) = \sigma_L(\Delta) + \sigma_L(R\overline{R}) = |\xi|^2 I_V. \tag{34}$$

On manifolds with boundary one has to impose boundary conditions to make these operators self-adjoint and elliptic. They read

$$B_L\psi(\epsilon) = 0, \tag{35}$$

$$B_F\psi(\varphi) = 0, \tag{36}$$

where $\psi(\epsilon)$ and $\psi(\varphi)$ are the boundary data (see Appendix) for the bundles G and V, respectively, and B_L and B_F are the corresponding boundary operators. In gauge theories one tries to choose the boundary operators B_L and B_F in a gauge-invariant way so that the condition

$$B_F\psi(R\epsilon) = 0 \tag{37}$$

is satisfied identically for any ϵ subject to the boundary conditions (35). This means that the boundary operators B_L and B_F satisfy the identity

$$B_F[\psi, R](I_G - B_L) \equiv 0, \tag{38}$$

where $[\psi, R] \equiv \psi R - R\psi$.

We will see that this requirement fixes completely the form of the as-yet-unknown boundary operator B_L. Indeed, the most natural way to satisfy the condition of gauge invariance is as follows. We decompose the cotangent bundle $T^*(M)$ in such a way that $\xi = (N, \zeta) \in T^*(M)$, where N is the inward-pointing unit normal to the boundary and $\zeta \in T^*(\partial M)$ is a cotangent vector on the boundary. Consider the restriction W_0 of the vector bundle V to the boundary. We define restrictions of the leading symbols of the operators R and Δ to the boundary, i.e.,

$$\Pi \equiv \sigma_L(\Delta; N)\Big|_{\partial M}, \tag{39}$$

$$\nu \equiv \sigma_L(R; N)\Big|_{\partial M}, \tag{40}$$

$$\mu \equiv \sigma_L(R; \zeta)\Big|_{\partial M}. \tag{41}$$

From Eq. (32) we thus have the identity

$$\Pi\nu = 0, \tag{42}$$

Moreover, from (33) and (34) we also have

$$\overline{\nu}\nu = I_G, \tag{43}$$

$$\overline{\nu}\mu + \overline{\mu}\nu = 0, \tag{44}$$

$$\overline{\mu}\mu = |\zeta|^2 I_G, \tag{45}$$

$$\Pi = I_V - \nu\overline{\nu}. \tag{46}$$

From (42) and (43) we find that $\Pi : W_0 \rightarrow W_0$ is a self-adjoint projector orthogonal to ν,

$$\Pi^2 = \Pi, \quad \Pi\nu = 0, \quad \overline{\Pi} = \Pi. \tag{47}$$

Then part of the boundary conditions for the operator F reads

$$\Pi\varphi\Big|_{\partial M} = 0. \tag{48}$$

The gauge transformation of this equation is

$$\Pi R\epsilon\Big|_{\partial M} = 0. \tag{49}$$

The normal derivative does not contribute to this equation and, hence, if Dirichlet boundary conditions are imposed on ϵ,

$$\epsilon\Big|_{\partial M} = 0, \tag{50}$$

Equation (49) is identically satisfied.

The easiest way to get the other part of the boundary conditions is just to set

$$\overline{R}\varphi\Big|_{\partial M} = 0. \tag{51}$$

Bearing in mind Eq. (5) we find that, under the gauge transformations (1), this is transformed into

$$L\epsilon\Big|_{\partial M} = 0. \tag{52}$$

If some ϵ is a zero-mode of the operator L, i.e., $\epsilon \in \operatorname{Ker} L$, this is identically zero. For all $\epsilon \notin \operatorname{Ker} L$ this identically vanishes when the Dirichlet boundary conditions (50) are imposed. In other words, the requirement of gauge invariance of the boundary conditions (36) determines in an almost unique way (up to zero-modes) that the ghost boundary operator B_L should be of Dirichlet type. In any case, Dirichlet boundary conditions for the operator L are *sufficient* to achieve gauge invariance of the boundary conditions for the operator F.

Since the operator $\overline{R}$ in the boundary conditions (51) is a first-order operator, the set of boundary conditions (48) and (51) follows the general scheme formulated in [1]. Separating the normal derivative in the operator $\overline{R}$ we find exactly the Gilkey–Smith boundary conditions [2] for operators of Laplace type:

$$\begin{pmatrix} \Pi & 0 \\ \Lambda & I_V - \Pi \end{pmatrix} \begin{pmatrix} [\varphi]_{\partial M} \\ [\varphi_{;N}]_{\partial M} \end{pmatrix} = 0, \tag{53}$$

where

$$\Lambda \equiv (I_V - \Pi)\left[\frac{1}{2}\left(\Gamma^i \widehat{\nabla}_i + \widehat{\nabla}_i \Gamma^i\right) + S\right](I_V - \Pi), \tag{54}$$

the matrices Γ^i having the form

$$\Gamma^i = -\nu\overline{\nu}\mu^i\overline{\nu}, \tag{55}$$

and S being an endomorphism. The matrices Γ^i are anti-self-adjoint, $\overline{\Gamma}^i = -\Gamma^i$, and satisfy the relations

$$\Pi\Gamma^i = \Gamma^i\Pi = 0. \tag{56}$$

Thus, one can now define the matrix

$$T \equiv \Gamma \cdot \zeta = -\nu\overline{\nu}\mu\overline{\nu}, \tag{57}$$

where $\mu \equiv \mu^j\zeta_j$, and study the condition of strong ellipticity $|\zeta| I_V - iT > 0$ obtained in [1]. Such a condition now reads

$$|\zeta| I_V - iT = |\zeta| I_V + i\nu\overline{\nu}\mu\overline{\nu} > 0. \tag{58}$$

Moreover, using Eqs. (55), (44) and (45), we evaluate

$$\Gamma^{(i}\Gamma^{j)} = -(I_V - \Pi)\mu^{(i}\overline{\mu}^{j)}(I_V - \Pi). \tag{59}$$

Therefore

$$T^2 = \Gamma^i\Gamma^j\zeta_i\zeta_j = -(I_V - \Pi)\mu\overline{\mu}(I_V - \Pi), \tag{60}$$

and

$$T^2 + |\zeta|^2 I_V = |\zeta|^2\Pi + (I_V - \Pi)[|\zeta|^2 I_V - \mu\overline{\mu}](I_V - \Pi). \tag{61}$$

Since, for non-vanishing ζ, the part proportional to Π is positive-definite, the condition of strong ellipticity for bosonic gauge theory means that

$$(I_V - \Pi)[|\zeta|^2 I - \mu\overline{\mu}](I_V - \Pi) > 0. \tag{62}$$

We have thus proved a theorem:

Theorem. *Let V and G be two vector bundles over a compact Riemannian manifold M with smooth boundary, such that* $\dim V > \dim G$. *Consider a bosonic gauge theory and let the first-order differential operator $R : C^\infty(G, M) \to C^\infty(V, M)$ be the generator of infinitesimal gauge transformations. Let $\Delta : C^\infty(V, M) \to C^\infty(V, M)$ be the gauge-invariant second-order differential operator of the linearized field equations. Let the operators $L \equiv \overline{R}R : C^\infty(G, M) \to C^\infty(G, M)$ and $F \equiv \Delta + R\overline{R}$ be of Laplace type and normalized by $\sigma_L(L) = |\xi|^2 I_G$ and $\sigma_L(F) = |\xi|^2 I_V$. Let $\sigma_L(R; N)\Big|_{\partial M} \equiv \nu$ and $\sigma_L(R; \zeta)\Big|_{\partial M} \equiv \mu$ be the restrictions of the leading symbol of the operator R to the boundary, N being the inward-pointing unit normal to the boundary and $\zeta \in T^*(\partial M)$ a cotangent vector, and $\Pi = I_V - \nu\overline{\nu}$.*

Then the generalized boundary-value problem (F, B_F) with the boundary operator B_F determined by the boundary conditions (48) and (51) is gauge-invariant provided that the ghost boundary operator B_L takes the Dirichlet form. Moreover, it is strongly elliptic with respect to the cone $\mathbf{C} - \mathbf{R}_+$ if and only if the matrix $[|\zeta| I_V + i\nu\overline{\nu}\mu\overline{\nu}]$ is positive definite. A sufficient condition for this is

$$(I_V - \Pi)[|\zeta|^2 I_V - \mu\overline{\mu}](I_V - \Pi) > 0. \tag{63}$$

In Euclidean quantum gravity, however, if a gauge-averaging functional of the de Donder type is chosen, with a gauge parameter such that the resulting operator on metric perturbations is of Laplace type, the boundary-value problem with boundary conditions (48) and (51) fails to be strongly elliptic [1]. Such a result raises deep interpretative issues, since it seems to imply that a Becchi–Rouet–Stora–Tyutin-invariant quantization of the gravitational field cannot be implemented on manifolds with boundary (cf. [3]). Two alternatives seem therefore to emerge:

(i) Use local boundary conditions which are not completely gauge-invariant because they do not involve tangential derivatives and hence preserve strong ellipticity. The corresponding quantum amplitudes should, however, be gauge-invariant [4].

(ii) Study non-local boundary conditions [5] and try to understand whether they are compatible with ellipticity, with self-adjointness and with the requirement of invariance under a suitable class of transformations.

It seems therefore that we are still at the very beginning of the process of understanding the interplay between the problems of Euclidean quantum gravity on the one hand, and the problems of global analysis on the other hand (different aspects of the same problem have been discussed in [6, 7]). Hopefully, new perspectives in fundamental physics and spectral geometry will be gained from such efforts.

Acknowledgments

The work of G. E. has been partially supported by PRIN97 'Sintesi'. Special thanks are due to Antonio Masiello for the warm hospitality in Monopoli, and to Emanuele Sorace for scientific advice.

Appendix

In the presence of a boundary for a Riemannian manifold M of dimension m, the local coordinates x^{μ}, $(\mu = 1, \ldots, m)$, are split into local coordinates $\widehat{x}^k$, $(k = 1, \ldots, m-1)$, on ∂M, and the geodesic distance to the boundary, r. Similarly, the coordinates ξ_{μ} on the cotangent bundle $T^*(M)$ are split into coordinates ζ_k on $T^*(\partial M)$, together with a real parameter ω. The notion of ellipticity in which we are interested then requires that the leading symbol $\sigma_L(F)$ of an operator F of Laplace type should be elliptic in the interior of M, and that a unique solution should exist of the ordinary differential equation:

$$\left[-\frac{d^2}{dr^2} + |\zeta|^2 - \lambda \right] \varphi(r) = 0, \tag{A.1}$$

subject to the boundary conditions and to a decay condition at infinity. A thorough formulation of boundary conditions demands some abstract thinking. For this purpose, one has to consider two vector bundles W_F and W'_F over the boundary of M, with a *boundary operator* B_F relating their sections, i.e.,

$$B_F : C^{\infty}(W_F, \partial M) \rightarrow C^{\infty}(W'_F, \partial M).$$

All information about normal derivatives of the fields is not encoded in B_F but in the *boundary data* $\psi_F(\varphi) \in C^{\infty}(W_F, \partial M)$. In our analysis one has

$$\psi_F(\varphi) = \begin{pmatrix} [\varphi]_{\partial M} \\ [\varphi_{;N}]_{\partial M} \end{pmatrix},$$

and the strong ellipticity condition demands that a unique solution of Eq. (A.1) should exist, subject to the boundary condition

$$\sigma_g(B_F)\psi_F(\varphi) = \psi'_F(\varphi) \qquad \text{for all } \psi'_F(\varphi) \in C^{\infty}(W'_F, \partial M), \tag{A.2}$$

and to the asymptotic condition

$$\lim_{r\to\infty} \varphi(r) = 0. \tag{A.3}$$

With standard notation, $\sigma_g(B_F)$ is the *graded leading symbol* of the boundary operator B_F. When the boundary conditions (53) are considered, one finds

$$\sigma_g(B_F) = \begin{pmatrix} \Pi & 0 \\ iT & I_V - \Pi \end{pmatrix}, \tag{A.4}$$

where T is the anti-self-adjoint matrix defined in (57).

When all the above conditions are satisfied for all $\zeta \in T^*(\partial M)$, for all $\lambda \in \mathbf{C}-\mathbf{R}_+$, for all $(\zeta, \lambda) \neq (0, 0)$, the boundary-value problem (F, B_F) is said to be strongly elliptic with respect to the cone $\mathbf{C} - \mathbf{R}_+$. This property is crucial for ensuring the existence of the L^2-trace, $\mathrm{Tr}_{L^2} \exp(-tF)$, which is frequently applied in quantum field theory and spectral geometry.

References

1. Avramidi I.G., Esposito G. (1999): Commun. Math. Phys. **200**, 495
2. Gilkey P.B., Smith L. (1983): J. Diff. Geom. **18**, 393
3. Moss I.G., Silva P.J. (1997): Phys. Rev. D **55**, 1072
4. Luckock H.C. (1991): J. Math. Phys. **32**, 1755
5. Esposito G. (1999): Class. Quantum Grav. **16**, 1113
6. Avramidi I.G., Esposito G. (1999): Heat-kernel asymptotics of the Gilkey–Smith boundary-value problem, in *Trends in Mathematical Physics, AMS/IP Studies in Advanced Mathematics*, ed. by V. Alexiades, G. Siopsis, American Mathematical Society, Vol. 13, Providence RI and International Press, Cambridge MA, pp. 15–34
7. Avramidi I.G., Esposito G. (1999): On ellipticity and gauge invariance in Euclidean quantum gravity, in *Trends in Mathematical Physics, AMS/IP Studies in Advanced Mathematics*, ed. by V. Alexiades, G. Siopsis, American Mathematical Society, Vol. 13, Providence RI and International Press, Cambridge MA, pp. 35–42

Maximal Acceleration Effects for Photons in Cavity Resonators

A. Feoli, G. Papini, G. Scarpetta

Abstract. Caianiello's model of relativistic dynamics with a maximal acceleration can, in principle, be tested by using photons in resonant cavities. The specific effect which is derived amounts to a shift of the cavity's resonant frequency.

The possibility that the proper acceleration of massive particles has an upper limit was first discussed by Caianiello [1] in the context of a geometrical unification of Quantum Mechanics and General Relativity in phase space [2]. Classical and quantum arguments supporting this possibility have also been adduced by several authors on different grounds [3]. The value of the maximal acceleration $A_m = 2mc^3/\hbar$ can be derived from quantum mechanical considerations [4] in which case m is the rest mass of the particle, or from the existence of a maximal temperature [5] through the Unruh relation [6]. In the latter case m is the Planck mass and the maximal acceleration becomes a universal constant. The extremely large value that A_m takes for all known particles makes a direct test very difficult. Nonetheless we discuss below a realistic experiment to detect the existence of a maximal acceleration using photons in a cylindrical cavity resonator of radius R and length l [7]. Our basic assumptions are:

1. the maximal acceleration depends on the rest mass of the particle and is given by A_m;
2. the model used to study the relativistic dynamics of a particle with maximal acceleration is the one originally proposed by Caianiello [8].

In Caianiello's approach, a particle of mass m accelerating along its worldline, in the absence of gravity, behaves dynamically as if embedded in a spacetime of metric

$$d\tilde{s}^2 = ds^2\left(1 - \frac{c^4|\ddot{x}|^2}{A_m^2}\right), \tag{1}$$

where ds^2 refers here to Minkowski space and $|\ddot{x}|^2 = |\eta_{\mu\nu}\ddot{x}^\mu\ddot{x}^\nu|$ is the squared length of its acceleration four-vector, also calculated in Minkowski space. This model has been successfully used, for instance, to solve the problems of the standard cosmological model [9], to smooth ultraviolet divergences in quantum field theory [10] and to improve the agreement between the predictions of quantum electrodynamics and the experimental data in the case of the Lamb shift for H, D and HE^+ [11].

As long as the photons are confined to the interior of the cylindrical cavity resonator (in this case the wave number is fixed by $k = p\pi/l$ with p null or a positive integer), they behave mechanically as massive particles because they acquire an effective mass. If the dielectric is vacuum, the spectrum of resonant frequencies [12] is

$$\frac{\omega_{mnp}^2}{c^2} = \frac{x_{mn}^2}{R^2} + \frac{p^2\pi^2}{l^2}, \tag{2}$$

where x_{mn} is the n-th root of the equation $J_m(x) = 0$. It follows from (3) that the effective mass for photons in the (m, n)-mode is given by $m_{\text{eff}} = \hbar x_{mn}/cR$. The lowest frequency TE-mode ($m = n = p = 1$) is represented by

$$\omega_0 \equiv \omega_{111} = 1.841\frac{c}{R}\left(1 + 2.912\frac{R^2}{l^2}\right)^{1/2} \tag{3}$$

and becomes the fundamental mode of the cavity for $l/R > 2.03$ [12]. In this case $m_{\text{eff}} = 1.841\hbar/cR$, while the momentum is $\hbar\pi/l$ because of the identity $(\pi/1.841)^2 = 2.912$. Purely indicative estimates can be obtained, e.g., for $R \simeq 10$ cm and $l \simeq 50$ cm. One then finds $m_{\text{eff}} \sim 6.5 \times 10^{-39} g$ and $A_m \sim 3.3 \times 10^{20}$ cm/s^2.

The application of Caianiello's model to the photons in the cavity requires that the two types of mass, real and effective, behave in entirely similar ways inertially and gravitationally. This is warranted, theoretically, by the fact that the solution of the Proca–Yukawa equation in the presence of inertial and gravitational fields does not distinguish between the two types of mass to first order in the metric deviation. To this order, which is certainly sufficient here, inertial and gravitational fields contribute only a phase factor to the wave function [13] which is then largely determined by the solution of the field-free equation. The equivalent behaviour of effective and real masses in the case of photons has also been confirmed experimentally by using rotating Fabry–Perot cavities [14].

The metric deviation $h_{\mu\nu} = -\left(\frac{\hbar|\ddot{x}|}{2cm_{\text{eff}}}\right)^2 \eta_{\mu\nu}$ must now be evaluated. The change in momentum for each reflection from the cavity walls results in an acceleration

$$c^2|\ddot{x}| \simeq \frac{\gamma}{m_{\text{eff}}}\frac{\Delta p}{\Delta t} = \frac{2\pi\hbar\gamma}{l m_{\text{eff}}\Delta t}, \tag{4}$$

where $m_{\text{eff}} v\gamma = \hbar\pi/l$ and γ is the usual Lorentz factor.

We obtain

$$h_{\mu\nu} = -\left(\frac{\pi R^3\omega_0}{(1.841)^3 l c^2 \Delta t}\right)^2 \eta_{\mu\nu}. \tag{5}$$

Equation (5) produces a change in the phase of the photon wave function which, on neglecting the spin, is seen to be given [15] by

$$\begin{aligned}\Delta\chi\phi_0 = &-\frac{1}{4}\oint dz^\lambda \left(\Gamma_{\alpha\lambda,\beta}(z) - \Gamma_{\beta,\lambda\alpha}(z)\right)\left[J^{\alpha\beta}(z), \phi_0(x)\right] \\ &+ \frac{1}{2}\oint dz^\lambda h_{\lambda\beta}\left[P^\beta, \phi_0(x)\right]\end{aligned} \tag{6}$$

where

$$\left[J^{\alpha\beta}(z), \phi_0(x)\right] \equiv i\left[(x^\alpha - z^\alpha)\partial^\beta\phi_0(x) - (x^\beta - z^\beta)\partial^\alpha\phi_0(x)\right],$$
$$\left[P^\beta, \phi_0(x)\right] \equiv i\partial^\beta\phi_0(x), \quad \Gamma_{\alpha,\lambda\beta} \equiv \frac{1}{2}\left(h_{\alpha\lambda,\beta} + h_{\alpha\beta,\lambda} - h_{\lambda\beta,\alpha}\right)$$

and $\phi_0(x)$ satisfies the free Klein–Gordon equation for a particle of mass m_{eff}. The phase change (6) must be evaluated along the closed path represented by the extremes of the cavity and the initial and final proper times. The boundary conditions require that ϕ_0 vanish at $x = 0$ and $x = l$; hence $\phi_0 \propto \sin\frac{\pi x}{l}$. If a reflection now takes place along the x-axis in a time Δt, from (5) and (6) we find that $\Delta\chi\phi_0 = \sin\frac{\pi x}{l} - h_{11}\frac{\pi}{l}v\Delta t\cos\frac{\pi x}{l} \sim \sin(\frac{\pi x}{l} - h_{11}\frac{\pi}{l}v\Delta t)$. The length of the cavity has therefore changed effectively from l to $l' = l + h_{11}v\Delta t$ and the conditions for the existence of stationary waves now require that $\sin(k'x) = 0$ at $x = 0$ and $x = l'$, or $k' = \frac{\pi p}{l'}$. This shift in wave number leads to the new fundamental mode frequency

$$\frac{\omega_0'^2}{c^2} \equiv \frac{\omega_{111}'^2}{c^2} = \frac{x_{11}^2}{R^2} + \frac{\pi^2}{l'^2} \simeq \frac{x_{11}^2}{R^2} + \frac{\pi^2}{l^2}\left(1 - 2h_{11}\frac{v\Delta t}{l}\right)$$

to first order in $h_{\mu\nu}$. Since the cavity is only three half-waveguide-lengths, a photon returns to its original position after three collisions with the walls. The process can be iterated ideally a number N of times. The resulting frequency shift is therefore

$$\Delta\nu = \frac{\omega_0' - \omega_0}{2\pi} \simeq -\frac{h_{11}Nv\Delta t\pi c^2}{2l^3\omega_0} \simeq \frac{\pi^4 N}{2\Delta t}\left(\frac{R}{1.841l}\right)^6 \tag{7}$$

which is the effect sought. In the case of perfectly reflecting walls one would choose $\Delta t \sim \omega_0^{-1}$ in (7). This is not the case in practice because losses take place in the cavity walls and in the coupling of the cavity to an EM detector. If $Q_T = \frac{Q_e Q_w}{Q_e + Q_w}$ is the total quality factor of the cavity and Q_e and Q_w those of the EM coupling and the walls respectively, then the losses broaden the resonance line whose width becomes $\Delta\omega = \frac{\omega_0}{Q_T}$ around the center frequency $\widetilde{\nu}_0 = \nu_0 - \frac{\nu_0}{2Q_T}$. The effect of (1) on $\widetilde{\nu}_0$ can be calculated readily from (7) as $\widetilde{\nu}_0' = \widetilde{\nu}_0 + \Delta\nu(1 - \frac{1}{2Q_T}) \simeq \widetilde{\nu}_0 + \Delta\nu$ for $Q_T \gg \frac{1}{2}$, while $\Delta t \sim \frac{Q_T}{\omega_0}$ is the typical time over which changes occur in the cavity. For the values of R and l selected we find that $\Delta\nu \sim 8 \times 10^{-5}\frac{N}{\Delta t}$ Hz.

When energy absorption takes place at the walls, one must also consider the corresponding effects in (5). To leading order, however, these amount to $\omega_0^2\epsilon^2/Q_T^2$ where ϵ is the energy loss at the walls, and are negligible for high values of Q_T. From $\widetilde{\nu}_0' \simeq \widetilde{\nu}_0 + \Delta\nu$, Eq. (7) and the fact that, over the observation time Q_e/ω_0, the photons are reflected a number of times $4N - 1 \simeq vQ_e/\omega_0 l$, we find that

$$\widetilde{\nu}_0' \simeq \widetilde{\nu}_0 + \frac{\pi^4\omega_0}{8Q_T}\left(\frac{R}{1.841l}\right)^6\left[1 + \pi Q_e\left(\frac{c}{l\omega_0}\right)^2\right], \tag{8}$$

while the condition $N \geq 1$ yields $Q_e \geq \frac{\omega_0^2}{\pi}\frac{l^2}{c^2}$. For the values of l and R selected we find that $Q_e \geq 30$ and

$$\widetilde{\nu}_0 \simeq \nu_0 - \frac{\nu_0}{4Q_T}[1 - 5 \times 10^{-4}(1 + 3.3 \times 10^{-2} Q_e)]. \tag{9}$$

The maximal acceleration correction in (1), represented by the second term on the r.h.s. of (8) or the second term in brackets in (9), appears, therefore, as a small decrease in the peak frequency shift of the wave packet. This shift increases for a generic (m, n, p)-mode because it depends strongly on p. On the contrary it disappears for $Q_e \sim 6.1 \times 10^4$. Relatively smaller values of Q_T seem to facilitate observation of the effect, while a high Q_e appears desirable. Very stable, low-loss superconducting cavities can be built to operate at 10^{10} Hz with $Q_T \sim 10^6$. Equation (9) yields in this case a frequency shift of the order of 10^{-10}. This is still two orders of magnitude below the resolution power of present low-noise electronics (see footnote on p. 560 of [16]). The value of $\Delta\nu$ could however be enhanced by 2π positive feedback, or by increasing the length of the observation time, if stability allows it.

References

1. Caianiello E.R. (1981): Lett. Nuovo Cimento **32**, 65;
   Caianiello E.R., De Filippo S., Marmo G., Vilasi G. (1982): Lett. Nuovo Cimento **34**, 112
2. Caianiello E.R. (1980): Il Nuovo Cimento B **59**, 350;
   Scarpetta G. (1984): Lett. Nuovo Cimento **41**, 51;
   Caianiello E.R., Scarpetta G., Marmo G. (1985): Il Nuovo Cimento A **86**, 337;
   Caianiello E.R. (1992): La Rivista del Nuovo Cimento **15** (4)
3. Caldirola P. (1981): Lett. Nuovo Cimento **32**, 264;
   Toller M. (1988): Nuovo Cimento B **102**, 261;
   Wood W.R., Papini G., Cai Y.Q. (1989): Nuovo Cimento B **104**, 653;
   Brandt H.E. (1989): Found. Phys. Lett. **2**, 39;
   Mashhoon B. (1990): Physics Letters A **143**, 176;
   Frolov V.P., Sanchez N. (1991): Nucl. Phys. B **349**, 815;
   Gasperini M. (1992): Gen. Rel. Grav. **24**, 219;
   Papini G., Wood W.R. (1992): Phys. Lett. A **170**, 409;
   Sanchez N. (1993): Conceptual aspects of string theory and the problem of maximal acceleration, in *Structure: from Physics to General Systems*, ed. by M. Marinaro, G. Scarpetta, Vol. 1, World Scientific, Singapore, p. 118
4. Caianiello E.R. (1984): Lett. Nuovo Cimento. **41**, 370;
   Wood W.R., Papini G., Cai Y.Q. (1989): Nuovo Cimento B **104**, 361 and (errata corrige) 727
5. Sakharov A.D. (1966): JETP Lett. **3**, 288
6. Brandt H.E. (1983): Lett. Nuovo Cimento **38**, 522
7. Papini G., Feoli A., Scarpetta G. (1995): Phys. Lett. A **50**
8. Caianiello E.R., Gasperini M., Predazzi E., Scarpetta G. (1988): Phys. Lett. A **132**, 82;
   Caianiello E.R., Feoli A., Gasperini M., Scarpetta G. (1990): Int. J. Theor. Phys. **29**, 131;
   Feoli A. (1993): Nucl. Phys. B **396**, 261

9. Gasperini M. (1987): Astrophys. Space Sci. **138**, 387;
Caianiello E.R., Gasperini M., Scarpetta G. (1991): Classical and Quantum Gravity **8**, 659;
Caianiello E.R., Feoli A., Gasperini M., Scarpetta G. (1999): Cosmological implications of maximal proper acceleration, in *Bogoliubovskie Ctenia, (Joint Institute Nuclear Research, Dubna, Russia 1994)*, p. 134;
Capozziello S., Lambiase G., Scarpetta G. (1999): Nuovo Cimento B **114**, 93
10. Nesterenko V.V., Feoli A., Lambiase G., Scarpetta G. (1999): Regularizing property of the maximal acceleration principle in quantum field theory. Phys. Rev. D **60**, 65001
11. Lambiase G., Papini G., Scarpetta G. (1997): Il Nuovo Cimento B **112**, 1003; (1998): Phys. Lett. A **244**, 349
12. Jackson J.D. (1975): Classical Electrodynamics. 2nd ed., Wiley, New York, p. 353
13. Cai Y.Q., Papini G. (1991): Phys. Rev. Lett. **66**, 1259; (1992): **68**, 3811
14. Heer C.V., Little J.A., Bupp J.R. (1968): Ann. Inst. Henri Poincare' A **8**, 311
15. Cai Y.Q., Papini G. (1989): Class. Quantum Gravity **6**, 407; (1989): Modern Phys. Lett. A **4**, 1143; (1990): Class. Quantum Gravity **7**, 269; (1990): Gen. Rel. Grav. **22**, 259;
Papini G. (1990): in: *Quantum Mechanics in Curved Space-Time*, ed. by J. Audretsch, V. de Sabbata, Plenum Press, New York, p. 473
16. Blair D.G., McClelland D.E., Bachor H.A., Sandeman R.J. (1991): Gravitational Wave Detectors, in *The Detection of Gravitational Waves*, ed. by D.G. Blair, Cambridge University Press, Cambridge

Review of Four-Dimensional Supersymmetric AdS Black Holes

D. Klemm

Abstract. We review four-dimensional asymptotically anti-de Sitter (AdS) dyonic black holes preserving some amount of supersymmetry. The implications of the spacetime topology on the supersymmetry properties are studied. In particular, we show that, for the spherical Kerr–Newman-AdS black hole to admit Killing spinors, it must carry nonzero angular momentum and electric charge only.

Black holes in anti-de Sitter space represent a subject of current interest, which is based, on the one hand, on Maldacena's conjecture of AdS/Conformal field theory-correspondence [1–4], and, on the other hand, on the fact that AdS space admits black holes with unusual topologies [5–10]. These so-called topological black holes have some intriguing properties [11]. Among them are the uncommon behaviour of the luminosity, another version of the information-loss paradox due to the boundary conditions necessary in an asymptotically AdS space, and a mass spectrum that seems to be difficult to reconcile with string theory. In fact, the argument of Horowitz-Polchinski [12], which describes the transition from a highly excited string state to a black hole, and provides a microscopic interpretation of black hole entropy, does not seem to work for topological black holes, at least not in a naive manner [11]. The reason for this is that the mass levels of a string in AdS space are $M \approx \ell^{-1}n$ for large quantum numbers n [13], ℓ being related to the cosmological constant via $\Lambda = -3/\ell^2$. This yields a black hole entropy proportional to n, whereas the string entropy is like $\sqrt{n}$ (cf. [11] for details). So it would seem that a string does not have enough degrees of freedom to account for black hole entropy. However, the correspondence principle of Horowitz–Polchinski, which has turned out to be very successful up to now, can not be rejected on the above, rather naive, arguments; for example, there are other mass spectrum regimes for a string in AdS (for $\ell/l_s \gg 1$, where l_s is the string length, the mass spectrum is as it is in flat space), or there could be a transition from a configuration of D-branes to a black hole. In short, it remains to be seen how exactly the argument of Horowitz–Polchinski works for topological black holes, and it seems quite challenging to try to give a microscopic description of the entropy of these objects within string theory, e.g., by using D-brane techniques. A first step in this enterprise is to find supersymmetric configurations, as, for BPS states, we know that the degeneracy at weak string coupling constant g_s does not change if one increases g_s.

Supersymmetry of AdS black holes with spherical event horizons have been studied before in the literature. Romans [14] showed that the Reissner-Nordström-AdS black hole is supersymmetric in two cases. The first appears for $q_m = 0$ and $q_e^2 = m^2$;

q_m, q_e and m being the magnetic charge, electric charge and mass parameter, respectively. The second emerges for $m = 0$ and $q_m = \pm\ell/2$ (we recall that $\Lambda = -3/\ell^2$ is the cosmological constant). However, all these supersymmetric configurations represent naked singularities. The situation is similar to the asymptotically flat Kerr–Newman black hole, which reaches the extreme limit $m^2 = a^2 + q_e^2$ (a denoting the rotation parameter) before the supersymmetry condition $m^2 = q_e^2$ is satisfied [15].

As far as the spherical Kerr–Newman-AdS solution is concerned, it was shown by Kostelecký and Perry [16] that, only for nonvanishing rotation parameter a, is it possible to obtain supersymmetric extremal black holes. This means that, in an AdS background, rotating black holes are the analogue of the asymptotically flat extreme Reissner-Nordström solution. This result was obtained by considering the Bogomol'nyi bound arising from the supersymmetry algebra osp(4|2) of $N = 2$, $d = 4$ AdS supergravity. It is given by $m = \sqrt{q_e^2 + q_m^2}(1 \pm a/\ell)$. In the present review we will show, that in this equation, the magnetic charge must be zero, thus making the Kostelecký-Perry result more precise. Besides, we will see that there also exists a supersymmetric configuration with vanishing mass parameter, but nonvanishing magnetic charge. This represents a naked singularity, and it was not obtained in [16].

Apart from a consideration of the Kerr–Newman-AdS black hole, in our review we shall also present the implications of other-than-spherical-event-horizon topology on the BPS properties of AdS black holes.

A natural candidate to address the issue of supersymmetry of topological AdS black holes is the model used in [16], i.e., $N = 2$, $d = 4$ gauged supergravity [17, 18]. In this theory, the rigid SO(2) symmetry, rotating the two independent Majorana supersymmetries present in the ungauged theory, is made local by introduction of a minimal gauge coupling between the graviphotons and the gravitinos. Local supersymmetry then requires a negative cosmological constant and a gravitino mass term. The theory has four bosonic and four fermionic degrees of freedom; it describes a graviton e_m^a, two real gravitinos ψ_m^i ($i = 1, 2$), and a Maxwell gauge field A_m. As mentioned above, the latter is minimally coupled to the gravitinos, with coupling constant ℓ^{-1}. In the following, we shall combine the two ψ_m^i as a single complex spinor $\psi_m = \psi_m^1 + i\psi_m^2$. The bosonic Lagrangian reads [17, 18] [1]

$$e^{-1}L = -\frac{1}{4}R + \frac{1}{4}F_{mn}F^{mn} - \frac{3}{2\ell^2}, \tag{1}$$

which is just the Lagrangian of Einstein-Maxwell gravity with negative cosmological constant $\Lambda = -3\ell^{-2}$. The amount of supersymmetry preserved by a purely bosonic background is determined by the dimension of the solution space of the Killing spinor equation

$$\widehat{\nabla}_m \epsilon = 0, \tag{2}$$

where $\widehat{\nabla}_m$ is the supercovariant derivative defined by

$$\widehat{\nabla}_m = \partial_m + \frac{1}{4}\omega_m{}^{ab}\gamma_{ab} - \frac{i}{\ell}A_m + \frac{1}{2\ell}\gamma_m + \frac{i}{4}F_{ab}\gamma^{ab}\gamma_m. \tag{3}$$

[1] We use the notation of [19].

The integrability conditions for (2) read

$$\widehat{R}_{mn}\epsilon = 0, \tag{4}$$

where

$$\widehat{R}_{mn} = \left[\widehat{\nabla}_m, \widehat{\nabla}_n\right] \tag{5}$$

is the supercurvature. (4) is necessary, but not sufficient, for the existence of Killing spinors. It guarantees that they exist locally, but globally there may exist no Killing spinor due to topological reasons.

We first study (2) and (4) for static topological black hole spacetimes

$$ds^2 = -V(r)dt^2 + V^{-1}(r)dr^2 + r^2 d\sigma^2, \tag{6}$$
$$F = q_e *\epsilon + q_m\epsilon, \tag{7}$$

where $d\sigma^2$ is the metric of a two-manifold Σ with volume form ϵ, while q_e and q_m denote the electric and magnetic charge parameters respectively. The equations of motion following from (1) require Σ to be a surface of constant curvature κ, and

$$V(r) = \kappa - \frac{2\eta}{r} + \frac{z^2}{r^2} + \frac{r^2}{\ell^2}, \quad z^2 = q_e^2 + q_m^2. \tag{8}$$

Without loss of generality, we can restrict ourselves to the cases $\kappa = 0, \pm 1$. If we assume Σ to be compact and orientable, it is diffeomorphic to the two-sphere S^2 for $\kappa = 1$, to a flat torus $\mathbb{R}^2/\mathbb{Z}^2$ for $\kappa = 0$, or to a Riemann surface of genus $g > 1$ for $\kappa = -1$. The topology of four-dimensional spacetime is $\mathbb{R}^2 \times \Sigma$. Black holes with metric (6) have an event horizon at the largest root r_+ of $V(r) = 0$. For $\eta = z = 0$, the manifold (6) is locally AdS; for $\kappa = 1$ we obtain AdS space, for $\kappa = 0$ a quotient of AdS with a naked singularity, and finally, for $\kappa = -1$, we obtain a quotient of AdS space with a black hole interpretation [5], a four-dimensional analogue of the Bañados-Teitelboim-Zanelli (BTZ) black hole [20].

For the metric (6) and the field strength (7), the supercurvature (5) can be written as a product

$$\widehat{R}_{mn} = \mathcal{P}\mathcal{G}_{mn}(r)\mathcal{O}, \tag{9}$$

where $\mathcal{G}_{mn}(r)$ is γ_{mn} times a scalar-valued function of r,

$$\mathcal{P} \equiv \frac{r^2}{2z} i F_{ab}\gamma^{ab}\gamma_1 \tag{10}$$

is an idempotent (and hence non-singular) operator, and $\mathcal{O}$ is given by

$$\mathcal{O} = \sqrt{V(r)} + \frac{r}{\ell}\gamma_1 + \left(\frac{z}{r} - \frac{\eta}{z}\right)\mathcal{P}. \tag{11}$$

As $\mathcal{G}_{mn}(r)$ and $\mathcal{P}$ are non-singular for $z \neq 0$ (the case $z = 0$ has to be considered separately), the integrability conditions for Killing spinors are equivalent to the vanishing of det $\mathcal{O}$. We obtain for the determinant

$$\det \mathcal{O} = \left\{ \delta_{g,1} - 1 - \frac{2q_m}{\ell} - \frac{(\eta^2 - 2q_m\eta r\ell^{-1})}{z^2} \right\} \times \left\{ \delta_{g,1} - 1 + \frac{2q_m}{\ell} - \frac{(\eta^2 + 2q_m\eta r\ell^{-1})}{z^2} \right\}. \tag{12}$$

Now det $\mathcal{O}$ is a function of r, and, for supersymmetric configurations, this function must vanish identically. For genus $g = 1$, this is satisfied when

$$q_m = 0 = \eta, \quad \Rightarrow \quad V(r) = \frac{r^2}{\ell^2} + \frac{q_e^2}{r^2}. \tag{13}$$

We observe that the lapse function is always positive in that case, i.e., the corresponding spacetimes represent naked singularities. This result is very similar to the Reissner-Nordström-AdS black hole considered in [14], where the supersymmetric configurations are also naked singularities.

For genus $g > 1$, (6) vanishes for $\eta = 0$ and $q_m = \pm\ell/2$. This yields

$$V(r) = \left(\frac{r}{\ell} - \frac{\ell}{2r}\right)^2 + \frac{q_e^2}{r^2}. \tag{14}$$

For vanishing electric charge, spacetime (6), with $V(r)$ given by (14), describes an extremal black hole. Thus, unlike in the case of spherical or toroidal event horizons, we now can hope to get a supersymmetric static extremal black hole. Clearly, this is not obvious, because it could happen that the Killing spinors (which exist locally, as (4) is satisfied), are not compatible with the identifications which must be carried out to get a compact event horizon. However, we shall see below that the Killing spinors depend only on the radial coordinate r, and consequently they do respect the identifications performed on Σ.

The extremal black hole found above is a solitonic object in the sense that the limit $\ell^{-1} \to 0$ does not exist (we recall that ℓ^{-1} is the coupling constant of the gauged theory, coupling the graviphoton to the gravitinos).

For genus $g > 1$, there is yet another case in which the integrability conditions (4) are satisfied, namely, for $\eta = 0 = q_m = q_e$. The spacetime is then a quotient space of AdS, and therefore admits Killing spinors locally. However, we will see that they do not exist globally, as the above-mentioned identifications are not respected. Thus the corresponding black hole is not supersymmetric.

Using the integrability condition $\mathcal{O}\epsilon = 0$, the Killing spinor equations (2) can be solved straightforwardly in the various cases found above (for details cf. [19]). As we saw, for genus $g = 1$, we only obtain naked singularities, in which we have no further interest. For genus $g > 1$, $V(r) = (r\ell^{-1} - \ell(2r)^{-1})^2 + q_e^2 r^{-2}$, we have a dyonic naked singularity if $q_e \neq 0$; for $q_e = 0$, however, we get an extremal magnetically charged black hole (a magnetic monopole hidden by an event horizon having the topology of a Riemann surface). The Killing spinors read

$$\epsilon(r) = \left(\sqrt{\sqrt{V(r)} + \frac{r}{\ell} - \frac{\ell}{2r}} - \sqrt{\sqrt{V(r)} - \frac{r}{\ell} + \frac{\ell}{2r}} i\gamma_0\right) P(-\gamma_1)P(i\gamma_{23})\epsilon_0, \tag{15}$$

where ϵ_0 is a constant spinor, and where we have defined the projection operator $P(\mathcal{L}) = (1 + \mathcal{L})/2$ for an idempotent $\mathcal{L}$. We see that ϵ_0 is subject to a double projection, so the solution space is one-(complex)-dimensional. The Killing spinor does not depend on the coordinates on Σ; thus it respects the identifications we have made on Σ to obtain a Riemann surface. Hence, for zero electric charge, we have obtained a supersymmetric extremal static black hole. This was not possible for spherical event horizons, i.e., for the Reissner-Nordström-AdS black hole, where all supersymmetric configurations were naked singularities [14]. So we see that admitting other spacetime topologies changes the supersymmetry properties.

For genus $g > 1$, $V(r) = -1 + r^2\ell^{-2}$, we have a quotient space of AdS describing an uncharged black hole. Without identifications on Σ, the spacetime is simply AdS viewed by a uniformly accelerated observer, the (noncompact) horizon being its acceleration horizon [10]. Only the compactification of the surfaces of constant r and t makes the spacetime become a black hole, with the singularity at $r = 0$ being a causal one, i.e., the manifold cannot be continued beyond this singularity, otherwise one would have closed timelike curves [5]. It is clear that locally this spacetime admits as many Killing spinors as AdS, but we have to check if they respect the identifications. One finds that the Killing spinors depend on the coordinates on Σ, so that they are not invariant under the transformations of the discrete group used in the identifications, and the black hole is not supersymmetric. Clearly this was to be expected, as a supersymmetric black hole necessarily must have zero temperature (note, however, that the converse is not true in general), whereas the hole considered above has nonvanishing Hawking temperature $T = 1/2\pi\ell$.

We now extend our discussion to several rotating generalizations of the static black holes considered above. As the corresponding metrics are rather complicated, the investigation of supersymmetry properties of these spacetimes is quite a formidable task. However, as we shall see below, it is still possible to solve the integrability conditions explicitly, which yields some interesting results.

First, we consider the spherical Kerr–Newman-AdS black hole with metric

$$\begin{aligned} ds^2 = &-\frac{\Delta_r}{\Xi^2\varrho^2}\left[dt - a\sin^2\theta d\phi\right]^2 + \frac{\varrho^2}{\Delta_r}dr^2 + \frac{\varrho^2}{\Delta_\theta}d\theta^2 \\ &+ \frac{\Delta_\theta \sin^2\theta}{\Xi^2\varrho^2}\left[a dt - (r^2 + a^2)d\phi\right]^2, \end{aligned} \tag{16}$$

where

$$\varrho^2 = r^2 + a^2\cos^2\theta, \quad \Xi = 1 - \frac{a^2}{\ell^2}, \tag{17}$$

$$\Delta_r = (r^2 + a^2)\left(1 + \frac{r^2}{\ell^2}\right) - 2mr + z^2, \quad \Delta_\theta = 1 - \frac{a^2}{\ell^2}\cos^2\theta. \tag{18}$$

Here a denotes the rotational and m the mass parameter. Furthermore we recall that $z^2 = q_e^2 + q_m^2$. The gauge field is given by

$$A = -\frac{q_e r}{\Xi\varrho^2}\left[dt - a\sin^2\theta d\phi\right] - \frac{q_m\cos\theta}{\Xi\varrho^2}\left[adt - (r^2 + a^2)d\phi\right]. \tag{19}$$

Let us define the critical mass parameter m_{extr},

$$\begin{aligned} m_{\text{extr}}(a, z) = {} & \frac{\ell}{3\sqrt{6}}\left(\sqrt{\left(1 + \frac{a^2}{\ell^2}\right)^2 + \frac{12}{\ell^2}(a^2 + z^2)} + \frac{2a^2}{\ell^2} + 2\right) \\ & \times \left(\sqrt{\left(1 + \frac{a^2}{\ell^2}\right)^2 + \frac{12}{\ell^2}(a^2 + z^2)} - \frac{a^2}{\ell^2} - 1\right)^{\frac{1}{2}}. \end{aligned} \tag{20}$$

A study of the positive zeroes of the lapse function Δ_r shows that the metric (16) describes a naked singularity for $m < m_{\text{extr}}$ and a black hole with an outer event horizon and an inner Cauchy horizon for $m > m_{\text{extr}}$. Finally, for $m = m_{\text{extr}}$, the lapse function has a double root and (16) represents an extremal black hole.

The Killing spinor integrability conditions (4) are satisfied in two cases. The first is

$$m = 0, \quad q_m^2 = \frac{\ell^2}{4}\left(1 - \frac{a^2}{\ell^2}\right)^2, \tag{21}$$

which describes supersymmetric naked singularities, as can be seen from the function

$$\Delta_r = \left[\frac{r^2}{\ell} + \frac{\ell}{2}\left(1 + \frac{a^2}{\ell^2}\right)\right]^2 + q_e^2. \tag{22}$$

The second case emerges for

$$q_m = 0, \quad m^2 = \left(1 \pm \frac{a}{\ell}\right)^2 q_e^2, \tag{23}$$

in which we have electrically charged possibly supersymmetric configurations. In the limit case $a = 0$ we recover the usual condition $m^2 = q_e^2$.

Now one may wonder why the constraint (23) on the electric charge cannot be rotated into a similar constraint on the magnetic charge or a combination of the two by an electromagnetic duality transformation. The reason for this is that the minimal coupling between the graviphoton and the gravitinos in the gauged supergravity theory breaks duality invariance. The bosonic action remains invariant upon gauging, but the Killing spinor equations do not.

Inserting (23) into the extremality condition (20), we obtain the relation

$$m^2 = \ell a \left(1 + \frac{a}{\ell}\right)^4 ; \tag{24}$$

the configurations that satisfy the latter equation are hence possibly supersymmetric extremal black holes, which carry electric charge and rotate. Expression (24) essentially coincides with the Bogomol'nyi bound found by Kostelecký and Perry [16], emerging from the susy algebra. However, we stress the fact that supersymmetry requires a vanishing magnetic charge in this case; the authors of [16] missed this condition in their paper. In summary, we can state that, in order to get extremal supersymmetric Kerr–Newman-AdS black holes, we must allow the holes to carry angular momentum, for, in the static case, all the supersymmetric configurations are naked singularities. Besides, the extremal supersymmetric holes carry only electric charge, and so we have made the result of Kostelecký and Perry more precise.

There also exist rotating generalizations of the toroidal and higher genus static black holes [21]. The former represents a stationary cylindrical AdS black hole, whereas the latter is a rotating black membrane in AdS space[2]. As far as the cylindrical black hole is concerned, one finds a behaviour similar to that of the Kerr–Newman-AdS solution, i.e., in order to have extremal supersymmetric configurations without naked singularities, one must allow the holes to rotate. Furthermore, the BPS states carry only electric charge. For the rotating membrane, one finds a one-parameter stationary generalization of the static higher genus BPS black hole with $\eta = 0$ and $q_m = \pm\ell/2$ [19].

To sum up, in the present review we considered four-dimensional asymptotically AdS dyonic black holes with various topologies in the context of gauged $N = 2$ supergravity. For toroidal or cylindrical topology, all static configurations preserving some amount of supersymmetry, are naked singularities, a behaviour familiar from the spherical Reissner-Nordström-AdS case studied previously by Romans. However, for black holes whose event horizons are Riemann surfaces of genus $g > 1$, we found an extremal supersymmetric black hole carrying purely magnetic charge, and admitting a one-(complex)-dimensional solution space of Killing spinors. As we said, this solitonic object also possesses a rotating generalization. However, for cylindrical or spherical topology, extremal supersymmetric black holes carrying only electric charge, can appear for nonvanishing angular momentum. Hence, in these cases, solitonic black holes must rotate. This is in agreement with the Kerr–Newman-AdS result of [16], emerging from considerations of the superalgebra. Yet, the authors of [16] did not obtain the condition of vanishing magnetic charge for these BPS states, so we have made more precise the Bogomol'nyi bound found in [16]. The rotating supersymmetric states with purely electric charge, appearing for cylindrical or spherical topology, have no analogue for the rotating generalizations of the higher genus solutions.

In conclusion, we can state that admitting unusual black hole topologies, and allowing the holes also to carry angular momentum, can lead to a new variety of

[2] In the rotating case, it is no longer possible to compactify the membrane to a Riemann surface [22].

states preserving some supersymmetry. It would be very interesting to understand the Bogomol'nyi bounds, found for unusual topologies, in terms of the superalgebra. However, this requires a careful definition of the mass and angular momentum of these rotating black configurations. As such a definition is a rather delicate question [21], such an enterprise becomes a nontrivial task which we leave for future investigations.

Acknowledgements

This work has been supported by a research grant within the scope of the *Common Special Academic Program III* of the Federal Republic of Germany and its Federal States, mediated by the DAAD.

References

1. Maldacena J.M. (1998): Adv. Theor. Math. Phys. **2**, 231
2. Witten E. (1998): Adv. Theor. Math. Phys. **2**, 253
3. Witten E. (1998): Adv. Theor. Math. Phys. **2**, 505
4. Gubser S., Klebanov I., Polyakov A. (1998): Phys. Lett. B **428**, 105
5. Åminneborg S., Bengtsson I., Holst S., Peldán P. (1996): Class. Quantum Grav. **13**, 2707
6. Mann R.B. (1997): Class. and Quantum Grav. **14**, 2927;
   Mann R.B. (1997): Topological black holes: Outside looking in, in *Workshop on the internal structure of black holes and spacetime singularities, Haifa (Israel) 29 Jun–3 Jul 1997*, ed. by L.M. Burko, A. Ori, Bristol, IOP, Isr. Phys. Soc., p. 534;
   Mann R.B., Smith W.L. (1997): Phys. Rev. D **56**, 4942
7. Lemos J.P.S. (1995): Class. Quantum Grav. **12**, 1081;
   Lemos J.P.S. (1996): Phys. Lett. B **353**, 46;
   Lemos J.P.S., Zanchin V.T. (1996): Phys. Rev. D **54**, 3840
8. Cai R.G., Zhang Y.Z. (1996): Phys. Rev. D **54**, 4891
9. Huang C.G., Liang C.B. (1995): Phys. Lett. A **201**, 27;
   Huang C.G. (1996): Acta Phys. Sin. **4**, 617
10. Vanzo L. (1997): Phys. Rev. D **56**, 6475
11. Klemm D., Vanzo L. (1998): Phys. Rev. D **58**, 104025
12. Horowitz G.T., Polchinski J. (1997): Phys. Rev. D **55**, 6189
13. Larsen A.L., Sánchez N. (1995): Phys. Rev. D **52**, 1051
14. Romans L.J. (1992): Nucl. Phys. B **383**, 395
15. Tod K.P. (1981): Phys. Lett. B **121**, 241
16. Kostelecký V.A., Perry M.J. (1996): Phys. Lett. B **371**, 191
17. Das A., Freedman D.Z. (1977): Nucl. Phys. B **120**, 221
18. Fradkin E.S., Vasiliev M.A. (1976): Lebedev Institute preprint N 197, unpublished
19. Caldarelli M.M., Klemm D. (1999): Nucl. Phys. B **545**, 434–460
20. Bañados M., Teitelboim C., Zanelli J. (1992): Phys. Rev. Lett. **69**, 1849
21. Klemm D., Moretti V., Vanzo L. (1998): Phys. Rev. D **57**, 6127
22. Klemm D. (1998): JHEP **11**, 019

Invariants of Spin Networks with Boundary in Quantum Gravity and TQFTs

G. Carbone, M. Carfora, A. Marzuoli

Abstract. The search for classical or quantum combinatorial invariants of compact n-dimensional manifolds ($n = 3, 4$) plays a key role both in topological field theories and in lattice quantum gravity (see, e.g., [1–4]). We present here a generalization of the partition function proposed by Ponzano and Regge to the case of a compact 3-dimensional simplicial pair $(M^3, \partial M^3)$. The resulting state sum $Z[(M^3, \partial M^3)]$ contains both Racah–Wigner $6j$ symbols associated with tetrahedra and Wigner $3jm$ symbols associated with triangular faces lying in ∂M^3. The analysis of the algebraic identities associated with the combinatorial transformations involved in the proof of the topological invariance makes manifest a common structure underlying the 3-dimensional models with empty and non-empty boundaries respectively. The techniques developed in the 3-dimensional case can be further extended in order to deal with combinatorial models for $n = 2, 4$ and possibly to establish a hierarchy among such models. As an example we derive here a 2-dimensional closed state sum model including suitable sums of products of double $3jm$ symbols, each of them being associated with a triangle in the surface.

Recall that a closed PL-manifold of dimension n is a polyhedron $M \cong |T|$, each point of which has a neighborhood, in M, PL-homeomorphic to an open set in $\mathbb{R}^n$. The symbol $\cong$ denotes homeomorphism, T is the underlying (finite) simplicial complex and $|T|$ denotes the associated topological space, namely, the set-theoretic union of all simplices of T endowed with its natural topology. PL-manifolds are realized by simplicial manifolds under the equivalence relation generated by PL-homeomorphisms. In particular, two n-dimensional closed PL-manifolds $M_1 \cong |T_1|$ and $M_2 \cong |T_2|$ are PL-homeomorphic, or $M_1 \cong_{PL} M_2$, if there exists a map $g : M_1 \to M_2$ which is both a homeomorphism and a simplicial isomorphism (see, e.g., [5] for more details on both this definition and other issues in PL-topology which will be used in what follows). We shall use the notation

$$T \longrightarrow M \cong |T| \tag{1}$$

to denote a *particular triangulation* of the closed n- dimensional PL-manifold M. In order to extend the previous notation to the case of a PL-pair $(M, \partial M)$ of dimension n, recall that a simplicial complex is *pure* provided that all its facets (i.e., its faces of maximal dimension) have the same dimension. Moreover, the *boundary complex* of a pure simplicial n-complex T is denoted by ∂T and is the subcomplex of T the facets of which are the $(n-1)$-faces of T which are contained in only one facet of T. The set of the interior faces of T is denoted by $\mathrm{int}\,(T) \doteq T \setminus \partial T$. Then

$$(T, \partial T) \longrightarrow (M, \partial M) \cong (|T|, |\partial T|) \tag{2}$$

denotes a triangulation on $(M, \partial M)$, where ∂T is the unique triangulation induced on the $(n-1)$-dimensional boundary PL-manifold ∂M by the chosen triangulation T in M.

Following [6], the connection between a recoupling scheme of SU(2) angular momenta and the combinatorial structure of a compact, 3-dimensional simplicial pair $(M^3, \partial M^3)$ can be established by considering *colored* triangulations which allow us to specialize the map (2) according to

$$(T^3(j), \partial T^3(j', m)) \longrightarrow (M^3, \partial M^3). \tag{3}$$

This map represents a triangulation associated with an admissible assignment of both spin variables to the collection of edges in $(T^3, \partial T^3)$ and of momentum projections to the subset of edges lying in ∂T^3. The collective variable $j \equiv \{j_A\}$, $A = 1, 2, \ldots, N_1$, denotes all the spin variables, n'_1 of which are associated with the edges in the boundary (for each A: $j_A = 0, 1/2, 1, 3/2, \ldots$ in $\hbar$ units). Note that the last subset is labelled both by $j' \equiv \{j'_C\}$, $C = 1, 2, \ldots, n'_1$, and by $m \equiv \{m_C\}$, where m_C is the projection of j'_C along the fixed reference axis (of course, for each m, $-j \leq m \leq j$, in integer steps). Consistency in the assignment of the j, j', m variables is ensured if we require that:

- each 3-simplex σ^3_B, $B = 1, 2, \ldots, N_3$, in $(T^3, \partial T^3)$ must be associated, apart from a phase factor, with a $6j$ symbol of SU(2), namely,

$$\sigma^3_B \longleftrightarrow (-1)^{\sum_{p=1}^{6} j_p} \begin{Bmatrix} j_1 & j_2 & j_3 \\ j_4 & j_5 & j_6 \end{Bmatrix}_B ; \tag{4}$$

- each 2-simplex σ^2_D, $D = 1, 2, \ldots, n'_2$, in ∂T^3 must be associated with a Wigner $3jm$ symbol of SU(2) according to

$$\sigma^2_D \longleftrightarrow (-1)^{(\sum_{s=1}^{3} m_s)/2} \begin{pmatrix} j'_1 & j'_2 & j'_3 \\ m_1 & m_2 & -m_3 \end{pmatrix}_D . \tag{5}$$

Then the following state sum can be defined:

$$Z[(M^3, \partial M^3)] =$$

$$= \lim_{L\to\infty} \sum_{\left\{ \begin{smallmatrix} (T^3(j), \partial T^3(j', m)) \\ j, j', m \leq L \end{smallmatrix} \right\}} Z[(T^3(j), \partial T^3(j', m)) \to (M^3, \partial M^3); L] \tag{6}$$

where

$$Z[(T^3(j), \partial T^3(j', m)) \to (M^3, \partial M^3); L] =$$

$$= \Lambda(L)^{-N_0} \prod_{A=1}^{N_1} (-1)^{2j_A}(2j_A+1) \prod_{B=1}^{N_3} (-1)^{\sum_{p=1}^{6} j_p} \begin{Bmatrix} j_1 & j_2 & j_3 \\ j_4 & j_5 & j_6 \end{Bmatrix}_B \cdot$$

$$\cdot \prod_{D=1}^{n'_2} (-1)^{(\sum_{s=1}^{3} m_s)/2} \begin{pmatrix} j'_1 & j'_2 & j'_3 \\ m_1 & m_2 & -m_3 \end{pmatrix}_D . \tag{7}$$

N_0, N_1, N_3 denote respectively the total number of vertices, edges and tetrahedra in $(T^3(j), \partial T^3(j', m))$, while n'_2 is the number of 2-simplices lying in $\partial T^3(j', m)$. Note that there is a factor $\Lambda(L)^{-1}$ for each vertex in $\partial T^3(j', m)$, with $\Lambda(L) \equiv 4L^3/3C$, C an arbitrary constant. It is also worthwhile to remark that products of $6j$ and $3jm$ coefficients of the kind which appear in (7) are known as *jm coefficients* in the quantum theory of angular momentum (see e.g., [7]). Their semiclassical limit can be defined in a consistent way by requiring that, simultaneously, $j, j' \to \infty$ and $m \to \infty$ with the constraint $-j' \leq m \leq j'$. The summation in (6) has precisely this meaning, apart from the introduction of the cut–off L. The state sum given in (6) and (7) when $\partial M^3 = \emptyset$ reduces to the usual Ponzano–Regge partition function for the closed manifold M^3 (see [1]); in such a case, it can be rewritten here as

$$Z[M^3] = \lim_{L\to\infty} \sum_{\{T^3(j), j \leq L\}} Z[T^3(j) \to M^3; L] \tag{8}$$

where the sum is extended to all assignments of spin variables such that each is not greater than the cut–off L, and each term under the sum is given by

$$Z[T^3(j) \to M^3; L] = \Lambda(L)^{-N_0} \prod_{A=1}^{N_1} (2j_A+1)(-1)^{2j_A}$$

$$\cdot \prod_{B=1}^{N_3} (-1)^{\sum_{p=1}^{6} j_p} \begin{Bmatrix} j_1 & j_2 & j_3 \\ j_4 & j_5 & j_6 \end{Bmatrix}_B . \tag{9}$$

As is well-known, the above state sum gives the semiclassical partition function of Euclidean gravity with an action discretized according to Regge's prescription [8].

The state sum given in (8) and (9) is formally invariant under a set of topological transformations performed on 3-simplices in $T^3(j)$: following Pachner [9], they are commonly known as *bistellar elementary operations* or *bistellar moves*. It is a classical result (see, e.g., [1] and [10]) that such moves can be expressed algebraically in terms of the Biedenharn–Elliott identity (representing the moves (2 tetrahedra) $\leftrightarrow$ (3 tetrahedra)) and of both the B-E identity and the orthogonality conditions (representing the moves (1 tetrahedron) $\leftrightarrow$ (4 tetrahedra)). The expression of the

Biedenharn–Elliott identity reads:

$$\sum_X (2X+1)(-1)^{\Theta+X} \begin{Bmatrix} a & b & X \\ c & d & p \end{Bmatrix} \begin{Bmatrix} c & d & X \\ e & f & q \end{Bmatrix} \begin{Bmatrix} e & f & X \\ b & a & r \end{Bmatrix} =$$

$$= \begin{Bmatrix} p & q & r \\ e & a & d \end{Bmatrix} \begin{Bmatrix} p & q & r \\ f & b & c \end{Bmatrix} \tag{10}$$

where $a, b, c, \ldots$ now denote angular momentum variables, $(2X+1)$ is the dimension of the irreducible representation of SU(2) labelled by X, and $\Theta = a+b+c+d+e+f+p+q+r$. The orthogonality conditions amount to

$$\sum_X (2X+1) \begin{Bmatrix} a & b & X \\ c & d & e \end{Bmatrix} \begin{Bmatrix} a & b & X \\ c & d & f \end{Bmatrix} = (2e+1)^{-1}\, \delta_{ef}\, \{ade\}\, \{bce\} \tag{11}$$

where the notation $\{ade\}$ stands for the triangular delta (viz.), $\{ade\}$ is equal to 1 if its three arguments satisfy triangular inequalities, and is zero otherwise) and $\delta_{ef} \equiv \delta(e, f)$.

The invariance under bistellar moves is related to the PL-equivalence class of the manifold involved. Indeed, Pachner proved in [9] that two closed n-dimensional PL-manifolds are PL-homeomorphic if, and only if, their underlying triangulations are related to each other by a finite sequence of bistellar moves. Thus, in particular, the state sum (8) is an invariant of the PL-structure of M^3.

Turning now to the case with boundary, we observe that (7) is manifestly invariant under bistellar moves which involve 3-simplices in int (T^3) (this is consistent with the remark that (6) reduces to (8) when $\partial M^3 = \emptyset$). In the non-trivial case $\partial M^3 \neq \emptyset$ new types of topological transformations have to be taken into account. Indeed Pachner introduced moves which are suitable in the case of compact n-dimensional PL-manifolds with a non–empty boundary, the *elementary shellings* (see [11]). As the term "elementary shelling" suggests, this operation involves the cancellation of one n-simplex (facet) at a time in a given triangulation $(T, \partial T) \to (M, \partial M)$ of a compact PL-pair of dimension n. In order to be deleted, the facet must have some of its faces lying in the boundary ∂T. It is possible to classify these moves according to the dimension of the components of the facet in ∂T, and it turns out that there are just n different types of elementary shellings in dimension n. Moreover, for each elementary shelling there exists an inverse move which corresponds to the attachment of a new facet to a suitable component in ∂T.

In [6] identities representing the three types of elementary shelling (and their inverse moves) for a 3-dimensional triangulation (3) were established. Following the notation of [12] one of the identities is displayed below:

$$\sum_{c\gamma}(2c+1)(-1)^{2c-\gamma}\begin{pmatrix} a & b & c \\ \alpha & \beta & \gamma \end{pmatrix}\begin{pmatrix} c & r & p \\ -\gamma & \varrho' & \psi \end{pmatrix}(-1)^{\Phi}\begin{Bmatrix} a & b & c \\ r & p & q \end{Bmatrix} =$$

$$= (-1)^{-2\varrho}\sum_{\kappa}(-1)^{-\kappa}\begin{pmatrix} p & a & q \\ \psi & \alpha & -\kappa \end{pmatrix}\begin{pmatrix} q & b & r \\ \kappa & \beta & -\varrho' \end{pmatrix} \qquad (12)$$

where Latin letters $a, b, c, r, p, q, \ldots$ denote angular momentum variables, Greek letters $\alpha, \beta, \gamma, \varrho, \psi, \kappa, \ldots$ the corresponding momentum projections and $\Phi \equiv a + b + c + r + p + q$. The topological content of the above identity is the following: on the left–hand side there appears a tetrahedron, two faces of which (lying in ∂T^3) share the edge labelled by c and its projection γ; the summation over c, γ provides for, on the right–hand side, the appearance of the other two faces which survive after the shelling. The inverse move, namely, the attachment of a new facet to a pair of triangles in ∂T^3, is obtained by reading the same identity backward.

Note that the complete set of identities can, in fact, be derived (up to suitable regularization factors) just from (12) and from both the orthogonality conditions for the $6j$'s (given in (11)) and the completeness conditions for the $3jm$ symbols which read

$$\sum_{c\gamma}(-1)^{2c-\gamma}(2c+1)\begin{pmatrix} a & b & c \\ -\alpha & -\beta & \gamma \end{pmatrix}\begin{pmatrix} b & a & c \\ -\beta' & -\alpha' & \gamma \end{pmatrix} = (-1)^{\alpha+\beta}\delta_{\alpha\alpha'}\,\delta_{\beta\beta'}. \qquad (13)$$

Recognition of the identities representing the elementary shellings and their inverse moves, together with a comparison with the expression given in (7), allows us to conclude that the state sum $Z[(M^3, \partial M^3)]$ is formally invariant both under (a finite number of) bistellar moves in the interior of $(M^3, \partial M^3)$ and under (a finite number of) elementary boundary operations, namely, shellings and inverse shellings. As regards PL-equivalence, we can exploit another result proved by Pachner in [11] which states that, if $(T_1, \partial T_1) \to (M_1, \partial M_1)$ and $(T_2, \partial T_2) \to (M_2, \partial M_2)$ are triangulations of PL, compact n-dimensional pairs, then $|(T_1, \partial T_1)| \cong_{PL} |(T_2, \partial T_2)| \iff (T_1, \partial T_1) \approx_{sh,bst} (T_2, \partial T_2)$, where the equivalence $\approx_{sh,bst}$ is both under elementary shellings and under bistellar elementary operations on n-simplices in $\mathrm{int}(T_1)$ or $\mathrm{int}(T_2)$. Thus (6) turns out to be an invariant of the PL-structure (actually it is a topological invariant, since we are dealing with 3-dimensional PL-manifolds).

As already stressed, the complete set of elementary shellings can be derived from a single identity, namely (12), together with conditions (11) and (13) on the symbols. The structure of such an identity strongly rensembles the Biedenharn–Elliott identity (10), both as regards the number of symbols involved and because the presence of a single sum over a j-variable. Recall also that the complete set of bistellar moves is actually derived from the B-E identity + (orthogonality conditions for the $6j$), apart from regularization. This similarity in the algebraic structure of the topological invariance in the two cases (closed and with boundary) is quite remarkable.

However, our previous analysis gives rise to other possible developments, in particular, as regards the search for state sum models of the same kind in dimension different from three. As we have already noted, the different types of elementary shelling acting on a simplicial n-dimensional pair $(T, \partial T)$ amount exactly to n, while the number of bistellar moves in the interior of $(T, \partial T)$ (or in T itself if it is closed) is $(n+1)$. Moreover, the central projection of each elementary shelling onto ∂T gives a particular bistellar move in dimension $(n-1)$ (with ∂T a triangulation of a closed $(n-1)$-dimensional manifold). It is also easy to check that the same kind of projection of the complete set of boundary operations reproduces the complete set of bistellar moves in the lower dimensional case. Starting from these remarks, we show in what follows how a consistent 2-dimensional state sum invariant arises as a *projection* of the invariant given in (6) and (7).

Since the structure of a local arrangement of 2-simplices in the state sum (7) is naturally encoded in (12), it turns out that a state sum for a 2-dimensional closed triangulation $T^2(j; m, m') \to M^2$ can be defined by associating with each 2-simplex $\sigma^2 \in T^2$ the following product of two $3jm$ symbols (a *double* $3jm$ symbol, for short):

$$\sigma^2 \longleftrightarrow (-1)^{\sum_{s=1}^{3}(m_s+m'_s)/2} \begin{pmatrix} j_1 & j_2 & j_3 \\ m_1 & m_2 & -m_3 \end{pmatrix} \begin{pmatrix} j_1 & j_2 & j_3 \\ m'_1 & m'_2 & -m'_3 \end{pmatrix} \tag{14}$$

where $\{m_s\}$ and $\{m'_s\}$ are two different sets of momentum projections associated with the same angular momentum variables $\{j_s\}$, $-j \leq m_s, m'_s \leq j$ $\forall s = 1, 2, 3$. The expression of the state sum of the given triangulation reads

$$Z[T^2(j; m, m') \to M^2; L] =$$

$$= \Lambda(L)^{-N_0} \prod_{A=1}^{N_1} (2j_A + 1)(-1)^{2j_A}(-1)^{-m_A - m'_A}$$

$$\prod_{B=1}^{N_2} \begin{pmatrix} j_1 & j_2 & j_3 \\ m_1 & m_2 & -m_3 \end{pmatrix}_B \begin{pmatrix} j_1 & j_2 & j_3 \\ m'_1 & m'_2 & -m'_3 \end{pmatrix}_B \tag{15}$$

where N_0, N_1, N_2 are the numbers of vertices, edges and triangles in T^2, respectively. Summing over all of the admissible assignments of $\{j; m, m'\}$ we get

$$Z[M^2] = \lim_{L\to\infty} \sum_{\{T^2(j;m,m'), j\leq L\}} Z[T^2(j; m.m') \to M^2; L] \tag{16}$$

where the regularization is carried out according to the usual prescription. In order to address the invariance of (15) we may start from a configuration representing two triangles glued together along a common edge q with momentum projections κ, κ'; taking into account (12) and (11) we get

$$\sum_q \sum_{\kappa,\kappa'} (2q+1)(-1)^{2q}(-1)^{-\kappa-\kappa'} \begin{pmatrix} p & a & q \\ \psi & \alpha & -\kappa \end{pmatrix} \begin{pmatrix} q & b & r \\ \kappa & \beta & \varrho \end{pmatrix} \cdot$$

$$\cdot \begin{pmatrix} p & a & q \\ \psi' & \alpha' & -\kappa' \end{pmatrix} \begin{pmatrix} q & b & r \\ \kappa' & \beta' & \varrho' \end{pmatrix} = \sum_c \sum_{\gamma,\gamma'} (2c+1)(-1)^{2c}(-1)^{-\gamma-\gamma'} \cdot$$

$$\cdot \begin{pmatrix} a & b & c \\ \alpha & \beta & \gamma \end{pmatrix} \begin{pmatrix} r & p & c \\ \varrho & \psi & -\gamma \end{pmatrix} \begin{pmatrix} a & b & c \\ \alpha' & \beta' & \gamma' \end{pmatrix} \begin{pmatrix} r & p & c \\ \varrho' & \psi' & -\gamma' \end{pmatrix}. \tag{17}$$

The geometrical meaning of the above identity should be clear: it represents the so-called *flip* in dimension 2, namely, the bistellar move $(2 \to 2)$ involving a pair of triangles.

In order to obtain the remaining bistellar moves, we start from a configuration of three triangles glued along their edges q, r, p in such a way that they share a common vertex. Using (12), (11) and (13) again we find that

$$\sum_{q,r,p} (2q+1)(2r+1)(2p+1)\,(-1)^{2q+2r+2p}\,. \tag{18}$$

$$\cdot \sum_{\kappa,\kappa'} \sum_{\varrho,\varrho'} \sum_{\psi,\psi'} (-1)^{-\kappa-\kappa'} (-1)^{-\varrho-\varrho'}\,.$$

$$\cdot (-1)^{-\psi-\psi'} \begin{pmatrix} p & a & q \\ \psi & \alpha & -\kappa \end{pmatrix} \begin{pmatrix} q & b & r \\ \kappa & \beta & -\varrho \end{pmatrix} \begin{pmatrix} r & c & p \\ \varrho & \gamma & -\psi \end{pmatrix}.$$

$$\cdot \begin{pmatrix} p & a & q \\ \psi' & \alpha' & -\kappa' \end{pmatrix} \begin{pmatrix} q & b & r \\ \kappa' & \beta' & -\varrho' \end{pmatrix} \begin{pmatrix} r & c & p \\ \varrho' & \gamma' & -\psi' \end{pmatrix} =$$

$$= \Lambda(L)^{-1} \begin{pmatrix} a & b & c \\ \alpha & \beta & \gamma \end{pmatrix} \begin{pmatrix} a & b & c \\ \alpha' & \beta' & \gamma' \end{pmatrix}. \tag{19}$$

This identity represents the barycentric subdivision (and its inverse operation) of a triangle of edges a, b, c, namely, the bistellar moves denoted by $(1 \leftrightarrow 3)$. As a matter of fact, the state sum given in (15) and (16) is formally invariant under (a finite number of) topological operations represented by (17) and (19). Thus, again from Pachner's theorem proved in [9], we conclude that it is a PL-invariant (indeed, a topological invariant).

The procedure outlined above for the 3-dimensional model can be further generalized to the case of the quantum enveloping algebra $U_{\mathbf{q}}(sl(2,\mathbb{C}))$ for $\mathbf{q}$ a root of unity. The corresponding *quantum* invariant $Z_{\mathbf{q}}[M^3, \partial M^3]$ is the counterpart of the Turaev–Viro invariant for a closed 3-dimensional PL-manifold M^3 (see [2]) and details on its derivation can be found in [6]. The same remark holds true for the 2-dimensional closed model as well. Moreover, this extension provides us with a *finite* topological invariant, which is easily evaluated and turns out to be obviously related to the Euler characteristic of the closed suface.

Without entering into technicalities, the quantum state sum which replaces (15) contains the $\mathbf{q}$-analog of the double Wigner $3jm$ symbols (14) (with a suitable choice of normalization), while the spin variables j take their values in a finite set $I = (0, 1/2, 1, \dots, \mathbf{k})$ where $\exp(\pi i/\mathbf{k}) = \mathbf{q}$. For each $j \in I$ a function $w^2(j) \equiv w_j^2 \doteq (-1)^{2x_j}[2x_j+1]_{\mathbf{q}} \in K^*$ is defined, where $K^* \equiv K \setminus \{0\}$ (K a commutative ring with unity). Recall that the notation $[.]_{\mathbf{q}}$ stands for a $\mathbf{q}$-integer, namely, $[n]_{\mathbf{q}}$

$= (\mathbf{q}^n - \mathbf{q}^{-n})/(\mathbf{q} - \mathbf{q}^{-1})$ and that, for each admissible triple (j, k, l), we have $w_j^{-2} \sum_{k,l} w_k^2 w_l^2 = w^2$, with $w^2 = -2\mathbf{k}/(\mathbf{q} - \mathbf{q}^{-1})^2$. Coming back to the terms of the quantum state sum, we see that the classical weights $(-1)^{2j}(2j+1)$ in (15) are replaced by w_j^2, while each of the factors Λ^{-1} becomes w^{-2}. Then the state sum $Z_{\mathbf{q}}[T^2(j; m, m') \to M^2]$ for a given triangulation can be easily evaluated, and it turns out that the associated quantum invariant amounts to

$$Z_{\mathbf{q}}[M^2] = w^2 \, w^{-2\chi(M^2)} \tag{20}$$

where $\chi(M^2)$ is the Euler characteristic of the manifold M^2. The above result shows that the 2-dimensional closed model derived from our 3-dimensional model with non-empty boundary is not trivial: we actually recover the only topological invariant which is significant for a closed surface.

Following the same line of reasoning which gives rise to the closed 2-dimensional model, we are currently addressing the analysis of the structure of the *projected* counterpart of the algebraic identities representing topological moves in dimension four. We think that the approach outlined in this paper can provide us with a hierarchy of models (with and without boundary) from dimension four to dimension two.

References

1. Ponzano G., Regge T. (1968): Semiclassical Limit of Racah Coefficients, in *Spectroscopic and Group Theoretical Methods*, ed. by F. Bloch, F. in Physics, pp. 1–58. North-Holland, Amsterdam, pp. 1–58
2. Turaev V., Viro O.Ya. (1992): State sum invariants of 3-manifolds and quantum $6j$-symbols. Topology **31**, 865–902
3. Ooguri H. (1992): Topological lattice models in four dimensions. Mod. Phys. Lett. A **7**, 2799–2810
4. Carter J.S., Kauffman L.H., Saito M. (1998): Structure and diagrammatics of four dimensional topological lattice field theories. Preprint, math. GT/9806 023
5. Rourke C., Sanderson B. (1982): Introduction to Piecewise Linear Topology. Springer, New York
6. Carbone G., Carfora M., Marzuoli A.: Wigner symbols and combinatorial invariants of three-manifolds with boundary. Preprint DFNT-T 14/98 and SISSA 118/98/FM
7. Yutsis A.P., Levinson I.B., Vanagas V.V. (1962): The Mathematical Apparatus of the Theory of Angular Momentum. Israel Program for Sci. Transl. Ltd. Jerusalem
8. Regge T. (1961): General Relativity without coordinates. Nuovo Cimento **19**, 558–571
9. Pachner U. (1987): Ein Henkeltheorem für geschlossene semilineare Mannigfaltigkeiten. Result. Math. **12**, 386–394
10. Carter J.S., Flath D.E., Saito M. (1995): The Classical and Quantum *6j*-Symbols. Math. Notes **43**. Princeton University Press, Princeton
11. Pachner U. (1990): Shellings of simplicial balls and p.l. manifolds with boundary. Discr. Math. **81**, 37–47
12. Varshalovich D.A., Moskalev A.N., Khersonski V.K. (1988): Quantum Theory of Angular Momentum. World Scientific, Singapore

A Review of Recent Results of the ζ-Function Regularization Procedure in Curved Spacetime

V. Moretti

Abstract. Some recent (1997–1998) theoretical results concerning the ζ-function regularization procedure used to renormalize, at one-loop, the effective Lagrangian, field fluctuations and the stress-tensor are reviewed along with some applications.

1 Quasifree QFT in curved static manifolds, the Euclidean approach and the ζ-function technique

Preliminaries

Let $(L, g^L_{\mu\nu})$ be a Lorentzian manifold which is assumed to be *globally hyperbolic*. This manifold is called (locally) *static* whenever it admits a (local) time-like Killing field ∂_t normal to a Cauchy surface Σ. In other words, the manifold admits a (local) coordinate frame $(x^0, x^1, x^2, x^3) \equiv (t, \mathbf{x})$, where $g_{0i} = 0$ $(i = 1, 2, 3)$ and $\partial_t g_{\mu\nu} = 0$. We consider a real scalar field ϕ propagating in L; its evolution equation can be expressed as

$$A'_L \phi = 0, \tag{1}$$

where $A'_L := -\nabla_\mu \nabla^\mu + V$, with V a smooth scalar field of the form

$$V(x) := \xi R + m^2 + V'(x). \tag{2}$$

Here V' is another smooth scalar field satisfying $\partial_t V = 0$, while ξ is a constant, $\xi \in \mathbb{R}$, R is the scalar curvature and m^2 the squared mass of the particles associated to the field. The operator A'_L acts on a space of real-valued C^∞ functions. A straightforward way to build up the QFT is by defining suitable Green functions of the operator A'_L [1], in particular, the Feynman propagator $G_F(x, x')$ or, equivalently, the Wightman functions $W_\pm(x, x')$. Then, generalizations of GNS theorems [2] in curved spacetime allow one to build up a corresponding Fock space and a quasifree QFT, with G_F corresponding to a (not necessarily pure) vacuum state. In a globally hyperbolic region where the static coordinates above are defined, it must be possible to choose Green functions which are invariant under translations of the local static Killing time $x^0 = t$. These Green functions should determine a static vacuum which may represent a thermal state. Moreover these coordinates allow one to perform the Wick rotation to get a Euclidean formulation of the QFT. This means that (locally) one can pass from the Lorentzian manifold $(L, g^L_{\mu\nu})$ to a Riemannian manifold (M, g_{ab}) by the analytic continuation $t \to i\tau$ where $t, \tau \in \mathbb{R}$. This defines a (local) Killing vector ∂_τ in the Riemannian manifold and a corresponding (local) "static" coordinate frame $(\tau, \mathbf{x})$ therein. As is well-known [1], when the Riemannian manifold has been made compact along the Euclidean time τ with a period β, $T = 1/\beta$ has to be interpreted as the temperature of the quantum state. The corresponding Feynman propagator, whenever continued in the Riemannian manifold, admits β as Euclidean temporal period. In this approach, the Feynman propagator $G_F(t - t', \mathbf{x}, \mathbf{x}')$ determines, and, generally speaking [1], itself is completely determined by, a proper Green function (in the spectral theory sense) $S_\beta(\tau - \tau', \mathbf{x}, \mathbf{x}')$ of a self-adjoint extension A of the operator

$$A' := -\nabla_a \nabla^a + V(\mathbf{x}) \; : \; C_0^\infty(M) \to L^2(M, d\mu_g). \tag{3}$$

(Here M can be restricted to a suitable region where the metric is static.) $S_\beta(\tau - \tau', \mathbf{x}, \mathbf{x}')$, called the Schwinger function, is the integral kernel of A^{-1}, assuming $A > 0$.

The *partition function* of the quantum state can be computed as the functional integral evaluated over the field configurations which are periodic with period β in Euclidean time

$$Z_\beta = \int \mathcal{D}\phi \, e^{-S_E[\phi]}, \tag{4}$$

the Euclidean action S_E being ($d\mu_g := \sqrt{g} d^4x$)

$$S_E[\phi] = \frac{1}{2} \int_M d\mu_g(x) \, \phi(x) A_x \phi(x). \tag{5}$$

Thus, formally speaking, one has

$$Z_\beta = \left\{ \det \left(\frac{A}{\mu^2} \right) \right\}^{-1/2}, \tag{6}$$

where μ is a mass scale which is necessary for dimensional reasons.

The ζ-function technique

The most intriguing problem related to the applications of (6) concerns the interpretation of the determinant of an operator. A interesting suggestion is given by the ζ-function procedure. Suppose A is an $n \times n$ positive-definite Hermitian matrix with eigenvalues $0 < \lambda_1 \leq \lambda_2 \leq \cdots \leq \lambda_n$. Then one can define the complex-valued function

$$\zeta(s|A) = \sum_{j=1}^{n} \lambda_j^{-s}, \tag{7}$$

where $s \in \mathbb{C}$. (Note that λ_j^{-s} is well-defined since $\lambda_j > 0$.) It is now a trivial task to prove that

$$\det A = e^{-\frac{d\zeta(s|A)}{ds}|_{s=0}}. \tag{8}$$

When M is a D-dimensional Riemannian compact manifold and A' is bounded below by a constant $b \geq 0$, this procedure can be generalized to operators. In this case, A' admits the Friedrichs self-adjoint extension A which is also bounded below by the same bound as for A', moreover the spectrum of A is discrete and each eigenspace has a finite dimension. Then one can consider the following series with $s \in \mathbb{C}$ (the prime on the sum means that any possible null eigenvalues are omitted):

$$\zeta(s|A/\mu^2) := \sum_j{}' \left(\frac{\lambda_j}{\mu^2}\right)^{-s}. \tag{9}$$

As is well-known, the series above converges provided $Re\, s > D/2$, where D is the dimension of M. Moreover, it is possible to continue the right-hand side above into a meromorphic function of s which is regular at $s = 0$ (see [3] for a short review of the principal properties of the convergence of the above series and for the corresponding bibliography). Following (6) and (8), the idea [4] is to define

$$Z_\beta := e^{\frac{1}{2}\frac{d}{ds}|_{s=0}\zeta(s|A/\mu^2)}, \tag{10}$$

where the function ζ on the right-hand side is the analytic continuation of that defined in (9). It is possible to define the ζ function in terms of the heat kernel of the operator A, $K(t, x, y|A)$. This is the smooth integral kernel of the (Hilbert-Schmidt, trace-class) operator e^{-tA}, $t > 0$. One has, for $Re\, s > D/2$,

$$\zeta(s|A/\mu^2) = \int_M d\mu_g(x) \int_0^{+\infty} dt\, \frac{\mu^{2s} t^{s-1}}{\Gamma(s)} \left[K(t, x, x|A) - P_0(x, x|A)\right], \tag{11}$$

where $P_0(x, y|A)$ is the integral kernel of the projector on the null-eigenvalues eigenspace of A.

If the manifold M is not compact, A also has a continuous-spectrum part; however, it is still possible to generalize the definitions and the above results by considering suitable integrals on the spectrum of the operator A provided that A is strictly positive (e.g, see [5]).

Another very useful tool is the *local* ζ function [5], which can be defined in two different but equivalent ways:

$$\zeta(s, x|A/\mu^2) = \int_0^{+\infty} dt\, \frac{\mu^{2s} t^{s-1}}{\Gamma(s)} \left[K(t, x, x|A) - P_0(x, x|A)\right], \tag{12}$$

and, with ϕ_j being the smooth eigenvector of the eigenvalue λ_j,

$$\zeta(s, x|A/\mu^2) = \sum_j{}' \left(\frac{\lambda_j}{\mu^2}\right)^{-s} \phi_j(x)\phi_j^*(x). \tag{13}$$

Both the integral and the series converge for $Re\, s > D/2$. The local zeta function enjoys the same analyticity properties as the integrated ζ function ([3, 5]). For future convenience it is also useful to define, in the sense of analytic continuation,

$$\zeta(s, x, y|A/\mu^2) = \int_0^{+\infty} dt\, \frac{\mu^{2s} t^{s-1}}{\Gamma(s)} \left[K(t, x, y|A) - P_0(x, y|A)\right] \tag{14}$$

(see [3, 6] for the properties of this off-diagonal ζ-function). In the framework of ζ-function regularization, the *effective Lagrangian* is defined as

$$\mathcal{L}(x|A)_{\mu^2} := \frac{1}{2}\frac{d}{ds}|_{s=0}\zeta(s, x|A/\mu^2), \tag{15}$$

and thus, in a thermal theory, $Z_\beta = e^{-S_\beta}$ where $S_\beta = \int d\mu_g \mathcal{L}_{\beta\mu^2}$. One of the first recent results which, for compact manifolds at least, generalizes to any dimension earlier results of Wald [5], has been obtained in [3]. This result proves how the effective Lagrangian can be obtained by a *point-splitting procedure* (see below); for D even, it reads

$$\mathcal{L}(y|A)_{\mu^2} = \lim_{x\to y}\left\{-\int_0^{+\infty}\frac{dt}{2t}K(t,x,y|A) - \frac{a_{D/2}(x,y)}{2(4\pi)^{D/2}}\ln\frac{\mu^2\sigma(x,y)}{2}\right.$$
$$\left.+\sum_{j=0}^{D/2-1}(\frac{D}{2}-j-1)!\frac{a_j(x,y|A)}{2(4\pi)^{D/2}}\left(\frac{2}{\sigma(x,y)}\right)^{D/2-j}\right\} - 2\gamma\frac{a_{D/2}(y,y)}{2(4\pi)^{D/2}}, \tag{16}$$

and, for D odd (note that μ disappears from the final result),

$$\mathcal{L}(y|A)_{\mu^2} = \lim_{x\to y}\left\{-\int_0^{+\infty}\frac{dt}{2t}K(t,x,y|A) - \sqrt{\frac{2}{\sigma(x,y)}}\frac{a_{(D-1)/2}(x,y)}{2(4\pi)^{D/2}}\right.$$
$$\left.+\sum_{j=0}^{(D-3)/2}\frac{(D-2j-2)!!}{2^{(D+1)/2-j}}\frac{a_j(x,y|A)}{2(4\pi)^{D/2}}\left(\frac{2}{\sigma(x,y)}\right)^{D/2-j}\right\}. \tag{17}$$

Here $\sigma(x, y)$ is one half the square of the geodesic distance of x from y and the coefficients a_j are the well-known off-diagonal coefficients of the small-t expansion of the heat-kernel (see [3] for a short review of the properties of these coefficients).

2 Improvements to the local ζ-function technique

Generalizations of the local ζ function technique

Besides the effective Lagrangian and the effective action, further important one-loop quantities are the *(quantum) field fluctuation* and the *averaged (quantum) stress tensor*. These quantities are given, in terms of the Euclidean path integral, by

$$\langle\phi^2(x)\rangle = \frac{\delta}{\delta J(x)}|_{J\equiv 0}\ln\int\mathcal{D}\phi\, e^{-S_E+\int d\mu_g\phi^2 J}, \tag{18}$$

$$\langle T_{ab}(x)\rangle = \frac{2}{\sqrt{g(x)}}\frac{\delta}{\delta g^{ab}(x)}\ln\int\mathcal{D}\phi\, e^{-S_E[g]}. \tag{19}$$

A very popular method for computing the quantities above, which in practice diverge, is the so-called *point-splitting procedure* [7–9]; however, it is possible to generalize the ζ-function method in order to construct suitable ζ-functions which regularize the quantities above directly, similar to the procedure for the effective Lagrangian [3, 6, 10, 11]. Let us consider the stress tensor; our approach to a direct ζ-function regularization procedure is based on the following chain of formal identities [10],

$$\sqrt{g(x)}\langle T_{ab}(x)\rangle \text{ “=” } 2\frac{\delta}{\delta g^{ab}(x)} \ln Z_\beta \text{ “=” } \frac{\delta}{\delta g^{ab}(x)} \frac{d}{ds}|_{s=0}\zeta(s|A/\mu^2)$$
$$\text{ “=” } \frac{\delta}{\delta g^{ab}(x)} \frac{d}{ds}|_{s=0}\sum_j{}' \left(\frac{\lambda_j}{\mu^2}\right)^{-s} \text{ “=” } \frac{d}{ds}|_{s=0}\mu^{-2s}\sum_j{}' \frac{\delta\lambda_j^{-s}}{\delta g^{ab}(x)}. \quad (20)$$

Thus, one *defines* the ζ-regularized (or renormalized) stress tensor as

$$\langle T_{ab}(x|A)\rangle_{\mu^2} := \frac{1}{2}\frac{d}{ds}|_{s=0}Z_{ab}(s, x|A/\mu^2), \quad (21)$$

where, *in the sense of the analytic continuation of the left-hand side*,

$$Z_{ab}(s, x|A/\mu^2) := 2\sum_j{}' \mu^{-2s}\frac{\delta\lambda_j^{-s}}{\delta g^{ab}(x)}. \quad (22)$$

The mathematical problem is whether the right-hand side above can be computed in practice and whether it defines an analytic function of s in a neighborhood of $s = 0$. We have the following result [6, 10].

Theorem 1. *If M is compact, $A \geq 0$ and $\mu^2 > 0$, then $Z_{ab}(s, x|A/\mu^2)$ is well-defined and is a C^∞ function of x which is also meromorphic in $s \in \mathbb{C}$. In particular, it is analytic in a neighborhood of $s = 0$.*

In practice, the above result also has been checked for several noncompact manifolds also containing singularities [10]. In these cases the summation on the right-hand side of (22) has to be changed into a spectral integration. The form of the series on the right-hand side of (22) is [6, 10]

$$s\sum_j{}' \left\{\frac{2}{\mu^2}\left(\frac{\lambda_j}{\mu^2}\right)^{-s-1} T_{ab}[\phi_j, \phi_j^*](x) + g_{ab}(x)\left(\frac{\lambda_j}{\mu^2}\right)^{-s}\right\}.$$

Here $T_{ab}[\phi_j, \phi_j^*](x)$ is the classical stress tensor evaluated on the modes of the Euclidean-motion operator $A = -\Delta + \xi R + m^2 + V(x)'$ (see [6, 10] for details). The series converges for $Re\, s > 3D/2 + 2$.

Similarly, it is possible to construct a ζ-function for the field fluctuation [11], [3]. One has

$$\langle\phi^2(x|A)\rangle_{\mu^2} := \frac{d}{ds}|_{s=0}\Phi(s, x|A/\mu^2),$$

where

$$\Phi(s, x|A/\mu^2) := \frac{s}{\mu^2}\zeta(s+1, x|A/\mu^2). \quad (23)$$

The properties of these functions have been studied in [3, 11] and several applications in concrete cases are considered (e.g., cosmic-string spacetime and homogeneous spacetimes). In particular, in [3], the problem of the change of the parameter m^2 in the field fluctuations has been studied.

Physical correctness of the given regularization procedures

We next consider the physical interest of these regularization techniques. To this end, the following quite general results are relevant and prove that the proposed techniques are physically good candidates [6, 10].

Theorem 2. *If M is compact, $A \geq 0$ and $\mu^2 > 0$, and the above averaged quantities are those defined above in terms of local ζ-function regularization, then:*

a) $< T_{ab}(x|A) >_{\mu^2}$ *defines a C^∞ symmetric tensorial field.*
b) *As in the classical result,*

$$\nabla^b \langle T_{bc}(x|A)\rangle_{\mu^2} = -\frac{1}{2} < \phi^2(x|A) >_{\mu^2} \nabla_c V'(x). \tag{24}$$

c) *The trace of the stress tensor is naturally decomposed into the classical and the (correct) anomalous part*

$$\begin{aligned} g^{ab}(x)\langle T_{ab}(x|A)\rangle_{\mu^2} = &\left(\frac{\xi_D - \xi}{4\xi_D - 1}\Delta - m^2 - V'(x)\right)\langle\phi^2(x|A)\rangle_{\mu^2} \\ &+ \delta_D \frac{a_{D/2}(x, x|A)}{(4\pi)^{D/2}} - P_0(x, x|A), \end{aligned} \tag{25}$$

where $\delta_D = 0$ if D is odd, $\delta_D = 1$ if D is even and $\xi_D = (D-2)/[4(D-1)]$.
d) *For any $\alpha > 0$*

$$\langle T_{ab}(x|A)\rangle_{\alpha\mu^2} = \langle T_{ab}(x|A)\rangle_{\mu^2} + t_{ab}(x)\ln\alpha, \tag{26}$$

where the form of $t_{ab}(x)$, which only depends on the geometry and is in agreement with Wald's axioms [9], has been given in [6].
e) *If $\partial_0 = \partial_\tau$ is a global Killing vector, the manifold admits periodicity β along the lines tangent to ∂_0 and M is smooth (near any fixed points of the Killing orbits) for $\beta \in (2\pi - \delta, 2\pi + \delta)$, $\delta > 0$ and, finally, Σ is a hyper surface normal to ∂_0 everawhere, then*

$$\frac{\partial}{\partial\beta}\ln Z(\beta)_{\mu^2} = \int_\Sigma d\mathbf{x}\sqrt{g(\mathbf{x})} < T_0^0(x, \beta|A) >_{\mu^2}. \tag{27}$$

Another general result, which concerns the possibility of obtaining a Lorentzian theory from a Euclidean one, is the following [6].

Theorem 3. *Let M be compact, $A \geq 0$, $\mu^2 > 0$, let M also be globally static with global Killing time ∂_τ and (orthogonal) global spatial section Σ and finally, let $\partial_\tau V' \equiv 0$. Then:*

a) $\partial_\tau < \phi^2(x|A) >_{\mu^2} \equiv 0$;
b) $\partial_\tau \mathcal{L}(x|A)_{\mu^2} \equiv 0$;
c) $\partial_\tau < T_{ab}(x|A) >_{\mu^2} \equiv 0$;
e) $\langle T_{0i}(x|A)\rangle_{\mu^2} \equiv 0$ *for* $i = 1, 2, 3, ..., D-1$,

where the above averaged quantities are those defined above in terms of local ζ-function regularization and where coordinates $\tau \equiv x^0, \mathbf{x} \in \Sigma$, are employed.

These properties allow one to continue the Euclidean quantities considered into imaginary values of the coordinate $\tau \mapsto it$, obtaining *real* functions of the Lorentzian time t.

Some of the properties above (i.e., in Theorems 1–3) also have been checked for noncompact and symmetric manifolds (Rindler spacetime, cosmic string spacetime, Einstein's open spacetime, H^N spaces, Gödel spacetime, BTZ spacetime) [10–14]. In particular, the theory of the regularization of the stress tensor and field fluctuations via local ζ-functions has been successfully employed to compute the back reaction on the three-dimensional BTZ metric [14] in the case where the singular ground state contains a naked singularity, and a semiclassical implementation of the cosmic censorship conjecture has been found.

Relation with the point-splitting technique

The procedure of point-splitting for renormalizing field fluctuations as well as the stress tensor wirh which we are concerned [7, 9], can be summarized as

$$\langle \phi^2(y)\rangle_{\text{ps}} = \lim_{x\to y} \{G(x, y) - H(x, y)\} \tag{28}$$

$$\langle T_{ab}(y)\rangle_{\text{ps}} = \lim_{x\to y} \mathcal{D}_{ab}(x, y)\,\{G(x, y) - H(x, y)\} + g_{ab}(y)Q(y)/D, \tag{29}$$

where $G(x, y)$ is one half of the Hadamard function (i.e., one half of the sum of the two Wightman functions) of the quantum state considered or, in the Euclidean approach, the corresponding Green (Schwinger) function. $H(x, y)$ is the *Hadamard local fundamental solution*, a *parametrix* for the Green function which depends only on the *local* geometry and which takes the short-distance singularity into account. $H(x, y)$ is represented in terms of a truncated series of functions of $\sigma(x, y)$. The operator $\mathcal{D}_{ab}(x, y)$ is a bi-tensorial operator obtained by "splitting" the argument of the classical expression of the stress tensor (see [6] for a quite general expression of this operator). Finally $Q(y)$ is a scalar obtained by imposing several physical conditions (essentially, the appearance of the conformal anomaly, the conservation of the stress tensor and the triviality of the Minkowskian limit) [9] on the left-hand side of (29) (see [6–9] for details). The expression for $H(x, y)$ is the following, in a geodesically convex neighborhood containing both x and y,

$$
\begin{aligned}
H(x, y) = & \frac{\sum_{j=0}^{L} u_j(x, y)\sigma(x, y)^j}{(4\pi)^{D/2}(\sigma(x, y)/2)^{D/2-1}} \\
& + \delta_D \left[\sum_{j=0}^{M} v_j(x, y)\sigma(x, y)^j \ln\left(\frac{\sigma(x, y)}{2}\right) \right] \\
& + \delta_D \sum_{j=0}^{N} w_j(x, y)\sigma(x, y)^j .
\end{aligned} \tag{30}
$$

Here L, M, N are fixed integers (see [6] for details), $\delta_D = 0$ if D is odd and $\delta_D = 1$ otherwise. The coefficients u_j and v_j are smooth functions of (x, y) which are completely determined by the local geometry. Conversely, the coefficients w_j are determined once one has fixed w_0. In dealing with Euclidean approaches, it is possible to make u_j and v_j explicit in terms of heat-kernel coefficients [3, 6]. The problem is the determination of the coefficient w_0. This coefficient should determine the form of the term Q above. However, there is no guarantee that *any* choice of w_0 determine a Q such that the stress tensor obtained fulfills the requisite physical conditions. (Nevertheless, as is well-known [9], a possible choice is $w_0 \equiv 0$).

We have the following result [3, 6].

Theorem 4. *If M is compact, $A \geq 0$ and $\mu^2 > 0$, and the above averaged quantities are those defined above in terms of local ζ-function regularization, then*

$$
\langle \phi^2(y|A) \rangle_{\mu^2} = \lim_{x \to y} \{G(x, y) - H(x, y)\}, \tag{31}
$$

$$
\langle T_{ab}(y|A) \rangle_{\mu^2} = \lim_{x \to y} \mathcal{D}_{ab}(x, y) \cdot \left\{G(x, y) - H_{\mu^2}(x, y)\right\} + g_{ab}(y)Q(y)/D, \tag{32}
$$

where $G(x, y) = \zeta(1, x, y|A/\mu^2)$ as given in (14), H_{μ^2} is completely determined by setting

$$
w_0(x, y) := -\frac{a_{D/2-1}(x, y|A)}{(4\pi)^{D/2}}[2\gamma + \ln \mu^2], \tag{33}
$$

and the term Q is

$$
Q(y) = -P_0(y, y|A) + \delta_D \frac{a_{D/2}(y, y|A)}{(4\pi)^{D/2}}. \tag{34}
$$

All the requirements on the stress tensor necessary to determine Q in the point-splitting approach are now satisfied by the left-hand side of (32). Moreover, independent of the ζ-function approach, the above choice for w_0 and Q should also work in Lorentzian and noncompact manifolds as pointed out in [6], where this conjecture has been checked for Minkowski spacetime.

References

1. Fulling S.A., Ruijsenaars S.N.M. (1987): Phys. Rep. **152**, 135
2. Kay B.S., Wald R.M. (1991): Phys. Rep. **207**, 49
3. Moretti V. (1999): Local ζ-function techniques vs point-splitting procedures: a few rigorous results. Commun. Math. Phys. **201**, 327
4. Hawking S.W. (1977): Commun. Math. Phys. **55**, 133
5. Wald R.M. (1979): Commun. Math. Phys. **70**, 226
6. Moretti V. (1999): One-loop stress-tensor renormalization in curved background: the relation between ζ-function and point-splitting approaches, and an improved point-splitting procedure. J. Math. Phys. **40**, 3843
7. Birrel N.D., Davies P.C.W. (1982): Quantum Fields in Curved Space. Cambridge University Press, Cambridge
8. Fulling S.A. (1991): Aspects of Quantum Field Theory in Curved Space-Time. Cambridge University Press, Cambridge
9. Wald R.M. (1994): Quantum Field theory and Black Hole Thermodynamics in Curved Spacetime. The University of Chicago Press, Chicago
10. Moretti V. (1997): Phys. Rev. D **56**, 7797
11. Iellici D., Moretti V. (1998): Phys. Lett. B **425**, 33
12. Caldarelli M. (1999): Nucl. Phys. B **549**, 499–515
13. Radu E. (1998): Phys. Lett. A **247**, 207
14. Binosi D., Moretti V., Vanzo L., Zerbini S. (1999): Quantum scalar field on the massless $(2+1)$-dimensional black-hole background. gr-qc/9809041, Phys. Rev. D **59**, 104017

Solitons in Three Space Dimensions as a Model for Relativistic Particles

L. Pisani

Abstract. This paper gives a short report of research carried out in the last five years on the existence of solitons in three space dimensions.
Benci, Fortunato and Pisani have developed a model equation proposed in 1964 by C. H. Derrick. Using some recent techniques of nonlinear functional analysis, the existence of a non-trivial solitary wave, with a topological constraint, has been proved.
Subsequent research has demonstrated the relativistic behaviour of these waves, multiplicity results and the interaction with the electromagnetic field.

1 A problem posed in the 1960s

In an attempt to find a model for extended elementary particles, as opposed to singular point particles, in the 1960s, some authors considered the following model equation

$$\frac{\partial^2 \psi}{\partial t^2} - c^2 \Delta \psi + V'(\psi) = 0. \tag{1}$$

In the one-dimensional case Enz (see [1]) showed that

$$\frac{\partial^2 \psi}{\partial t^2} - c^2 \frac{\partial^2 \psi}{\partial x^2} + V'(\psi) = 0 \tag{2}$$

has time-independent solutions, where the energy is localized about a point in the x-axis; if we further require stability, then only certain discrete energy values are permitted. In addition these one-dimensional solutions possess certain symmetry and topological properties which Enz suggests might correspond in the three-dimensional case to such discrete quantum numbers as charge or parity.

The *sine-Gordon* equation

$$\frac{\partial^2 \psi}{\partial t^2} - c^2 \frac{\partial^2 \psi}{\partial x^2} + \sin \psi = 0 \tag{3}$$

gives an example of this kind of property: it is well-known that the static solutions of (3), *kink* and *anti-kink*, give rise to solitons, that is, solutions with finite energy which behave, and interact, as particles (see, e.g., [2]).

These suggestive results led Derrick (see [3]) to the following problem: "*Can (1) have stable, time-independent, localized solutions in three dimensions?*"

The answer given by Derrick is "no". Indeed, if $V(\xi) \geq 0$, then (1) has no localized, nontrivial solutions at all; otherwise, he deduces that the static solutions are unstable, since the second variation of the energy is not positive.

Derrick proposes several ways out of this difficulty; one of them is the following. Equation (1) is the Euler–Lagrange equation relative to the action

$$\mathcal{S} = \int \mathcal{L}\, dx\, dt$$

where the Lagrangian density $\mathcal{L}$ is given by

$$\mathcal{L} = -\left(c^2 |\nabla \psi|^2 - \left(\frac{\partial \psi}{\partial t}\right)^2\right) - V(\psi). \tag{4}$$

Derrick suggests taking a Lagrangian density of the form

$$\mathcal{L} = -\alpha \left(c^2 |\nabla \psi|^2 - \left(\frac{\partial \psi}{\partial t}\right)^2\right) - V(\psi) \tag{5}$$

with α a polynomial; indeed (5), as well as (4), gives rise to Lorentz-invariant equations. In particular for every static solution $u = u(x)$, we have also a one-parameter family of travelling solutions

$$\psi(x, t) = u\left(\frac{1}{\gamma}(x_1 - vt), x_2, x_3\right), \tag{6}$$

where $|v| < c$ and $\gamma = \sqrt{1 - v^2/c^2}$.

Derrick concludes: "Such a Lagrangian, however, leads to a very complicated differential equation." In fact, in the 1960s, the methods of nonlinear analysis were not sufficiently developed to study this kind of equation.

2 A model proposed in the 1990s

In [4], Benci, Fortunato and Pisani considered model (5); they assume that

$$\alpha(\sigma) = \sigma + \frac{\epsilon}{3}\sigma^3$$

where ε is an arbitrary positive constant. Note that, for sake of simplicity, in the following, we set $c = 1$.

The equation for the static solutions $u = u(x)$ becomes

$$-\operatorname{div}\left(\left(1 + \varepsilon |\nabla u|^2\right) \nabla u\right) + V'(u) = 0.$$

Since they are searching for waves with a topological constraint, in [4] the authors consider fields

$$\psi = (\psi^1, \psi^2, \psi^3, \psi^4).$$

So, in the static case, we have a system of 4 quasilinear elliptic equations

$$-\text{div}\left(\left(1+\varepsilon\left|\nabla u\right|^2\right)\nabla u^j\right)+\frac{\partial V}{\partial\xi^j}(u)=0,\qquad(1\leqslant j\leqslant 4). \tag{7}$$

As target space for the field ψ, we consider $\mathbf{R}^4\setminus\{\overline{\xi}\}$, with $\overline{\xi}\neq 0$ (such a space might be called an *internal parameters space*). Then, as to the function $V\in C^2(\mathbf{R}^4\setminus\{\overline{\xi}\},\mathbf{R})$, we make the following assumptions

$$V(\xi)\geqslant V(0)=0; \tag{8}$$

there exist $a, r>0$ such that

$$\left|\xi-\overline{\xi}\right|<r\Rightarrow V(\xi)>\frac{a}{\left|\xi-\overline{\xi}\right|^6}; \tag{9}$$

$$V''(0)[\xi,\xi]=\omega_0^2\left|\xi\right|^2,\ \text{with}\ \omega_0>0. \tag{10}$$

Remark 1. The linearization of Eq. (7) at 0 gives a system of Klein–Gordon equations

$$\frac{\partial^2\psi^j}{\partial t^2}-\Delta\psi^j+\omega_0\psi^j=0.$$

Equations (7) have a variational structure: the solutions can be obtained as critical points of the energy functional

$$E(u)=\int_{\mathbb{R}^3}\left[\frac{1}{2}\left|\nabla u\right|^2+\frac{\varepsilon}{6}\left|\nabla u\right|^6+V(u)\right]dx. \tag{11}$$

This functional is real-valued and continuous on

$$\Lambda=\left\{u\in H\mid\forall x\in\mathbf{R}^3:u(x)\neq\overline{\xi}\right\},$$

which is an open subset of the function space

$$H=H^1\left(\mathbf{R}^3,\mathbf{R}^4\right)W^{1,6}\left(\mathbf{R}^3,\mathbf{R}^4\right).$$

The functions in H are continuous and decay to zero at infinity.

Now we recall that, up to a multiplicative constant, on $\mathbf{R}^4\setminus\{\overline{\xi}\}$ there exists a unique 3-form η which is closed but non-exact. For every $u\in\Lambda$, we can consider the pull-back of η with respect to the function u:

$$\varrho(u,\nabla u)=u^*(\eta)=\frac{1}{\left|S^3\right|}\frac{\det(u,\nabla u)}{\left|u-\overline{\xi}\right|^4},$$

where $\left|S^3\right|$ is the measure of the sphere S^3. Thus we define the *charge* of a vector field $u\in\Lambda$ as

$$ch(u)\overset{\text{def}}{=}\int_{\mathbf{R}^3}\varrho(u,\nabla u)\,dx\in\mathbf{Z}. \tag{12}$$

Now we can state the result proved in [4].

Theorem 2. *There exists a weak solution of (7) which is obtained as a minimum of the energy functional (11) in*

$$\Lambda^* = \{u \in \Lambda \mid ch(u) \neq 0\}.$$

From the mathematical point of view, we have the following difficulties.

1. The functional is defined on an open subset, so we have to check the behavior of E on the boundary of Λ; the order of singularity in (9) is chosen so that the energy diverges for every $u \in H$ that touches $\overline{\xi}$.
2. The domain for the independent variable x is unbounded; so, in order to find local minima, the authors use a splitting lemma in the spirit of the Concentration-Compactness Lemma.
3. The connected components of Λ, which are characterized by the charge and are also called *sectors*, are not weakly closed. So an open question is whether the solution which realizes the minimum in Λ^* has charge equal to 1.
4. At present, we are not able to prove that these minima solutions are stable and give rise to solitons, so we call them *solitary waves*.

3 Interaction with an electromagnetic field

Since the topological invariant can be interpreted as the electric charge, Benci, Fortunato, Masiello and Pisani (see [5]) have studied the interaction between the solution ψ and the electromagnetic field (E, H). This field is described by the electromagnetic 4-vector potential (A, ϕ):

$$E = -(A_t + \nabla\phi) \tag{13}$$

$$H = \nabla \times A. \tag{14}$$

For the system "soliton-electromagnetic field", the total action is given by

$$\mathcal{S} = \mathcal{S}(\psi, A, \phi) = \int \mathcal{L}_1(\psi)\, dx dt + \int \mathcal{L}_2(A, \phi)\, dx dt + \int \mathcal{L}_3(\psi, A, \phi)\, dx dt$$

where $\mathcal{L}_1$, $\mathcal{L}_2$ and $\mathcal{L}_3$ denote the Lagrangian density of the field ψ, of the electromagnetic field and of the interaction between ψ and (E, H), respectively.

The Lagrangian density of the electromagnetic field is defined in the standard way as

$$\mathcal{L}_2 = \frac{1}{8\pi}\left(|E|^2 - |H|^2\right).$$

The Lagrangian density relative to the interaction is given by

$$\mathcal{L}_3 = \psi^*(\eta) \wedge \omega$$

where $\psi^*(\eta)$ is the pullback of η, a 3-form in $\mathbf{R}^4\setminus\{\overline{\xi}\}$ closed but not exact, and

$$\omega = A_1 dx^1 + A_2 dx^2 + A_3 dx^3 - \phi dx^4,$$

where $x_4 = t$.

If we denote by $(-1)^i J_i(\psi, \nabla\psi, \psi_t)$ the first three components of $\psi^*(\eta)$, then we obtain the usual form

$$\mathcal{L}_3 = (J(\psi, \nabla\psi, \psi_t) \mid A) - \varrho(\psi, \nabla\psi)\phi.$$

Remark 3. Since η is closed, $\psi^*(\eta)$ is also closed and then the continuity equation holds true:

$$\nabla J + \frac{\partial \varrho}{\partial t} = 0. \tag{15}$$

Equation (15) has several consequences: first, we can interpret ϱ as charge density and J as current density; second, the energy

$$\int \mathcal{L}_3(\psi, A, \phi)\, dx dt \tag{16}$$

is gauge invariant.

Remark 4. From the definition of J, it follows that

$$\frac{\partial \psi}{\partial t} = 0 \Rightarrow J = 0,$$

namely, the current density vanishes whenever the charge carried out by the soliton does not move.

If we consider the variation of $\mathcal{S}$ with respect to A and ϕ, then we get

$$\nabla \times (\nabla \times A) = 4\pi J - \frac{\partial}{\partial t}(A_t + \nabla\phi) \tag{17}$$

$$-\nabla(A_t + \nabla\phi) = 4\pi\varrho. \tag{18}$$

Using (13) and (14), we can write (17) and (18) as

$$\nabla \times H = 4\pi J(\psi, \nabla\psi, \psi_t) + E_t$$
$$\nabla \cdot E = 4\pi\varrho(\psi, \nabla\psi),$$

which complete the Maxwell equations.

In [5] the following result has been proved.

Theorem 5. *There exist*

$$u : \mathbf{R}^3 \to \mathbf{R}^4\setminus\{\overline{\xi}\}$$
$$\phi : \mathbf{R}^3 \to \mathbf{R}$$

such that $ch(u) \neq 0$ *and* $(u, 0, \phi)$ *is a (weak) static solution of the Euler–Lagrange equation*

$$d\mathcal{S} = 0.$$

Note that, for this problem, which again has a variational structure, we have to find critical points of a functional which is strongly unbounded from below and from above. So the solution (u, ϕ) we find is a saddle point.

4 Variations and developments

In the last section of [4], by studying the relations between energy and momentum of the traveling solutions (6), it is shown that:

1. the rest mass is a scalar and not a tensor;
2. the mass equals the energy (Einstein's equation) and increases with the velocity by the factor γ.

In a subsequent paper [6], the same authors have studied the case of a vector field

$$\psi = (\psi^1, \psi^2, \ldots, \psi^k)$$

that takes values in a suitable subset $\Omega \subset \mathbf{R}^k$. Now the sectors of Λ are classified by means of a global topological invariant $u^\# \in \pi_3(\Omega)$ and then we have a multiplicity result related to the structure of $\pi_3(\Omega)$.

In the case $\Omega = \mathbf{R}^4 \backslash \{\xi_1, \ldots, \xi_k\}$, where the fields are classified by k integers (*quantic numbers*), we get the existence of k distinct local minima.

Under suitable symmetry assumptions on V, the result of [6] can be applied also in the case of only one singularity $\{\overline{\xi}\}$: in [7] the authors prove the existence of another non-trivial solution, which is different from that found in [4], since it has void charge.

Another interesting change of target space is the case of a manifold $\mathcal{M}$ (due to Benci, Giannoni and Piccione; see [8]). Here it is not necessary to have a singular function V in order to get a non-trivial classification of static fields; thus we see some similarity with the model for elementary particles due to Skyrme (see [9]).

Using this framework, in [10], Benci and Fortunato have proved the existence of string-like solitons, in the same spirit as that of some recent studies of Faddeev and Niemi (see [11]).

The most recent variation on our model is contained in [12]. In this paper, the authors study a field equation which is related to the Double Solution Theory of de Broglie (causal interpretation of Quantum Theory).

This class of results shows that the use of nonlinear interactions plays an essential role in the description of relativistic and quantic phenomena by means of classical fields (solutions of PDE in a Newtonian space-time).

References

1. Enz U. (1963): Discrete mass, elementary length and a topological invariant as a consequence of a relativistic invariant variational principle. Phys. Rev. **131**, 1932–1934
2. Lamb G.L. (1980): Elements of soliton theory. Wiley, New York
3. Derrick C.H. (1964): Comments on nonlinear wave equations as model for elementary particles. Jour. Math. Phys. **5**, 1252–1254

4. Benci V., Fortunato D., Pisani L. (1998): Solitons like solutions of a Lorentz invariant equation in dimension 3. Reviews in Mathematical Physics **3**, 315–344
5. Benci V., Fortunato D., Masiello A., Pisani L. (1999): Solitons 1 and electromagnetic field. Math. Z., 73–102
6. Benci V., Fortunato D., Pisani L. (1996): Remarks on topological solitons. Topological Methods in Nonlinear Analysis **7**, 349–367
7. D'Avenia P., Pisani L.: Remarks on the topological invariants of a class of solitary waves. Nonlinear Analysis T.M.A., in press
8. Benci V., Giannoni F., Piccione P. (2000): Solitons on manifolds. Advances in Differential Equations, in press
9. Skyrme T.H. (1961): A non linear field theory. Proc. Roy. Soc. A **260**, 127–138
10. Benci V., Fortunato D. (1999): Existence of string-like soitons. Ricerche di Matematica, **48**, Suppl., pp. 399–406
11. Faddeev L., Niemi A.J. (1997): Stable knot-like structures in classical field theory. Nature **387**, 58–61
12. Benci V., Fortunato D.: On the Existence of the Impossible Pilot Wave, in *Proc. of Conference "Calculus of Variations and Related Topics", March 1998, Haifa, Israel*

Relativistic Astrophysics and Cosmology

Relativistic Astrophysics and Cosmology

F. Occhionero

Without pretending to represent the vast extent of the scientific interests of our national community, this year's Workshop has been dedicated to four sharp wedges of active research in the discipline:

- Recent Results about Gamma Ray Bursters (GRB's);
- Accretion onto Black Holes (BH's);
- Backgrounds of Gravitational Waves (GW's);
- Large and Very Large Scale Structure of the Universe (LSS and VLSS).

On GRB's, we heard the enthusiastic talk of Dr. E. Costa, (IAS–CNR), one of the leaders among the group of scientists of the Dutch-Italian enterprise, about the satellite "Beppo–SAX", named after the Italian physicist Giuseppe Occhialini. Rapid location of GRB's in the sky by the advanced tecnology of Beppo–SAX made it possible to alert optical telescopes around the world immediately and to catch several of the GRB's optical afterglows with them and so to collect the basic spectroscopic information: redshift. One cosmic record after the other paved the way to the cosmological scenario (leaving the galactic one in the shadow) and to a fascinating energy problem: thanks to Beppo–SAX, after struggling in the mud for thirty years, GRB's are on the move again. It is strongly regretted that the talk could not be written up for these Proceedings.

On BH's J. Miller (SISSA and Oxford) presented new results on non-stationary disk accretion for potential application both to the stellar mass BH's of our Galaxy and to the huge monsters lurking in the cores of active galactic nuclei. Seeking to explain the observed (and very different) time scales, the author concentrates on instabilities of different origin and follows them numerically, under reasonable approximations, with a Lagrangian code. Beautiful examples of inward and outward waves are displayed for relevant values of the parameters.

On stochastic backgrounds of GW's, while most of the present interest in interferometric detectors concentrates on inflationary tensor perturbations, R. Schneider ("La Sapienza" and INFN) sheds light on the GW component generated by an early burst of supernovae explosions and the ensuing core collapse and/or neutron star spin–down radiation. The author then proceeds to evaluate the detectability of this less known background.

Supernovae explosions have always enjoyed the center stage of the astrophysical play: this will become even more true with the strengthening connection with GRB's.

On LSS and VLSS: firstly, in the last decade, dramatic observations of the cosmic density of clustered matter have forced cosmologists to face a conflict with the inflationary paradigm in favor of an open model. (It must be said at the very start, however, that the early nucleation of primordial bubbles in an inflationary phase transition would make a superhorizon, but finite bubbles look like an infinite open Universe and would thus reconcile openness with inflation, or better with "open" inflation.) Secondly, the famous high redshift supernovae observations permitted the recovery of flatness thanks to a mysterious dark energy capable of accelerating the cosmic expansion close to the present epoch. At the modest cost of a cosmic coincidence or concordance, all this points to a cosmological constant or to a frozen scalar field. Here, as alternative explanations of the cosmic acceleration seem still possible, I am taking the point of view that the primordial phase transition generating the superhorizon bubbles and the VLSS is repeated inside each bubble Universe and generates an LSS with fractal dimension up to scales of the order of hundreds of Mpc. A sort of "foam within a foam" might describe what happens inside and outside the horizon.

Voids in Open Universes

F. Occhionero, L. Amendola

Abstract. The universe may be open and its galaxies may be distributed on spherical shells around the large scale voids that are detected in redshift surveys. To account for both features at once, we devise and present a new inflationary model that harbors two episodes of phase transitions. Of these transitions the former nucleates super-horizon-sized bubbles, which are seen from inside as open universes, while the latter generates a distribution of sub-horizon voids, of sizes typically around 10–100 Mpc/h, inside the open universes. The model realizes this sequence along the same slow-rolling path, by modulating the energy difference between vacuum states. Thus, with one physical process, it is possible to reconcile inflation with the recent observations of a low Ω_0 and to seed the present large scale voids.

A non-Gaussian signal on the microwave background might be the smoking gun.

1 Introduction

The properties of the global space-time geometry of the universe and the properties of the small inhomogeneities it contains are at the center of modern cosmology. In fact, these are two aspects of the same general question: the solution of the general relativistic field equations and the reconstruction of the initial conditions from present observations.

The global space-time geometry, under the hypothesis of homogeneity and isotropy, can be embodied in a small set of unknown parameters, like Ω_0, H_0, Λ, which express the basic dynamics of the model. As observations are extended and made more precise, the cosmological parameters are being constrained more and more tightly. At the moment we are witnessing the breakthrough of Λ (for a review see [1]), on the basis of the high redshift SNe Ia results [2, 3]. However, the assessment is far from definitive: for example, in [4] the authors claim that an explanation of the SNe results does not warrant necessarily a cosmological constant, but may be achieved as well in the scenario of a local underdense bubble, hundreds of Mpc in size, in an open universe.

Thus, we may be back in the familiar situation in which Ω_0 is smaller than unity, $\Omega_0 \approx 0.3$ [5], in clear contrast with the naive inflationary expectation that $\Omega_0 = 1$. Even this low value of Ω_0 is certainly far from conclusive. First, it refers essentially only to the matter contained within clusters; secondly, even for clusters there still may be observational biases that produce a value smaller that the real one. However, it is necessary to start speculating on the possibility that we live in an open universe, and to try to reconcile the low-Ω observations with the inflationary paradigm. The most promising hypothesis so far is probably that the local open horizon volume is but a bubble embedded in a globally flat universe still undergoing inflation, [6–10].

The walls of the bubble are then invisible because expanding at the velocity of light: from within, our horizon volume looks like an infinite open space.

As regards the small inhomogeneties superimposed on the global background, we are still far from an accurate and uncontroversial modeling of the observations. Partly due to still sparse data (the volume of universe accurately sampled by redshift surveys is less than 5%), partly to the creative power of theorists, there are now several models of large scale clustering that compete to become the new standard model, after the dismissal of the simplest model of CDM. In this context, however, a trend that seems to emerge from redshift surveys is that of an ever more structured distribution of matter, that is hardly possible to characterize simply by two-point descriptors. In particular, several groups reported the existence of huge regions either totally devoid of luminous matter or simply underdense [11–13]. These voids can be described in a first approximation as spherical bubbles, with an empty or underdense core and a thin shell. Typical sizes range from several Mpc to $40\,h^{-1}$ Mpc or more in diameter. Although these voids are not necessarily in contrast with standard models of structure formation, in particular, with models in which small and Gaussian initial inhomogeneities grow by pure gravitational instability into non-linear and non-Gaussian objects, it is worth considering them as additional ingredients on their own and studying their possible independent role in the process of structure formation. Dynamical models have been given in terms of the Tolman solution [14, 15]. A distribution of large scale voids has been studied by several authors, in order to assess their dynamics and their ability to correctly reproduce the statistics observed [16–19]. Their impact on the cosmic microwave background has also been studied [20–22]. Since their signal is typically ring-like, it might be possible to identify them on high resolution maps of the CMB.

To summarize, the status of recent cosmological observations and theoretical requirements is consistent with an inflationary model in which sub-horizon scale voids are produced within super-horizon-sized bubbles. This paper presents a simple realization of this model.

The physical scheme for such a model is very easy to understand in a two-field inflationary potential $U(\omega, \psi)$, such that the slow-rolling inflation is driven in the ω direction. Let the potential in the ψ direction have two minima (actually, two channels directed along ω) with an energy difference ΔU that depends on ω. As ω evolves, a transition occurs along ψ from the higher to the lower level. Now, since the tunneling rate depends on the energy difference, i.e., on ω, and thus ultimately on time, we can find a potential for which the transition rate is modulated in a way which produces *two* epochs of intense tunneling. To be more detailed, let us label time with the number N of e-foldings before the end of inflation, putting $N_H = 60$ as the beginning of the observable part of inflation. We can then find a potential with two channels along which the following phases take place: a) first, much earlier than the observable 60 e-foldings, the energy difference is vanishingly small, so that no transition occurs; b) second, as we approach N_H, the tunneling rate from the minimum A to the minimum B increases, and reaches a maximum before N_H so that bubbles of open universes nucleate in B; c) third, at some $N_* \geq N_H$, the energy difference reaches zero so that tunneling halts abruptly without achieving completion; d) fourth, below $N = 60$, the rate increases again, but this time the transition occurs from B to A, since the energy difference has the opposite sign from before; e) fifth, the whole space in phase B is converted into phase A, and the transition effectively stops again, after completion. At this point we find in our universe a number of open universes (A→B) the inside of which is filled by sub-horizon bubbles (B→A). Outside the open-universe bubbles, space is still expanding inflationarily so that the bubbles do not percolate. Finally, inflation ends in both channels, leaving behind a non-negligible fraction of space in super-horizon-sized open-universe bubbles, filled with a distribution of sub-horizon voids that are the seeds voids of the present-day.

The essential point in the mechanism is the change of sign of the energy difference around N_H. As we will show, all the other features arise almost automatically once this is realized. It is through this sign inversion that two distinct epochs of bubble formation are possible, on cosmological scales ($> 6000\,\mathrm{h}^{-1}$ Mpc) and on the astrophysical scale ($\approx$10–100 h^{-1} Mpc). The observed value of Ω_0 and the typical scale for the voids in the galaxy distribution fix the times at which the two epochs occur.

The rest of the paper is organized as follows. In Sect. 2 we review the basics of tunneling in two-field potentials, and propose a first simple potential that arises in quadratic gravity. As this potential fails however, although by a narrow margin, in Sect. 3 we identify a modified version of it in GR that implements our scheme. In Sect. 4 we present our conclusions.

2 Bubble spectra

As regards the mechanism that could trigger the formation of bubbles, La [23] suggested a pivotal cosmogonical role for the primordial phase transition (PT) consisting of the nucleation of bubbles of true vacuum (TV) in a false vacuum (FV) background via the quantum tunneling (QT) of a scalar field ψ across a potential barrier.

We learn from the pre-inflationary, but cosmologically minded, works of Coleman and collaborators, [24–26], how to evaluate the fundamental quantities, at least in the semi-classical limit and in the thin wall approximation, namely,

- the Euclidean action S on the famous bounce solution (explicit expression in (16) below);
- the tunneling rate,

$$\Gamma = \mathcal{M}^4 \exp(-S), \tag{1}$$

  where $\mathcal{M}$ is of the order of the energy scale of SSB, and finally,
- the number of bubbles nucleated per unit vol. and time,

$$\frac{dn_B}{dt} = \Gamma V_{\mathrm{FV}}, \tag{2}$$

$$V_{\mathrm{FV}} = \frac{4\pi}{3}\frac{1}{H_{\mathrm{in}}^3}\left(\frac{a}{a_{\mathrm{in}}}\right)^3 \exp(-I),$$

$$I(t) = \frac{4\pi}{3}\int_{-\infty}^{t} dt'\Gamma(t')\left(a(t')\int_{t'}^{t}\frac{dt''}{a(t'')}\right)^3,$$

where $a(t)$ is the scale factor of a flat Friedmann metric,

$$a(t)/a_{\rm in} = \exp\alpha(t), \quad N_{\rm in} = \alpha + N, \tag{3}$$

and the subscript *in* refers to the epoch when the nucleation process begins, $N_{\rm in} > N_H$.

In order to pin down the physics, however, the most interesting quantity is the dimensionless rate

$$Q = \frac{4\pi}{9}\frac{\Gamma}{H^4}, \tag{4}$$

which, on attaining unity, signals directly the achievement of percolation: the nucleation of one bubble in each horizon four volume.

For cosmological applications to a de Sitter background (fixed H) and for fixed Γ it was immediately recognized [27] that of this process one has either too much ($Q >> 1$) or too little ($Q << 1$), but hardly the right amount, due to the presence of the exp $-I$ factor in (2); see, e.g., the "graceful exit" problem of Old Inflation. It became clear with extended inflation [28] and better still with two-field inflation [29, 30] that the process does work if Γ, and/or H, are variable and if, with their variation, an epoch $t_{\rm nucl}$ occurs where

$$Q(t_{\rm nucl}) \approx 1. \tag{5}$$

Furthermore, this timing must be such that the bubbles are able to inflate to the right size: for instance, if they all were too small, say $\leq 10, h^{-1}$ Mpc in comoving size, they would be already thermalized in the Radiation Dominated Era (RDE) [31] and would not be of any use in the Matter Dominated Era (MDE). In practice, the burst of nucleation must happen early enough, i.e., the number of e-folds from the PT to the end of inflation, $N(\omega)$, the convenient time parameter, must be 50 or 60 for the bubbles to attain astrononomical sizes (tens of Mpc) or cosmological sizes (thousands of Mpc), respectively. For this to be possible, it is necessary that the potential $U(\omega)$ of the TV remain > 0 and capable of driving the required N's of slow-roll after ψ has tunneled.

After changing variables from t to N we may extract from (2) an expression for the fraction of TV, $F^{\rm TV}$, nucleated around N:

$$\frac{dF^{\rm TV}}{dN} = \frac{dn_B}{dN}\times\left(\frac{L(N)}{L(N_{\rm in})}\right)^3 = -3Q(N)\exp(-I), \tag{6}$$

$$L(N) = \frac{1}{H(N)}\times\exp N.$$

In the following we will treat the I-term in either of the two approximations relevant to the present problem:

i) on larger cosmological scales $I \approx 0$ corresponding to an incipient PT far from completion, or
ii) on smaller astronomical scales I grows quickly, completing the PT.

Quadratic gravity

In general relativity (GR) it takes a second field ω, driving inflation with its potential $U(\omega)$, to make $Q(\omega)$ astronomically interesting. A way to design such a configuration is suggested by quadratic gravity (QG), a theory derived from the quadratic Lagrangian

$$\mathcal{L}_{\text{grav}} = -R + \frac{R^2}{6M^2}, \tag{7}$$

(where $M < 10^{-5} M_{\text{Planck}}$ from CMB isotropy; see the pioneering work of Starobinsky [32]). Actually, in this theory, inflationary solutions are attractors for a vast region of the PS of the initial conditions [33].

The conformal transformation which brings QG to canonical GR [34],

$$\widetilde{g}_{\alpha\beta} = e^{2\omega} g_{\alpha\beta}, \quad e^{2\omega} = \left| \frac{\partial \mathcal{L}_{\text{grav}}}{\partial R} \right| = 1 - \frac{R}{3M^2}, \tag{8}$$

on a Friedmann background yields the Hubble expansion

$$H^2 = \frac{8\pi}{3} U(\omega), \tag{9}$$

under the conformal potential

$$\begin{aligned} U(\omega) &= \frac{R\mathcal{L}' - \mathcal{L}}{16\pi |\mathcal{L}'|^2} \\ &= \frac{3M^2}{32\pi} e^{-4\omega} (e^{2\omega} - 1)^2. \end{aligned} \tag{10}$$

As the latter approaches a constant as $\omega \to \infty$,

$$U(\omega) \simeq \frac{3M^2}{32\pi}, \quad H^2 = \left(\frac{M}{2}\right)^2, \tag{11}$$

QG's inflation is de Sitter. Finally a convenient way of writing the slow-rolling solution is

$$\frac{4}{3} N(\omega) = e^{2\omega} - 1, \tag{12}$$

which links the time-keeping N to the slow-roll driving ω.

When adding the ingredient of matter,

$$\mathcal{L}_{\text{tot}} = \mathcal{L}_{\text{grav}} + \mathcal{L}_{\text{mat}},$$

in the form of the scalar field ψ,

$$\mathcal{L}_{\text{mat}} = 16\pi \left[\frac{1}{2} \psi_{,\sigma} \psi^{,\sigma} - V(\psi) \right],$$

a coupling to gravity through R^2 (not R as usual),

$$\mathcal{L}_{\rm grav} = -R + \frac{R^2}{6M^2 W(\psi)}, \tag{13}$$

gives convenient shape to the conformal potential of (10). In fact, under (8), the latter takes the form

$$U(\omega, \psi) = e^{-4\omega}\left[V(\psi) + \frac{3M^2}{32\pi} W(\psi)(e^{2\omega} - 1)^2\right], \tag{14}$$

which is easily carved for the purpose of QT [35]. If the dimensionless W is double-welled and degenerate, like the quartic

$$W(\psi) = 1 + \frac{\lambda}{\psi_0^4}\psi^2(\psi - \psi_0)^2,$$

and, if V is single-welled, like the parabola

$$V(\psi) = \frac{1}{2}m^2\psi^2,$$

then $U(\omega, \psi)$ will have two non-degenerate, almost parallel channels, in which QT can take place:

- the TV channel is at $\psi = 0$;
- the FV channel is at $\psi \simeq \psi_0$,, and in between;
- the barrier of height $\propto \lambda$ is at $\psi = \psi_0/2$.

Thus QG allows the implementation of what is needed by splitting the potential in

- a slow-roller, and
- a symmetry breaker:

$$\begin{aligned} U(\omega, \psi) &= U_{\rm slowroll}(\omega, \psi) + \Delta U_{\rm symm.b.}(\omega, \psi), \\ U_{\rm slowroll}(\omega, \psi) &= \frac{3M^2}{32\pi}\left(\frac{e^{2\omega} - 1}{e^{2\omega}}\right)^2 W(\psi) \simeq \frac{3M^2}{32\pi} W(\psi), \\ \Delta U_{\rm symm.b.}(\omega, \psi) &= e^{-4\omega} V(\psi) \simeq \frac{1}{N^2} V(\psi), \end{aligned} \tag{15}$$

with the latter, $\Delta U_{\rm symm.b.}$, removing the degeneracy and allowing bubble production only for finite $N >> 1$, but not as $N \to \infty$.

Turning now to Coleman's action, with (15) its evaluation becomes straightforward:

$$S = \frac{27\pi^2}{2(\Delta U_{\rm symm.b.})^3}\left[\int_0^{\psi_0} d\psi \sqrt{2 \times \frac{3M^2}{32\pi} \times (W(\psi) - 1)},\right]^4 \tag{16}$$

$$= \frac{27\pi^2}{2(\Delta U_{\text{symm.b.}})^3} \left(\frac{3M^2 M_{Pl}^2}{16\pi} \right)^2 \frac{\lambda^2 \psi_0^4}{6^4}.$$

It displays a crucial dependence on $(\Delta U_{\text{symm.b.}})^3$ in the denominator (and on a λ^2 in the numerator) and for a quartic $W(\psi)$ becomes $S \propto e^{8\omega} \propto N^4$ (we have lost a factor $e^{4\omega}$ due to the conformal transformation [30]).

For the purpose of tuning the process, we now introduce two new parameters, N_0 and N_1, combinations of the five fundamental parameters $(\mathcal{M}, M, m, \lambda, \psi_0)$: in detail, if we write (16) as

$$S(N) = \left(\frac{N}{N_1} \right)^4, \quad \frac{1}{N_1^4} = \frac{M^4 M_{Pl}^4}{108 m^6 \psi_0^2}, \tag{17}$$

then N_1 signals the end of the nucleation of spherical bubbles, since $S(N_1) = 1$, in favor of spinodal decomposition (occurring for $N < N_1$). Likewise, if we write (4) as

$$Q(N) = \exp\left[\left(\frac{N_0}{N_1} \right)^4 - \left(\frac{N}{N_1} \right)^4 \right], \quad \exp\left(\frac{N_0}{N_1} \right)^4 = \frac{64\pi}{9} \left(\frac{\mathcal{M}}{M} \right)^4, \tag{18}$$

then N_0 signals percolation and marks the end of the PT, since $Q(N_0) = 1$. Now, from (17) and (18) we can compute bubble spectra [35, 36]: thus astronomical information encoded in the bubble spectrum in principle translates into information about the fundamental parameters [37].

Swapping roles

The potential (15) has the useful property of a variable energy difference between the two vacuum states, but lacks the decisive feature of a sign inversion to swap the role of the vacuum channels. This feature is easy to introduce with a new factor

$$\Delta U_{\text{symm.b.}}(\omega, \psi) = e^{-4\omega} V(\psi) \times \left(1 - \frac{N(\omega)}{N_*} \right). \tag{19}$$

Under its action, the vacuum channels exchange their roles at N_*, i.e., if QT occurs from $\psi = 0$ to $\psi \approx \psi_0$ before N_*, it will occur from $\psi \approx \psi_0$ to $\psi = 0$ thereafter.This is the simplest realization of the mechanism able to produce two transitions, one inside the other. However, it is not yet what we need, because, owing to the de Sitter nature of QG's inflation, this change of sign is accompanied in the channel by minima and maxima that stop the slow-roll. For this reason we will go back to GR, carrying over this suggestion in an *ad hoc* potential, and, of course, we will choose $N_* \simeq 60 >> 1$, in order to separate

- early nucleation at (large) cosmological scales ($N > N_*$, label "C"), from
- later nucleation at (smaller) astronomical ones ($N < N_*$, label "A").

3 Bimodal spectra

Seen from inside, a finite bubble mimics an open Universe [6–8]. Hence, observations [5] claiming that $\Omega_0, \approx 0.3, <, 1$, are not in conflict with inflation if we live inside a bubble nucleated at time suitable for its Ω_0 to have grown exactly to the value observed, namely [9], at an $N(\Omega_0)$ given by (see Tab. 1)

$$\Omega_0 = \frac{1}{1 + 4 \exp 2\left[N_H - N(\Omega_0)\right]}. \tag{20}$$

We must now produce two generations of bubbles, one inside the other, and make sure at the same time that the first PT does not run to completion (in earlier work [36], we achieved this by introducing a ramp in the potential barrier). Assuming GR as the underlying gravity, two fields ω and ψ, and the simplest chaotic potential,

$$U(\omega) = \frac{1}{2} M^2 \omega^2 = \frac{M^2}{4\pi} N, \tag{21}$$

where we made use of $N = N(\omega) = 2\pi\omega^2$, we forge our phenomenological *Ansatz* as in (15):

$$\begin{aligned} U_{\text{slowroll}}(N, \psi) &= \frac{M^2}{4\pi} N W(\psi), \\ \Delta U_{\text{symm.b.}}(N, \psi) &= V(\psi) \frac{N^{2/3}}{1 + N^q} \left(1 - \frac{N}{N_*}\right), \quad q >> 1, \end{aligned} \tag{22}$$

where the two episodes of nucleation are limited by $\Delta U_{\text{symm.b.}}(N \to \infty) \to 0$ on the large scales, on small scales by the completion itself of the PT, and are separated on intermediate scales by $\Delta U_{\text{symm.b.}}(N_*) = 0$. The factor $N^{2/3}$ is relatively unimportant as it is meant only as a formal simplification; instead, q will be used to constrain the position and the width of the peak of the bubble spectrum at the C-scales (see (26) below). (The lack of such a free parameter is another reason why QG cannot be used for this toy model).

In place of (16) we end up now with two different expressions for the action on the A- and the C-scales due to the two signs of $\Delta U_{\text{symm.b.}}$, which, upon defining as

Table 1. Number of e-foldings at nucleation and comoving size of open Universe bubbles for different Ω_0. The two values N_+ and N_- that bracket $\Omega_0 = .3$ will be used below

Ω_0	$N(\Omega_0)$	$L(N(\Omega_0))$ in h^{-1} Mpc
.2	60.0=N_-	6,000
.3	60.3	8,000
.4	60.5=N_+	10,000
1	∞	∞

in (17) a suitable $N_1 = g(M, m, \lambda, \psi_0)$ where the action on the A-scales becomes unity and the spinodal decomposition begins, are written as

$$S_A(N) = \left(\frac{N}{N_1}\right)^{3q} \left(\frac{1 - N_1/N_*}{1 - N/N_*}\right)^3, \quad N < N_*, \tag{23}$$
$$S_C(N) = \left(\frac{N}{N_1}\right)^{3q} \left(\frac{1 - N_1/N_*}{N/N_* - 1}\right)^3, \quad N > N_*.$$
$$\frac{(1 - N_1/N_*)^3}{N_1^{3q}} = \frac{M^4 M_{Pl}^4}{48 m^6 \psi_0^2},$$
$$S_A(N_1) = 1.$$

Similarly, we have two expressions for Q defined in (4),

$$Q_{A,C}(N) = \left(\frac{N_0}{N}\right)^2 \times \exp\left[S_A(N_0) - S_{A,C}(N)\right], \tag{24}$$
$$Q_A(N_0) = 1,$$
$$\pi \left(\frac{\mathcal{M}}{M}\right)^4 = N_0^2 \exp\left[\left(\frac{N_0}{N_1}\right)^{3q} \left(\frac{1 - N_1/N_*}{1 - N_0/N_*}\right)^3\right],$$

and two expressions for the fraction of TV, (6),

$$\frac{dF_A^{TV}}{dN} = -3Q_A(N), \exp -I, \quad \frac{dF_C^{TV}}{dN} = -3Q_C(N). \tag{25}$$

The definitions of (23) and (24) parallel respectively the definitions of (17) and (18). We see then from the second equation of (23) that the vanishing of $\Delta U_{\text{symm.b.}}$ and the ensuing singularity in S at an N_* introduces an intermediate minimum in the action at an $N_{\min} > N_*$: in our example

$$S_{\min} = \left(\frac{N_{\min}}{N_*}\right)^{3q} \left(\frac{1 - N_1/N_*}{N_{\min}/N_* - 1}\right)^3, \tag{26}$$
$$N_{\min} = N_* + \Delta N_{\min} = \frac{q}{q-1} N_*, \quad q = \frac{N_{\min}}{\Delta N_{\min}} \simeq \mathcal{O}(10).$$

Hence, if $S_{\min}$ is large and $\exp(-S_{\min})$ is small, the C-scale PT does not complete – as required. This is the interesting case in which we can ask that some 10% of the FV is turned into TV.

For this purpose, we explore the parameter space which is 4-dimensional, N_0, N_1, N_*, q; note however that there are 5 microphysical parameters and that N_* and q are only phenomenological. First, we choose a pair (N_0, N_1) starting for simplicity with $N_1 = N_0 \in [53, 57]$ to have poweron comoving radial scales $\simeq$10–20 h^{-1} Mpc, prior to the MDE overcomoving expansion. The latter is really unknown and ranges between a factor of 4 – as in Bertschinger's 1985 [38] self–similar solutions – and a factor of 2 if the bubbles are not totally empty (see the figure in [15]); also the loss of energy for expanding into the cosmic medium may be strong and bring this factor down to unity. Second, we replace q by $\Delta N_{\min} = N_{\min} - N_*$ given by (26) and explore numerically $N_* \in [55, 60]$ and $\Delta N_{\min} \in [1, 6]$ with a 50×50 grid to search for acceptable solutions. We display in Fig. 1 an example of bimodal spectrum.

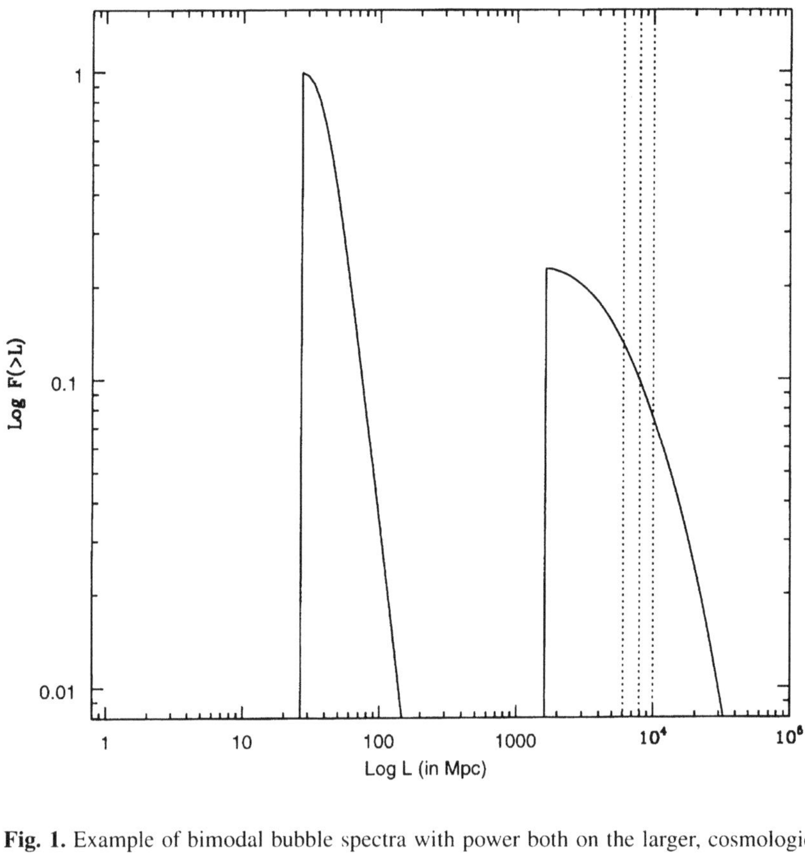

Fig. 1. Example of bimodal bubble spectra with power both on the larger, cosmological scales (right peak) and, separately, on the smaller, astronomical scales (left peak). The first phase transition does not achieve percolation, but yields only 20–25% of true vacuum, while the second phase transition does achieve percolation. Vertical dotted lines mark superhorizon bubble sizes of 6,000, 8,000, and 10,000 h^{-1} Mpc, corresponding to Ω_0=.2.3, and .4 according to Table I. The given spectrum has $N_0 = N_1 = 56$, $N_* = 57.1$, $N_{\min} = 60.0$ and $\Delta N_{\min} = 2.9$. From the unobservable part of the spectrum we find that the probability of finding ourselves in an open $.2 <, \Omega_0, < .4$ bubble is of the order of 5%. The observable part of the spectrum has power all the way up to hundreds of Mpc [4]

4 Conclusions

"We need inflation more than ever if the observations prove that $\Omega_0 < 1$" [8]. In fact, two quite distinct chains of observational evidence in favor of the openness of the Universe, and in favor of the large scale voids, can be explained within the realm of canonical inflation and by a single physical mechanism, quantum tunneling *à la Coleman* in a primordial first-order phase transition. With two scalar fields – one slow-rolling and the other tunneling – and assuming General Relativity as the underlying gravity, we have designed a toy model for such a process by carefully carving into the potential two vacuum channels that swap their roles of true and false vacua. Essentially only one phenomenological assumption is needed for this purpose, that the potential difference between the two channels vanishes at some scale intermediate between the cosmological and the astronomical scales: this introduces a singularity in the Euclidean action and stops the first phase transition (the one on the larger scales) as required in open cosmology when only a minor fraction ($\geq 10\%$) of true vacuum has been generated.

This true vacuum phase shows up in the form of large bubbles of sizes both larger and smaller than the horizon: for some combinations of the parameters the bubbles with the right Ω_0, $.2 < \Omega_0 < .4$, (larger than the horizon and sized up to 10,000 h^{-1} Mpc) occupy a fraction of a few percent of the available space, implying only a modest use of the anthropic principle.

The present work is an extension of previous ones aimed at explaining either the astronomical bubbles or the cosmological ones and contains phenomelogical macrophysical parameters in order to fit the bubble spectra as they become available. An attempt is made to relate microphysical or fundamental parameters to astronomical observables.

The galaxy formation *loci* in this scenario should be the mildly overcomoving spherical shells surrounding the large voids, shells in front and inside of which (most of) the material content of the observable Universe will have been processed by gas-astrophysics. For this reason, although N-body work and numerical hydro will be needed ultimately to assess the implications for galaxy formation and clustering, it seems that the LSS will appear sheet-like, with an effective fractal dimension $D = 2$ up to 50–100 h^{-1} Mpc, not in disagreement with the accepted views [39, 40].

Thus, from knowledge of the (relatively) local environment, it may become possible to speculate about the Universe beyond our horizon and about fundamental physics at energies beyond the reach of our terrestrial experiments.

References

1. Bahcall N.A., Ostriker J.P., Perlmutter S., Steinhardt P.S. (1999): Science **284**, 1481
2. Riess A.G., Filipponko A.V., Challis P. (1998): Astron. J. **116**, 1009
3. Perlmutter S., Alderimg G., Goldhaber G. (1999): Astrophys. J. **517**, 565
4. Goodwin S.P., Thomas P.A., Barber A.J.: The local to global H_0 ratio and the S Ne results. astro-ph/9906187
5. Bahcall N.A., Fan X. (1998): Astrophys. J. **504**, 1

6. Gott J.R. (1982): Nature **295**, 304
7. Gott J.R., Statler T.S. (1984): Phys. Lett. B **136**, 157
8. Gott J.R. (1997): Open, CDM Inflationary Universes, in *Critical Dialogues in Cosmology*, ed. by N. Turok, World Scientific, Singapore, p. 519 ff.
9. Bucher M., Goldhaber A.S., Turok N. (1995): Phys. Rev. D **52**, 3314 and 5538
10. Linde A., Mezhlumian A. (1995): Phys. Rev. D **52**, 6789
11. Kirshner R.P., Oemler A., Schecter P.L., Shectman S.A. (1981): Astrophys. J. Lett. L **248**, 57
12. deLapparent V., Geller M., Huchra J.P. (1989): Astrophys. J. **343**, 1
13. El-Ad H., Piran T., da Costa L.N. (1997): MNRAS **287**, 790
14. Occhionero F., Santangelo P., Vittorio N. (1983): Astron. Astrophys. **117**, 365
15. Occhionero F., Baccigalupi C., Amendola L., Monastra S. (1997): Phys. Rev. D **56**, 7588
16. Blumenthal G.R., da Costa L.N., Goldwirth D.S., Lecar M., Piran T. (1991): Astrophys. J. **388**, 234
17. Liddle A.R., Wands D. (1991): MNRAS **253**, 637
18. Amendola L., Occhionero F. (1993): Astrophys. J. **413**, 39
19. Amendola L., Borgani S. (1994): Mon. Not. R. Astr. Soc. **266**, 191
20. Baccigalupi C. (1998): Astrophys. J. **496**, 615
21. Amendola L., Baccigalupi C., Occhionero F. (1998): Astrophys. J. Lett. L **492**, 5
22. Sakai N., Sugijama N., Yokoyama J. (1999): Astrophys. J. **510**, 1
23. La D. (1991): Phys. Lett. B **265**, 232
24. Coleman S. (1977): Phys. Rev. D **15**, 2929
25. Callan C.G. Jr., Coleman S. (1977): Phys. Rev. D **16**, 1762
26. Coleman S., De Luccia F. (1980): Phys. Rev. D **21**, 3305
27. Guth A.H., Weinberg E.J. (1983): Nucl. Phys. B **212**, 321
28. La D., Steinhardt P.J. (1989): Phys. Rev. Lett. **62**, 376
29. Adams F.C., Freese K. (1991): Phys.Rev. D **43**, 353
30. Kolb E.W. (1991): Physica Scripta T **36**, 199
31. Vadas S.L. (1993): Phys. Rev. D **48**, 4582
32. Starobinsky A.A. (1979): Sov. Phys. JETP Letters **30**, 682
33. Capozziello S., Occhionero F., Amendola L. (1993): Intern. J. Mod. Phys. D **1**, 615
34. Whitt B. (1984): Phys. Lett. B **145**, 176
35. Occhionero F., Amendola L. (1994): Phys. Rev. D **50**, 4846
36. Amendola L., Baccigalupi C., Occhionero F. (1996): Phys. Rev. D **54**, 4760
37. Amendola L., Baccigalupi C., Konoplich R., Occhionero F., Rubin S. (1996): Phys. Rev. D **54**, 7199
38. Bertschinger E. (1985): Astrophys. J. Suppl. Ser. **58**, 1
39. Davis M. (1997): Is the Universe Homogeneous on Large Scales?, in *Critical Dialogues in Cosmology*, ed. by N. Turok, World Scientific, Singapore, p. 13 ff.
40. Wu K.K.S., Lahav O., Rees M.J. (1999): Nature **397**, 225

Non-Stationary Accretion onto Black Holes

J.C. Miller, E. Szuszkiewicz

Abstract. Results are presented from an ongoing research programme to study non-stationary disc accretion onto black holes using a time-dependent Lagrangian computer code and working within the "slim disc" approach. Phenomena considered include limit cycle behaviour, instabilities at the sonic point and disco-seismic modes.

1 Introduction

Accretion discs can be subject to instabilities related to thermal, viscous and acoustic mechanisms. If an instability arises in an initially stationary flow, this implies that a significant time-dependent change will occur in it, but only in extreme cases would this lead to the disc being disrupted. We are carrying out a programme of numerical calculations to investigate the time-dependent behaviour produced as a result of the onset of various types of instability, working within the slim disc approach and using a Lagrangian hydrodynamic code [1,2].

In the present progress report, we focus on work which we have carried out to study the thermal instability, driven by radiation pressure, which can occur in fully ionized plasma in the inner part of the disc. It is interesting to make a comparison between this and the well-known "dwarf nova" type of instability which is connected with motion of the boundary between ionized and non-ionized matter further out in the disc. The present thermal instability is also related to motion of a boundary, in this case the boundary between regions of radiation pressure domination and gas pressure domination. It has potential applications both on the scale of stellar-mass black holes and also on the scale of the much larger black holes probably present in active galactic nuclei. In the first case, the applications are to QPOs in black-hole candidate systems and to rapid bursters; in the second case, to AGN variability on timescales from a thousand days to a thousand years and to possible mechanisms for the launching of jets.

In disc accretion, rotation in the flow gives rise to centrifugal forces which then cause the material to form a flattened disc. We will here be assuming that the disc is *axisymmetric* and *suitably thin*. In the *thin disc model* of Shakura & Sunyaev (1973) [3] (which became an important standard reference point for subsequent work), the disc is taken to be *extremely* geometrically thin with centrifugal forces essentially balancing gravity at each point so that the disc is said to be "Keplerian". While the rate at which material spirals inwards under the action of viscosity is calculated within this model, the *effects* of the inward advection are not included. The vertical structure is approximated by means of averaging or vertical integration, giving rise to an essentially one-dimensional problem.

The *slim disc model*, introduced by Abramowicz et al. in 1988 [4], retains the assumption that the disc is geometrically thin but the centrifugal force is no longer taken to balance gravity exactly. There is a link between the thickness of the disc and the values of pressure and viscosity: the thicker the disc, the higher must be the pressure and (in general) the viscosity. This means that, in going from extremely thin discs to ones which are thicker but still not very thick (i.e., the "slim" discs,) the effects of advection become progressively more important. The slim disc approach includes the effects of advection in a correct way while continuing to treat the vertical structure by means of averaging or vertical integration. Including the effects of advection is particularly important in the inner parts of a black-hole accretion disc; a Keplerian disc ends at the location of the marginally stable orbit whereas a real accretion flow continues inwards and must pass through a sonic point (where the radial infall velocity becomes greater than the local sound speed) before reaching the event horizon. For stationary models, this changes the nature of the solution because of the presence of a critical point in the flow at which regularity conditions must be satisfied and the system of equations must be solved as an eigenvalue problem.

2 Strategy of the calculations

As outlined above, our objective is to study time-dependent, non-stationary behaviour of accretion discs around black holes. For doing this, we have constructed a Lagrangian hydrodynamics code with suitable grid structure and re-mapping procedures to allow satisfactory coverage of the computational domain without requiring an exorbitant number of grid points. Our strategy has been to start from the simplest picture and then to add successive improvements one at a time so that the effects of each could be clearly seen.

Our starting approximations were as follows:

- Gravity has been represented by the *pseudo-Newtonian potential* introduced by Paczyński & Wiita (1980) [5]:

$$\Phi = -\frac{GM}{\sqrt{r^2+z^2} - r_G} \tag{1}$$

  where G is the gravitational constant, M is the mass of the black hole, r and z are cylindrical-polar radial and vertical coordinates centred on the black hole (with the equatorial plane of the disc being at $z = 0$) and r_G is the Schwarzschild radius defined by $r_G = 2GM/c^2$, with c being the velocity of light. Using this form for the gravitational potential but otherwise proceeding using Newtonian equations, incorporates many essential features of the flow onto a non-rotating black hole (e.g., the locations of the marginally-stable and marginally-bound orbits are given correctly) without needing to make fully relativistic calculations. The pseudo-Newtonian potential has played an important role in the development of accretion theory even if it is now something which one would like to proceed beyond (as we plan to do in the course of our programme of calculations).

- Vertical hydrostatic equilibrium was assumed.
- The action of viscosity was described by means of an α law. The azimuthal equation of motion is:

$$\varrho \frac{Dl}{Dt} = \frac{1}{r}\frac{\partial}{\partial r}\left(r^2 \tau_{\varphi r}\right) \tag{2}$$

where ϱ is the density, D/Dt denotes a radially-comoving Lagrangian time derivative, l is the specific angular momentum and $\tau_{\varphi r}$ is the shear component of the viscous stress tensor given by

$$\tau_{\varphi r} = \mu r \frac{\partial \Omega}{\partial r} \tag{3}$$

with μ being the dynamical viscosity coefficient and Ω being the angular velocity. For a *stationary, Newtonian, Keplerian* disc, this expression can be rewritten as

$$\tau_{\varphi r} = -\alpha p \tag{4}$$

where α is a (constant) viscosity parameter, and this form is also frequently used in other contexts.

- The disc material was treated as being *optically thick* with the radiative flux emitted from the surface being given by

$$F^- = \frac{16\sigma T^4}{\tau_R} \tag{5}$$

where σ is the Stefan-Boltzman constant, T is the temperature and τ_R is the Rosseland mean optical depth. The corresponding expression for the radiation pressure is

$$p_{\rm rad} = \frac{4\sigma T^4}{3c}. \tag{6}$$

If the medium is not optically thick, it is convenient to use the grey atmosphere approximation given by Hubeny (1990) [6]. The expressions for F^- and $p_{\rm rad}$ are then

$$F^- = 6\frac{4\sigma T^4}{\frac{3\tau_R}{2} + \sqrt{3} + \frac{1}{\tau_P}} \tag{7}$$

where τ_P is the Planck mean optical depth, and

$$p_{\rm rad} = \frac{F^-}{12c}\left(\tau_R + \frac{2}{\sqrt{3}}\right). \tag{8}$$

We have used this prescription in later calculations.

3 System of equations

The complete system of dynamical equations used in this work can be written as follows:

- Mass conservation equation

$$\frac{D\Sigma}{Dt} = -\frac{\Sigma}{r}\frac{\partial}{\partial r}(r v_r) \tag{9}$$

where Σ is the surface density (the integral of ϱ over the thickness of the disc) and v_r is the radial velocity (negative for an inflow). It is often convenient to write v_r in units of the local sound speed c_s.

- Radial equation of motion

$$\frac{Dv_r}{Dt} = -\frac{1}{\varrho}\frac{\partial p}{\partial r} - \frac{\partial \Phi}{\partial r} + \frac{l^2}{r^3} \tag{10}$$

where p is the total pressure (gas pressure plus radiation pressure).

- Azimuthal equation of motion

$$\frac{Dl}{Dt} = -\frac{\alpha}{r\Sigma}\frac{\partial}{\partial r}\left(r^2 P\right) \tag{11}$$

where P is the vertically-integrated pressure.

- Vertical equation of motion

$$\frac{Dv_z}{Dt} = -\frac{1}{\varrho}\frac{\partial p}{\partial z} - \frac{\partial \Phi}{\partial z}. \tag{12}$$

If hydrostatic equilibrium is assumed in the vertical direction, the acceleration term vanishes and the resulting equation can be treated analytically to give an expression for the height H of the disc.

- Energy equation

$$\varrho T \frac{DS}{Dt} = Q_{\rm vis} + Q_{\rm rad} \tag{13}$$

where S is the specific entropy, $Q_{\rm vis}$ is the rate (per unit volume) at which heat is generated by viscous friction and $Q_{\rm rad}$ (a negative quantity) is the equivalent rate at which energy is radiated away ($Q_{\rm rad} = -F^-/H$). Equation (13) is solved implicitly for the temperature T.

4 Results

Our procedure for the calculations is to calculate first an initial stationary model using the techniques normally adopted for that, and then to transfer the data onto the finite-difference grid used for the time dependent code. For unstable initial models, naturally-occurring perturbations are then sufficient to set off time-varying behaviour. Here, we present results obtained for two different representative cases using mainly the set of starting assumptions outlined above but including the grey atmosphere approach when the medium was not optically thick.

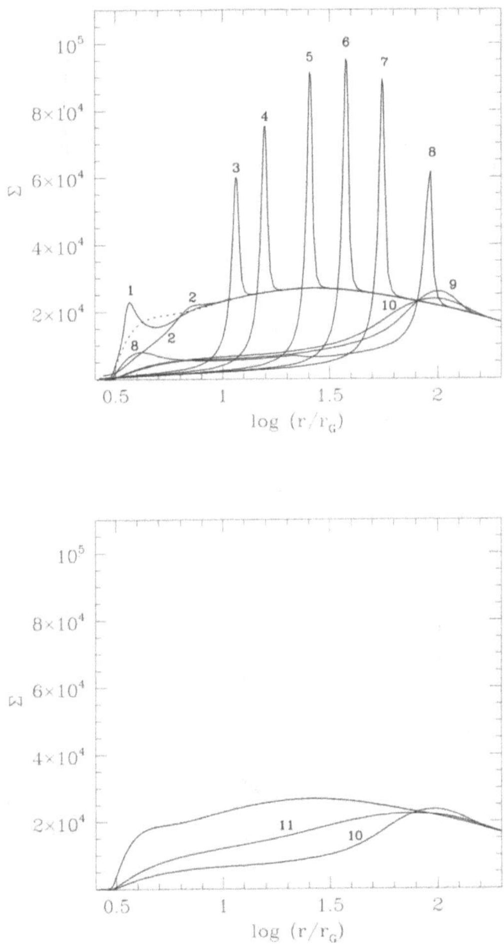

Fig. 1. Time evolution of a thermally-unstable disc with $\alpha = 10^{-1}$. The surface density Σ (measured in cgs units) is plotted against $\log(r/r_G)$ at a succession of times

The first case has $\alpha = 10^{-1}$ and starts from an initial stationary model having a thermally unstable region between $4.5\, r_G$ and $17.5\, r_G$. The mass of the black hole is $10\, M_{\odot}$. After less than one second of the evolution, two density waves are sent out from around $6\, r_G$ (within the unstable region), one moving inwards and the other outwards. This can be seen in Fig. 1, where the surface density Σ is plotted against r/r_G at a succession of times (marked in sequence from 1 to 11 with the dotted curve corresponding to the initial time and the unmarked solid curve in the lower panel being the last one). The ingoing wave (see the curve labelled 1) propagates quickly, passes through the sound horizon (where the inward velocity is equal to the sound speed) and disappears into the black hole, taking with it a significant amount of matter. The outgoing wave is launched on its progress out through the disc only when the ingoing wave has fully passed in through the sound horizon. At the onset of the instability, the temperature rises significantly in the unstable region, increasing the contribution of the radiation pressure by nearly an order of magnitude. The first effect of this is to push the ingoing density wave into the black hole leaving behind an underdensity while causing just a small increase of density at the place from which the outgoing wave will be launched. As the outgoing wave progresses outwards (with the Mach number of the outflow v_r/c_s reaching values as high as 0.2) it partially evacuates the inner part of the disc but also heats the material through which it passes, causing it to expand vertically as shown in Fig. 2. In order for the outgoing wave to continue its propagation, it is necessary that the material behind it should remain in a hot state. Once the wavefront has moved beyond the linearly-unstable region of the initial model, it becomes progressively harder for the temperature to be maintained in the high state and eventually the front starts to weaken (see curve 8) and the temperature falls dramatically down to a low state well below that of the initial model. The front dies after about 20 seconds when it has reached about $100\,r_G$ and a slow filling up of the under-dense region then begins, proceeding on the viscous timescale, with an associated progressive rise in temperature. Eventually, when the configuration has returned near to its initial state (876 seconds after the start), a second cycle begins which is very similar to the first. (There is a slight difference because the configuration does not pass exactly through the stationary state at the end of the cycle.) After this, there is a succession of similar cycles (with period $\sim$ 780 seconds) constituting a limit cycle behaviour. These results are in substantial agreement with ones obtained by Honma, Matsumoto & Kato (1991) [7] using a different numerical method.

The second case has $\alpha = 10^{-3}$ and starts from an initial stationary model with an almost identical thermally unstable region to that above. The mass of the black hole is again $10\,M_{\odot}$. For this case, no limit cycle behaviour was observed and the run ended with the formation of a velocity spike adjacent to the sonic point which appeared to be catastrophic (see Fig. 3). A key difference between the two cases is that, for high α, the sonic point in the stationary models is outside the location of the marginally stable orbit (at $3\,r_G$) whereas, for low α, it is inside it. As the evolution proceeds, the inward-moving density wave passes essentially unimpeded through the sonic point in the case of high alpha whereas, for low alpha, it is stopped outside the sonic point and pushes it inwards towards the black hole eventually giving rise to the formation of the velocity spike which led to termination of the run. (Note that the sound speed c_s used in Fig. 3 is the local *isothermal* sound speed $(p/\varrho)^{1/2}$ which is not exactly the quantity which is relevant for considerations of causal connection. The effective "sonic point" for these purposes comes *near* to the place where $v_r/c_s = -1$

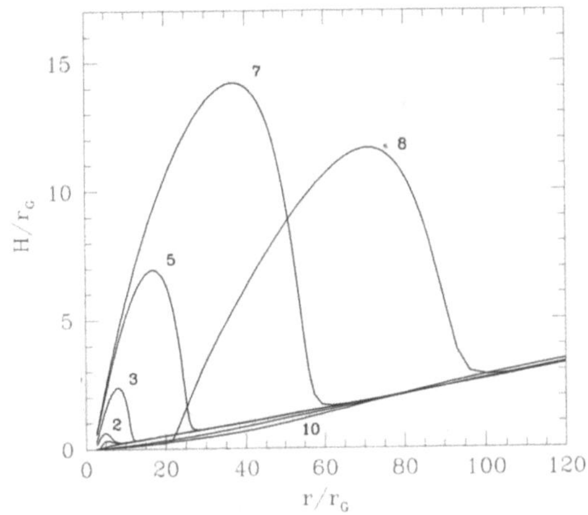

Fig. 2. The half-thickness of the disc H, measured in units of r_G, is plotted as a function of r/r_G for some of the times shown in Fig. 1

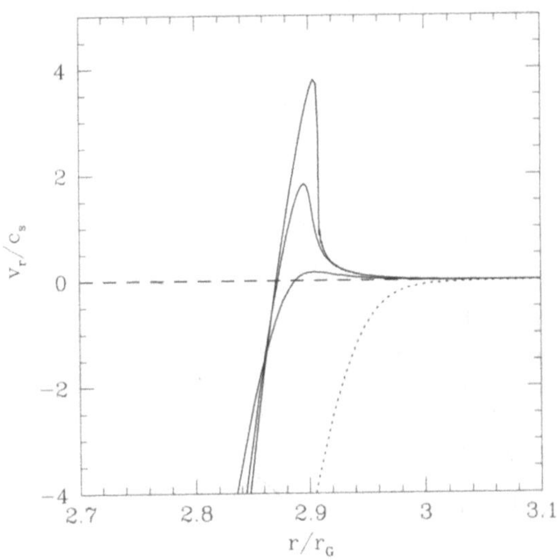

Fig. 3. The Mach number v_r/c_s is plotted as a function of r/r_G at a succession of times during the growth of the instability near to the sonic point for the case with $\alpha = 10^{-3}$. The dotted line shows the initial Mach number profile and the rapid growth of the sharp-peaked structure is shown by profiles for three later output times

but does not precisely coincide with it.) Experiments with this case have indicated that further improvements in the numerical treatment and representation of the input physics would certainly enable us to follow the evolution beyond the present point of breakdown and it will be interesting to see what emerges from this. However, for now we limit ourselves to saying that importantly different behaviour has been seen for the low and high α cases within the present treatment.

5 Conclusions

In this paper, we have presented results for the evolution of thermally unstable discs with both low and high α, mainly following the line of our starting assumptions apart from the fact that the grey atmosphere approach was used when appropriate. The work is continuing. An additional feature which we have observed in some subsequent calculations, is the occurrence of regular pulsations in the inner part of the disc which then propagate outwards and modulate the amplitude of the outward-moving density wave shown in Fig. 1. These regular pulsations may be disco-seismic *p-waves*. This is still under investigation. Various further improvements are currently being made to the code including implementation of a more physical viscosity treatment with $\tau_{\varphi r} \propto \partial\Omega/\partial r$, removal of the assumption of vertical hydrostatic equilibrium (solving a dynamical equation for H) and introducing the capacity to treat two-temperature discs (where ions and electrons can have different temperatures). Results from this will be presented in due course.

References

1. Szuszkiewicz E., Miller J.C. (1997): MNRAS **287**, 165
2. Szuszkiewicz E., Miller J.C. (1998): MNRAS **298**, 888
3. Shakura N.I., Sunyaev R.A. (1973): Astron. Astrophys. **24**, 337
4. Abramowicz M.A., Czerny B., Lasota J.-P., Szuszkiewicz E. (1988): Ap. J. **332**, 646
5. Paczyński B., Wiita P.J. (1980): Astron. Astrophys. **88**, 23
6. Hubeny I. (1990): Ap. J. **351**, 632
7. Honma F., Matsumoto R., Kato S. (1991): PASJ **43**, 147

Stochastic Backgrounds of Gravitational Waves from Compact Sources

R. Schneider, V. Ferrari, S. Matarrese

Abstract. We estimate the gravitational wave background generated by a cosmological population of core-collapse supernovae. In particular, both the radiation emitted during the collapse to black holes and the spin-down radiation emitted by hot, young rapidly rotating neutron stars have been considered. Within their peculiar emission properties, these astrophysical sources of gravitational radiation are likely to have been formed since the beginning of star formation. In order to estimate their rate of formation throughout the Universe we have reconstructed the supernova rate as a function of redshift from an observation-based determination of star formation rate density evolution. We have investigated the overall signal produced by the ensemble of sources exploring the parameter space and discussing its possible detectability.

1 Introduction

The last few years have seen an enormous increase of interest in gravitational wave sources in view of their possible detection with several terrestrial interferometers now under construction.

Most of the gravitational radiation sources that contribute to the frequency bandwidth of terrestrial and space interferometers, are related to the fate of very massive stars which, having reached the final stages of their nuclear burning, explode leaving behind compact remnants (see [1] for a recent review).

As a consequence of the strong energy loss to gravitational waves, these sources cannot remain highly dynamical for long and thus we expect them to be transient sources with low event rates. As an example, the observed local rate of Type II and Type Ib/c supernovae is on the order of 1–2 events every 100 yr [2]. Thus, meeting with a reasonable event rate for these objects will require reaching out to a substantial fraction of the Universe and it is plausible to expect that the strongest sources will be extra-galactic [1].

All these sources of gravitational waves are likely to have been formed since the onset of galaxy formation ($z \lesssim 5$). Thus, we expect that their collective effect might be that of creating a background of gravitational waves.

These astrophysical backgrounds might superimpose or eventually conceal signals of primordial origin, the production of which is a robust prediction of any model which attempts to describe physics at Planck scale. Some of these models have been thoroughly investigated, (see [3] for a recent review). Even though some scenarios are believed to be illustrative rather than realistic, significant research activity is devoted to deriving from the theory, in a self-consistent way, the parameters which describe the resulting gravitational spectra [4].

Conversely, gravitational backgrounds generated by astrophysical sources have not yet met with a systematic study which correctly accounts for the source redshift evolution.

Recent investigations of gravitational backgrounds generated by extra-galactic populations of sources have proved that the main contributions to the resulting signal come from sources at redshifts $z \leq 1-2$, and their formation rate can not be simply extrapolated from its local value but must account for the evolution of the overall galactic population [5–7].

The model we have adopted for the redshift evolution of the source rate of formation is described in Sect. 1 and it is based on the star formation history derived by UV-optical observations of star forming galaxies out to redshifts of $\sim 4-5$ (see, e.g., [8, 9]).

The gravitational wave sources for which the extra-galactic background contributions have been investigated so far are white dwarf binary systems during the early in-spiral phase [7] and core-collapse SNae. In particular, we have considered the gravitational waves emitted during core-collapse to a black hole [5] and the gravitational radiation emitted by newly formed, rapidly rotating, hot neutron stars with an instability in their r-modes [6]. The first choice was motivated by the results of numerical simulations of core-collapses: unlike the case of core-collapse to a neutron star, the gravitational wave emission spectrum produced during core-collapse to a black hole is rather generic, in the sense that it is sufficiently independent of the initial conditions and of the equation of state of the collapsing star (see [10] for a recent review). The second kind of sourc was considered because of its high efficiency in producing gravitational signals: although preliminary, the investigations of r-mode instabilities in highly rotating young neutron stars have proved that almost $2/3$ of the stars initial rotational energy is converted to gravitational waves, making the process very interesting for gravitational wave detection [11–15].

A brief description of the characteristics of the source emission spectra is given in Sect. 2. Finally, in Sect. 3, we derive the spectra of the corresponding backgrounds, explore the parameter space and discuss their detectability.

2 Star formation history

In the last few years, the extraordinary advances attained in observational cosmology have led to the possibility of identifying actively star forming galaxies at increasing cosmological look-back times (see e.g. [16]). Using the rest-frame UV-optical luminosity as an indicator of the star formation rate and integrating on the overall galaxy population, the data obtained with the *Hubble Space Telescope* (HST [17, 18]) Keck, and other large telescopes [9, 19], together with the completion of several large redshift surveys [20–22], have enabled us, for the first time, to derive coherent models for the star formation rate evolution throughout the Universe.

A collection of some of the data obtained at different redshifts together with a proposed fit is shown in Fig. 1. Because dust extinction can lead to an underestimate of the real UV-optical emission and, ultimately, of the real star formation activity, the

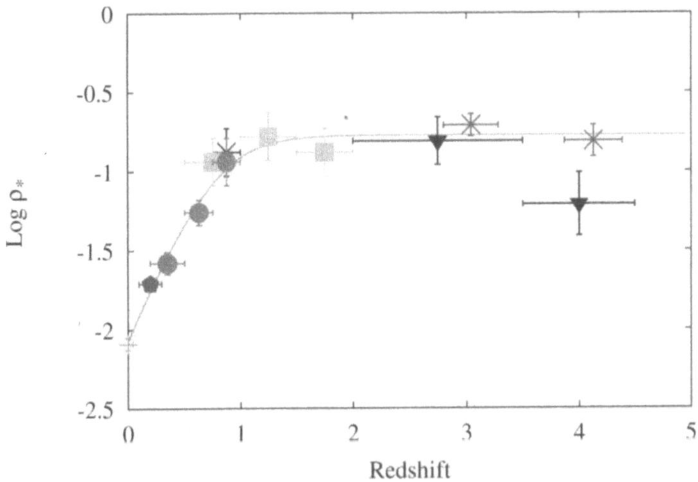

Fig. 1. The Log of the star formation rate density in units of $M_{\odot}\mathrm{yr}^{-1}\mathrm{Mpc}^{-3}$ as a function of redshift for a cosmological background model with $\Omega_M = 1$, $\Omega_\Lambda = 0$, $H_0 = 50\,\mathrm{kms}^{-1}\mathrm{Mpc}^{-1}$ and a Salpeter IMF (see text). The data points correspond to [20] (cross), [21] (filled pentagon), [22] (filled circles), [18] (filled squares), HDF ([17], filled triangles), [9, 19] (asterisks)

data shown in Fig. 1 have been corrected upwards according to factors implied by the Calzetti dust extinction law (see [9]). Although the strong luminosity evolution observed between redshift 0 and 1–2 is believed to be quite firmly established, the amount of dust correction to be applied at intermediate redshift (thus the amplitude of the curve at $z \sim 1-2$) as well as the behaviour of the star formation rate at high redshift is still relatively uncertain. In particular, the decline of the star formation rate density implied by the $\langle z \rangle = 4$ point of the *Hubble Deep Field* (HDF, see Fig. 1) is now contradicted by the star formation rate density derived from a new sample of Lyman break galaxies with $\langle z \rangle = 4.13$ [9] which, instead, seems to indicate that the star formation rate density remains substantially constant at $z > 1-2$. It has been suggested that this discrepancy might be caused by problems of sample variance in the HDF point at $\langle z \rangle = 4$ [9]. Thus, we have up-dated the star formation rate model that we have previously considered in the analysis even though gravitational wave backgrounds are almost insensitive to the behaviour of the star formation rate at $z > 1-2$ because the contribution of very distant sources is very weak [5, 6]. Conversely, if a larger dust correction factor should be applied at intermediate redshifts, this would result in a similar amplification of the gravitational background spectra. From the star formation history plotted in Fig. 1, it is possible to infer the formation rate (number of objects formed per unit time) of a particular population of gravitational wave sources (remnants) by integrating the star formation rate density over the comoving volume element out to redshift z and considering only those progenitors with masses falling in the correct dynamical range for the remnant to

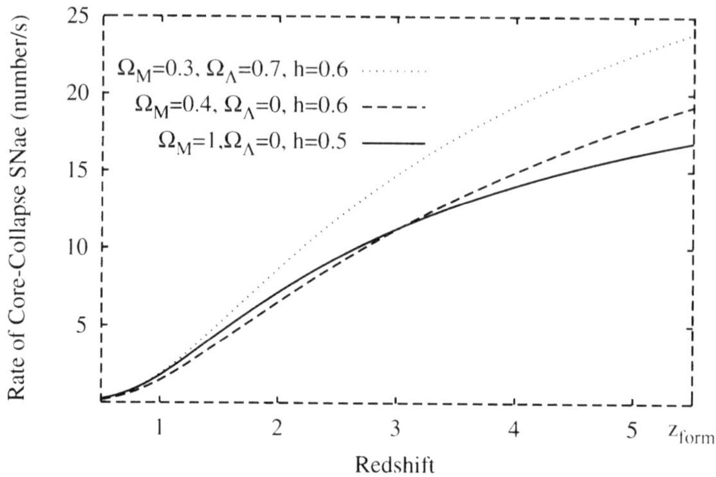

Fig. 2. The rate of core-collapse supernovae (progenitor masses in the range $8M_\odot \lesssim M \lesssim 125M_\odot$) for three different cosmological background models. When varying the cosmological parameters, both the comoving volume element and the star formation rate density are properly modified (see [5])

form, i.e.,

$$R(z) = \int_0^z dz' \frac{\dot{\varrho}_\star(z')}{1+z'} \frac{dV}{dz} \int_{\Delta M} dM' \, \Phi(M'), \tag{1}$$

where the factor $(1+z')^{-1}$ takes into account dilution due to cosmic expansion and $\Phi(M)$ is the initial mass function (IMF) chosen to be of Salpeter type, $\Phi(M) \propto M^{-(1+x)}$ with $x = 1.35$.

Stellar evolution models have shown that single stars with masses $\geq 8M_\odot$ pass through all phases of nuclear burning and end up as core-collapse supernovae leading to a neutron star or a black hole remnant. While there seems to be general agreement that progenitors with masses in the range $8M_\odot \lesssim M \lesssim 20M_\odot$ leave neutron star remnants, the value of the minimum progenitor mass which leads to a black hole remnant is still uncertain, mainly because of the unknown amount of fall-back of material during the supernova explosion [23, 24]. In our analysis, a reference interval of $25M_\odot \lesssim M \lesssim 125M_\odot$ was considered but we have also investigated the effects of choosing lower limits of $20M_\odot$ and $30M_\odot$ as well as an upper limit of $60M_\odot$ [5]. The rate of core-collapse SNae predicted for three cosmological background models is shown in Fig. 2 as a function of redshift. The main difference between the three cosmologies is introduced by the geometrical effect of the comoving volume and is significant at $z \gtrsim 1-2$. This implies that the gravitational backgrounds, which are mainly contributed by sources at $z \lesssim 1-2$, are almost insensitive to the cosmological parameters.

The total black hole formation rate R_{BH} and neutron star formation rate R_{NS} predicted by our model are,

$$R_{BH} = 3.3-4.7\,\mathrm{s}^{-1} \qquad R_{NS} = 13.6\text{–}19.3\,\mathrm{s}^{-1} \tag{2}$$

depending on the cosmological background model considered. The value predicted by our model for the local core-collapse SNa rate is in good agreement with available observations [2].

3 The single source emission spectra

The emission spectrum that we have adopted as our model for the gravitational wave radiated from a core which is collapsing to a black hole was that obtained from a fully non-linear numerical simulation of Einstein+hydrodynamic equations of an axisymmetric core collapse [25, 26].

The main properties of the spectrum are shown in Fig. 3 for the collapse to a $1.5 M_\odot$ black hole at a distance of 15 Mpc and for three assigned values of the angular momentum.

The relevant quantity is the rotational parameter $a = J/(G M_{\rm core}^2/c)$. In fact, there is a maximum in the emission at a frequency which depends on the value of a and whose amplitude, for values of a in the range $0.2 < a < 0.8$, scales as a^4. This peak is located at a frequency which is very close to the frequency of the lowest $m = 0$ quasi-normal mode. This means that a substantial fraction of the energy will

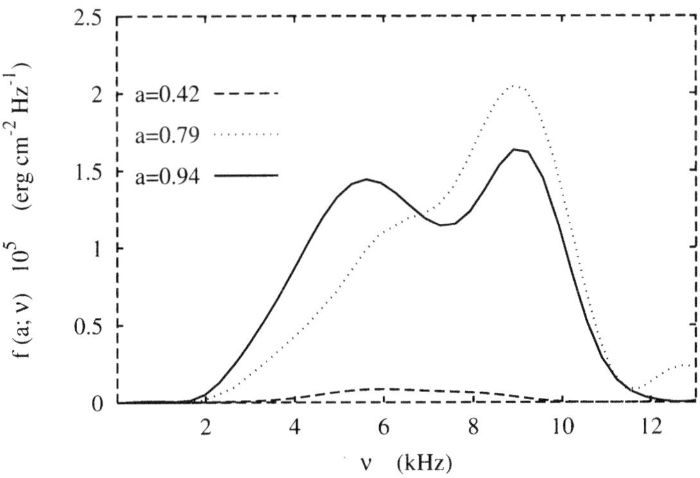

Fig. 3. The average energy flux emitted during the axisymmetric collapse of a rotating, polytropic star to a black hole of $M_{\rm core} = 1.5\, M_\odot$ at a distance of 15 Mpc. The three curves correspond to assigned values of the rotational parameter

be emitted after the black hole has formed: it will oscillate in its quasi-normal modes until its residual mechanical energy is radiated away in gravitational waves.

For high values of the rotational parameter, the geometry of the collapse is different as the star becomes flattened into the equatorial plane and then bounces vertically, but still continues to collapse inward until the black hole is formed. In this case, a low frequency component appears, with an amplitude which may become comparable to that of the peak corresponding to the quasi-normal modes.

In general, the efficiency of this axisymmetric core-collapse to a black hole is $\Delta E_{GW}/M_{\rm core}c^2 \leq 7 \times 10^{-4}$. It should be remembered that less symmetric configurations may result in a more efficient production of gravitational waves.

A number of investigations of relativistic rotating stars has recently led to the discovery of a new class of instability modes, called the r-modes [11–15].

These modes are characterized by having the Coriolis force as the restoring force and thus they are relevant only for rotating stars. Even though the analyses carried out so far are still preliminary and are based on several approximations, these modes, whose instability is driven by gravitational radiation, appear to radiate efficiently in gravitational waves a large part of the initial rotational energy in a relatively small time interval.

A preliminary estimate of the corresponding emission spectrum was recently obtained in [14] for a polytropic neutron star model with a $1.4 M_{\odot}$ mass and a radius of 12.53 km. We have adopted their proposed spectrum as our model for single source emission in order to estimate the gravitational background produced by young, hot, rapidly rotating neutron stars through the r-mode instability [6].

The evolution of the angular momentum of the star is determined by the emission of gravitational waves, which couple to the r-modes through the current multipoles, primarily that with $l = m = 2$. For this mode, the frequency of the emitted gravitational radiation is $\nu = (2/3\pi)\,\Omega$. The star is assumed to be initially rotating at its maximum spin rate, i.e., at its Keplerian value Ω_K, which corresponds to a gravitational wave frequency of $\sim$ 1400 km, for the star model considered [14]. The evolution of Ω during the phase in which the amplitude of the mode is small can be determined from the standard multipole expression for angular momentum loss, and from the energy loss due to gravitational emission and to dissipative effects induced by the bulk and shear viscosity. In this phase, Ω is nearly constant and the instability grows exponentially. After a short time, the amplitude of the mode becomes close to unity and non-linear effects saturate and halt further growth of the mode. This phase lasts for approximately 1 yr, during which the star loses angular momentum radiating approximately 2/3 of its initial rotational energy in gravitational waves, up to a point where the angular velocity reaches a critical value, Ω_c. This value can be determined by solving the equation $1/\tau(\Omega_c) = 0$, where τ is the total dissipation time-scale which can be decomposed as a sum of the damping times associated to gravitational emission, to shear and to bulk viscosity. $\tau(\Omega_c)$ is clearly a function of the temperature of the star, and it has been shown that the r-mode instability operates only in hot neutron stars ($T > 10^9$ K) [14]. Below 10^9 K, superfluidity and other non-perfect fluid effects become important and damping due to viscosity dominates with respect to the destabilizing effect of gravitational radiation. For the star model considered, $\Omega_c \approx 566\,\text{Hz}$, which corresponds to a final spin period of $\approx 11\,\text{ms}$ and to $\nu_{\text{min}} \approx 120\,\text{Hz}$. Below this critical value, viscous forces and gravitational radiation damp out the energy left in the mode, and the star slowly reaches its final equilibrium configuration.

The qualitative picture that arises from this simple model is believed to be sufficiently reliable, even though various uncertainties and approximations might affect the quantitative results for the initial rotation of the star after collapse, for the spin-down time-scales as well as for the final rotation period [15]. However, in this framework the expression for the energy spectrum can be approximated as follows:

$$\frac{dE_{GW}}{d\nu} \approx \frac{4}{3} E_K \frac{\nu}{\nu_{\text{max}}^2} \quad \text{for} \quad \nu_{\text{min}} \leq \nu \leq \nu_{\text{max}} \tag{3}$$

where E_K indicates the initial rotational energy [14].

Thus, the mean flux emitted by this source can be written as

$$f(\nu) = \frac{1}{4\pi d^2} \left(\frac{dE_{GW}}{d\nu} \right). \tag{4}$$

4 The stochastic backgrounds

In order to evaluate the spectral energy density, $dE/dtdSd\nu$, of the stochastic backgrounds produced by the radiation emitted during an axisymmetric black hole collapse and by the spin-down radiation from newly born neutron stars, we need to convolve the differential rate of sources, $dR(z)$, with the flux emitted by a single source at redshift z as it would be observed today (see [5, 6]). This means that we account for the luminosity distance damping on the flux emitted by a single source and we redshift the emission frequencies.

The corresponding values of the closure energy densities of gravitational waves can be obtained as follows:

$$\Omega_{GW}(\nu_{\text{obs}}) = \frac{\nu_{\text{obs}}}{c^3 \varrho_{cr}} \frac{dE}{dtdSd\nu}, \tag{5}$$

where $\varrho_{cr} = 3H_0^2 / 8\pi G$ and are shown in Fig. 4. These figures have been obtained for a flat cosmological background model with zero cosmological constant and with a Hubble constant of $H_0 = h\,100 = 50\,\text{km/s}^{-1}\,\text{Mpc}^{-1}$.

As previously mentioned, the effect of a varying cosmological background is negligible on the final properties of the stochastic backgrounds. In fact, the amplification of the rate at high redshifts shown in Fig. 2 for an open model and a model with a cosmological constant is mostly suppressed by the inverse squared luminosity distance dependence of the single source spectrum for the same models.

The closure density of the black hole collapse background is shown in the top panel of Fig. 4 for three values of the rotational parameter. Since we do not know the distribution of angular momenta, for each curve all the sources of the ensemble were

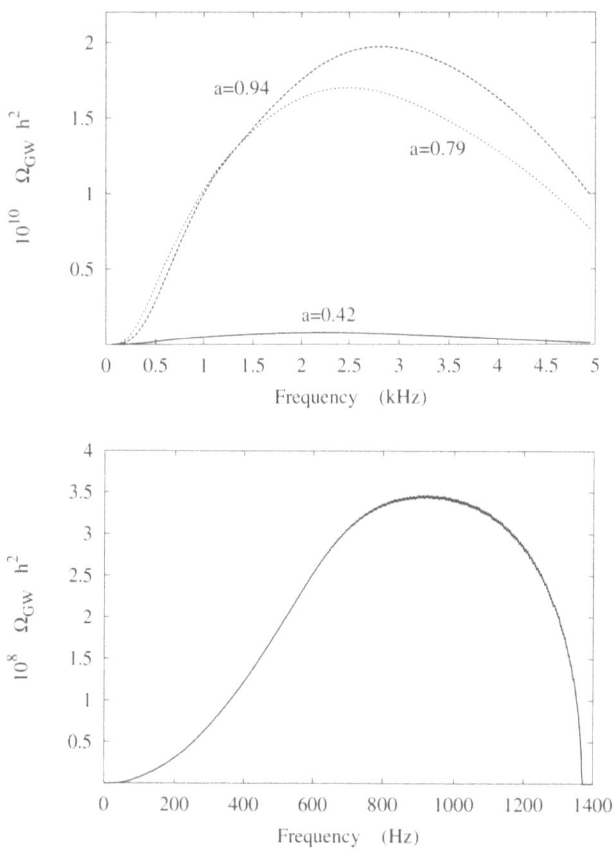

Fig. 4. The functions $h^2\,\Omega_{GW}$ plotted against observational frequency corresponding to the background produced by the radiation emitted by an ensemble of axisymmetric black hole collapses (top panel) and by the spin-down radiation emitted by young rapidly rotating neutron stars (bottom panel). The three curves in the top panel correspond to assigned values of the rotational parameter. A flat cosmological background model with zero cosmological constant and $h = H_0/100 = 0.5$ is assumed (see text)

assumed to have the same value of a. Depending on this value, the closure density has a maximum amplitude in the range $\sim 10^{-9}$–10^{-10} at frequencies between $\sim$ 2–3 kHz. Even though the final properties of the background depend on the model that we have assumed as being representative of the process of gravitational collapse to a black hole, the relevant features of the energy spectrum which we use to model each single event are likely to represent a generic situation reasonably, (see the discussion in [5]). As for the dependence on the formation rate of black holes, the uncertainties which affect the evolution of the star formation rate at high redshifts are completely irrelevant whereas variations induced by different lower and upper mass cut-offs of the progenitor mass range are limited to a factor $\lesssim 2$ [5].

As shown in the bottom panel of Fig. 4, the closure density for the neutron star background has a larger amplitude than that in the previous case and the main part of the signal is concentrated at lower frequencies. In fact, it is characterized by a wide maximum, ranging from ~ 0.7–1 kHz, with an amplitude of a few 10^{-8}. Allowing for variations in $\nu_{\rm min}$ and $\nu_{\rm max}$ does not substantially alter the main features of the background although some quantitative differences appear both in the small and large frequency part of the signal (see Figs. 6 and 8 in [6]).

Neutron star background allows a clear inspection of the impact of star formation rate evolution on its final properties. In fact, in this case all the sources have the same mass and, thus, elements of the ensemble at the same redshift have exactly the same emission properties. Therefore, it is easier to distinguish the effects of the source rate evolution from that of the spectrum of each single event. The bottom panel of Fig. 5 shows the spectrum of the neutron star background. The maximum amplitude occurs around ~ 700 Hz. This means that the most significant contribution to the background signal comes from neutron stars at their maximum spin rate (~ 1400 Hz, for our model) which are formed at redshifts $z \sim 1$–2 where the star formation rate reaches its maximum value before entering its high redshift plateau. Similarly, if one takes into account the fact that the mean value of the core mass which collapses is the fact ≈ 4–$5 M_{\odot}$, the corresponding maximum in the contribution of a mean single source occurs at rest-frame frequencies in the range 2–3 kHz. From the top panel of Fig. 4 it is possible to see that the maximum amplitude in the black hole background spectra corresponds to frequencies ≈ 1–2 kHz, depending on the value of the rotational parameter. Thus, the relevant contribution to the final black hole background signal comes from those sources which are formed around $z \sim 1$–2.

Moreover, it is important to note that, for sources, such as those we have described, which emit gravitational waves at rest-frame frequencies $\nu \gtrsim 100$ Hz, at frequencies 1–100 Hz, where cross-correlation between terrestrial interferometers can be accomplished, the stochastic background signal is entirely produced at $0 < z < 1$–2.

We can conclude that a reliable estimate of astrophysical backgrounds can not set the important effect of star formation rate evolution.

Finally, it is possible to show that the first generation of interferometers will not reach the sensitivities required to observe these backgrounds. In fact, the relevant part of the signal is at relatively high frequencies where, at their actual sites, the interferometers that will soon start to operate can not be cross-correlated.

For the first generation of interferometers, the best signal-to-noise ratio is obtained by cross-correlating VIRGO and GEO600 optimally oriented. Assuming one year of integration, $S/N \sim 2 \times 10^{-3}$. For the same integration time, two LIGO interferometers with advanced sensitivities give $S/N \sim 1.23$ at their actual sites and $S/N \sim 15$ if they were at a distance of ~ 300 km.

Though signal-to-noise ratios calculated for interferometer-bar pairs, such as VIRGO-NAUTILUS or GEO600-NAUTILUS, are still very low, two hollow spheres with $\sqrt{S_n(200\,\mathrm{Hz})} \sim 10^{-24}$ placed at the same site would reach, in one year of in-

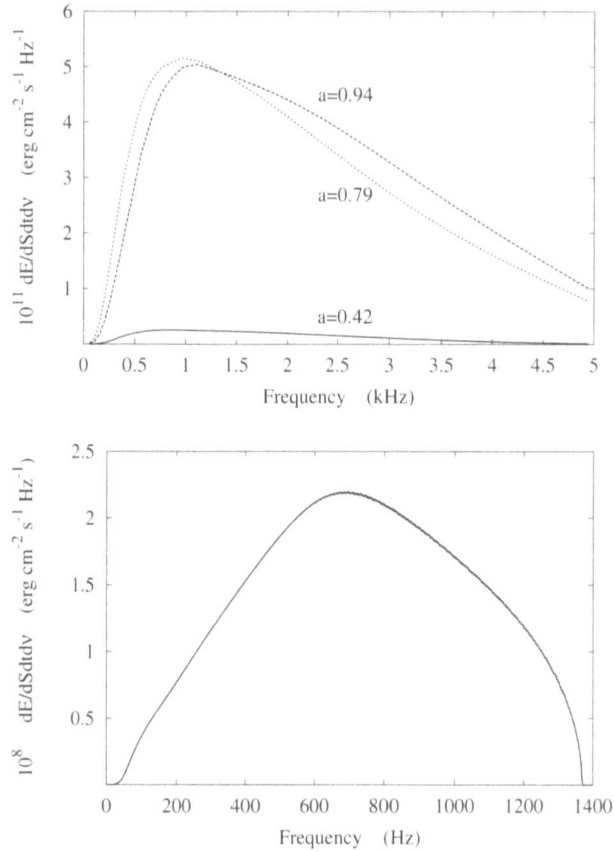

Fig. 5. The spectra plotted against the observational frequency corresponding to the background produced by the radiation emitted by an ensemble of axisymmetric black hole collapses (top panel) and by the spin-down radiation emitted by an ensemble of rapidly rotating neutron star (bottom panel). The three curves in the top panel in correspond to assigned values of the rotational parameter. A flat cosmological background model with zero cosmological constant and $h = H_0/100 = 0.5$ is assumed (see text)

tegration, a signal-to-noise ratio $S/N \sim 1$ [27]. So far, the stochastic backgrounds we have described were considered to be continuous. This is always the case for the background produced by the spin-down radiation emitted by rapidly rotating neutron stars, as the signal from each single source is emitted over a relatively long time interval, on the order of 1 yr (see Sect. 3). Thus, these signals can superimpose and do form a continuous background [5]. Conversely, the background produced by core collapses to black holes can be described as a continuous signal only if an observational run is much longer than the mean temporal interval between two successive bursts. In fact, the typical duration of the gravitational signal emitted by each source is much shorter than that of the previous case, on the order of a ms. Thus, the contributions from the elements of the ensemble do not superimpose but rather generate a shot-noise background, characterized by a succession of isolated bursts with a mean separation of a few seconds, much longer than the typical duration of each burst [5]. The peculiar statistical character of this background might be exploited in order to design a specific algorithm which may favour its detection.

References

1. Thorne K.S. (1997): Probing Black Holes and Relativistic Stars with Gravitational Waves, in *Proceedings of a Conference in Memory of S. Chandrasekhar*, ed. by R.M. Wald, University of Chicago Press, Chicago, pp. 41–77
2. Cappellaro E., Turatto M., Tsvetkov D.Y., Bartunov O.S., Pollas C., Evans R., Hamuy M. (1997): A&A **322**, 431
3. Allen B. (1997): The stochastic gavity-wave background: sources and detection, in *Relativistic Gravitation and Gravitational Radiation*, ed. by J.P. Lasota, J.A. Marck, Cambridge University Press, Cambridge, p. 373 ff.
4. Brustein R. (1998): Cosmic Gravitational Wave Background in String Cosmology, in *The Second Amaldi Conference on Gravitational Waves*, ed. by E. Coccia, G. Pizzella, G. Veneziano, World Scientific, Singapore, p. 149
5. Ferrari V., Matarrese S., Schneider R. (1999): MNRAS **303**, 247
6. Ferrari V., Matarrese S., Schneider R. (1999): MNRAS **303**, 258
7. Kosenko D.I., Postnov K.A. (1998): A & A **336**, 786
8. Madau P., Pozzetti L., Dickinson M. (1998): ApJ **498**, 106
9. Steidel C.C., Adelberger K.L., Giavalisco M., Dickinson M., Pettini M. (1999): ApJ **519**, 1
10. Ferrari V., Palomba C. (1998): Int. J. Mod. Phys. **D**
11. Andersson N. (1998): ApJ **502**, 708
12. Friedman J.L., Morsink S.M. (1998): ApJ **502**, 714
13. Lindblom L., Owen B.J., Morsink M. (1998): Phys. Rev. Lett. **80**, 4843
14. Owen B.J., Lindblom L., Cutler C., Schutz B.F., Vecchio A., Andersson N. (1998): Phys. Rev. D **58**, 084020
15. Andersson N., Kokkotas K., Schutz B.F. (1998): ApJ in press, astro-ph/9805225
16. Ellis R.S. (1997): A.R.A.A. **35**, 389
17. Madau P., Ferguson H.C., Dickinson M.E., Giavalisco M., Steidel C.C., Fruchter A. (1996): MNRAS **283**, 1388
18. Connolly A.J., Szalay A.S., Dickinson M., SubbaRao M.U., Brunner R.J.(1997): ApJ L **11**
19. Steidel C.C., Giavalisco M., Pettini M., Dickinson M., Adelberger K.L. (1996): ApJ L **462**, 17
20. Gallego J., Zamorano J., Arágon-Salamanca A., Rego M. (1995): ApJ L **455**, 1
21. Treyer M.A., Ellis R.S., Milliard B., Donas J., Bridges T.J. (1998): MNRAS **300**, 303
22. Lilly S.J., Le Févre O., Hammer F., Crampton D. (1996): ApJ L **460**, 1
23. Woosley S.E., Weaver T.A. (1995): ApJS **101**, 181
24. Woosley S.E., Timmes F.X. (1996): Nucl. Phys. A **606**, 137
25. Stark R.F., Piran T. (1985): Phys. Rev. Lett. **55**, 891
26. Stark R.F., Piran T. (1986): A numerical computation of the gravitational radiation from rotating gravitational collapse, in *Proc. Fourth Marcel Grossmann Meeting on General Relativity*, ed. by R. Ruffini, Elsevier, Amsterdam, p. 327
27. Coccia E., Fafone V., Frossati G., Lobo J.A., Ortega J.A. (1998): Phys. Rev. D **57**, 2051

Experimental Gravitation

Sights of Experimental Gravity

E. Coccia

The current status of experimental gravitational physics in Italy is very good. As I write this introduction, two advanced detectors of gravitational waves, NAUTILUS and AURIGA, located at laboratories of Frascati and Legnaro respectively, are monitoring the strongest sources in our galaxy and in the local Group. Together with the EXPLORER detector at CERN, operated by the Rome group, they form an Italian network of detectors operating in tandem, which is the heart of the recently formed International Gravitational Event Collaboration, including two more detectors, one in the USA and another in Australia.

The construction of the French–Italian large interferometer, VIRGO, near Pisa, one of the biggest and most ambitious projects ever realized, completes the picture of the Italian effort in this field. The possibility of investigating a stochastic gravitational wave background by correlating the outputs of these detectors, as envisaged, is exciting and will give, for the first time, a probe for physics at the Planck scale.

These sophisticated detectors require an extreme level of isolation from all ambient noises. The contributions contained in the workshop report on the influence of two important sources: the seismoacoustic and the newtonian.

A fundamental topic in gravitational physics is the validity of Newton's gravitational law. Many experiments were stimulated by doubts rised by Fishback. A possible non-result result would indicate the existence of mediators with non-zero mass, predicted by some superstring theories.

Measurement of deviations from the inverse square law in the geophysical region are, however, very difficult and are the subject of another contribution.

We shall see the progress at the next SIGRAV Conference.

Mechanical Isolation of Gravitational Wave Antennae: Present and Future

E. Coccia, V. Fafone, S. Merkowitz, G. Raffone, M. Schipilliti, M. Visco

Abstract. Efficient mechanical isolation from environmental noise is one of the main characteristics required for a gravitational wave antenna. We discuss general design optimization criteria for a multistage mechanical filter. Recent experimental results obtained by a prototype to be used for a spherical detector are presented.

1 Introduction

Resonant gravitational wave antennae are among the most sensitive motion detectors ever made. Present experiments measure displacements whose spectral amplitude is of the order of $10^{-22}\frac{\mathrm{m}}{\sqrt{\mathrm{Hz}}}$ at frequencies near 1 kHz [1–5]. The goal of detecting events produced by supernovae explosions in the Virgo Cluster, which would ensure an appreciable rate, requires an improvement in sensitivity of at least one order of magnitude. Considerable rejection of external vibrational noise is mandatory to obtain such a good performance.

One can identify two categories of mechanical noise acting on the apparatus. The first is due to human activity around the laboratory and can be avoided in principle by a careful choice of site and control of the environment around the experiment. The second kind of noise, such as seismic noise, can be reduced only by using vibration isolation systems. Seismic noise can be modeled, in the frequency range of interest, by the function of frequency $x(\nu) = \frac{A}{\nu^2}\left(\frac{\mathrm{m}}{\sqrt{\mathrm{Hz}}}\right)$ where $A = 10^{-6} \div 10^{-8}$ depending on the site characteristics. Therefore the expected seismic noise contribution at 1 kHz is $(10^{-12} \div 10^{-14})\frac{\mathrm{m}}{\sqrt{\mathrm{Hz}}}$, and filters able to attenuate displacement of 10 orders of magnitude are required to obtain the target sensitivity of $10^{-23}\frac{\mathrm{m}}{\sqrt{\mathrm{Hz}}}$.

2 Ideal elementary attenuator

The simplest example of a low-pass mechanical filter, for a single degree of freedom, is an elementary harmonic oscillator made of a mass M attached to a spring having stiffness $K = \omega_0^2 M$ and mechanical quality factor Q. The displacement response S of the system to a sinusoidal displacement at frequency ω applied to the base is given by

$$S(\omega) = \frac{\sqrt{1 + Q^2\left(\frac{\omega_0}{\omega}\right)^2}}{\sqrt{\left(\frac{\omega_0}{\omega} - \frac{\omega}{\omega_0}\right)^2 Q^2 + 1}}. \qquad (1)$$

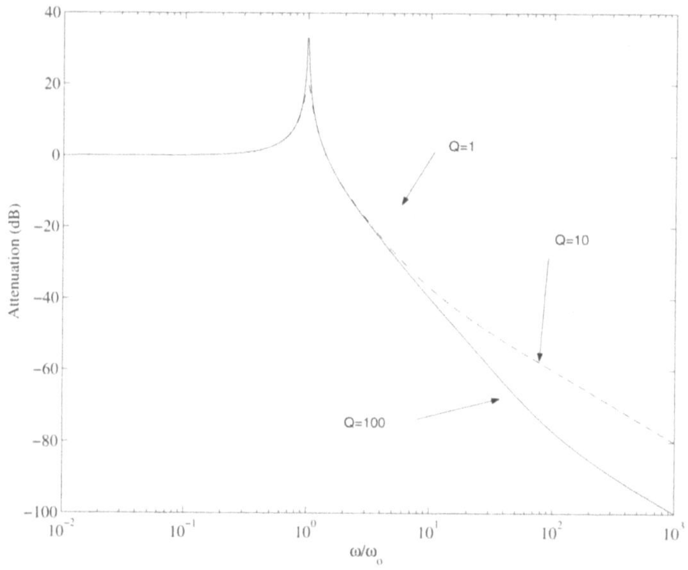

Fig. 1. Displacement response of a harmonic oscillator

In Fig. 1 the function S is plotted against ω/ω_0 for three different values of Q. One can identify different regions in the function curves. When $\omega << \omega_0$ the filter does not attenuate. At $\omega = \omega_0$ the amplitude is amplified by a factor Q. When $\omega >> \omega_0$ there are two possibilities: if $Q >> \frac{\omega}{\omega_0}$ the attenuation is given by $\left(\frac{\omega}{\omega_0}\right)^2$; otherwise it is $\frac{\omega}{\omega_0}\frac{1}{Q}$.

Even if, from (1), it appears feasible, by the use of a single harmonic oscillator with resonance $\nu_0 = \frac{\omega_0}{2\pi} \simeq 10^{-3}$ Hz and $Q >> 10^5$, to obtain an attenuation of 200 dB at 1 kHz, required by a gravitational wave antenna, in the real world this is impossible for two main reasons. Firstly it is impossible to built a spring "soft" enough to have a low frequency resonance, but "strong" enough to support the antenna mass. Secondly a real spring has not only its fundamental resonance, but also an infinite number of self-resonances at higher frequencies.

In practice the required attenuation is attained by using a stack of several single filters, which provide a total attenuation given by the product of the attenuations of the single stages.

3 Real elementary attenuator

We begin the analysis of the behaviour of a real attenuator by discussing the properties of a single stage; later we will analyze the problem of the design of a multistage attenuator.

The springs are made, at room temperature, from air dampers and layers of rubber. At low temperature metallic cables or cantilever beams are used. The springs are loaded by masses whose eigenfrequencies must be far from the working frequency of the antenna.

We will limit our analysis to low temperature filters, whose design is more critical.

The behaviour of the metallic rods used in each stage is analyzed by considering the displacement along the beam axis and transverse to it. Torsional vibrations, for our purposes, can be neglected.

Longitudinal vibrations

Longitudinal vibrations in a thin rod or in a cantilever beam, if dissipations are negligible, are described by the following equation:

$$\frac{\partial^2 z(x,t)}{\partial x^2} - \frac{1}{v_l^2}\frac{\partial^2 z(x,t)}{\partial t^2} = 0 \tag{2}$$

where $z(x, t)$ is the longitudinal displacement of the rod at time t, v_l is the sound velocity given by $\sqrt{\frac{E}{\varrho}}$, E and ϱ being respectively the Young modulus and the density of the material used.

The boundary conditions corresponding to a rod having cross section A, loaded at the extreme $x = L$ by a mass M and moved at the attaching point $x = 0$ by a sinusoidal signal of amplitude B and frequency ω, are:

$$\begin{aligned} &z(0,t) = B\cos(\omega t), \\ &E \cdot A \left.\frac{\partial z(x,t)}{\partial x}\right|_{x=L} = -M\frac{\partial^2 z(x,t)}{\partial t^2}. \end{aligned} \tag{3}$$

From the analytical solution of the equation, one can get the attenuation as a function of the frequency:

$$S_l(\omega) = \frac{z(L,t)}{z(0,t)} = \frac{1}{\frac{\omega}{\omega_l}\frac{M}{M_s}\sin\left(\frac{\omega}{\omega_l}\right) - \cos\left(\frac{\omega}{\omega_l}\right)} \tag{4}$$

where $\omega_l = \frac{v_l}{L}$ and $M_s = A \cdot L \cdot \varrho$ is the mass of the rod.

The function S_l of $\frac{\omega}{\omega_l}$ is reported in Fig. 2. The frequency response is similar to that of a single oscillator (Eq. (1) and Fig. 1), the attenuation for frequencies higher than the cut frequency is still given by $\left(\frac{\omega}{\omega_0}\right)^2$, but many self-resonances appear.

An analytical expression for the cut frequency can be calculated from Eq. (4) by using the approximation $\frac{\omega}{\omega_l} << 1$

$$\omega_{0l} = \sqrt{2}\frac{\omega_l}{\sqrt{1 + 2\frac{M}{M_s}}}. \tag{5}$$

The analytical expression for the higher self-resonaces can be calculated from Eq. (4) with $\frac{\omega}{\omega_l}\frac{M}{M_s} >> 1$:

$$\omega_{rl}^n = n\pi\omega_l \quad \text{with } n > 1. \tag{6}$$

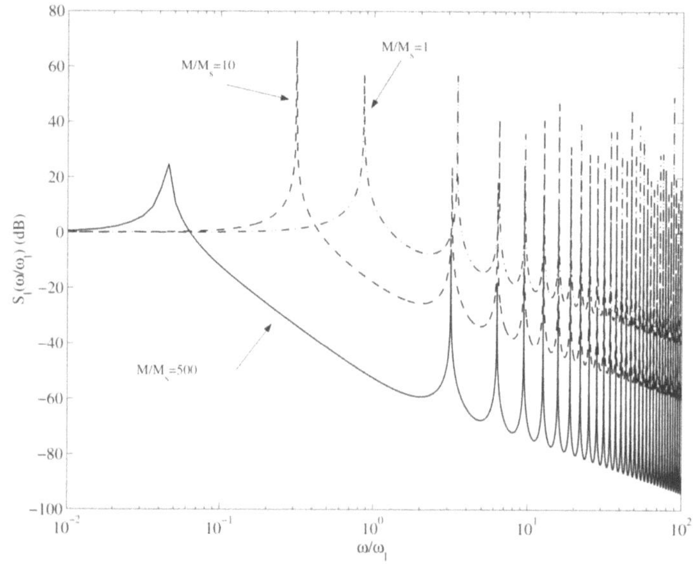

Fig. 2. Longitudinal attenuation

Transverse vibrations

The transverse displacement $u(x,t)$ of the bar satisfies the following equation:

$$E \cdot I \frac{\partial^4 u(x,t)}{\partial x^4} - P \frac{\partial^2 u(x,t)}{\partial t^2} + \varrho \cdot A \frac{\partial^2 u(x,t)}{\partial t^2} = 0 \tag{7}$$

where I is the second moment of inertia and $P = M \cdot g \cos\theta$ is the axial tensile force ($\theta = 0°$ for a cable and $\theta = 90°$ for a cantilever; g is the acceleration due to gravity). The boundary conditions for a beam rigidly attached to a base oscillating at frequency ω with amplitude B and loaded at $x = L$ by a mass M having a moment of inertia J are the following:

$$\begin{aligned} u(0,t) &= B\cos(p \cdot t) \quad & E \cdot I \left.\frac{\partial^2 u(x,t)}{\partial x^2}\right|_{x=L} &= -J \left.\frac{\partial^3 u(x,t)}{\partial t^2 \partial x}\right|_{x=L} \\ \left.\frac{\partial u(x,t)}{\partial x}\right|_{x=0} &= 0 & E \cdot I \left.\frac{\partial^3 u(x,t)}{\partial x^3}\right|_{x=L} &= M \frac{\partial^2 u(x,t)}{\partial t^2}. \end{aligned} \tag{8}$$

A general solution of Eq. (7) can be readily found if $P = 0$; in this case the transfer function is given by

$$S_t(\omega) = \frac{u(L,t)}{u(0,t)} = \frac{\frac{J}{J_s}\alpha^3 (\sin\alpha + \sinh\alpha) - (\cos\alpha + \cosh\alpha)}{D} \tag{9}$$

where

$$D = \frac{M}{M_s}\alpha\left(\sin\alpha\cosh\alpha - \cos\alpha\sinh\alpha\right) + \frac{J}{J_s}\alpha^3\left(\sin\alpha\cosh\alpha + \cos\alpha\sinh\alpha\right)$$
$$+ \frac{M}{M_s}\frac{J}{J_s}\alpha^4\left(\cos\alpha\cosh\alpha - 1\right) - \left(\cos\alpha\cosh\alpha + 1\right)$$

and $\alpha = \sqrt{\frac{\omega}{\omega_t}}$ and $\omega_t = \sqrt{\frac{E \cdot I}{\varrho A}\frac{1}{L^2}}$.

In Fig. 3 attenuation against ω/ω_t is reported for different values of J/J_s.

The loading mass rotary inertia produces a splitting of the cut frequency. An approximate expression for these two frequencies can be calculated from (9) with $\frac{\omega}{\omega_t} << 1$:

$$\omega_{0t\mp} = \sqrt{210}\,\omega_t \frac{\sqrt{\left(3\frac{J}{J_s} + \frac{M}{M_s}\right) \mp \sqrt{9\left(\frac{J}{J_s}\right)^2 + 3\frac{J}{J_s}\frac{M}{M_s} + \left(\frac{M}{M_s}\right)^2}}}{\sqrt{105\frac{J}{J_s} + 2\sqrt{\frac{M}{M_s}}}}. \tag{10}$$

An analytical expression for the higher self-frequencies can be obtained from (9) with $\frac{\omega}{\omega_t} >> 1$:

$$\omega_{rt}^n = n \cdot \frac{9}{4}\pi^2\omega_t \quad \text{for } n > 1. \tag{11}$$

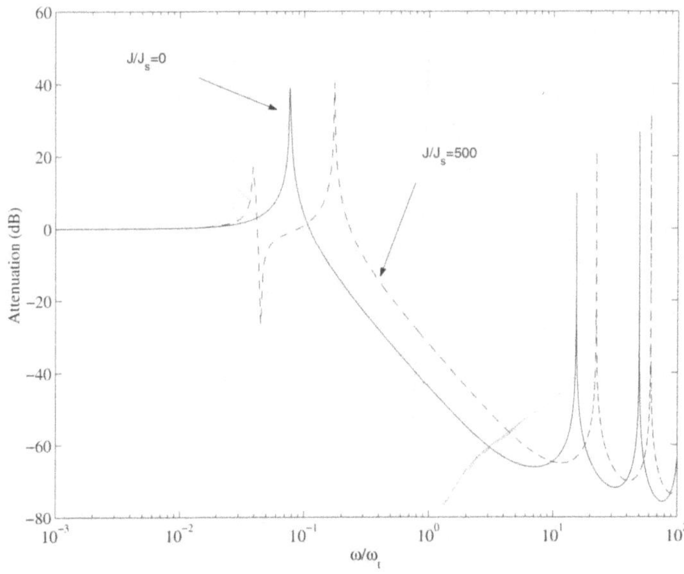

Fig. 3. Transverse attenuation for two different values of J/J_s

4 Optimization criteria

In the previous paragraphs we have shown how a cable or a cantilever beam loaded by a mass acts as a single stage low pass filter; now we discuss a way of building more complicated structures.

The main characteristic required by a reliable mechanical filter for a gravitational wave experiment are the following:

1. it must provide the required attenuation in all directions,
2. no self-resonances of the filter must be near to the antenna frequency,
3. the response must be linear, in order to avoid up-conversion phenomena that could shift low frequency disturbances to the working frequency of the antenna,
4. the system must be free from acoustic emission.

We begin by discussing points 3 and 4 which imply some easy constraints. To maintain the apparatus in linear working conditions, displacements must be small. This can be achieved by avoiding the choice of cut frequencies which are too low. In fact, when the mechanical Q is high, the decay time of an oscillation at very low frequency can be so long that the effects of subsequent disturbances can overlap.

It is beyond the scope of this paper to discuss the process by which acoustic emissions are produced, but it is well-known that they strongly depend on the stress applied to the material. As a consequence, acoustic emissions are negligible when the maximum stress is kept, within any part of the isolator, below 10% of the yield stress.

We next discuss how to satisfy the first two points required in the project for a filter. We will refer to filters made of cables and cantilevers such as those presented earlier.

To obtain the best attenuation in each stage of a filter, the cut frequencies, both for longitudinal and transverse vibrations, must be as low as possible. At the same time self-resonances must be far from the working frequency of the antenna ω_a. Past experience has show that, very often, due to model errors, estimation of self-frequencies is quite approximate and therefore the probability of having a self-resonance near to the working frequency is not negligible, especially if the self-resonances are not widely spaced. To avoid this critical condition it is convenient to require that all the self-resonances of the system are well above the working frequency ω_a. As in most practical situations the transverse cut frequency is much lower than the longitudinal one $\omega_{0t} < \omega_{0l}$ and, further, the first longitudinal self-resonance is much higher than the transverse self-resonance $\omega_{rt} < \omega_{rl}$, it is sufficient to require that

$$\begin{aligned} \omega_{0l} &<< \omega_a, &(12)\\ \omega_{rt} &> \omega_a. &(13) \end{aligned}$$

These conditions, on using (5) and (11) and choosing the minimum cable section $\left(A = \sqrt{\frac{M \cdot g}{k_f}}\right)$ with respect to the maximum stress applicable to the beam k_f, become

$$\sqrt{g}\sqrt{\frac{E}{k_f}\frac{1}{\sqrt{L}}} << \omega_a, \tag{14}$$

$$\frac{9\pi^2}{4}\sqrt{\frac{E}{\varrho}\frac{1}{L^2}}\sqrt{\frac{I}{A}} > \omega_a. \tag{15}$$

The left-hand terms of these two expressions are the product of factors: a numerical part, followed by an expression related to the material used, and a term depending on the length. It easy to see that a material like titanium, having high yield strength and low density, is to be preferred. In expression (15) there is also present a factor given by the square-root of the ratio between the second moment of inertia and the cross-section area. This factor is always greater for hollow cross-section, which should be preferred over bulky ones. Once the material and the cross-section are set, optimization is very simple: the length is chosen in such a way as to satisfy condition (15) and consequently all the parameters are fixed.

We stress that the optimization criteria do not include any consideration of the loading mass; therefore, the cut frequency and attenuation of a single stage do not depend on it.

We limited our discussion to a peculiar geometry which is easy to model but the results are quite general and suggest common design criteria. However, more complicated structures of attenuation, made of many stages, require numerical methods in order to be analyzed adequately. Nowadays, many powerful finite element analysis packages are available for this kind of study.

5 Multistage filters

The filters used for gravitational wave antennae must be cooled to a low temperature; therefore, the number of stages and their masses should be minimized so as not to increase the cost and complexity of the experimental apparatus.

Thus two important questions concerning multistage filters are: is there a minimum value for the masses to be used in a filter? Is there a number of stages that optimize filter performances?

The general solution of these problems is not trivial and therefore we will limit our analysis to a special case that is easy to solve but gives an idea of the general behaviour.

We consider a multistage filter made of n cables, with the same lengths and cross-sections equal to the minimum allowed by the yield strength, suspending $(n-1)$ stages of identical mass $M_f/(n-1)$ and an antenna of mass M_a. If the total attenuation can be expressed as the product of the attenuations of each of the stages and if the attenuation of a single stage can be expressed as the square of the ratio between the longitudinal cut frequency ω_{0l} and the working frequency ω_a, then the total attenuation attainable at frequency ω_a by using n stages is given by

$$S_{\text{tot}} = \left(\frac{\omega_{0l}}{\omega_a}\right)^{2n} \prod_{i=2}^{n}\left[\frac{M_a}{M_f}(n-1)+i-1\right]. \tag{16}$$

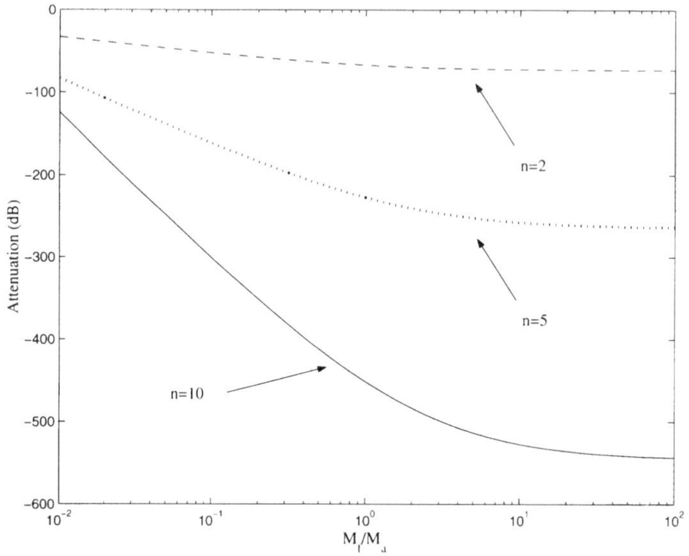

Fig. 4. Total attenuation of a n-stage filter as a function of the ratio between the total mass used in the stages and the antenna mass, calculated for $\frac{\omega_{0l}}{\omega_a} = 1.5 \cdot 10^{-2}$

In Fig. 4 we report, for different values of the number of stages, the total attenuation S_{tot} against the ratio $\frac{M_f}{M_a}$ for a typical value of $\frac{\omega_{0l}}{\omega_a}$. We note that the total mass used for the filter is fixed to the value M_f.

Expression (16) suggests some general considerations.

- The attenuation of a multistage filter is not independent of the loading mass, as in a single stage filter, since the spring at each stage must be shaped by taking into account all the stages that are hung upon it.
- The total attenuation depends on the ratio between the total mass used in the filter and the antenna mass. In an efficient filter this value must be greater then 0.1.
- It is always helpful to increase the number of stages.

6 Filters in present and future antennae

We briefly review the general criteria adopted in present detectors [6–8], showing which solutions are most suitable to use in the future.

Vibration isolation in present antennae is accomplished in two stages: a room temperature isolation stack, made of multiple alternating layers of metal and rubber; a low temperature stage made of multimode metallic isolators. Different solutions were adopted for the connection between low temperature filters and the cooling system. The groups that designed the two antennae ALLEGRO [2] and NIOBE [3] decided to separate the mechanical filters, which suspend the antenna, from the cooling apparatus. The antenna is cooled by a helium reservoir independently supported. On the other hand, in the antennae made by the Italian groups (EXPLORER [1], NAUTILUS [4] and AURIGA [5]), the mechanical filters also act as radiation shields.

Both of the technical solutions worked properly, allowing long periods of operation of the detectors, but we consider the first technical choice more suitable for adopting in the new generation of detectors. In fact, even if a larger experimental volume is required, there are less constraints on the geometry of the filters and possible future modifications are less difficult.

In recent years the development of a new generation of resonant gravitational wave antennae of spherical shape has begun. This kind of detector should bring an increase in sensitivity due to the possibility of having a larger mass with the same working frequency as that of present antennae (around 1 kHz) and to the fact that it is omni-directional having sensitivity which is constant over a solid angle).

Our group has started the design of a prototype of spherical antenna made of aluminum alloy 5056, 0.8 m in diameter and having quadrupole modes resonating around 3.2 kHz [9]. We have designed a filter for this prototype using the general criteria introduced in the previous paragraphs. The isolator will consist of six stages to achieve an attenuation of 200 dB. The mass of each stage is a steel cylinder (120 kg in mass, 304 mm in diameter and 200 mm thick); the springs three for each stage, are made of steel cables (of diameter 13.4 mm and length 80 mm). Some preliminary tests have been made for two stages. As an example, in Fig. 5, longitudinal attenuation against frequency is reported for the first stage. The transfer function measurements are drawn with a solid line while the dashed line is the result of the

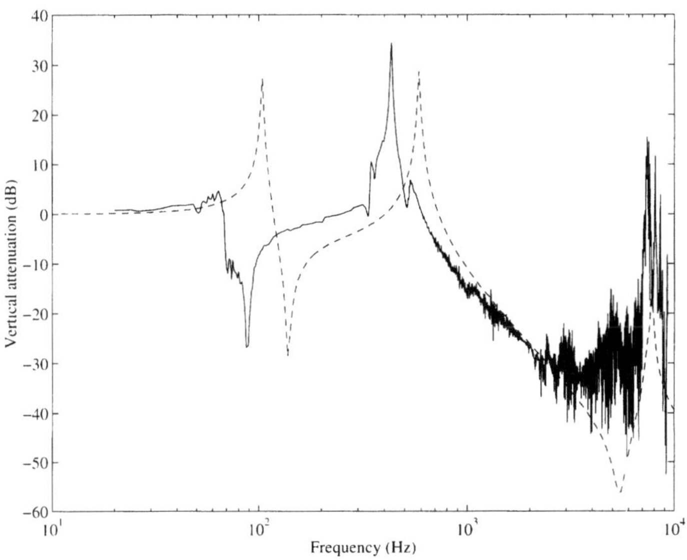

Fig. 5. Vertical attenuation of one stage, experimental value (solid line) and numerical model (dashed line)

response simulated by a finite element program (COSMOS /M). This single stage provides an attenuation of around 30 dB; the agreement between the model and the experimental measurements is extremely satisfactory.

References

1. Astone P., Bassan M., Bonifazi P. (1993): Phys. Rev. D **47**, 2
2. Geng Z.K., Hamilton W.O., Johnson W.W. (1995): Operation of the Allegro Detector at LSU, in *Proc. First Edoardo Amaldi Conference, Frascati, 1994*, ed. by E. Coccia, G. Pizzella, F. Ronga, World Scientific, Singapore, p. 128
3. Blair D.G., Ivanov E.N., Tobar M.E. (1995): Phys. Rev. D **74**, 1908
4. Astone P., Bassan M., Bonifazi P. (1997): Astrop. Phys. **7**, 231
5. Cerdonio M., Franceschini L., Fontana G. (1995): Status of the Auriga Gravitational Wave Antenna and Perspectives for the Gravitational Wave Search with Ultracryogenic Resonant Detector, in *Proc. First Edoardo Amaldi Conference, Frascati, 1994*, ed. by E. Coccia, G. Pizzella, F. Ronga, World Scientific, Singapore, p. 176
6. Aldcroft T.L., Michelson P.F., Taber R.C., McLoughlin F.A. (1992): Rev. Sci. Instrum. **63**, 8
7. Veitch P.J. (1991): Rev. Sci. Instrum. **62**, 1
8. Coccia E. (1982): Rev. Sci. Instrum. **53**, 148
9. Merkowitz S., Coccia E., Fafone V. (1999): Rev. Sci. Instrum. **70**, 1553

Off-Line Subtraction of Seismic Newtonian Noise

G. Cella

Abstract. A fundamental limit for the sensitivity of an Interferometric Gravitational Wave detector in the low frequency region is imposed by the effect of environmental mass density fluctuations. These fluctuations generate stochastic gravitational fields which couple directly to the apparatus test masses, bypassing seismic isolation systems. In this paper the results of a preliminary investigation on the possibility of reducing this kind of noise are reported. We focus on the feasibility of off-line noise subtraction. In this approach the mass density fluctuations are monitored by an appropriate set of measurement devices. The resulting signals are combined linearly with the output of the interferometer to obtain partial cancellation of the noise.

1 Introduction

Density mass fluctuations are generated continuously in the environment by a variety of mechanisms. Examples include atmospheric pressure fluctuations, infrastructure movements, human activities and seismic fluctuations of the ground.

These mass fluctuations are the sources of a gravitational field which, though very weak, couples directly to the test masses of an interferometric gravitational wave detector. The effect of this coupling on the sensitivity curve of a gravitational wave detector was estimated in a series of works [1–5]. The main result is that this source of noise could be relevant, given the planned sensitivity of the current generation interferometers, in the frequency band between 1 and 10 Hz. Below this range it is overwhelmed rapidly by seismic noise, above by thermal noise. For a concrete example see Fig. 1, where the more important source of noise for the VIRGO interferometer are compared.

All these estimates are normalized to the power spectrum amplitude of seismic motion, which, in some cases, is not known very well and probably overestimated [6]. An attempt to explore the feasibility of the reduction of this "Newtonian noise" is in our opinion worth doing, especially in the perspective of second generation cryogenic detectors. For these Newtonian noise could become the fundamental limitation in the low frequency range.

In the following we will focus on seismic generated Newtonian noise. Preliminary, unpublished results show that atmospheric generated Newtonian noise could also be relevant, in fact, the most relevant. We expect that it will be possible to carry out a similar analysis in this case as well.

Different strategies can be elaborated to reduce the effect of seismic originated Newtonian noise. The more obvious is the construction of a mold of appropriate depth around test masses, in order to isolate a sufficiently large ground volume from seismic motion. Accurate modelling is required to estimate the effectiveness of this

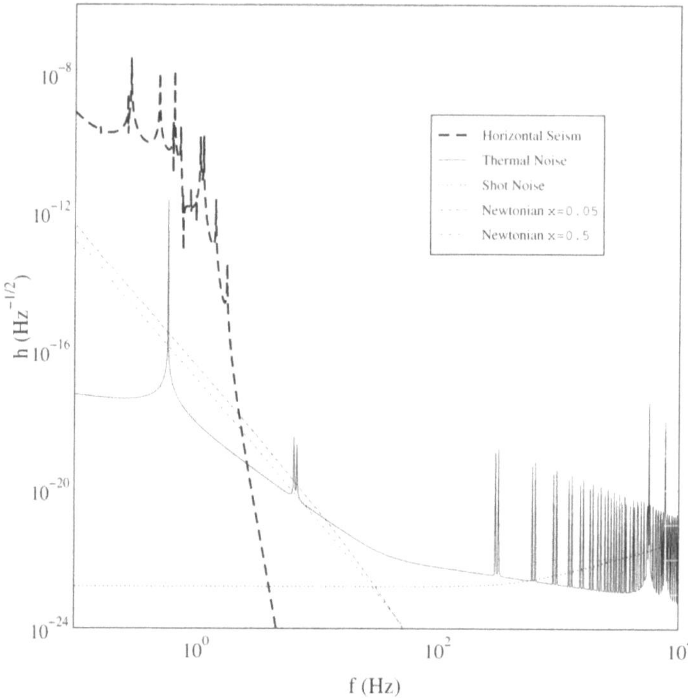

Fig. 1. Sensitivity curve for the VIRGO interferometer. The two different extimates for the Newtonian noise correspond to two different values of the ratio between transverse and longitudinal sound speed in the terrain considered

approach. This must take into account, for example, the effect of the diffraction of elastic seismic waves on the mold.

It is also possible to contemplate the construction of dynamic structures which could provide a screening effect. However this approach appears to be very tricky and, to obtain relevant effects, these dynamical structures would probably have to be very finely tuned.

Another important point is that these solutions require us to introduce permanent modifications in the apparatus, which should be avoided as much as possible in order not to add non-controlled, systematic effects.

In this preliminary stage of investigation it seems wiser to concentrate our attention on a third approach, which does not require apparatus modifications. This is based on the possibility of monitoring the environmental sources of Newtonian noise and correcting "off-line" the output of the experiment using the information obtained.

2 Off-line subtraction: generalities

Newtonian noise, seen as a random process in the time domain, is a linear function of the mass density fluctuations in the environment. Suppose that we know, at each point in time, the displacement of each point in the ground from its equilibrium position $\mathbf{u}(\mathbf{x}, t)$ The force experienced by an isolated test mass in $\mathbf{x}_0$ can be written as

$$\mathbf{F}(t) = G \int \varrho(\mathbf{x}) \frac{\mathbf{x} - \mathbf{x}_0}{|\mathbf{x} - \mathbf{x}_0|^3} \nabla \cdot \mathbf{u}(\mathbf{x}, t) dV. \tag{2.1}$$

In the general case we can write, for the Newtonian part of the output signal of an interferometer, $N(t) = \widetilde{(N|B)}$, where $|B)$ is the state of the ground (explicitly, the displacement field $\mathbf{u}$) and $(N|$ some linear operator which depends on the geometry of the interferometer. Suppose now that we place a measure instrument, for example, an accelerometer, at a point $\mathbf{x}$ of the ground. The part of the instrument output correlated with seismic motion will also be a linear function of the ground fluctuations, which we can write in the form $\widetilde{(a(x)|B)}(t)$.

More generally, we write the output of the interferometer in the form

$$\widetilde{H}(t) = \widetilde{(N|B)}(t) + \widetilde{h}(t), \quad \text{with} \quad E[\widetilde{(N|B)}\widetilde{h}] = 0 \tag{2.2}$$

with the assumption that both $\widetilde{(N|B)}$ and $\widetilde{h}$ are stationary, zero-mean stochastic processes. The $E[\cdots]$ average must be appropriately defined. For example, it could be the usual $T \to \infty$ limit of the ensemble average over strips of length T. In Eq. (2.2) $\widetilde{h}$ represents the fraction of the output uncorrelated with seismic motion, as, for example, thermal noise, shot noise, gravitational wave signals. In the same way we write the output of an accelerometer as

$$\widetilde{A}_i(t) = \widetilde{(a_i|B)}(t) + \widetilde{\eta}_i, \quad \text{with} \quad E[\widetilde{(a_i|N)}\widetilde{\eta}_i] = 0 \tag{2.3}$$

and, in this case, $\widetilde{\eta}$ represents the intrinsic instrument noise. We assume that, if the output of n different accelerometers is known, the maximum reduction of Newtonian noise could be obtained by constructing a "subtracted signal" H_S that can be written as

$$\widetilde{H}_S(t) = \widetilde{H}(t) - \sum_i \int_{-\infty}^{t} \widetilde{w}_i(t - t')(N|a_i)\widetilde{A}_i(t')dt' \tag{2.4}$$

or, in the frequency domain,

$$\begin{aligned} H_S(\omega) &= H(\omega) - \sum_i w_i(\omega)(N|a_i)A_i(\omega) \\ &= H(\omega) - \sum_i w_i(\omega)(N|a_i)\left[(a_i|B)(\omega) + \eta_i(\omega)\right]. \end{aligned} \tag{2.5}$$

Here the $w_i(\omega)$ are a set of functions (one for each accelerometer) which must be calculated. The factor $(N|a_i)$ is not a stochastic process, and was introduced for convenience. It represents the coupling of the fluctuations measured by a given accelerometer to the Newtonian noise, and depends only on the interferometer geometry and on the accelerometer position relative to it. With its aid Eq. (2.5) suggest that what we are doing is subtracting from H the Newtonian noise signal due to the density fluctuation fraction controlled by our acceleration measurements.

It is also evident that, in the limit of an infinite number of accelerometers, it is always possible to write $(N|B) = \sum_i w_i(N|a_i)(a_i|B)$, so that, in this case, the subtraction of Newtonian noise should be complete in the absence of instrumental noise.

If the noise is stationary, and if the processes $(a_i|B)$ are Gaussian, the subtraction procedure (2.5) is optimal. The ω dependency of the weights w_i is due to the fact that the optimal signal to subtract encodes the information about seismic dynamics, and this means that it must have memory.

In order to determine the functions w_i we have to fix a well-defined quantity that we want to minimize. In the stationary case it is natural to use the frequency integral of the interferometer output power spectrum density, weighted by a function $g(\omega)$ which selects the frequency band of interest,

$$\Gamma[w_1, \cdots, w_n] = \int g(\omega)\overline{|H_S|^2}d\omega. \tag{2.6}$$

As indicated, this is a (quadratic) functional of the unknown functions $w_i(\omega)$. In order to simplify our expression we now specialize to the case $g(\omega) = \delta(\omega - \omega_0)$, so that the expression to be minimized is

$$\begin{aligned}\Gamma &= f_i f_j^\star (C_{ij} + R_{ij}) w_i(\omega_0) w_j^\star \\ &\quad - f_i^\star (N_i + Z_i) w_i^\star(\omega_0) - f_i (N_i + Z_i)^\star w_i(\omega_0).\end{aligned} \tag{2.7}$$

Summation over a repeated index, which labels the accelerometers, is understood. Each quantity which appears in this expression admits a simple interpretation. The factor $f_i = (N|a_i)$ has been described earlier. The array $C_{ij} = E[(a_i|B)(B|a_j)]$ is simply the statistical correlation between the outputs of i-th and j-th accelerometers, in the absence of instrumental noise. To obtain the correlation between the outputs of two real instruments we must add the $R_{ij} = E[\eta_i \eta_j^\star]$ array, which is the correlation of the intrinsic noises of accelerometers i and j.

The vector $N_i = E[(N|B)(B|a_i)]$ is the statistical correlation between the fraction of the interferometer output and of the i-th accelerometer output of seismic origin. Finally the vector $Z_i = E[\widehat{H}\eta_i^\star]$ is the correlation between the intrinsic noise of the i-th instrument and the fraction of the interferometer output uncorrelated with seismic origins.

Minimizing Γ we easily find the optimal weights

$$w_i^{\text{opt}}(\omega_0) = f_i^{-1}(C_{ij} + R_{ij})^{-1}(N_i + Z_i)^\star \tag{2.8}$$

and the reduction of noise power spectrum in ω_0

$$|H|^2 - |H_S|^2 = (C_{ij} + R_{ij})^{-1}(N_i + Z_i)(N_j + Z_j)^\star. \tag{2.9}$$

This last expression is a nonlinear function of the position and orientation of each accelerometer. Our objective is to optimize these parameters for a given number of instruments, to understand what is the best result we can achieve.

The quantities C_{ij} and N_i depend on the properties of seismic noise, which are connected in turn to the dynamics of the background motion. While it is relatively easy to measure C_{ij}, to obtain experimental information on N_i we need the sensitivity of the complete interferometer. However both C_{ij} and N_i can be calculated starting from a theoretical model of the ground seismic motion. Some of these calculations can be found in [3], and a detailed account will be published elsewhere [7, 8]. For what concerns the terms connected with instrumental noise we note that the non-diagonal part of the R_{ij} matrix and the vector Z_i should be negligible.

An intuitive understanding of the result of the optimization procedure can be obtained by looking at Eq. (2.9). In order to maximize this expression we can try to move all the accelerometers to the position which is maximally coupled to Newtonian noise. In this way the N_i terms grow. However the accelerometers cannot become too near to each other, because in this case the C_{ij} terms grow as well. The optimal configuration is obtained by balancing these two factors. Note that if, the intrinsic noise R_{ij} is large, the C_{ij} terms become less important, and all the accelerometers will try to cluster in similar positions and orientations, in such a way as to improve the statistics.

3 Some results

We studied Eq. (2.9) numerically, finding the optimal configurations for a fixed number of accelerometers. The starting point was knowing C_{ij} and N_i as functions of the positions of the accelerometers. Next we minimized numerically the optimal weight subtracted power spectrum by adjusting the accelerometer positions. In Fig. 3 we show the convergence of the mininization algorithm, plotting the relative power spectrum reduction against the iterative procedure step. The three different curves are the results for three different initial conditions. In particular the case with the best initial condition corresponds to a regularly spaced lattice of accelerometers. It is apparent that this configuration is also quite suboptimal. It is also evident that the final result is independent on the initial condition. In some cases (large number of accelerometers) this result can be achieved, avoiding local minima, with a simulated annealing procedure.

In Fig. 3 we plot the results obtained for two simple models. The first is the model used by Saulson to obtain an early estimate of Newtonian noise [1]. Its main assumption is that seismic motion can be modelled by a partition of the ground into cubic cells which oscillate coherently, but are completely uncorrelated to each other. The second is a slight modification of Saulson's model (which we will call the coherent model) which takes mass conservation into account. From our point of view this means that the motions of nearby cells are no longer uncorrelated [3].

A first observation is that the subtraction procedure is less effective at higher frequencies. This fact has a simple explanation, because the coherence length of the seismic motion (the cell dimension of Saulson's model) is proportional to the inverse of the frequency. In order to control a given volume at some specified level, we need a number of accelerometers which grows as the third power of the frequency.

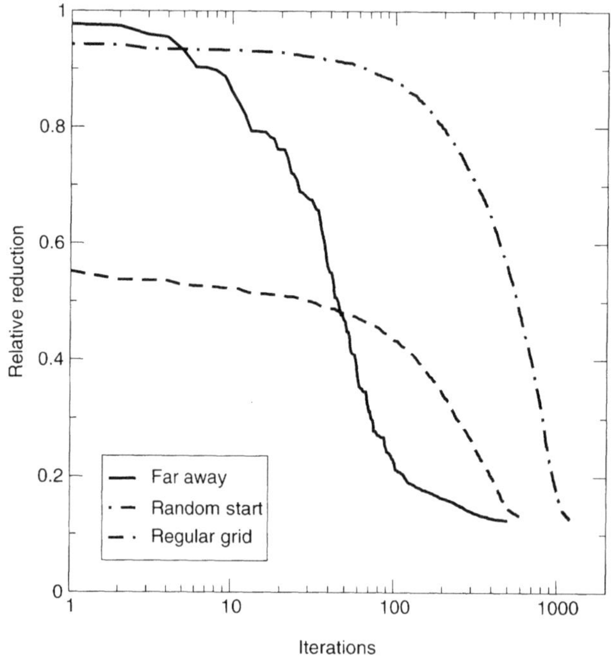

Fig. 2. Convergence of the algorithm for the minimization of the best weight subtracted power spectrum. The position of the accelerometers is adjusted, starting from some initial conditions, until the relative reduction of power spectrum converges. Organization of the accelerometers into a somewhat regular configuration can be observed. This configuration is model dependent and also frequency dependent

Another point is that we can use only accelerometers located on the surface without making worse the overall performance of our procedure, as long as the number of accelerometers is not too big (less than 40 for the Saulson's model). Note that the importance of underground measurements is greater for Saulson's model. This is a consequence of the fact that, in this case, surface measurements give us no information at all on the dynamics below the level of the first cell.

The performance of the subtraction procedure is not exceptional: for a reasonable number of accelerometers we can expect a relative reduction of an order of magnitude. We can see that a more coherent seismic dynamics improves the result, and that the two simple models we have used probably underestimate the coherence of the real dynamics. In addition they do not model the contributions to Newtonian noise due to the surface discontinuity, and it turns out that these are the most relevant ones.

We expect to obtain better results with more realistic models. We have evaluated C_{ij} and N_i for the elastic wave model used in [3] to predict Newtonian noise amplitude and extensive numerical simulations are currently in progress. However we do not expect that a relative reduction of Newtonian noise amplitude of greater than two orders of magnitude can be obtained.

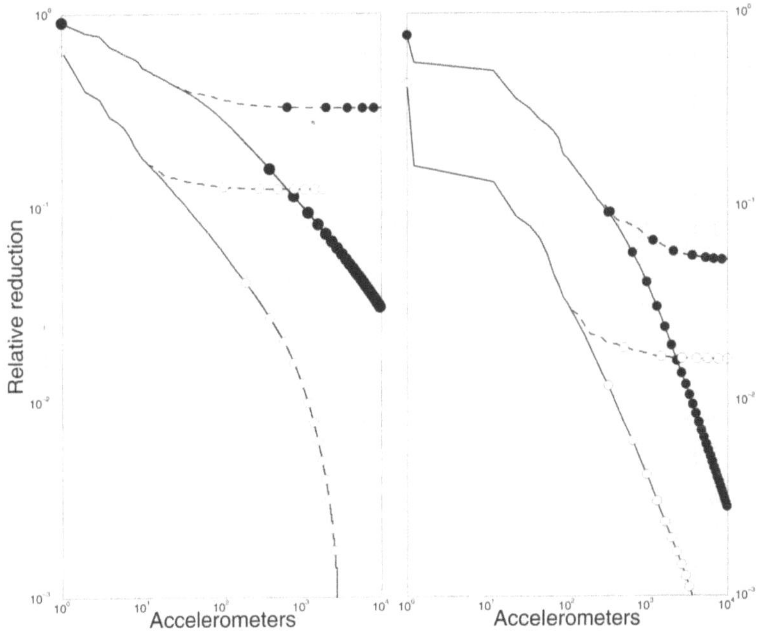

Fig. 3. Relative reduction of noise amplitude for Saulson's model (left) and the coherent model (right). The empty circles are for $f_0 = 1\,\text{Hz}$, the filled for $f_0 = 4\,\text{Hz}$. The plots with dotted lines were obtained by constraining the accelerometers to the surface

There is a general comment that can be made about the calculation of the functions C_{ij} and N_i. For any reasonable model of background seismic motion in the frequency range considered, it is important to consider the effect of dissipation on elastic modes. The main consequence is a modification of the correlation functions C_{ij} which becomes exponentially damped for large separation between points, with a finite coherence length ξ. Experimental measurements lead to the conclusion that ξ could also be of the order of magnitude of wavelength. This means that the local seismic noise level in the frequency band investigated is strongly influenced by local sources (wind, human activities, scattering processes) and an important assumption is that the frequency spectrum of these sources is wide-band.

In other words the typical expression for the amplitude of a given elastic mode, in the presence of dissipative effects, can be written as the convolution between a model dependent kernel and a source strength function. We can normalize all quantities to this unknown source strength if the convolution integral is dominated by the poles of the kernel. If this is not the case, then the structure of the excitation process becomes important (imagine, for example, a very narrow peak in the spectral distribution of the forces connected to the wind) and must be modeled.

By using this assumption the effect of dissipation can be included in the model for seismic dynamics and the seismic correlation functions can be evaluated in closed form for the case of a homogeneous ground. Detailed results will be presented elsewhere [8].

4 Conclusions and perspectives

The estimation of the feasibility of off-line subtraction of Newtonian noise is based on modellling the seismic dynamics. From a practical point of view an important issue is that we understand the effect of a wrong model on the performance of the subtraction procedure. The functions w_i optimized for a given model can give poor performance if used in a different situation. Some effects are not easy to model, in particular, the dissipative nature of the terrain and the seismic wave scattering generated by the presence of inhomogeneities. Both could be relevant, as they alter the coherence length of seismic motion and can couple otherwise independent normal modes.

A particular model can be a good approximation only in special conditions, for example, only in a particular frequency band, or in a particular atmospheric condition, or when the level of human activity is low. The validation of a model is a very important point that can be achieved with an extensive set of seismic correlation measurements.

Another possible approach is that of "model–independent" subtraction. The main point is the optimization of the functions w_i by the direct experimental estimation and minimization of the Newtonian noise power spectrum. This can be seen as the training problem for a M–adaline network with delays [9], and can be solved with standard adaptive techniques. Numerical simulations are in progress to test the effectiveness of this method. Note that experimentally it is easy to adapt the functions w_i, but it is not easy to adapt the positions and the orientations of the accelerometers. This means that a good theoretical model is in any case important in order to guess a good configuration for the instruments.

Our method for the determination of the optimal subtraction procedure is based on the assumption that the stochastic processes in which we are interested are stationary. If this is not true, then the main point is the precise determination of the quantity which we want to minimize, as the noise power spectrum is longer a useful concept. In some simple cases our formalism requires only minor modifications, for example, the redefinition of the $E[\cdots]$ average. Further investigations are needed.

As a final comment we stress that our method is independent on the particular instrument used to monitor density fluctuations. We have used accelerometers, but there are many alternative possibilities which can be investigated using the same formalism. A good instrument has a strong overlap N_i with the seismic modes which are maximally coupled to Newtonian noise, and an accurate choice can allow us to obtain major improvements.

References

1. Saulson P. (1984): Terrestrial gravitational noise on a gravitational wave antenna. Phys. Rev. D **30**, 732
2. Saulson P. (1994): Fundamental of Interferometric Gravitational Wave Detectors. World Scientific, Singapore
3. Beccaria M., Bernardini M, Braccini S. (The VIRGO Pisa collaboration) (1998): Relevance of Newtonian seismic for the VIRGO interferometer sensitivity. Class. Quant. Grav. **15**, 3339
4. Hugues S.A., Thorne K. (1998): Seismic gravity-gradient noise in interferometric gravitational wave detectors. Phys. Rev. D **58**, 122002
5. Thorne K.S., Winstein C.J. (1999): Phys. Rev. D **60**, 082001
6. Verkindt D. (1993): Gravitational noise on Virgo mirrors. Virgo note 93/019
7. Cella G. (2000): Off line subtraction of Newtonian noise, in *Proc. XXXIVth Rencontres de Moriond, Gravitational Waves and Experimental Gravity*
8. Cella G. (2000): Off line subtraction of Newtonian Noise. Preprint IFUP-TH 10/00, to be submitted to Phys. Rev. D
9. Widrow B., Stearns S.D. (1985): Adaptive Signal Processing. Prentice-Hall, Englewood Cliffs

de Broglie-like Oscillations of Spacetime

A. Feoli, G. Scarpetta

Abstract. We give two examples of de Broglie-like solutions of the linearized field equations of Einstein General Relativity. Hence a geometric interpretation of the wave associated with a quantum particle is possible. The limits and the advantages of this geometrical model are discussed.

1 Introduction

From 1926, in the "theory of the double solution" [1], de Broglie proposed an alternative to Born's probabilistic interpretation of the wave function. He supposed that the "associated waves have physical reality and must be expressed as functions of the three space coordinates and of the time". All particles are accompanied in spacetime by a real "pilot wave" $\psi = R(t, x, y, z) \exp\left[\frac{i}{\hbar} S(t, x, y, z)\right]$, and the momentum P^ν of a particle is related to the phase S of the wave field in such a way that the guiding formula

$$P^\mu = -\partial^\mu S \tag{1}$$

holds. A sufficiently complete formulation of this idea was realized in 1952 in Bohm's hidden variable theory [2]. Many efforts have been made to improve this approach, both from an experimental point of view in detecting the effects of real "de Broglie pilot waves" [3] and from a theoretical point of view in the Stochastic Interpretation of Quantum Mechanics [4]. Vigier and his co-workers showed, in fact, that waves represent collective motions on the top of a chaotic medium (the covariant subquantal Dirac "aether") which induces random stochastic jumps at the velocity of light. These real soliton waves can be considered as describing the particle's immediate neighbourhood, i.e., the aether's reaction to the particle's motion. In the framework of a geometric interpretation of quantum mechanics, de Broglie waves are either found as solutions of geodesic equations in phase space [5] or derived from a Weyl generalization of Einstein Relativity [6] or inserted in more complex models of geometric quantization [7]. In this approach, the aims of this paper are:

- to propose a geometrical interpretation of the wave associated with a quantum particle, using a simpler way, by considering perturbations to the flat metric in the *standard* linearized Einstein field equations;
- to show how some aspects of quantum mechanics can be obtained from the mathematical formalism of General Relativity without any generalization of standard Riemannian spacetime.

2 The geometric model

Our goals can be obtained by starting from the well-known theory of gravitational waves in which the metric tensor is that of Minkowski plus a perturbation

$$g_{\mu\nu} = \eta_{\mu\nu} + h_{\mu\nu}(x), \tag{2}$$

where $|h_{\mu\nu}| << 1$. From the standard theory we know that the linearized Einstein field equations in vacuo are

$$\Box h_{\mu\nu} = 0 \tag{3}$$

if the Lanczos conditions

$$\partial_\mu \left(h^\mu_\nu - \frac{1}{2}\delta^\mu_\nu h \right) = 0 \tag{4}$$

are satisfied. The solution of the classical theory is

$$h_{\mu\nu} = e_{\mu\nu}\psi = \mathrm{Re}\,[e_{\mu\nu} A \exp(-i K_\mu x^\mu)] \tag{5}$$

where $e_{\mu\nu}$ is the polarization tensor, A is a constant such that $A << 1$ and $K^\nu \equiv (\omega/c, \mathbf{k})$ satisfies $K_\nu K^\nu = 0$.

In this paper we seek de Broglie-like solutions and we start with the ansatz

$$h_{\mu\nu} = e_{\mu\nu}\psi = e_{\mu\nu} R(t, x, y, z) \exp\left(\frac{i}{\hbar} S(t, x, y, z) \right), \tag{6}$$

where R and S are functions to be determined. We consider the case of a free particle moving with the classical constant momentum $(P^\nu)_{cl} = m\gamma(c, \mathbf{v}) = (E/c, \mathbf{p})$ in flat spacetime with an associated wave ψ. We neglect the back-reaction of $h_{\mu\nu}$ on the motion of the source.

The field equation (3) can be split into its real and imaginary parts

$$\eta^{\alpha\beta}(\partial_\alpha S)(\partial_\beta S) = \hbar^2 \frac{\Box R}{R} \tag{7}$$

$$\partial_\mu J^\mu = 0, \tag{8}$$

where

$$J^\mu = \varrho u^\mu = R^2 \left(-\frac{1}{m} \partial^\mu S \right) \tag{9}$$

and (8) represents the conservation of the probability current.

Equation (1), typical of a de Broglie wave, is obtained if, in Eq. (7), we look for solutions that satisfy the equation

$$\frac{\Box R}{R} = \left(\frac{mc}{\hbar} \right)^2. \tag{10}$$

In this way Eq. (7) becomes

$$\eta^{\alpha\beta}(\partial_\alpha S)(\partial_\beta S) = m^2 c^2 = P_\mu P^\mu \tag{11}$$

which is recognized as the Hamilton–Jacobi equation for a relativistic particle in flat spacetime. As the velocity is constant, we have from (1) that

$$S = -P^{\mu}x_{\mu}, \tag{12}$$

where P^{ν} is the classical relativistic four-momentum $(P^{\nu})_{cl}$.

We can look for a solution of the metric perturbation in the form of cylindrical waves considering a world-tube around the x-axis and imposing the condition that $R(x, y, z, t) = f(\sqrt{y^2 + z^2})$. By defining $\beta \equiv \sqrt{y^2 + z^2}$ we found [10] a solution of Eq. (10) which is the following

$$R = AJ_o(g\beta) \tag{13}$$

in which the Bessel function of the zero-th order has the form of a wave packet but only for the spatial directions orthogonal to that of propagation. The gauge conditions (4) become

$$\left(e^{\mu}_{\nu} - \frac{1}{2}\delta^{\mu}_{\nu}e\right)\partial_{\mu}R = 0 \tag{14}$$

$$\left(e^{\mu}_{\nu} - \frac{1}{2}\delta^{\mu}_{\nu}e\right)\partial_{\mu}S = 0$$

from which we calculate the nonvanishing components of the polarization tensor: $e_{00} = e_{11} = -(v^2 + c^2)/c^2$, $e_{22} = e_{33} = (c^2 - v^2)/c^2$ and $e_{01} = e_{10} = 2v/c$. *Thus we have shown that it is possible to obtain de Broglie waves as real as gravitational waves.*

A typical effect due to the passage of a gravitational wave is the variation of the relative distance between two test particles. This can be studied also in the case of our de Broglie waves. We consider the wave packet

$$h_{\mu\nu} = \mathrm{Re}\left\{e_{\mu\nu}AJ_o(g\beta)\exp\left[-\frac{i}{\hbar}(Et - px)\right]\right\}$$

propagating along the x-axis and two masses initially separated by a distance ℓ along the y-axis of a suitable reference frame. The geodesic deviation caused by a gravitational field is

$$\frac{D^2\xi^{\mu}}{c^2 d\tau^2} = -R^{\mu}_{\nu\gamma\delta}\sigma^{\nu}\sigma^{\delta}\xi^{\gamma},$$

where τ is the proper time along the world-line of the two masses, σ^{μ} is their four-velocity and ξ^{μ} their relative distance. We can use two approximations:

- the velocity of the two test particles is very small compared with the speed of light so $\sigma^{\mu} \simeq (1, 0, 0, 0)$;
- the displacement δ^{μ} of the two masses due to the wave is very small so, in $\xi^{\mu} = (0, \delta^1, \delta^2 + \ell, \delta^3)$, we can consider $\ell >> \delta^i$.

In this framework the maximal acceleration of geodesic deviation is

$$\ddot{\xi}_2 \simeq -c^2 R_{2020}\ell \simeq -\frac{A\ell c^2}{2}\left[\left(1 + \frac{v^2}{c^2}\right)\frac{\partial^2 J_0}{\partial y^2} + g^2 J_0\right],$$

where v is the velocity of the source of the de Broglie wave.

3 Spherical symmetry

If we want to obtain a solution of the linearized Einstein equations with spherical symmetry, we must start with a different ansatz

$$h_{\mu\nu} = N\partial_\mu \partial_\nu \psi. \tag{15}$$

Using (15), Eqs. (3) and (4) are automatically satisfied by the solution of $\Box\psi = 0$ if Schwarz' theorem on the commutation of partial derivatives is satisfied.

In this case the condition (10) is verified by the non-dispersive de Broglie wave packet found first by de Broglie [8] and then thoroughly studied by L. Mackinnon [9]:

$$R = A\frac{\sin gr}{gr}, \tag{16}$$

where $g = mc/\hbar$ and $r = \sqrt{\gamma^2(x - vt)^2 + y^2 + z^2}$ is the distance from the particle position, so that

$$\psi = A\frac{\sin gr}{gr} \exp\left[-\frac{i}{\hbar}(Et - px)\right]. \tag{17}$$

In the rest-frame of the source we have

$$\phi = A\frac{\sin gr'}{gr'} \exp(-i\omega t'), \tag{18}$$

where $r' = \sqrt{x^2 + y^2 + z^2}$ and $\omega = cg$.

Hence we can consider the Mackinnon solution as ψ in (15), just outside the radius of the source so that we can avoid the singularity at $r = 0$. The choice of the constant N in (15) in the form $N = g^{-2}$ is dictated by dimensional constraints. Using (17) in (15), we see that the real part of the metric perturbation becomes, to first order in $\hbar$,

$$\begin{aligned} h_{\mu\nu} = & - A\frac{\sin gr}{gr}\left(\partial_\mu r \partial_\nu r + \frac{P_\mu P_\nu}{m^2c^2}\right)\cos(S/\hbar) \\ & - A\frac{\cos gr}{gr}\left[\frac{(\partial_\nu r)P_\mu + (\partial_\mu r)P_\nu}{mc}\right]\sin(S/\hbar). \end{aligned} \tag{19}$$

This is a non-dispersive de Broglie wave packet that acquires a clear geometric meaning as it is a solution of the linearized Einstein field equations. As in the Schwarzschild case, the metric holds in the vacuum outside the source that is localized in a spacetime region of the order of its Compton wavelength, on the peak of a de Broglie wave packet.

The tensor (19) can also be obtained by using (18) in (15) and then performing the coordinate transformation

$$h'_{\mu\nu} = h_{\alpha\beta} \frac{\partial \xi^\alpha}{\partial x^\mu} \frac{\partial \xi^\beta}{\partial x^\nu} \tag{20}$$

on $h_{\mu\nu}$, where ξ^μ and x^μ are related by a Lorentz boost in the x-direction.

We can also perform, provided that $|h'_{\mu\nu}| << 1$, a transformation of the rest-frame metric tensor in spherical coordinates (t, r, θ, φ). The only nonzero components are

$$h_{oo} = h_{11} = -A \frac{\sin gr'}{gr'} \cos(\omega t'), \tag{21}$$

$$h_{o1} = h_{1o} = -A \frac{\cos gr'}{gr'} \sin(\omega t'). \tag{22}$$

Note also that (19, 21, 22) preserve the classical limit because, when $\hbar \to 0$, the quantum perturbations disappear and $g_{\mu\nu} = \eta_{\mu\nu}$. The same result is obtained also in the limits $m \to \infty$ or $r \to \infty$.

4 Conclusions

Finally we briefly discuss the limits and the advantages of our model. It is well-known [11] that, for a Klein–Gordon particle, the equation of motion can be written as a geodesic equation in the metric

$$g_{\mu\nu} = [1 + (\hbar/mc)^2 \Box R/R]\eta_{\mu\nu} \tag{23}$$

where $\eta_{\mu\nu} = \text{diag}\,(1, -1, -1, -1)$.˙but this geometric model has some problems as pointed out by Holland [12]:

1. there is no well-defined "source" of the metrical field in the physical sense that a localized mass may generate a gravitational field;
2. the metrical coefficients are independent of the wave intensity and do not, in general, diminish in magnitude with distance in what is otherwise empty space;
3. in the framework of de Broglie–Bohm theory the J^0 component of the current is of indefinite sign and hence cannot be treated as a probability density and J^μ is not generically timelike.

In our particular case the geometrical model solves all these problems and furthermore we have the advantage of deducing our metric directly from the Einstein field equations and not from the Klein–Gordon equation. The geometric interpretation of de Broglie waves is now possible in the standard framework of General Relativity but this model works in the free particle case. Now the main problem is how to introduce interactions without generalizing Riemannian space to Finsler space, as Holland proposes when he generalizes the metric (23) to include an electromagnetic field.

Furthermore the Mackinnon solution has to be treated carefully. In fact the ansatz (15) leads to a solution that can be gauged away by a suitable coordinate transformation of the form $x^{\mu'} = x^\mu + \xi^\mu(x)$, leading to

$$h'_{\mu\nu} = h_{\mu\nu} - \partial_\nu \xi_\mu - \partial_\mu \xi_\nu = 0 \tag{24}$$

and such that $\Box \xi^\mu(x) = 0$. According to Einstein [13]: "If an h-field can be made to vanish by the addition of terms like those in (24), i.e., by means of an infinitesimal transformation, then the gravitational field being described is only an apparent field". But "the geometrical point of view shows that this is too narrow an interpretation. The connection is a geometric object and so the potential should be regarded as physical", Chap. I in [14]. In our case connections and metrics assume non-trivial values in the rest-frame of the particle and in all the frames connected by means of a Lorentz transformation, so a geometric interpretation is still possible even in the spherical case.

Acknowledgements

The research was supported by MURST funds ex 40% and ex 60%.

References

1. de Broglie L. (1927): Jour. de Phys. **5**, 225; (1926): C.R. Acad. Sci. **183**, 447; (1927): **184**, 273; (1927): **185**, 380
2. Bohm D. (1952): Phys. Rev. **85**, 166 and 180;
   de Broglie L. (1960): Nonlinear Wave Mechanics. Elsevier, Amsterdam;
   Vigier J.P. (1991): Found. Phys. **21**, 125
3. Wang L.J., Zou X.Y., Mandel L. (1991): Phys. Rev. Lett. **66**, 1111;
   Zou X.Y., Grayson T., Wang L.J., Mandel L. (1992): Phys. Rev. Lett. **68**, 3667
4. Bohm D., Vigier J.P. (1954): Phys. Rev. **96**, 208;
   Nelson E. (1966): Phys. Rev. **150**, 1079;
   Lehr W., Park J. (1977): Jour. Math. Phys. **18**, 1235;
   Vigier J.P. (1979): Lett. Nuovo Cimento **24**, 258 and 265
5. Caianiello E.R. (1982): Lett. Nuovo Cimento **35**, 381; (1992): La Rivista del Nuovo Cimento **15** (4);
   Caianiello E.R., Marmo G., Scarpetta G. (1983): Lett. Nuovo Cimento **36**, 487; (1983): Lett. Nuovo Cimento **37**, 361
6. Santamato E.(1984): Phys. Rev. D **29**, 216; (1985): D **32**, 2615
7. Woodhouse N.M.J. (1980): Geometric Quantization. Clarendon Press, Oxford;
   Prugovecky E. (1992): Quantum Geometry. Kluwer Academic Publishers, Dordrecht
8. de Broglie L. (1925): C.R. Acad. Sci. **180**, 498
9. Mackinnon L. (1978): Found.Phys. **8**, 157; (1981): Lett. Nuovo Cimento **31**, 37; (1981): **32**, 311
10. Feoli A., Scarpetta G. (1998): Found. Phys. Lett. **11**, 395
11. Synge J.L. (1954): Geometrical Mechanics and de Broglie Waves. University Press, Cambridge;
    Vigier J.P. (1968): C.R. Acad. Sci. Paris **266**, 598

12. Holland P.R. (1987): Found. Phys. **17**, 345; (1993): Phys. Rep. **224**, 95
13. Einstein A., Rosen N. (1937): J. Franklin Inst. **223**, 43
14. Atiyah M.F. (1988): Collected Works. Vol. 5, *Gauge Theories*. Clarendon Press, Oxford, p. 11

Testing Newton's Inverse Square Law

S. Focardi

Abstract. We briefly describe a recent experiment aimed at verifying the dependence on the distance in Newton's inverse square law. Then we present two new experiments for: (*i*) testing the previous result; (*ii*) making a measurement of the G constant by a method different from those normally employed.

1 Introduction

About ten years ago, doubts about the validity of Newton's gravitational law were raised by Fischbach et al. [1] as a consequence of their re-analysis of the classical Eötwös experiment. The validity of this law has been carefully tested in laboratory experiments for distances that are less than 1 m [2]. Above 10 km, the law has been verified by comparing values for G obtained by measuring the motions of the Moon and the satellite LAGEOS [3]. However, in the geophysical region, i.e., for distances in the range 1 m $\div$ 10^4 m, measurements cannot be made with high precision because of difficulties both in moving large masses and in detecting small effects.

Deviations from the inverse square law would indicate the existence of gravitational mediators having a mass different from zero. Their existence is indeed predicted in several superstring theories that are highly accredited today in the attempt to unify the gravitational force with the other fundamental interactions. The absence of theoretical suggestions on the mass values, and, consequently, on the interaction range of such hypothetical particles, makes very important any effort at exploring the validity of Newton's law, especially in the geophysical region.

Many experiments were stimulated by Fischbach's hypothesis [1]. A common characteristic in them is a comparison of the earth's gravitational field at different heights with that calculated starting from the Newton's law. The latter depends on mass distribution; this can be evaluated by mapping the values of the gravitational acceleration g at a reference level. For this reason, all the experiments of this type depend on the model used to deduce the mass distribution from the values of g. Another important characteristic of such experiments is the fact that they use La Coste–Romberg mechanical gravimeters to measure g at different heights. These gravimeters can be moved easily to different heights but they are characterized by a continuous drift superimposed on the tidal signal. They also need calibration obtained by a comparison with absolute gravimeters. As a consequence, even under the best conditions, gravity intervals cannot be measured by La Coste–Romberg gravimeters with an accuracy better than 5 μGal (1 Gal $= 1$ cm s^{-2}) [4].

In the next section we describe the experiment performed. Since the experimental results obtained in these measurements show a deviation from the inverse square law, a new experiment, described in Sect. 3, is scheduled. Its aim is to check the result previously obtained.

Despite its importance, the constant G, which plays the role of the constant coupling in gravitational interactions, is not well determined. Moreover, recent determinations of G reported in the literature point to the existence of systematic errors. In Sect. 4, we describe our proposal for a measurement of G by a method different from the traditional ones.

2 Our experiment

The main characteristic of our experiment [5] is that the measurements are independent of the terrestrial gravitational field. Another important feature is seen in the sensitivity of the superconducting gravimeter used. It can detect variations in g of the order of 10 nGal. Therefore it has a sensitivity at least 100 times greater than that of a La Coste–Romberg mechanical gravimeter.

The basic idea of the experiment consists in using as a source of displacement of large masses a lake, whose level changes in the daytime, and in measuring the variations in g yielded with the superconducting gravimeter. The lake we used is Brasimone, 845 m above sea level, approximately midway between Bologna and Florence. Brasimone, together with Suviana Lake, acts as a power storage device used by ENEL: water is pumped from Suviana up to Brasimone during the night, then the inverse process is used to obtain electric power at hours of maximum need.

On the shore of Brasimone, in a ENEA research centre, a 100 m long tunnel extends toward the centre of the lake, 30 m below ground level. As shown in Fig. 1, we positioned the gravimeter at the end of the tunnel, in order to measure gravity variations induced by the water masses moved. The resulting gravitational signal differs only by 8% from that produced by an infinite sheet. This makes its dependence on the shape of the shore not particularly critical.

The shore of the lake has been measured with great accuracy in order to improve the precision of the calculated gravitational effect. In particular, we have installed a

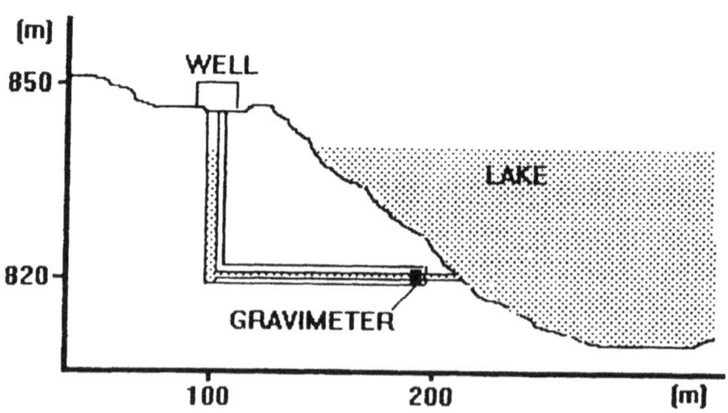

Fig. 1. Sketch of the measurement site

22 station reference network along the border of the lake. Then we performed high-precision geometric leveling, electromagnetic distance measurements and angular measurements, which allows us to know the distances (hundreds of meters) between the stations to an accuracy of 1 cm. An aerophotogrammetric survey, a photogrammetric survey and a GPS (Global Positioning System) static observation campaign, together with a numerical model, have permitted us to reduce to 1.4×10^{-4} the relative error on the calculated Newtonian effect.

The great precision of the superconducting gravimeter (GWR type D60/T015, from GWR, San Diego (CA), USA) is due to the stability in the time of its supercurrents. A niobium superconducting sphere is maintained against its gravity by the magnetic field gradient generated by two superconducting coils. Its height is kept constant by changing the current circulating in a third coil. The voltage applied by the system to the feedback coil is the instrument signal output.

The superconducting gravimeter is not an absolute instrument since it measures only variations of g. For this reason, it needs to be calibrated. This has been done by moving a well-known annular mass (273.40 ± 0.01 kg) up and down around the gravimeter [6]. The maximum Δg variation, calculated from the geometry of the ring, and assuming for G the CODATA value $G = 6.67259\times10^{-11}\,\mathrm{m^3kg^{-1}s^{-2}}$, is

$$\Delta g_{\max} = 6.731(1)\ \mu\mathrm{Gal} \tag{1}$$

with a total relative uncertainty of about 0.01%.

All calibration measurements were performed in an ENEA center laboratory far from the lake. In this laboratory we also recorded, for five months, the gravimeter signal, which is known to depend on the local elastic characteristic. This permitted us to obtain a good determination of the short period tidal components [7]. The same measurements were used to obtain the pressure admittance ($-0.201\pm 0.007\ \mu$Gal/mbar) and the instrumental drift ($2.709 \pm 0.001\ \mu$Gal/day).

The gravimeter was installed at the end of the tunnel twice and was run there for a total of 327 days. During such time intervals, we continuously recorded many other parameters besides the gravity signal. In particular, among others, we recorded the level and temperature of the lake, barometric pressure, air temperature and humidity. The lake temperature was measured at two different depths (-1 m and -5 m) along the vertical, above the gravimeter position. In fact, we verified that the maximum temperature difference measured at any depth, in various places and in different seasons, never exceeded ±0.8 K. Such great thermal stability is due to the continuous water exchange between the two basins which prevents the formation of thermal gradients.

The level lake was continuously monitored (every 20 seconds) by measuring the length of a stainless-steel wire fixed to a buoy floating in the well (see Fig. 1) which communicates with the lake. The wire was maintained at a constant tension and wound on a precision wheel whose angular position permitted us to detect 0.1 mm variations in the length of the wire . The whole system was calibrated by comparison with measurements obtained by an electromagnetic distance-measuring instrument. The relative error associated with the level of the lake measurements was 2×10^{-4}.

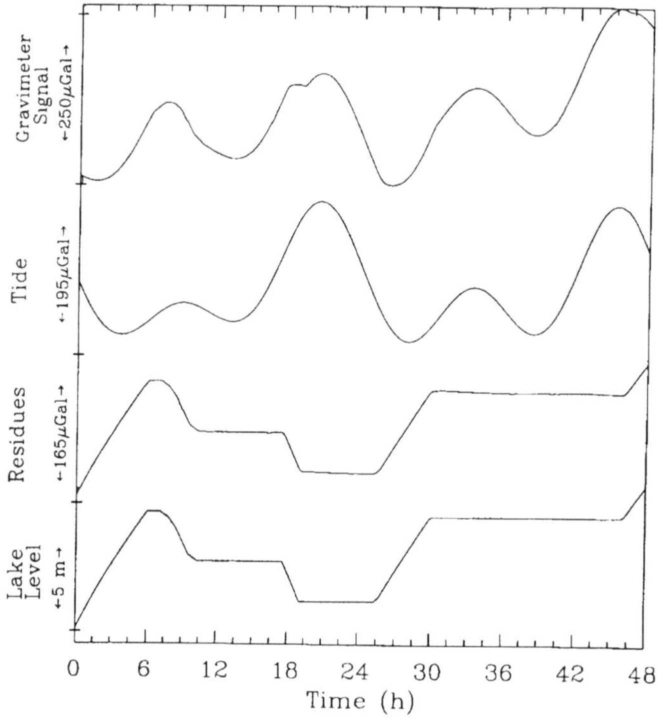

Fig. 2. Gravimetric signal, tidal signal, water mass effect and lake level

Figure 2 is relative to a period of two days and from top to bottom shows the gravity signal recorded, the tidal effect calculated, the lake gravity effect obtained as a difference and the measured level of the lake. Comparison of the last two curves allows us to obtain the experimental effect due to the water mass moved. Our analysis was made by comparing the theoretical (Newtonian) and the experimental effects associated with 10 cm lake level variations. Such changes of level, which occur in brief periods on the order of 15 minutes, allowed us to disregard any effect due to temperature changes of the water, to water table variations and to the lack of knowledge of the long period tidal components.

The calculated Newtonian effect utilized both water and air densities: the water density was obtained of the extrapolating at the temperature of the lake the value measured at 21 °C with an absolute densitometer of 10^{-6} g,cm^{-3} sensitivity. This allowed a specific mass determination with a relative error of 2.6×10^{-5}. The air density was calculated starting from the measured values of air temperature, atmospheric pressure and vapor pressure. The relative uncertainty of the correction which takes into account subtracting the value for air from that for water is 1×10^{-5}. For these reasons the relative precision of the calculated effect is determined by that due to the level of the lake, i.e., 10^{-4}.

The main correction to the experimental data depends on the subsidence and uplift produced by the load variations on the gravimeter station located at the end of the tunnel. In order to obtain the total effect we employed three different methods: we used: *(i)* a clinometer in the first 100 m (inside the tunnel); *(ii)* a high-precision spirit survey from 100 m to 1.5 km; *(iii)* a numerical model from 1.5 km to infinity. We obtained in total

$$\Delta h = (0.57 \pm 0.11)\ \text{mm for 1 m of water.} \tag{2}$$

The measured gravitational signal was therefore corrected by a term

$$\Delta g = (0.08 \pm 0.03)\ \mu\text{Gal for 1 m of water} \tag{3}$$

which takes into account the Bouguer correction and the density variation of the surrounding medium. This corresponds to a relative error of 8×10^{-4} for the gravitational signal.

From the continuous set of data, we extracted the gravitational effect corresponding to a fixed (10 cm) lake level variation. The data we obtained were grouped in intervals characterized by having the same height with respect to the gravimeter station. Fig. 3 shows the experimental and the calculated (Newtonian) values against distance from the gravimeter. The ratio between experimental and theoretical effects is

$$R = 1.0127 \pm 0.0013. \tag{4}$$

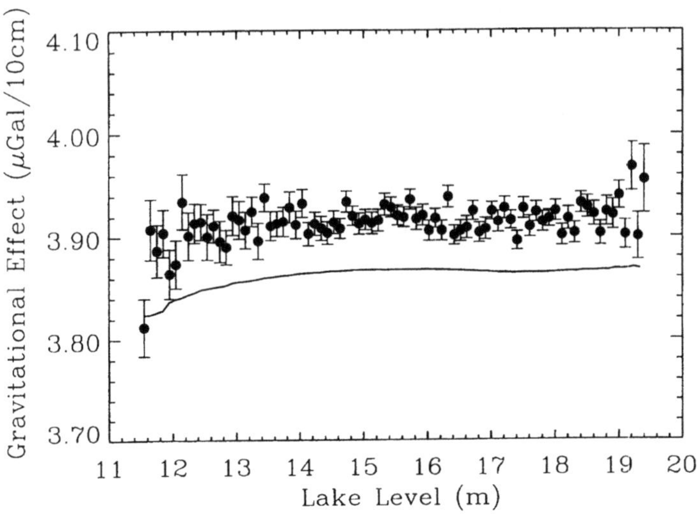

Fig. 3. Experimental and theoretical (continuous curve) gravitational effect for 10 cms of water level variation against vertical distance from the gravimeter

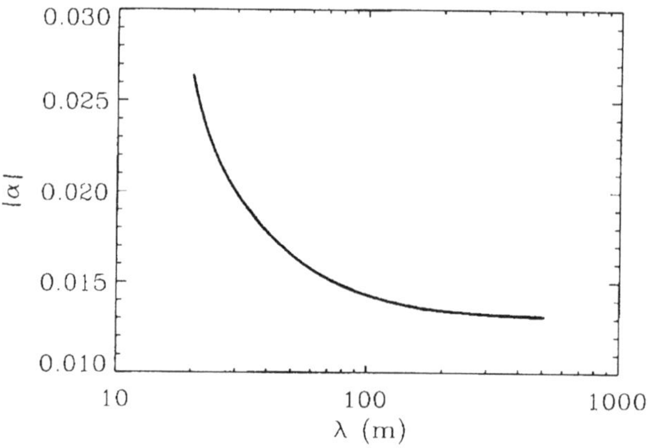

Fig. 4. Relationship between α and λ in the range $20 \div 500\,\text{m}$

By assuming the validity of Newton's law and the dependence of G on distance, R can be written as

$$R = \frac{G_{\text{far}}}{G_{\text{near}}}. \tag{5}$$

Since the $G(r)$ hypothesis disagrees with experimental results [2, 3], a more realistic position is to assume the existence of non-Newtonian terms in the gravitational law.

By assuming a gravitational potential energy of

$$U = -G\frac{m_1 m_2}{r}(1 + \alpha e^{-r/\lambda}) \tag{6}$$

our experimental results allow us to exclude λ values smaller than 20 m. The observed effect indicates a negative λ value which corresponds to a repulsive term (due to a spin 1 particle: graviphoton). Using relation (6), our experimental data are compatible with values of (λ, α) pairs which lie on the curve shown in Fig. 4. Since a large effect would have already been observed in other experiments, we consider $\lambda > 100\,\text{m}$ and $\alpha \approx 0.013$ as the most probable solution.

3 A check of the experiment

The error quoted in (4), which corresponds to a relative sensitivity of 1.3×10^{-3}, essentially derives from two effects [5]: gravimeter calibration which contributes 9.2×10^{-4} and vertical deformation which contributes 8×10^{-4}. The latter practically establishes a lower bound of $\approx 10^{-3}$ to the precision in the determination of the ratio R.

Almost all the measurements, and also the data analysis, were checked by different approaches as described in detail in [5]. The only parameter which has not been verified at the maximum level of precision is gravimeter calibration. Furthermore, the gravimeter was calibrated by using a well-known annular mass in a laboratory far from the lake. The calibration coefficients relative to the runs in the tunnel were determined by means of an electrostatic calibration. This consists in measuring the output signal (in volts) produced electrostatically by applying a known voltage to the gravimeter capacitive bridge which controls the position of the niobium sphere.

For these reasons we think that a new measurement aimed at checking the previous one should solve the problem of calibrating the instrument in the tunnel directly.

Absolute gravimeters measure the gravity acceleration g by controlling the fall of a test mass in a vacuum. The most recent absolute gravimeters have an important characteristic: a high operation rate (up to a single drop every 15 s). This allows us to obtain a consistent number of drops (of the order of 10^4) in a few days, reducing the standard deviation to a minimum value which depends on the instrument itself and on the geophysical noise (imprecise knowledge of tidal effects, non-local changes in atmospheric pressure, water table variations).

We have compared our superconducting gravimeter with an FG5 [8] absolute gravimeter owned by the Strasbourg EOST (Ecole et Observatoire des Sciences de la Terre) Institute. We operated the two instruments simultaneously in a laboratory far from the lake and obtained a superconducting constant calibration (in volt/μGal) with a relative precision of 1.4×10^{-3}.

We plan to use another gravimeter which is based on launching the test mass (doubling the path length) and thus giving a precision no worse than that of FG5. Such a gravimeter is presently operated manually, but will be modified in order to make it automatic.

Our experimental program can be summarized as follows:

(i) Initially the absolute gravimeter will be installed in the same laboratory (far from the lake) where the superconducting gravimeteris installed. The gravitational signal will be used to obtain the calibration coefficient. This run will take a few days.

(ii) Later, the absolute instrument will be moved to the end of the tunnel and will be operated there for about two months. During this period the superconducting gravimeter will allow us to obtain all the gravitational effects by direct measurement. These will be subtracted from the values measured by the absolute gravimeter in order to obtain the effect of the lake.

(iii) Subsequently the absolute gravimeter will be moved to the laboratory. A final comparison of the gravimeters will be performed to check the calibration coefficient value.

Starting from the precision measured in the calibration process, we think that the final result should be obtained with a total relative precision of about 2×10^{-3}. This semms to be sufficient to verify the previous experiment.

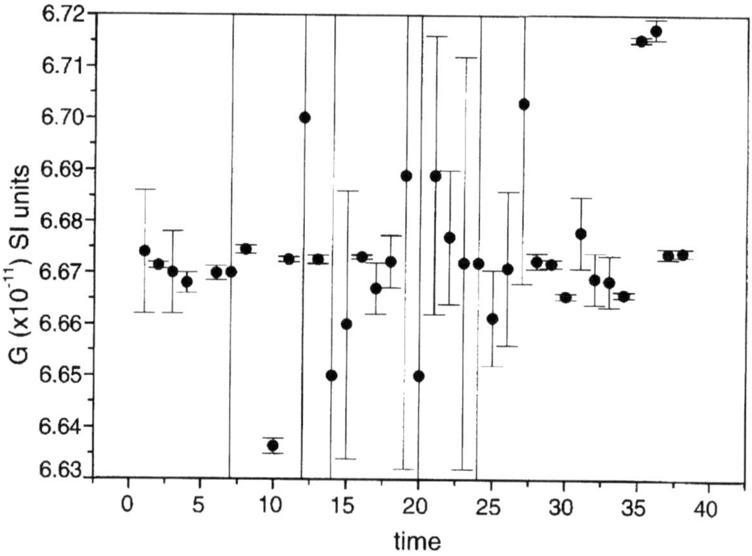

Fig. 5. G values obtained in the period 1969–1996

4 The value of the constant G

As pointed out by Gillies [9] in a report on the Newtonian gravitational constant, the value of G is not known with the precision we would like a fundamental constant to have. Indeed, the difficulties in measuring G with high precision are due to the extreme weakness of gravity.

In Fig. 5 the values of G measured from 1969 to 1996 are shown plotted against time. Data are taken from [9]. The value whose abscissa is equal to 13 is not the result of a measurement; it is the CODATA value established in 1986. As clearly shown in 5, some of the most recent results are in strong disagreement. This indicates the existence of systematic errors, one of which depends on the inelasticity of the fibres used in experiments based on a torsion pendulum [10].

For these reasons a measurement of G based on a method different from those typically used is of great interest. After the second calibration described in the previous section, we will use the annular mass moved around the superconducting gravimeter to obtain a value for G at a distance on the order of 1 meter. The relative precision of this determination can be estimated as on the order of 2×10^{-3}. Our experimental project will thus permit us to obtain, at this level of precision, the value of G in two different regions of distances.

References

1. Fischbach E., Sudarsky D., Szafer A., Talmadge C., Aronson S.H. (1986): Reanalysis of the Eötwös experiment. Phys. Rev. Lett. **56**, 3–6

2. Cook A.H. (1988): Experiments on gravitation. Rep. Prog. Phys. **51**, 707–757
3. Cook A.H. (1987): The inverse square law of gravitation. Contemp. Phys. **28**, 159–175
4. Speake C.C. (1992): Experiments on the Fifth Force, in *Proc. International School of Physics E. Fermi*, ed. by L. Crovini, J.J. Quinn, Course CX, North Holland, Amsterdam, pp. 154–177
5. Achilli V., Baldi P., Casula G., Errani M., Focardi S., Palmonari F.. Pedrielli F. (1997): A geophysical experiment on Newton's inverse-square law. Nuovo Cimento B **112**, 775–804
6. Achilli V., Baldi P., Casula G., Errani M., Focardi S., Guerzoni M.. Palmonari F., Raguní G. (1995): A calibration system for superconducting gravimeters. Bull. Geod. **69**, 73–80
7. Baldi P., Casula G., Focardi S., Palmonari F. (1995): Tidal analysis of data recorded by a superconducting gravimeter. Ann. Geofis. **38**, 161–166
8. Niebauer T.M., Sasagawa G.S., Faller J.E., Hilt R., Klopping F. (1995): A new generation of absolute gravimeters. Metrologia **32**, 159–180
9. Gillies G.T. (1997): The Newtonian gravitational constant: recent measurements and related studies. Rep. Prog. Phys. **60**, 151–225
10. Kuroda K. (1995): Does the time-of swing method give a correct value of the Newtonian gravitational constant? Phys Rev. Lett. **75**, 2796–2798

MIX
Papier aus verantwortungsvollen Quellen
Paper from responsible sources
FSC® C105338

If you have any concerns about our products,
you can contact us on
ProductSafety@springernature.com

In case Publisher is established outside the EU,
the EU authorized representative is:
Springer Nature Customer Service Center GmbH
Europaplatz 3, 69115 Heidelberg, Germany

Printed by Libri Plureos GmbH
in Hamburg, Germany